Oxidative Folding of Peptides and Proteins

RSC Biomolecular Sciences

Editorial Board:

This Series is devoted to coverage of the interface between the chemical and biological sciences, especially structural biology, chemical biology, bio- and chemo-informatics, drug discovery and development, chemical enzymology and biophysical chemistry. Ideal as reference and state-of-the-art guides at the graduate and post-graduate level.

Titles in the Series:

Visit our website at www.rsc.org/biomolecularsciences

For further information please contact:
Sales and Customer Care, Royal Society of Chemistry, Thomas Graham House, Science Park, Milton Road, Cambridge, CB4 0WF, UK
Telephone: +44 (0)1223 432360, Fax: +44 (0)1223 426017, Email: sales@rsc.org

Oxidative Folding of Peptides and Proteins

Edited by

Johannes Buchner
Department of Chemistry, Technical University of Munich,
Garching, Germany

Luis Moroder
Max Planck Institute, Martinsried, Germany

RSCPublishing

ISBN: 978-0-85404-148-0

A catalogue record for this book is available from the British Library

Published by The Royal Society of Chemistry,
Thomas Graham House, Science Park, Milton Road,
Cambridge CB4 0WF, UK

Registered Charity Number 207890

For further information see our website at www.rsc.org

Foreword

Proteins are marginally stable entities. There is only a small energetic difference between the native and the manifold of unfolded states. Their dynamic nature, however, is crucial for performing the diverse functions proteins exert. Especially in the harsh environment outside the cell, proteins and peptides are stabilized by covalent bonds between cysteines against stress of different kinds which may result in unfolding and proteolytic attack. It became evident only recently that the redox reactions involved in formation of these disulfide bonds are catalyzed by an enzyme machinery present in all kingdoms of life. The advent of recombinant DNA technology allowed us, using this trick of nature, to stabilize proteins of interest for biotechnology, widening its application beyond the classical production of disulfide-stabilized peptides by organosynthetic approaches. Furthermore, in the context of the recombinant production of disulfide-containing proteins efficient strategies for their oxidation and correct bond formation were developed.

Although disulfide bonds have been known for decades and their effects have been studied extensively in a variety of proteins, significant progress has been made especially in recent years. However, a comprehensive description of this field of research covering both disulfide-containing peptides and proteins together with the cellular machinery required for their formation was missing. This book was therefore long needed and overdue. The editors have assembled an impressive line-up of the world's leading experts reporting on the different topics related to disulfide bonds in proteins and peptides, ranging from cell biology and biochemistry to peptide chemistry and biotechnology. Each chapter is composed as an independent unit enabling the reader to pick the topic of immediate interest. Given the importance of disulfide-bonded proteins and peptides for the architecture and stability of these biomolecules, this book is an invaluable source of information combining for the first time the protein and peptide worlds.

Robert Huber

Max-Planck-Institut für Biochemie, D-82152 Martinsried, Germany
School of Biosciences, Cardiff University, Cardiff, UK
ZMB, Universität Duisburg-Essen, Essen, Germany

Preface

The formation of disulfide bonds is the major natural post-translational modification of proteins and peptides which adds additional cross-links into the polypeptide chain. Especially for secreted cysteine-containing peptides and proteins, these disulfide bonds are often crucial for their stability and the final three-dimensional structure. However, there is an intrinsic danger in this process, as mispairing of cysteine residues can prevent the polypeptides from reaching their native functional conformation. Furthermore, disulfide-bond formation is a slow process for which there is a cellular machinery to catalyze and guide the reaction. In this book we present an overview of the different aspects associated with this post-translational modification. For the first time, both the protein and peptide worlds are presented, together with the intent to highlight specific unique aspects and at the same time to emphasize common principles.

The often-used term oxidative folding refers to the complex process that combines native disulfide-bond formation and folding, *i.e.* formation of the native three-dimensional structure from the cysteine-containing unfolded polypeptide chain. It has been a long time since the first successful, but largely unnoticed, studies on the oxidative assembly of the insulin molecule from the two reduced peptide chains[1] and particularly the classic experiments by Anfinsen *et al.*[2] on the oxidative refolding of reduced unfolded ribonuclease A, which were of seminal importance in establishing the central principle of self-assembly of proteins,[3] to today's concepts of the folding of disulfide-bonded proteins and the knowledge gained on the cellular machinery assisting this process. Compared to the folding of proteins that lack disulfide bonds,

[1] Y. C. Du, Y. S. Zhang, Z. X. Lu and C. L. Tsou, *Sci. Sin.*, 1961, **10**, 84–104.
[2] C. B. Anfinsen, E. Haber, M. Sela and F. H. White, *Proc. Natl. Acad. Sci. USA*, 1961, **47**, 1309–1314.
[3] C. B. Anfinsen, *Science*, 1973, **181**, 223–230.

RSC Biomolecular Sciences
Oxidative Folding of Peptides and Proteins
Edited by Johannes Buchner and Luis Moroder
© Royal Society of Chemistry 2009
Published by the Royal Society of Chemistry, www.rsc.org

oxidative folding of cysteine-containing peptides and proteins is slow because of its dependence on redox reactions such as oxidation, reduction and reshuffling. For *in vivo* oxidative folding, organisms from bacteria to humans exploit highly specialized enzymes, which catalyze disulfide formation, reduction and reshuffling and drive the folding protein to its native structure. Great headway has been made in understanding the oxidative folding processes *in vivo* and the role of thiol protein oxidases and disulfide isomerases (PDIs) involved, which catalyze both formation and isomerization of disulfide bonds in the eukaryotic endoplasmic reticulum and the prokaryotic periplasm, respectively. Just recently a similar system has been discovered in the intermembrane space of mitochondria. Furthermore, specific protective systems turned on upon oxidative stress are under intensive investigation (Chapters 1.1 to 1.8). The presence of disulfide bonds has a pronounced and complex effect on the folding and stability of proteins with contributions affecting the unfolded state and the kinetics of folding. Furthermore, folding intermediates can readily be trapped in disulfide-rich proteins and peptides resulting in significant advances of our understanding of putative folding mechanisms (Chapter 2.1). Relying on improved knowledge of the *in vivo* and *in vitro* processes and the redox potentials involved (see Chapters 2.1 and 3) experimental procedures for *in vitro* oxidative folding of biochemically and pharmacologically relevant disulfide-rich proteins, but also of cystine-rich peptides, have been established. The experiences accumulated over the years have also been exploited in developing the biotechnology of oxidative folding in industry (Chapter 2.2).

Although the increasing number of disulfide-rich peptides detected and isolated from organisms of all kingdoms of life such as hormones, defensins, toxins, enzyme inhibitors, *etc.* are products of post-translational processing of larger proforms, optimized *in vitro* oxidative folding of the lower-sized mature forms is possible, even in high yields (Chapters 6.1, 6.4 and 6.5). For the few exceptions chemical approaches to regioselective disulfide formation still remains indispensable and thus the method of choice. The progress achieved in the synthesis of peptides and proteins as well as in the required chemistry for selective cysteine-pairings has advanced to a state of art that generally allows for synthetic access to such disulfide-rich peptides and small proteins with the correct disulfide framework and thus native three-dimensional structure (Chapter 6.2). The surprising observation that, with rather few exceptions, the sequence pattern of the cysteine residues in disulfide-rich peptides is dictating the native disulfide framework and thus three-dimensional structure, independently of the overall sequence homology, represents an additional biochemical puzzle, so far unsolved (Chapters 6.1 and 6.3). However, it opened the way to explore the usefulness of such native disulfide frameworks with the related spatial structures as scaffolds in the rational design and production of miniproteins with targeted functions as biocatalysts or biopharmaceuticals (Chapter 7). While selenocysteine as the cysteine chalcogen analog in native proteins exerts mostly catalytic functions (Chapter 5), its highly reducing redox potential and thus the great stability of diselenides towards reducing agents offers new promising perspectives in replacing native cystine frameworks with the robust selenocystine scaffolds for the design of bioactive peptides and proteins (Chapters 5 and 8). Such an approach could well represent a valid alternative to the use of engineered disulfides for stabilization of tertiary folds of *de novo* designed proteins (Chapter 4).

The topics selected for this monograph will provide the reader with in-depth insight into the present knowledge associated with the folding of disulfide-bonded proteins and peptides, an aspect of biochemistry which, according to a recent editorial in *Science*,[4] still represents one of the biggest unsolved problems in today's science.

Johannes Buchner
Department of Chemistry
Technische Universität München
Lichtenbergstr. 4
85747 Garching
Germany

Luis Moroder
Max Planck Institute of Biochemistry
Am Klopferspitz 18
82152 Martinsried
Germany

[4] Editorial, *Science*, 2005, **309**, 78–102.

Contents

RSC Biomolecular Sciences
Oxidative Folding of Peptides and Proteins
Edited by Johannes Buchner and Luis Moroder
© Royal Society of Chemistry 2009
Published by the Royal Society of Chemistry, www.rsc.org

**Chapter 1.7   Eukaryotic Protein Disulfide-isomerases and their Potential
               in the Production of Disulfide-bonded Protein Products:
               What We Need to Know but Do Not!**
Robert B. Freedman

CHAPTER 1

Oxidative Folding of Proteins in vivo

CHAPTER 1.1

Thioredoxins and the Regulation of Redox Conditions in Prokaryotes

CARSTEN BERNDT AND ARNE HOLMGREN

The Medical Nobel Institute for Biochemistry, Department of Medical Biochemistry and Biophysics, Karolinska Institutet, SE-17177 Stockholm, Sweden

The cellular redox state is a crucial mediator of several aspects of life, *e.g.* growth and apoptosis. It is based on low-molecular-weight thiols such as glutathione (GSH) and protein thiols, providing a reducing or an oxidizing environment. The cytoplasm with its low redox potential favors the reduction of cysteinyl residues, whereas the prokaryotic periplasm supports disulfide-bond formation. In principle thiol-disulfide pairs in proteins have two possible functions. First, disulfides often can contribute to the overall structure and stability of the protein; second, the redox state of the cysteinyl residues can regulate the activity of the protein. The modifications of cysteine residues and thereby the redox state of the particular compartment are

RSC Biomolecular Sciences
Oxidative Folding of Peptides and Proteins
Edited by Johannes Buchner and Luis Moroder
© Royal Society of Chemistry 2009
Published by the Royal Society of Chemistry, www.rsc.org

controlled by thiol-disulfide oxidoreductases, which mainly belong to the thioredoxin family of proteins.

1.1.1   The Thioredoxin Family of Proteins

Escherichia coli thioredoxin 1 (Trx1), the first member of the thioredoxin family of proteins, was discovered more than 40 years ago as an electron donor for ribonucleotide reductase (RNR).[1,2] In all organisms, this enzyme is essential for DNA synthesis during both replication and repair.[3] The second member of the Trx family, glutaredoxin (Grx), was discovered as a GSH-dependent electron donor for RNR in an *E. coli* mutant lacking Trx.[4] Today, the Trx protein family is first and foremost defined by a structural motif named the Trx fold.[5] In spite of considerable variation in overall structure, the Trx fold is present in a variety of functionally different proteins:[6,7] thiol-disulfide oxidoreductases, disulfide isomerases, glutathione S-transferases,[8] thiol-dependent peroxidases[9] and chloride intracellular channels.[10]

1.1.1.1   The Thioredoxin Fold

The Trx fold motif consists of a central four-stranded β-sheet surrounded by three α-helices[11,12] (Figure 1.1.1A). This basic βαβαββα topology can only be found in bacterial glutaredoxins, while thioredoxins contain an additional β-sheet and α-helix at the N-terminus[13] (Figure 1.1.1B,C). Variations of this motif have also been identified in domains of 723 proteins,[14] *e.g.* the Trx fold domains of *E. coli* DsbA, a protein necessary for disulfide-bond formation in the periplasm[15] (Figure 1.1.1D).

Hallmarks of the Trx motif are a *cis*-proline residue located before β-sheet three and the Cys-X-X-Cys active site motif located on the loop connecting

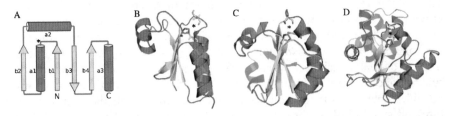

Figure 1.1.1   The Thioredoxin (Trx) fold. A) The topology of the basic Trx fold, which is present in bacterial Grxs (see B), consists of a four stranded β-sheet and three surrounding α-helices. Structures of Glutaredoxin 1 (B, PDB code: 1egr), Thioredoxin 1 (C, 1xob) and DsbA (D, 1a2l) were presented as Trx-fold oxidoreductases from *E. coli*. The secondary structural elements of the Trx fold are shown in yellow and red, additional elements are shown in magenta and blue. The *cis*-Pro residues are depicted in purple, the active site Cys residues in grey. In addition, the localization of the active site motif is marked by asterisks.

β-sheet one and α-helix one.[11] The nature and composition of these two amino acids dramatically affects the standard redox potential of the particular proteins. In *E. coli*, the strongest reductant, cytosolic Trx (Cys-Gly-Pro-Cys), has a redox potential of $\Delta E'_0 = -270$ mV,[16] the strongest oxidant, DsbA (Cys-Pro-His-Cys), has a redox potential of $\Delta E'_0 = -122$ mV.[17] Mutation of the Cys-Gly-Pro-Cys active site in Trx to the corresponding Cys-Pro-His-Cys active site of DsbA resulted in an increase of its standard midpoint potential to $\Delta E'_0 = -204$ mV.[18] Supportingly, the redox potential of a DsbA mutant harboring the Cys-Gly-Pro-Cys active site of Trx decreased by 92 mV.[19] Several other amino acids outside the active site motif have been determined as important for the redox potential of Trx fold oxidoreductases.[20-22]

1.1.1.2 Thioredoxins and the Thioredoxin System

In 1968 *E. coli* Trx1 was sequenced, revealing the characteristic Cys-Gly-Pro-Cys active site motif[2] and seven years later the Trx fold was described for the first time.[11] Since then, more than 200 structures of different Trxs were solved including structures of both oxidized and reduced Trxs. These structures are very similar, but reduction induces local conformational changes in the area of the active site (*e.g.* ref. 23). The two cysteinyl residues in the active site of Trx are utilized to reduce the protein disulfide formed during the catalytic cycle of RNR.[1,3] Today we know Trx as a general disulfide reductase[24] reducing disulfide bonds by a ping-pong mechanism[25] (Figure 1.1.2). The N-terminal active site thiol of Trxs possesses an unusual low pK_a value,[26] whereas the pK_a of the C-terminal active site is higher than that of cysteine in solution.[27] The low pK_a of the N-terminal cysteine of the *E. coli* Trx1 active site was shown to be related to the carboxyl group of Asp 26 and the ε-amino group of Lys 57.[27] Hence, the thiol group of the N-terminal active site cysteine is readily deprotonated under physiological conditions. Recently, it was shown by single molecule force-clamp spectroscopy that efficient catalysis requires a reorientation of the substrate disulfide bond.[28] This investigation demonstrated that the rate-limiting step of Trx activity is the orientation of the N-terminal active site cysteine of Trx and the two disulfide bridged cysteines of the substrate in a 180° angle. This reorientation provides the condition for the nucleophilic attack of the N-terminal Cys resulting in a covalent intermediate mixed disulfide between the Trx N-terminal active site and one of the substrate's cysteinyl side-chains.[26] The C-terminal active site cysteinyl side-chain reduces this disulfide yielding the reduced substrate and a disulfide in the active site of Trx. Subsequently, the disulfide in the active site of Trx is reduced by the dimeric flavo-enzyme thioredoxin reductase (TrxR) at the expense of NADPH (for a more detailed overview, see ref. 12).

E. coli contains two thioredoxins. Trx1, a protein with a molecular mass of 12 kD, and Trx2, a protein of 15.5 kDa, which contains an N-terminal domain of 32 amino acids including two additional Cys-X-X-Cys motifs. These four extra cysteines are able to coordinate zinc.[29]

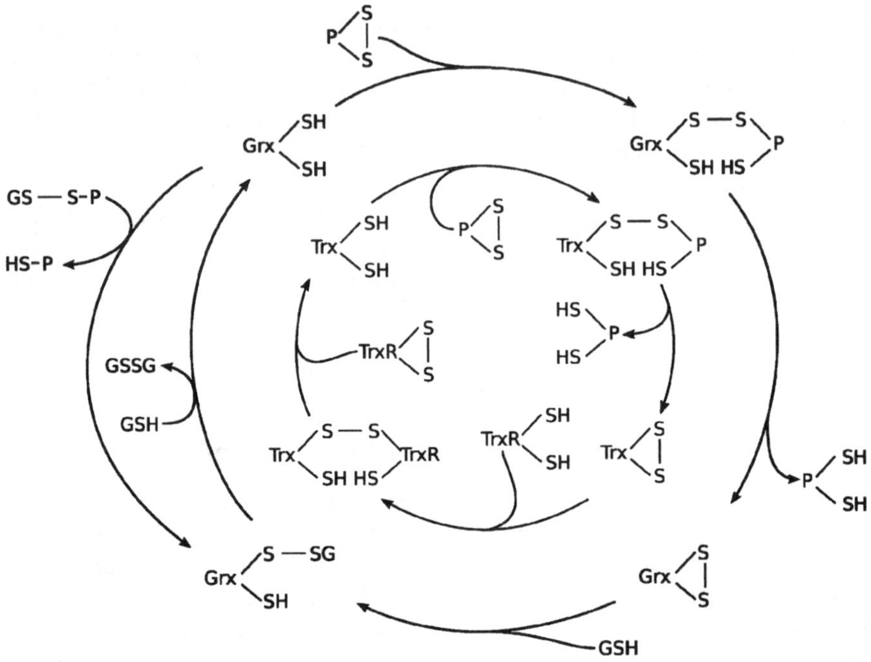

Figure 1.1.2   Reaction mechanisms and redox cycles of Thioredoxins and Glutaredoxins. Thioredoxin (Trx, inner circle) catalyzes the reversible reduction of protein disulfides (P-S-S) utilizing both active site cysteine
residues, reducing the target disulfide to form a covalent-mixed disulfide
intermediate, which in turn is reduced by the C-terminal active site
thiolate (-SH). In a similar manner Trx is afterwards reduced by Trx
reductase (TrxR). The reaction mechanism of Glutaredoxin (Grx, outer
circle) is similar to those of Trx. Grx is reduced by glutathione (GSH)
leading to a mixed disulfide, which is reduced by a second molecule of
GSH. Grx is also specifically able to reduce protein-GSH-mixed disulfides (P-S-SG).

1.1.1.3   Glutaredoxins and the Glutaredoxin System

Glutaredoxins (Grxs) exist in all glutathione (GSH)-containing life forms. As
described in Section 1.1.1.1 bacterial Grxs displayed the basic architecture of
the Trx fold. Similar to Trxs, the structural comparison of reduced[30] and
oxidized[31] *E. coli* Grx1 revealed more or less identical overall structures but
significant changes in and around the active site.[32]

Based on their active site motifs, Grxs can be divided into two major categories: the dithiol Grxs (active site: Cys-Pro-Tyr-Cys) and the monothiol Grxs
(active site: Cys-Gly-Phe-Ser). Dithiol Grxs are general thiol-disulfide oxidoreductases reducing some protein disulfides like that in *E. coli* ribonucleotide
reductase with a dithiol mechanism as described for Trxs above (Figure 1.1.2).
In addition, Grxs are able to reduce protein-GSH mixed disulfides

(de-glutathionylation) utilizing a mechanism that requires only the N-terminal active site residue (monothiol mechanism; for a more detailed overview, see ref. 33). Grxs use GSH as electron donor. The resulting glutathione disulfide (GSSG) is reduced by glutathione reductase (GR) with electrons from NADPH[34,35] (Figure 1.1.2). The molecular mechanism and the functions of monothiol glutaredoxins, which are prevalent inactive in (dithiol) Grx-specific activity assays, are only beginning to emerge (for a recent review see ref. 36).

E. coli contains the three dithiol Grxs 1–3,[37,38] and the monothiol Grx4.[39] Grxs 1, 3 and 4 are single domain proteins with molecular masses of 10, 9 and 13 kDa, respectively, whereas Grx2 is a larger protein of 24 kDa. Only the N-terminal part of Grx2 is a Trx-fold domain; the overall structure resembles that of GSH-S-transferases.[40] Recently, several monothiol Grxs including *E. coli* Grx4 were described as [FeS] proteins.[41] The cofactors are coordinated by the N-terminal active site cysteine of two Grx monomers and two non-covalently bound molecules of GSH as described before for human Grx2 and poplar GrxC1.[42–44] So far, *E. coli* Grx4 is the only essential member of the Trx family of proteins in *E. coli*.[39]

1.1.1.4 NrdH and Other Related Proteins

E. coli NrdH (9 kDa) shows similarities to Grxs in its secondary and tertiary structure.[45] However, it lacks the GSH binding site and is therefore not reduced by GSH. *In vitro* and *in vivo* studies identified TrxR as an electron donor of NrdH.[45,46] NrdH was shown to act as electron donor for RNRs.[45,47]

YbbN/Trxsc is a protein of 31 kDa expressed in the cytoplasm of *E. coli*.[48] Its N-terminus is homologous to Trxs, but without a Cys-X-X-Cys motif.

Peroxiredoxins (reviewed in ref. 49), a ubiquitous family of thiol-dependent proteins reducing peroxides like hydrogen peroxide or peroxinitrite, contain Trx-fold domains. In *E. coli*, four of these peroxiredoxins are present. The primary scavenger of endogenous hydrogen peroxide is alkyl hydroperoxide peroxidase C (AhpC),[50] which is reduced by AhpF. Thiol peroxidase (Tpx), the gene product of *btuE*, a GSH peroxidase homolog, and bacterioferritin comigratory protein (BCP) are reduced by Trx1, or at least are Trx dependent.[51,52]

GSH-S-transferases, as described for *E. coli* Grx2, contain a Trx domain. The *E. coli* genome encodes for eight of these proteins transferring GSH to electrophilic compounds and thereby detoxifying them.[53]

In the periplasm, oxidative folding takes place performed by proteins of the Dsb family (reviewed in ref. 54 and Chapter 1.2). DsbA introduces disulfides into substrates, and the protein disulfide isomerases DsbC and DsbG reduce incorrect formed disulfide bonds. All these proteins are Trx-fold oxido-reductases.[15,55,56] The eukaryotic counterparts to *E. coli* Dsbc/DsbG, for instance human and yeast PDIs (protein disulfide isomerases), are Trx-fold proteins as well.[57,58]

1.1.2   Functions of Thioredoxin and Glutaredoxin

As in the chapters above we will focus on *E. coli* as the best investigated pro-karyotic organism. Here, numerous functions have been described for Trxs and Grxs, both as electron donors as well as regulators of cellular function in response to oxidative stress (Figure 1.1.3).

Within the framework of this chapter, it is important to note that a number of functions described for these proteins are independent from their oxido-reductase activity, for instance *E. coli* Trx as subunit of the T7 DNA polymerase complex,[59] or its activity as molecular (co-)chaperone (ref. 60, see 1.1.3.).

1.1.2.1   Regulation of Redox Conditions

Trxs and Grxs keep a reduced environment in the cytoplasm by reducing protein disulfides and protein-GSH mixed disulfides.

By modulating the redox state of protein thiols Trxs and Grxs can function as regulators of transcription factors. OxyR is such a factor activating tran-scription of several *E. coli* genes encoding proteins defending against hydrogen peroxide induced oxidative stress, *e.g.* catalase 1, AhpCF, GSH reductase, Grx1 and Trx2.[61] The formed disulfide, which activates OxyR, is most likely *in vivo* reduced by Grx1.[61] The activity of the transcription factor SoxR is regulated by a [2Fe2S] cluster. If the cluster is oxidized, the transcription of proteins involved in defense against oxidative stress is activated.[62] *In vitro*, the reconstitution of the [FeS] cluster of SoxR is promoted by the Trx system. In

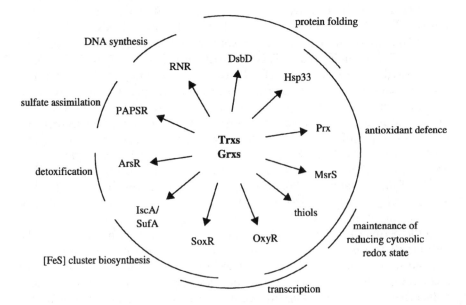

Figure 1.1.3   Substrates of Thioredoxins (Trxs) and Glutaredoxins (Grxs) in *E. coli* (for details see text).

addition, SoxR inactivation is inhibited in *E. coli* mutants lacking TrxR and GR.[63]

Trx and Grx have also been described as electron donors for antioxidant enzymes, for instance, peroxiredoxins as described in Section 1.1.1.4.

Methionine sulfoxides that may form during oxidative stress by reactive oxygen species are reduced by methionine sulfoxide reductases (Msrs).[64] In *E. coli* six Msrs are present,[65] which are most likely using Trx1 as the electron donor *in vivo*.[66] Since increased expression of Grx1 and Trx2 in an *E. coli* mutant lacking Trx1 allows growth using methionine sulfoxide as the sole source of methionine, these two oxidoreductants are also potential electron donors for Msrs *in vivo*.[66]

1.1.2.2   Regulation of Metabolic Enzymes

Trx and Grx were first described as electron donors for RNR.[1,4,67] RNRs catalyze the reduction of nucleotides to deoxynucleotides and are thus essential for DNA synthesis.[68,69] *E. coli* cells are equipped with three RNR enzymes. While RNR1a and RNRIII are essential for aerobic and anaerobic metabolism, respectively, RNR1b is not essential.[69,70] The rate-limiting step during enzymatic turn-over of ribonucleotides by RNR1a is the reduction of a disulfide performed by Trxs or Grxs.[1,4,37,71,72] In *E. coli*, Grx1 may be the primary electron donor for RNR1a *in vivo*.[73,74] Trx1 may be involved in activation of RNRIII[75] *via* reduction of a disulfide.[76]

Bacteria are able to satisfy their need for reduced sulfur by assimilation of inorganic sulfate. Reduction of sulfate (SO_4^{2-}) to sulfide (S^{2-}) requires eight electrons and takes place in two steps. First, sulfate is activated to adenylylsulfate (APS) by ATP sulfurylase and then to phoshoadenylylsulfate (PAPS) by APS kinase and subsequently reduced to sulfite (SO_3^{2-}) by PAPS reductase. Secondly, sulfite is reduced to sulfide using six electrons provided by NADPH (reviewed in ref. 139). Trx1 and Grx1 were identified as electron donors for APS kinase[77] and PAPS reductase in *E. coli*.[78–81]

Arsenate reductase (ArsC) detoxifies arsenate to arsenite with electrons provided by *E. coli* Grxs 1, 2 and 3.[82] The substrate for Grxs is a glutathionylated arsenate intermediate.

[FeS] clusters, found in all life forms, can undergo reversible redox reactions, determine protein structure, act as catalytic centers and as sensitive sensors of iron and various oxygen species.[83,84] The biosynthesis of these cofactors is therefore essential for catalytic function of several enzymes. During biosynthesis sulfide and iron are delivered to a scaffold protein, which coordinates the newly synthesized [FeS] cluster before transferring it to target proteins. In *E. coli*, two independent systems for [FeS] cluster biosynthesis are present: the Isc (iron sulfur cluster) and the Suf (mobilization of sulfur) systems.[85,86] The Trx system was shown to mediate iron binding of IscA[87] and SufA,[88] proteins that have been described as alternative scaffold proteins,[89] or as potential iron donors for the formation of [FeS] clusters in the scaffold IscU.[90,91] Monothiol

Grxs including *E. coli* Grx4 are crucially involved in iron–sulfur cluster biosynthesis and regulation of iron homeostasis.[92–94]

1.1.3   Thioredoxins, Glutaredoxins and Protein Folding

Trxs are by far the most well known members of the Trx family. The investigations on these proteins, as well as Grxs, provided general insights in functions and mechanisms of all members of the Trx family. Even though most of these initial general experiments were done *in vitro*, they provided important concepts for the *in vivo* situation. The different ways in which Trxs and Grxs can participate in prokaryotic but also eukaryotic protein folding are the subject of this chapter.

1.1.3.1   Regulation of Protein Folding *via* Electrons Provided by Thioredoxins and Glutaredoxins

Trxs and Grxs has been described as regulators of many proteins involved in folding *via* their redox activities.[95,96]

In prokaryotes, Trx and Grx catalyze the reduction of protein disulfides in *E. coli* Hsp33.[97] This reduction led to the formation of Zn coordinating inactive monomers. Under conditions of oxidative stress, Hsp33 forms dimers protecting unfolded proteins against aggregation (see also Chapter 1.8).

Yeast Ssp2[98] and human Hdj2,[99] two redox-regulated co-chaperones, are reduced by Trxs.

As described in Section 1.1.1.4 peroxiredoxins interact with the Trx system. Some eukaryotic peroxiredoxins were characterized as chaperones. H_2O_2-induced chaperone activity of yeast and human Prxs 2 requires the active site cysteinyl side-chains and the Trx system as cofactor.[100,101] In concert with yeast Prx2, the Trx system protects ribosomes against stress-induced aggregation.[102]

Electrons provided by Trx1 are also essential for the correct oxidative folding of proteins in the *E. coli* periplasm (Figure 1.1.4, black), where the two protein disulfide isomerases DsbC and DsbG reduce incorrect formed disulfide bonds for further isomerization.[54] A constant electron supply is guaranteed by the membrane protein DsbD, which itself is reduced by cytoplasmic Trx.[103]

Several proteomic approaches identified molecular chaperones as new targets of the Trx and Grx systems: human Trx1 was identified as interaction partner of 14-3-3 ζ proteins during interphase and mitosis in HeLa cells.[104] Synergistically with α-crystallin, the human Trx system can recover inactivated GR in human aged clear and cataract lenses.[105] Cyclophilin, a peptidyl-prolyl *cis-trans* isomerase essential for protein folding[106,107] is activated *via* reduction by spinach Trx-m.[108] Cyclophilin was also found in investigations aiming at the identification of new targets of *A. thaliana* cytosolic Trx-h3 and spinach plastidic Trx-m.[109,110] Using poplar Grx-C4 as bait Rouhier *et al.* identified a 14-3-3 protein (At5g6543), Hsp60, Hsp70 and cyclophilin as interacting proteins in extracts of different plants.[111] The following human chaperones were identified as targets

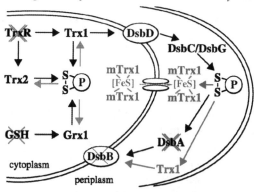

Figure 1.1.4　Thioredoxin (Trx) and Glutaredoxin (Grx) as reductants and oxidants in a prokaryotic cell. Under physiological conditions (black), the *E. coli* cytoplasm is maintained by the action of Thioredoxin (Trx) and Glutaredoxin (Grx) in a reduced state, whereas the periplasm provides a more oxidizing environment allowing the formation and isomeration of protein disulfides (P-S-S) by DsbA and DsbC/DsbG, respectively. Trxs1 and 2 and Grx1 are reduced by Thioredoxin reductase (TrxR) or GSH (arrows indicate electron flow). In *E. coli* mutants lacking TrxR and/or GSH (*via* depletion of GSH reductase, γ-glutamylcysteine synthase or GSH synthetase) oxidized Trxs and Grx1 form disulfides in the cytoplasm (red). Trx1, when exported to the periplasm, is able to complement *E. coli* cells deficient in DsbA (green). A mutated Trx1, which coordinates a [FeS] cluster in cytoplasm if TrxR is absent, can complement an *E. coli* strain lacking both DsbA and DsbB, the electron acceptor of DsbA (blue).

for reversible glutathionylation: Hsp10,[112] Hsp60,[113,114] Hsp70,[112–115] Hsp90,[113] Hsc70,[113,115] 14-3-3 proteins,[113] cyclophilin A,[112,114,115] and Cox17.[116] As mentioned in Section 1.1.1.3 Grxs are highly specific for protein-GSH mixed disulfides, which suggests that all the above-listed chaperone-GSH mixed disulfides are likely targets for Grxs *in vivo*. Indeed, it was shown that glutathionylated human Hsc70 is a substrate for Grx1, and that reduction of the mixed disulfide decreased chaperone activity.[117]

1.1.3.2　Thioredoxins and Glutaredoxins Acting as Protein Disulfide Isomerases or Molecular Chaperones

Trxs and Grxs are not only regulators of proteins involved in folding, they are also able to act directly in oxidative folding or as chaperones.

As described above (Section 1.1.1.2), a recent paper dealing with the characterization of the reaction mechanism of *E. coli* Trx1 using single molecule force-clamp spectroscopy suggests that a local refolding of the substrate is part of the catalytic activity.[28] Hence, chaperone activity is coupled to rate enhancements in the order of four to five orders of magnitude compared to small reductants like dithiothreitol in reduction of protein disulfides.

E. coli Trx1 fusion was shown to increase the levels of soluble proteins heterologously expressed in *E. coli*, for instance numerous mammalian proteins,[118–120] *Clostridium tetani* fragment C of tetanus toxin[121] or the single-chain variable fragment of antibodies.[122] Several plasmids were constructed to express Trx fusion proteins to avoid inclusion body formation during recombinant expression of proteins in *E. coli*.[118,119] The induction of correct folding was independent of Trx's redox activity since fusion with a Trx mutant harboring an Ala-Gly-Pro-Ala active site was as efficient as the wild type Trx fusion partner.[122] The fused Trx, covalently linked to the protein of interest, may act as a molecular chaperone preventing precipitation and aggregation of the fused partners until these reach a stable folding state.[118,119]

Direct activity as molecular chaperone was demonstrated *in vitro* by Richarme and coworkers. *E. coli* Trx1 and the Trx homolog YbbN/Trxsc were able to refold citrate synthase and α-glucosidase with an efficiency comparable to those of chaperones like DnaK and different heat shock proteins.[48,60] As observed for the chaperone activity of fused Trx neither the redox state of the Trxs nor the active site cysteines or other amino acids important for redox function are required for chaperone activity measured *in vitro*.[60,123] Corroboratively, Trx1 can stimulate the refolding of MglB, a protein without cysteines.[60] Unlike molecular chaperones, Trx and YbbN/Trxsc do not preferentially bind unfolded proteins and do not protect citrate synthase against thermal degradation.[48,60]

It is not only *E. coli* proteins that have been described as chaperones since also Trx1, but not Trx2 from *Helicobacter pylori*, promoted the renaturation of arginase.[123]

Eukaryotic mitochondrial monothiol Grx5 and several other monothiol Grxs including *E. coli* Grx4 have been implied in iron–sulfur cluster biosynthesis.[92–94] Depletion of Grx5 led to increased iron levels and decreased enzymatic function of iron-sulfur proteins such as aconitase and succinate dehydrogenase.[92] The transfer of [FeS] clusters from scaffold to target protein (see Section 1.1.2.2.) is dependent on a functional HscA/HscB chaperone system.[85,124] Grx5 deletion in yeast resulted in accumulation of iron–sulfur clusters on the scaffold protein.[93] This phenotype can be rescued by over-expression of the HscA-type chaperone Ssq1[92] indicating that Grx5 may act in concert with the HscA/Ssq1 HscB/Jac1 chaperone couple in [FeS] cluster biosynthesis.

Protein disulfide isomerases catalyze the isomerization of disulfide-bond formation in the oxidative environments of the prokaryotic periplasm and the eukaryotic endoplasmatic reticulum (overviews in refs. 54 and 125).

Trx1, when exported to the periplasm, is able to complement *E. coli* strains deficient in the periplasmatic thiol oxidase DsbA at concentrations that allow efficient re-oxidation of Trx by DsbB[126] (Figure 1.1.4, green). Mutated Trx1 variants mimicking active sites of other Trx-fold oxidoreductases resulting in a higher redox potentials (see Section 1.1.1.1) were more efficient in complementation.[126,127] Beyond that, some Trx1 mutations were able to perform disulfide-bond formation pathway independent of DsbA and DsbB.[128] The active site mutants of *E. coli* Trx1 were created by random mutagenesis and the most efficient mutants contained the active sites Cys-Ala-Cys-Cys and Cys-Ala-Cys-Ala. These Trx mutants coordinated an [2Fe2S] cluster that was essential for the catalysis of oxidative protein folding (Figure 1.1.4, blue).

As described in Section 1.1.2.1. Trxs and Grxs usually maintain a reducing environment in the cytosol by reduction of disulfide bonds. But when the Trx system is impaired, for instance by inhibition of TrxR, disulfide-bond formation can occur in the cytoplasm of *E. coli*.[129] This effect is enhanced if, in addition to the Trx system, the Grx system is disturbed as well.[130] It was shown that the resulting disulfide bridge formations in the cytoplasm are attributed not only to the lack of reductase activity of Trxs 1 and 2 and Grx1, but also to the accumulation of oxidized redoxins that serve as oxidants and catalysts of disulfide-bond formation[66] (Figure 1.1.4, red).

In vitro both *E. coli* Trx1 and Grx1 are able to refold ribonuclease A (RNase A). Oxidized and a mixture of oxidized and reduced Trx efficiently catalyzes the refolding of reduced, denatured RNase, or oxidized, scrambled RNase, respectively.[131] As shown by kinetic analyses Grx1 only needs the N-terminal active site cysteine for RNase A refolding.[132] The further investigation of the mechanism showed that Grx1 acted by the monothiol mechanism (Figure 1.1.2) and that ribonuclease-GSH mixed disulfide was the substrate.[133] The investigation of an *in vitro* model of the mechanism by which PDI catalyzes formation of disulfide bonds in the presence of GSSG indicated that PDI primarily catalyzes formation and breakage of GSH-protein mixed disulfides.[134] *E. coli* Grx1 displayed synergistic activity together with PDI when the redox potential of the GSH redox buffer was low enough to reduce the active site disulfide in Grx,[133] but was also able to refold RNase A without PDI.[135,136] Therefore, GSH mixed disulfides were believed to be important folding intermediates *in vivo*. Today it seems that in eukaryotic systems GSH is needed as a redox balancing system, providing reducing equivalents for the reduction of PDIs, and for the protection against reactive oxygen species, which are formed during oxidation of PDIs *via* Ero1 using oxygen as the ultimate electron acceptor (reviewed in ref. 137). In *E. coli* GSH is present in the periplasm but not essential for disulfide-bond formation or isomeration.[138]

1.1.4 Concluding Remarks

Over the last 40 years research has shown the essential roles of thiol-disulfide oxidoreductases of the Trx family of proteins in regulating the cellular redox conditions. In prokaryotes these proteins are crucial for both reduction of protein thiols in cytoplasm and oxidation of protein thiols in periplasm. The numerous interaction partners of Trx-fold proteins identified so far indicate the potential of this structural motif as protein binding scaffold promising more interaction partners detected by future research. Protein interaction is of course a condition for donating or accepting electrons but also for folding activity.

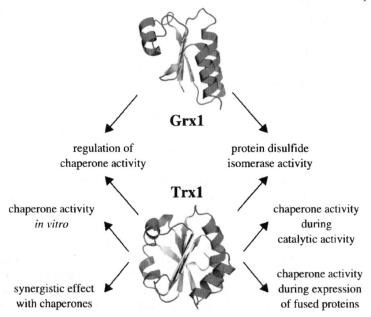

Figure 1.1.5 Summary of functions of Thioredoxin (Trx) and Glutaredoxin (Grx) in protein folding (for details see text).

Therefore it is not surprising that – in addition to regulation of different proteins involved in folding – Trxs and Grxs possess protein folding activity (Figure 1.1.5). *In vivo* an essential chaperone activity of Trx and Grx for the correct folding of a specific substrate is not yet described, but the *in vitro* investigations prepared the ground for the insights of the enzymatic mechanisms of proteins like DsbA or PDI.

Acknowledgments

This work was supported by the Deutsche Forschungsgemeinschaft, Karolinska Institutet, the Swedish Cancer Society, the Swedish Research Council Medicine and the Knut and Alice Wallenberg Foundation.

References

1. T. C. Laurent, E. C. Moore and P. Reichard, *J. Biol. Chem.*, 1964, **239**, 3436–3444.
2. A. Holmgren, *Eur. J. Biochem.*, 1968, **6**, 475–484.
3. P. Nordlund and P. Reichard, *Annu Rev. Biochem.*, 2006, **75**, 681–706.
4. A. Holmgren, *Proc. Natl. Acad. Sci. USA*, 1976, **73**, 2275–2279.
5. A. Holmgren, *Annu. Rev. Biochem.*, 1985, **54**, 237–271.
6. J. L. Pan and J. C. A. Bardwell, *Protein Sci.*, 2006, **15**, 2217–2227.

7. J. L. Martin, *Structure*, 1995, **3**, 245–250.
8. M. Nishida, S. Harada, S. Noguchi, Y. Satow, H. Inoue and K. Takahashi, *J. Mol. Biol.*, 1998, **281**, 135–147.
9. H. J. Choi, S. W. Kang, C. H. Yang, S. G. Rhee and S. E. Ryu, *Nat. Struct. Biol.*, 1998, **5**, 400–406.
10. R. H. Ashley, *Mol. Membr. Biol.*, 2003, **20**, 1–11.
11. A. Holmgren, B. O. Söderberg, H. Eklund and C. I. Brändén, *Proc. Natl. Acad. Sci. USA*, 1975, **72**, 2305–2309.
12. A. Holmgren, *Structure*, 1995, **15**, 15.
13. H. Eklund, C. Cambillau, B. M. Sjöberg, A. Holmgren, H. Jörnvall, J. O. Höög and C. I. Brändén, *EMBO J.*, 1984, **3**, 1443–1449.
14. N. V. G. Yuan Qi, *Proteins*, 2005, **58**, 376–388.
15. J. L. Martin, J. C. A. Bardwell and J. Kuriyan, *Nature*, 1993, **365**, 464–468.
16. G. Krause, J. Lundström, J. L. Barea, C. Pueyo de la Costa and A. Holmgren, *J. Biol. Chem.*, 1991, **266**, 9494–9500.
17. M. Wunderlich and R. Glockshuber, *Protein Sci.*, 1993, **2**, 717–726.
18. E. Mössner, M. Huber-Wunderlich and R. Glockshuber, *Protein Sci.*, 1998, **7**, 1233–1244.
19. M. Huber-Wunderlich and R. Glockshuber, *Fold. Des.*, 1998, **3**, 161–171.
20. P. J. Gane, R. B. Freedman and J. Warwicker, *J. Mol. Biol.*, 1995, **249**, 376–387.
21. J. Blank, T. Kupke, E. Lowe, P. Barth, R. B. Freedman and L. W. Ruddock, *Antioxid. Redox Signal.*, 2003, **5**, 359–366.
22. A. Porat, C. H. Lillig, C. Johansson, A. P. Fernandes, L. Nilsson, A. Holmgren and J. Beckwith, *Biochemistry*, 2007, **46**, 3366–3377.
23. M. F. Jeng, A. P. Campbell, T. Begley, A. Holmgren, D. A. Case, P. E. Wright and H. J. Dyson, *Structure*, 1994, **2**, 853–868.
24. A. Holmgren, *J. Biol. Chem.*, 1979, **254**, 9113–9119.
25. A. Holmgren, *J. Biol. Chem.*, 1979, **254**, 9627–9632.
26. G. B. Kallis and A. Holmgren, *J. Biol. Chem.*, 1980, **255**, 10261–10265.
27. H. J. Dyson, M. F. Jeng, L. L. Tennant, I. Slaby, M. Lindell, D. S. Cui, S. Kuprin and A. Holmgren, *Biochemistry*, 1997, **36**, 2622–2636.
28. A. P. Wiita, R. Perez-Jimenez, K. A. Walther, F. Gräter, B. J. Berne, A. Holmgren, J. M. Sanchez-Ruiz and J. M. Fernandez, *Nature*, 2007, **450**, 124–127.
29. J. F. Collet, J. C. D'Souza, U. Jakob and J. C. A. Bardwell, *J. Biol.Chem.*, 2003, **278**, 45325–45332.
30. P. Sodano, T. H. Xia, J. H. Bushweller, O. Björnberg, A. Holmgren, M. Billeter and K. Wüthrich, *J. Mol. Biol.*, 1991, **221**, 1311–1324.
31. T. H. Xia, J. H. Bushweller, P. Sodano, M. Billeter, O. Björnberg, A. Holmgren and K. Wüthrich, *Protein Sci.*, 1992, **1**, 310–321.
32. J. J. Kelley, T. M. Caputo, S. F. Eaton, T. M. Laue and J. H. Bushweller, *Biochemistry*, 1997, **36**, 5029–5044.
33. C. Berndt, C. H. Lillig and A. Holmgren, *Am. J. Physiol. Heart Circ. Physiol.*, 2007, **292**, 1227–1236.

34. A. Holmgren, *J. Biol. Chem.*, 1989, **264**, 13963–13966.
35. A. Holmgren, C. Johansson, C. Berndt, M. E. Lönn, C. Hudemann and C. H. Lillig, *Biochem. Soc. Trans.*, 2005, **33**, 1375–1377.
36. E. Herreo and M. A. de la Torre-Ruiz, *Cell Mol. Life Sci.*, 2007, **64**, 1518–1530.
37. A. Holmgren, *J. Biol. Chem.*, 1979, **254**, 3664–3671.
38. F. Aslund, B. Ehn, A. Miranda-Vizuete, C. Pueyo and A. Holmgren, *Proc. Natl. Acad. Sci. USA*, 1994, **91**, 9813–9817.
39. A. P. Fernandes, M. Fladvad, C. Berndt, C. Andresen, C. H. Lillig, P. Neubauer, M. Sunnerhagen, A. Holmgren and A. Vlamis-Gardikas, *J. Biol. Chem.*, 2005, **280**, 24544–24552.
40. B. Xia, A. Vlamis-Gardikas, A. Holmgren, P. H. Wright and H. J. Dyson, *J. Mol. Biol.*, 2001, **310**, 907–918.
41. A. Picciocchi, C. Saguez, A. Boussac, C. Cassier-Chauvat and F. Chauvat, *Biochemistry*, 2007, **46**, 15018–15026.
42. C. H. Lillig, C. Berndt, O. Vergnolle, M. E. Lönn, C. Hudemann, E. Bill and A. Holmgren, *Proc. Natl. Acad. Sci. USA*, 2005, **102**, 8168–8173.
43. C. Berndt, C. Hudemann, E. M. Hanschmann, R. Axelsson, A. Holmgren and C. H. Lillig, *Antioxid. Redox Signal.*, 2007, **9**, 151–157.
44. N. Rouhier, H. Unno, S. Bandyopadhyay, L. Masip, S. K. Kim, M. Hirasawa, J. M. Gualberto, V. Lattard, M. Kusunoki, D. B. Knaff, G. Georgiou, T. Hase, M. K. Johnson and J. P. Jacquot, *Proc. Natl. Acad. Sci. USA*, 2007, **104**, 7379–7378.
45. M. Stehr, G. Schneider, F. Aslund, A. Holmgren and Y. Lindqvist, *J. Biol. Chem.*, 2001, **276**, 35836–35841.
46. S. Gon, M. J. Faulkner and J. Beckwith, *Antioxid. Redox Signal.*, 2006, **8**, 735–742.
47. A. Jordan, F. Aslund, E. Pontis, P. Reichard and A. Holmgren, *J. Biol. Chem.*, 1997, **272**, 18044–18050.
48. T. Caldas, A. Malki, R. Kern, J. Abdallah and G. Richarme, *Biochem. Biophys. Res. Comm.*, 2006, **343**, 780–786.
49. Z. A. Wood, E. Schröder, J. Robin Harris and L. B. Poole, *Trends Biochem. Sci.*, 2003, **28**, 32–40.
50. L. C. Seaver and J. A. Imlay, *J. Bacteriol.*, 2001, **183**, 7173–7181.
51. L. M. S. Baker and L. B. Poole, *J. Biol. Chem.*, 2003, **278**, 9203–9211.
52. W. Jeong, M. K. Cha and I. H. Kim, *J. Biol. Chem.*, 2000, **275**, 2924–2930.
53. J. F. P. Chris, L. Rife, G. Xiao, G. L. Gilliland and R. N. Armstrong, *Proteins*, 2003, **53**, 777–782.
54. J.-F. Collet and J. C. A. Bardwell, *Mol. Microbiol.*, 2002, **44**, 1–8.
55. A. A. McCarthy, P. W. Haebel, A. Törrönen, V. Rybin, E. N. Baker and P. Metcalf, *Nat. Struct. Biol.*, 2000, **7**, 196–199.
56. B. Heras, M. A. Edeling, H. J. Schirra, S. Raina and J. L. Martin, *Proc. Natl. Acad. Sci. USA*, 2004, **101**, 8876–8881.
57. J. Kemmink, N. J. Darby, K. Dijkstra, M. Nilges and T. E. Creighton, *Biochemistry*, 1996, **35**, 7684–7691.

58. G. Tian, S. Xiang, R. Noiva, W. J. Lennarz and H. Schindelin, *Cell*, 2006, **124**, 1085–1088.
59. A. Holmgren, I. Ohlsson and M.-L. Grankvist, *J. Biol. Chem.*, 1978, **253**, 430–436.
60. R. Kern, A. Malki, A. Holmgren and G. Richarme, *Biochem. J.*, 2003, **371**, 965–972.
61. F. Aslund, M. Zheng, J. Beckwith and G. Storz, *Proc. Natl. Acad. Sci. USA*, 1999, **96**, 6161–6165.
62. M. Chander and B. Demple, *J. Biol. Chem.*, 2004, **279**, 41603–41610.
63. H. Ding and B. Demple, *Biochemistry*, 1998, **37**, 17280–17286.
64. N. Brot and H. Weissbach, *Arch. Biochem. Biophys.*, 1983, **223**, 271–281.
65. D. Spector, F. Etienne, N. Brot and H. Weissbach, *Biochem. Biophys. Res. Commun.*, 2003, **302**, 284–289.
66. E. J. Stewart, F. Aslund and J. Beckwith, *EMBO J.*, 1998, **17**, 5543–5550.
67. E. C. Moore, P. Reichard and L. Thelander, *J. Biol. Chem.*, 1964, **239**, 3445–3452.
68. L. Thelander and P. Reichard, *Annu. Rev. Biochem.*, 1979, **48**, 133–158.
69. A. Jordan and P. Reichard, *Annu. Rev. Biochem.*, 1998, **67**, 71–98.
70. R. Ortenberg, S. Gon, A. Porat and J. Beckwith, *Proc. Natl. Acad. Sci. USA*, 2004, **101**, 7439–7444.
71. A. Holmgren, *J. Biol. Chem.*, 1979, **327**, 3672–3678.
72. F. Aslund, K. D. Berndt and A. Holmgren, *J. Biol. Chem.*, 1997, **272**, 30780–30786.
73. A. Holmgren, *J. Biol. Chem.*, 1979, **327**, 3672–3678.
74. A. Potamitou, A. Holmgren and A. Vlamis-Gardikas, *J. Biol. Chem.*, 2002, **277**, 18561–18567.
75. D. Padovani, E. Mulliez and M. Fontecave, *J. Biol. Chem.*, 2001, **276**, 9587–9589.
76. J. Andersson, M. Westman, M. Sahlin and B. M. Sjöberg, *J. Biol. Chem.*, 1997, **275**, 19449–19455.
77. U. Schriek and J. D. Schwenn, *Arch. Microbiol.*, 1986, **145**, 32–38.
78. M. L. Tsang, *J. Bacteriol.*, 1981, **146**, 1059–1066.
79. J. D. Schwenn and U. Schriek, *Z. Naturforsch.*, 1987, **42**, 93–102.
80. M. Russel, P. Model and A. Holmgren, *J. Bacteriol.*, 1990, **172**, 1923–1929.
81. C. H. Lillig, A. Prior, J. D. Schwenn, F. Aslund, D. Ritz, A. Vlamis-Gardikas and A. Holmgren, *J. Biol. Chem.*, 1999, **274**, 7695–7698.
82. J. Shi, A. Vlamis-Gardikas, F. Aslund, A. Holmgren and B. P. Rosen, *J. Biol. Chem.*, 1999, **274**, 36039–36042.
83. H. Beinert, R. H. Holm and E. Münck, *Science*, 1997, **277**, 653–659.
84. H. Beinert, *J. Biol. Inorg. Chem.*, 2000, **5**, 2–15.
85. F. Barras, L. Loiseau and B. Py, *Adv. Microb. Physiol.*, 2005, **50**, 41–101.
86. C. Ayala-Castro, A. Saini and F. W. Outten, *Microbiol. Mol. Biol. Rev.*, 2008, **72**, 110–125.
87. H. Ding, J. Yang, L. C. Coleman and S. Yeung, *J. Biol. Chem.*, 2007, **282**, 7997–8004.
88. J. Lu, J. Yang, G. Tan and H. Ding, *Biochem. J.*, 2008, **409**, 535–543.

89. C. Krebs, J. N. Agar, A. D. Smith, J. Frazzon, D. R. Dean, B. H. Huynh and M. K. Johnson, *Biochemistry*, 2001, **40**, 14069–14080.
90. H. Ding, R. J. Clark and B. Ding, *J. Biol. Chem.*, 2004, **279**, 37499–37504.
91. J. Yang, J. P. Bitoun and H. Ding, *J. Biol. Chem.*, 2006, **281**, 27956–27963.
92. M. T. Rodriguez-Manzaneque, J. Tamarit, G. Belli, J. Ros and E. Herrero, *Mol. Biol. Cell*, 2002, **13**, 1109–1121.
93. U. Mühlenhoff, J. Gerber, N. Richhardt and R. Lill, *EMBO J.*, 2003, **22**, 4815–4825.
94. M. M. Molina-Navarro, C. Casas, L. Piedrafita, G. Belli and E. Herrero, *FEBS Lett.*, 2006, **580**, 2273–2280.
95. U. Jakob, W. Muse, M. Eser and J. C. A. Bardwell, *Cell*, 1999, **96**, 341–352.
96. R. C. Cumming, N. L. Andon, P. A. Haynes, M. Park, W. H. Fischer and D. Schubert, *J. Biol. Chem.*, 2004, **279**, 21749–21758.
97. J. H. Hoffmann, K. Linke, P. C. F. Graf, H. Lilie and U. Jacob, *EMBO J.*, 2004, **23**, 160–168.
98. F. Vignols, N. Mouaheb, D. Thomas and Y. Meyer, *J. Biol. Chem.*, 2003, **282**, 4516–4523.
99. H.-I. Choi, S. P. Lee, K. S. Kim, C. Y. Hwang, Y.-R. Lee, S.-K. Chae, Y.-S. Kim, H. Z. Chae and K.-S. Kwon, *Free Rad. Biol. Med.*, 2006, **40**, 651–659.
100. H. H. Jang, K. O. Lee, Y. H. Chi, B. G. Jung, S. K. Park, J. H. Park, J. R. Lee, S. S. Lee, J. C. Moon, J. W. Yun, Y. O. Choi, W. Y. Kim, J. S. Kang, G.-W. Cheong, D.-J. Yun, S. G. Rhee, M. J. Cho and S. Y. Lee, *Cell*, 2004, **117**, 625–635.
101. J. C. Moon, Y.-S. Hah, W. Y. Kim, B. G. Jung, H. H. Jang, J. R. Lee, S. Y. Kim, Y. M. Lee, M. G. Jeon, C. W. Kim, M. J. Cho and S. Y. Lee, *J. Biol. Chem.*, 2005, **280**, 28775–28784.
102. J. D. Rand and C. M. Grant, *Mol. Biol. Cell*, 2006, **17**, 387–401.
103. A. Rietsch, P. Bessette, G. Georgiou and J. Beckwith, *J. Bacteriol.*, 1997, **179**, 6602–6608.
104. S. E. M. Meek, W. S. Lane and H. Piwnica-Worms, *J. Biol. Chem.*, 2004, **279**, 32046–32054.
105. H. Yan, J. J. Harding, K. Xing and M. F. Lou, *Curr. Eye Res.*, 2007, **32**, 455–463.
106. G. Fischer, B. Wittmann-Liebold, K. Lang, T. Kiefhaber and F. X. Schmid, *Nature*, 1989, **337**, 476–478.
107. N. Takahashi, T. Hayano and M. Suzuki, *Nature*, 1989, **337**, 473–475.
108. K. Motohashi, F. Koyama, Y. Najanishi, H. Ueoka-Nakanishi and T. Hisabori, *J. Biol. Chem.*, 2003, **278**, 31848–31852.
109. C. Marchand, P. Le Marechal, Y. Meyer, M. Miginiac-Maslow, E. Issakidis-Bourguet and P. Decottignies, *Proteomics*, 2004, **4**, 2696–2706.
110. K. Motohashi, A. Kondoh, M. T. Stumpp and T. Hisabori, *Proc. Natl. Acad. Sci. USA*, 2001, **98**, 11224–11229.
111. N. Rouhier, A. Villarejo, M. Srivastava, E. Gelhaye, O. Keech, M. Droux, I. Finkemeier, G. Samuelsson, K.-J. Dietz, J.-P. Jacquot and G. Wingsle, *Antioxid. Redox Signal.*, 2005, **7**, 919–929.

112. M. Fratelli, E. Gianazza and P. Ghezzi, *Expert Rev. Proteomics*, 2004, **1**, 365–376.
113. C. Lind, R. Gerdes, Y. Hamnell, I. Schnuppe-Koistinen, H. Brockenhuus von Löwenhielm, A. Holmgren and I. A. Cotgreave, *Arch. Biochem. Biophys.*, 2002, **406**, 229–240.
114. M. Fratelli, H. Demol, M. Puype, S. Casagrande, I. Eberini, M. Salmona, V. Bonetto, M. Mengozzi, F. Duffieux, E. Miclet, A. Bachi, J. Vande-kerckhove, E. Gianazza and P. Ghezzi, *Proc. Natl. Acad. Sci. USA*, 2002, **99**, 3505–3510.
115. M. Fratelli, H. Demol, M. Puype, S. Casagrande, P. Villa, I. Eberini, J. Vanderkerckhove, E. Gianazza and P. Ghezzi, *Proteomics*, 2003, **3**, 1154–1161.
116. A. Voronova, J. Kazantseva, M. Tuuling, N. Sokolova, R. Sillard and P. Palumaa, *Prot. Exp. Purif.*, 2007, **53**, 138–144.
117. G. Hoppe, Y.-C. Chai, J. W. Crabb and J. Sears, *Exp. Eye Res.*, 2004, **78**, 1085–1092.
118. E. R. LaVallie, E. A. Diblasio-Smith, L. A. Collins-Racie and J. M. McCoy, *Methods Enzymol.*, 2000, **326**, 322–340.
119. E. R. La Vallie, E. A. Diblasio-Smith, L. A. Collins-Racie, Z. Lu and J. M. McCoy, *Methods Mol. Biol.*, 2003, **205**, 119–140.
120. E. R. LaVallie, E. A. DiBlasio, S. Kovacic, K. L. Grant, P. F. Schendel and J. M. McCoy, *Biotechnology (NY)*, 1993, **11**, 187–193.
121. A. V. Ribas, P. L. Ho, M. M. Tanizaki, I. Raw and A. L. Nascimento, *Biotechnol. Appl. Biochem.*, 2000, **31**, 91–94.
122. P. Jurado, V. de Lorenzo and L. A. Fernandez, *J. Mol. Biol.*, 2006, **357**, 49–61.
123. D. J. McGee, S. Kumar, R. J. Viator, J. R. Bolland, J. Ruiz, D. Spada-fora, T. L. Testerman, D. J. Kelly, L. K. Pannell and H. J. Windle, *J. Biol. Chem.*, 2006, **281**, 3290–3296.
124. L. E. Vickery and J. R. Cupp-Vickery, *Crit. Rev. Biochem. Mol. Biol.*, 2007, **42**, 95–111.
125. C. W. Gruber, M. Čemažar, B. Heras, J. L. Martin and D. J. Craik, *Trends Biochem. Sci.*, 2006, **31**, 455–464.
126. L. Debarbieux and J. Beckwith, *J. Bacteriol.*, 2000, **182**, 723–727.
127. S. Jonda, M. Huber-Wunderlich, R. Glockshuber and R. Mössner, *EMBO J.*, 1999, **18**, 3271–3281.
128. L. Masip, J. L. Pan, S. Halder, J. E. Penner-Hahn, M. P. DeLisa, G. Georgiou, J. C. A. Bardwell and J. F. Collet, *Science*, 2004, **303**, 1185–1189.
129. A. I. Derman, W. A. Prinz, D. Belin and J. Beckwith, *Science*, 1993, **262**, 1744–1747.
130. P. H. Bessette, F. Aslund, J. Beckwith and G. Georgiou, *Proc. Natl. Acad. Sci. USA*, 1999, **96**, 13703–13708.
131. V. P. Pigiet and B. J. Schuster, *Proc. Natl. Acad. Sci. USA*, 1986, **83**, 7643–7647.
132. K. W. Walker, M. M. Lyles and H. F. Gilbert, *Biochemistry*, 1996, **35**, 1972–1980.

133. J. Lundstrom-Ljung and A. Holmgren, *J. Biol. Chem.*, 1995, **270**, 7822–7828.
134. N. J. Darby and T. E. Creighton, *Biochemistry*, 1995, **34**, 11725–11735.
135. M. Ruoppolo, J. Lundstrom-Ljung, F. Talamo, P. Pucci and G. Marino, *Biochemistry* 1997, **36**, 12259–12267.
136. R. Xiao, J. Lundstrom-Ljung, A. Holmgren and H. F. Gilbert, *J. Biol. Chem.*, 2005, **280**, 21099–21106.
137. S. Chakravarthi, C. E. Jessop and N. J. Bulleid, *EMBO Rep.*, 2005, **7**, 271–275.
138. J. Messens, J. F. Collet, K. Van Belle, E. Brosens, R. Loris and L. Wyns, *J. Biol. Chem.*, 2007, **282**, 31302–31307.
139. J. D. Schwenn, in *Sulphur metabolism in higher plants*, W. J. Cram (ed.), Backhuys Publishers, Leiden, The Netherlands, 1997, 39–58.

CHAPTER 1.2

Disulfide-bond Formation and Isomerization in Prokaryotes

GORAN MALOJČIĆ[a] AND RUDI GLOCKSHUBER[b]

[a] Institute of Molecular Biology and Biophysics, ETH Zürich, HPK E14, CH-8093 Zürich, Switzerland; [b] Institute of Molecular Biology and Biophysics, ETH Zürich, HPK E17, CH-8093 Zürich, Switzerland

1.2.1 Introduction

Structural disulfide bonds are a typical feature of secretory proteins and are often required for protein folding and stability. Formation of a disulfide bond from a cysteine pair in a newly synthesized protein is a redox reaction that requires the interaction of the folding protein with an oxidant that accepts the two electrons generated (Scheme 1.2.1).

As the redox environment in the cellular cytoplasm is reducing, structural disulfide bonds are essentially only formed in oxidizing compartments of the secretory pathway, namely the periplasmic space in bacteria and the endoplasmic reticulum in eukaryotic cells. This chapter focuses on the mechanisms underlying the oxidative folding of proteins in the periplasm of Gram-negative bacteria, where more than half of the periplasmic and outer membrane proteins contain at least one structural disulfide bond.[1–3]

In vitro, disulfide bonds can be formed spontaneously by oxidation with molecular oxygen, but the rate of this reaction is several orders of magnitude lower than the rate of disulfide-bond formation in vivo.[4,5] Moreover, the number of combinations of possible disulfide cross-links increases by around one order of magnitude with each additional cysteine pair according to

RSC Biomolecular Sciences
Oxidative Folding of Peptides and Proteins
Edited by Johannes Buchner and Luis Moroder
© Royal Society of Chemistry 2009
Published by the Royal Society of Chemistry, www.rsc.org

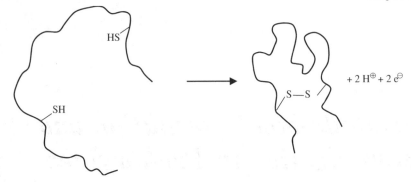

Scheme 1.2.1

Equation (1.2.1)

$$c = \frac{n!}{(n - 2k)! \cdot 2^k \cdot k!} \qquad (1.2.1)$$

where c is the number of possible configurations for formation of k disulfide bonds, k is the number of disulfide bonds and n is the number of all cysteines in the protein. Thus, not only disulfide-bond formation, but also the rearrangement of non-native disulfide bonds to the native configuration becomes a rate-limiting reaction for folding of proteins with multiple disulfide bonds. It is thus not surprising that both processes, formation and isomerization of disulfide bonds, are catalyzed reactions *in vivo*. Indeed, all cells that produce disulfide-bonded proteins contain thiol-disulfide oxidoreductases that catalyze these two rate-limiting reactions in oxidative folding. In *Escherichia coli*, both catalytic processes are separated and regulated *via* distinct pathways and catalyzed by Dsb proteins ("Dsb" stands for **d**i**s**ulfide-**b**ond formation): the disulfide-bond formation pathway is catalyzed by the dithiol oxidase DsbA and the quinone reductase DsbB, and the reductive disulfide isomerization pathway involves the transmembrane electron transfer catalyst DsbD, the disulfide isomerases DsbC and DsbG, and the cytochrome *c* maturation factor CcmG. In this chapter, we review the current knowledge and the open questions on oxidative protein folding in prokaryotic organisms.

1.2.2   Disulfide-bond Formation

1.2.2.1   The Periplasmic Dithiol Oxidase DsbA

The periplasmic dithiol oxidase DsbA was the first identified bacterial protein required for disulfide-bond formation *in vivo*.[6,7] The *dsbA*⁻ strains are hypersensitive to benzylpenicillin, dithiothreitol and metals[8,9] and show reduced levels of secreted proteins such as the outer-membrane protein OmpA and alkaline phosphatase,[6] which are apparently degraded when they cannot attain their native tertiary structure due to the lack of structural disulfides. Furthermore, *dsbA⁻* strains do not exhibit motility since the P-ring of the flagellar motor fails to assemble.[10] In addition, *dsbA⁻* mutants of many pathogenic Gram-negative bacteria lose virulence, since virulence factors such as adhesive pili or toxins fail to fold and assemble properly.[11]

DsbA is a 21.1 kDa monomeric periplasmic enzyme consisting of two domains: a core domain with a thioredoxin fold that is characteristic for cellular enzymes that catalyze disulfide-bond formation, reduction and isomerization, and an α helical domain that is inserted into the thioredoxin fold (Figure 1.2.1).[12,13] In common with other thiol-disulfide oxidoreductases with a thioredoxin fold, DsbA possesses a CXXC motif in the active site (X = any amino acid), with the active-site sequence CPHC. The active-site cysteine pair is found predominantly oxidized *in vivo*.[14,15] DsbA is a particularly reactive dithiol oxidase which rapidly and randomly introduces disulfide bonds into unfolded proteins *via* disulfide exchange *in vitro*[16–18] (Figure 1.2.1). *In vivo*, DsbA appears to be so reactive that it favors formation of sequential disulfide bonds in unfolded substrate proteins during translocation into the periplasm.[19] On the one hand, this favors formation of correct disulfide bonds in substrates with non-overlapping disulfide bonds. On the other hand, it also favors formation of non-native disulfide bonds in substrates with overlapping disulfide bonds, which then require the disulfide isomerases DsbC and DsbG for the rearrangement of non-native disulfides introduced by DsbA (see below).

DsbA has the most reactive and unstable disulfide bond in the family of thioredoxin-like oxidoreductases, which is reflected by the high intrinsic redox potential of the active-site cysteine pair ($E^{\circ\prime} = -122$ mV).[17] The high redox potential of DsbA is a direct consequence of the extremely low pK_a value of the more N-terminal active site cysteine (Cys30, $pK_a = 3.4$, which compares with pK_a values of 9–9.5 for normal cysteine residues).[28] The low pK_a of Cys30 guarantees that Cys30 in reduced DsbA is present in the thiolate anion form at physiological pH. The thiolate anion is not only stabilized by the tertiary structure of the protein, but in turn also stabilizes the tertiary structure of reduced DsbA. Formation of the active-site disulfide bond eliminates the stabilizing thiolate anion, which explains the observation that the oxidized form of DsbA is less stable than the reduced form.[17,18] Equation (1.2.2) describes the corresponding pH dependence of the difference in free energy of folding between the oxidized and reduced form of DsbA

$$\Delta\Delta G^0_{ox/red} = -RT \ln \left[\frac{1 + 10^{pH - pK_a^N}}{1 + 10^{pH - pK_a^U}} \right] \tag{1.2.2}$$

where pK_a^N is the pK_a of Cys30 in native, reduced DsbA, and pK_a^U is the normal pK_a of Cys30 in the unfolded protein. Assuming a value of pK_a^U of 9.5, Equation (1.2.2) predicts that the oxidized protein is 20.5 kJ mol⁻¹ less stable than the reduced form at pH 7.0, which is in good agreement with experimental data ($\Delta\Delta G^0_{ox/red} = 22$ kJ mol⁻¹).[20]

A

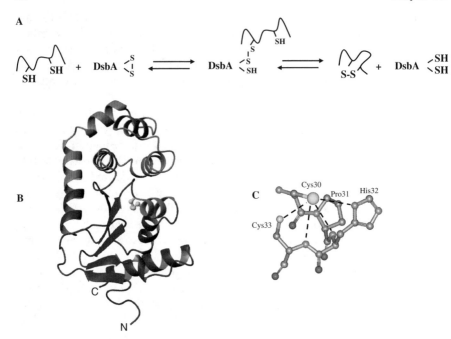

Figure 1.2.1   Introduction of disulfide bonds into substrate proteins by DsbA, and
three-dimensional structure of DsbA. **(A)** The disulfide exchange reaction between oxidized DsbA and a reduced polypeptide proceeds *via* a
mixed disulfide between the substrate and Cys30 of DsbA. **(B)** Ribbon
diagram of the X-ray structure of oxidized DsbA.[13] The thioredoxin
domain is shown in blue, the inserted α-helical domain is shown in red.
Active-site cysteines are shown in yellow, and the active-site helix
(residues 30–37) is depicted in green. **(C)** Stabilization of the thiolate
anion form of Cys30 in reduced DsbA by four alternative hydrogen
bonding possibilities.[30]

The low pK_a of Cys30 in DsbA also has direct consequences on the kinetics of
disulfide exchange reactions between DsbA and substrate proteins or organic
dithiols such as dithiothreitol. At pH 7.0, DsbA oxidizes unfolded substrate proteins
with rate constants of about $10^6 M^{-1} s^{-1}$, and the oxidation of small dithiols such as
dithiothreitol is about one order of magnitude slower.[21,22] Compared to the small
rate constants of disulfide exchange reactions between small organic thiols and
disulfides at pH 7.0,[23] the reaction with DsbA is 4–6 orders of magnitude faster.

As the rate constants of disulfide exchange reactions depend on the pK_a values
of all sulfur atoms involved,[23–28] the equilibrium constants of disulfide exchange
between thioredoxin-like enzymes and a thiol/disulfide redox pair, like reduced
and oxidized glutathione (GSH and GSSG, respectively), are predictable. The
deduced redox potentials are in surprisingly good agreement with the experimental data for known members of the thioredoxin family, which range between
–122 mV for DsbA as the most oxidizing member, and –270 mV for thioredoxin
as the most reducing member (for a review see ref. 29).

The Cys30/Cys33 pair of DsbA is located at the N-terminus of an α-helix. The negative charge on the Cys30 thiolate in reduced DsbA is stabilized electrostatically by the partial positive charge of the helix dipole, and alternative hydrogen bonding possibilities of the Cys33 thiol, the main chain amides of His32 or Cys33 and the NHδ1 of His32.[13,30,31] DsbA also possesses a hydrophobic groove flanking the active-site disulfide, which is supposed to increase the affinity of DsbA for unfolded polypeptide chains and may contribute to the fast disulfide exchange reactions of DsbA bringing the active-site disulfide bond into close proximity of free thiols in protein substrates.[13,30,31]

1.2.2.2 DsbB

1.2.2.2.1 Regeneration of Oxidized DsbA

Disulfide exchange between DsbA and the nascent proteins renders DsbA reduced and inactive. Its reoxidation is performed by the inner membrane protein DsbB. The observation that *E. coli* mutants lacking *dsbB* exhibit the same pleiotropic phenotype as those lacking *dsbA* suggested that DsbB and DsbA belong to the same pathway.[8,10,32] Furthermore, in *dsbB⁻* strains DsbA accumulates in the reduced state, while it is completely oxidized in wild type strains.[14,15] The isolation of a DsbA–DsbB mixed-disulfide complex confirmed their direct interaction.[15,33]

The reoxidation of DsbA *via* DsbB is coupled to the reduction of quinones from the inner bacterial membrane by DsbB.[10,14,15,32,33] Under aerobic conditions, the electrons are passed from DsbB onto ubiquinone-Q8 (Figure 1.2.2), which then donates the electrons to molecular oxygen *via* terminal oxidases.[34,35] Under anaerobic conditions, the electrons are passed from DsbB to menaquinone-Q8 (Figure 1.2.2) and then to fumarate or alternative electron

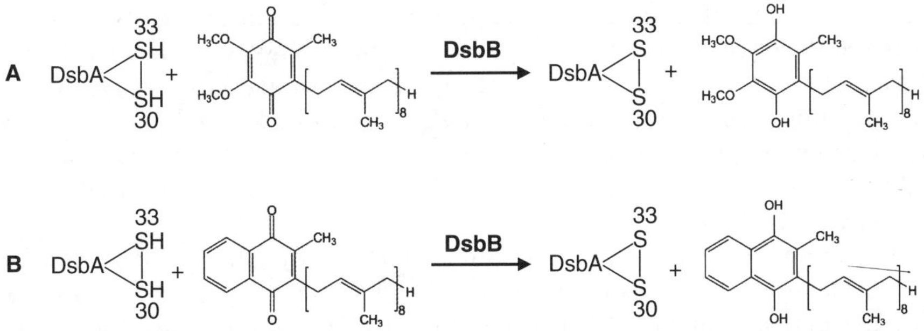

Figure 1.2.2 Reactions catalyzed by the quinone reductase DsbB. Active-site cysteine residues in DsbA are numbered. Under aerobic conditions (**A**), DsbA is oxidized by ubiquinone-Q8, while menaquinone-Q8 serves as the electron acceptor under anaerobic conditions (**B**).

acceptors.[34] *In vitro* experiments with purified components confirmed that quinones directly receive electrons from DsbB.[34] Thus, DsbB generates a disulfide bond in DsbA *de novo* by reduction of quinones and joins oxidative protein folding with the respiration pathways.[36,37]

1.2.2.2.2   Structure and Mechanism of DsbB

DsbB is a 20.1 kDa inner membrane protein that possesses four transmembrane helices, two periplasmic loop regions,[38,39] with both the amino- and carboxy-terminus located in the cytoplasm. Each of the two periplasmic loops contains one pair of invariant cysteines. Cys41 and Cys44 form a disulfide bond in the amino-terminal (first) periplasmic loop, whereas Cys104 and Cys130 form a disulfide bond in the carboxy-terminal (second) loop.[39] While the Cys41/Cys44 pair interacts with the quinone, the Cys104/Cys130 pair undergoes disulfide exchange with DsbA, in which Cys30 of DsbA is disulfide-linked with Cys104 of DsbB (Figure 1.2.3A).

The mechanistic details of the reaction of DsbB have been extensively studied by both biochemical[6,14,15,19,32,34–38,40–46] and theoretical methods.[41] In addition, the most recently elucidated crystal structure of the inactive variant DsbB(Cys130Ser)[38] (Figure 1.2.4) confirmed the proposed topology of DsbB, identified an additional alpha helix formed by the residues of the second periplasmic loop and provided further insights into the reaction catalyzed by this unusual membrane protein.

The aromatic ring of the quinone is bound in the proximity of the Cys41/Cys44 pair and Arg48, which is also essential for catalysis.[42] These residues are situated near the periplasmic end of the second transmembrane helix (Figure 1.2.3A). Upon reaction with DsbA, the mixed disulfide bond (DsbB-Cys104)–(Cys33-DsbA) is formed, together with an inter-loop disulfide Cys130–Cys41 within DsbB. The resulting, unpaired Cys44 is engaged in a charge transfer interaction with the quinone ring, giving rise to the pink color characteristic for this complex. The data suggest that *de novo* disulfide-bond generation at the active site of DsbB proceeds *via* the transient formation of a covalent bond between the Cys44 sulfur and the quinone (Figure 1.2.5). Subsequently, Cys41 attacks Cys44, which regenerates the Cys41/44 disulfide bond in DsbB and releases ubiquinol. This mechanism of *de novo* disulfide-bond generation by DsbB resembles the mechanism in the active site of the FAD-dependent enzyme dihydrolipoyl dehydrogenase,[47] where the analogous charge transfer intermediate forms (red color).

The thermodynamics of the oxidation of DsbA by quinones catalyzed by DsbB has been a matter of controversy.[48,49] Various redox potentials for the reactive cysteine pairs of DsbB have been measured, and the situation has not yet been clarified. The bottom line, however, is that the overall process, *i.e.* the oxidation of DsbA by ubiquinone, is thermodynamically favorable. In common with other (bio)chemical reactions, the catalysis by DsbB conceivably involves high-energy intermediates.

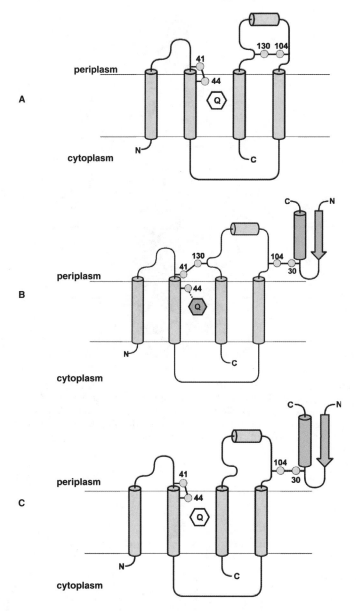

Figure 1.2.3 Membrane topology of DsbB, alone and in different complexes with DsbA. DsbB is depicted in yellow, the part of DsbA interacting with DsbB is indicated in blue. Cysteines are represented as yellow circles and numbered, and the bound quinone is represented as a hexagon. **(A)** DsbB consists of four transmembrane helices and two periplasmic loops each containing a pair of essential cysteines.[39] **(B)** The complex of the wild type DsbB with the single-cysteine variant DsbA(Cys33Ala) contains a rearranged intramolecular disulfide bond (Cys130-Cys41) and the unpaired Cys44 thiolate of DsbB, which forms a charge-transfer interaction with quinone, giving rise to the pink color of the complex.[40,41] **(C)** Disulfide-bonding pattern in the X-ray structure of the mixed disulfide complex between the inactive DsbB(Cys130Ser) variant and DsbA(Cys33Ala).[38]

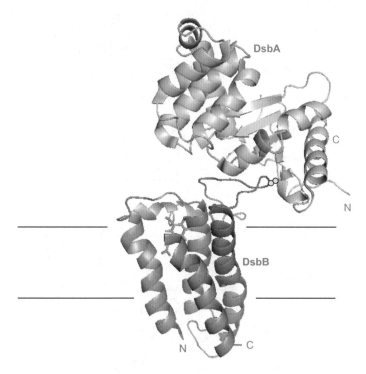

Figure 1.2.4   Ribbon diagram of the 3.7-Å crystal structure of the mixed disulfide complex between DsbB(Cys130Ser) and DsbA(Cys33Ala).[38] The quinone ring of DsbB bound ubiquinone Q8 is indicated in pink, and the position of the intermolecular disulfide bond between Cys30(DsbA) and Cys104(DsbB) is indicated in yellow.

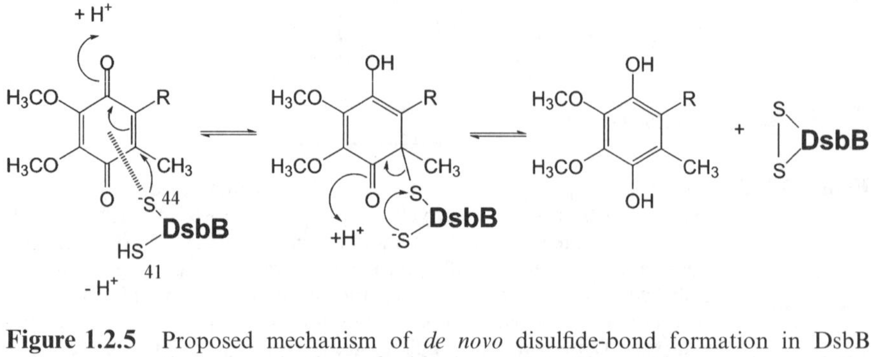

Figure 1.2.5   Proposed mechanism of *de novo* disulfide-bond formation in DsbB through reduction of ubiquinone Q8. The residue numbers of the reactive cysteines in DsbB that react with the quinone ring are indicated. R represents the hydrophobic Q8 tail with eight isoprene units.

1.2.3 Disulfide-bond Isomerization

1.2.3.1 Disulfide-bond Isomerase DsbC

The dithiol oxidase DsbA introduces non-native disulfide bonds into substrate proteins *in vivo* and *in vitro*.[17,20,22,35,50–53] This not only leads to non-native conformations in folding proteins that are aggregation and degradation prone, but also causes molecules with non-native disulfide configurations that can no longer isomerize spontaneously to the native configuration when the substrate protein i) is completely oxidized and ii) no longer possesses free cysteine thiols that could attack non-native disulfides intramolecularly. Consequently, proper folding of scrambled and completely oxidized substrates requires the transient reduction of at least one of the non-native disulfide bonds to allow disulfide rearrangement to the native configuration.

The first step in the discovery of the isomerization pathway in prokaryotes was the identification of the periplasmic protein disulfide isomerase DsbC.[54,55] The *E. coli dsbC* gene was isolated independently by screening for dithiothreitol-sensitive mutations[54] and as a suppressor of a *dsbA* null phenotype.[55] Like *dsbA⁻* and *dsbB⁻* strains, *dsbC* mutants show defects in disulfide-bond formation, but the effect is significantly milder.[56] For instance, mutations in *dsbC* affect neither cell motility nor disulfide-bond formation in OmpA and β-lactamase. The disulfide isomerase function of DsbC was subsequently established using *in vitro* folding assays with model substrates and confirmed in expression studies of heterologous proteins in various *dsb⁻* backgrounds.[19,52,56–60] It was found that DsbC is required for the correct folding of proteins with multiple disulfide bonds. Hiniker and Bardwell showed that two *E. coli* periplasmic proteins, RNaseI (four disulfide bonds) and MepA (six conserved cysteines), are *in vivo* substrates of DsbC.[2] The levels of RNaseI and MepA were strongly decreased in both *dsbA⁻* and *dsbC⁻* strains. In contrast, the levels of two other DsbA-dependent proteins with multiple disulfide bonds, PhoA and DppA, were unchanged in *dsbC⁻* strains. While the latter proteins contain only consecutive disulfide bonds, RNaseI and MepA contain both non-consecutive and consecutive disulfide bonds. These findings suggest that DsbC is required for the folding of proteins with non-consecutive disulfide bonds, whereas DsbA alone is sufficient for folding of periplasmic proteins with consecutive disulfide bonds. The hypothesis that disulfide-bond connectivity determines whether a protein requires DsbC for proper folding was further supported by the experiments demonstrating that the acid phosphatase phytase (AppA), a protein with one non-consecutive disulfide bond, requires DsbC for proper folding, while folding of an AppA variant lacking the non-consecutive bond was DsbC-independent. Conversely, a homolog of AppA that contains only consecutive bonds does not require DsbC, but becomes DsbC-dependent upon engineering of a non-consecutive bond.[50]

DsbC is a V-shaped homodimer of 25.6 kDa subunits and, like DsbA, belongs to the thioredoxin superfamily. Each DsbC monomer consists of an N-terminal dimerization domain and a C-terminal thioredoxin domain with the active site motif CGYC. Its redox potential (-140 mV)[61,62] is almost as oxidizing as that of DsbA, and the more N-terminal active site cysteine (Cys 98) also shows a very low pK_a value of 4.1.[62] The dimerization of DsbC is required for its function *in vivo*.[62,63] The crystal structure of oxidized DsbC (Figure 1.2.6A), together with biochemical experiments, revealed that DsbC combines two important features required for disulfide-bond isomerase activity: a reactive cysteine pair in the active site that can attack and reshuffle disulfide bonds in substrates, and a hydrophobic binding site for unfolded or partially folded substrate proteins.[64] The active sites in the DsbC homodimer are separated by ~40 Å and oriented towards the interior of the "V". The surface of the cleft, defined by the two "arms" of the "V", is lined with hydrophobic and uncharged residues, enabling hydrophobic interactions. The size of the cleft may vary between more open and more closed conformations according to the bound polypeptide.[1] Compared to the peptide binding site in the DsbA monomer, the binding site in the DsbC homodimer has about a 10^4-fold higher affinity[57] and is a prerequisite for the disulfide isomerase activity of DsbC, most likely due to strongly increased effective concentrations of the active sites relative to non-native disulfides in the substrate.

In contrast to DsbA, which is active in its oxidized form as dithiol oxidase, DsbC needs to be reduced for being active as an isomerase, because only the reduced form of the enzyme can attack non-native disulfides in substrates.[58,59] The low pK_a of Cys 98 guarantees that its nucleophilically active thiolate anion form is always populated at physiological pH values. Figure 1.2.7 shows the two potential mechanisms for the reaction catalyzed by DsbC, which both start with the formation of a DsbC-substrate mixed disulfide. It is, however, still unknown whether the substrate stays covalently bound to DsbC during disulfide-bond isomerization (mechanism *a* in Figure 1.2.7), or whether the substrate becomes partially reduced by DsbC, followed by spontaneous, intramolecular disulfide rearrangement and reoxidation by DsbC (mechanism *b*). The latter mechanism also agrees with the high redox potential of DsbC, which makes substrate oxidation thermodynamically favorable.

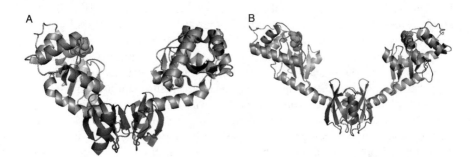

Figure 1.2.6   Ribbon diagrams of the structures of DsbC[64] **(A)** and DsbG[84] **(B)**. The dimerization domains in both proteins are indicated in dark green, the thioredoxin domains in light green and the linker helices connecting both domains in the monomer are depicted in grey. Orange balls represent the active site cysteine residues.

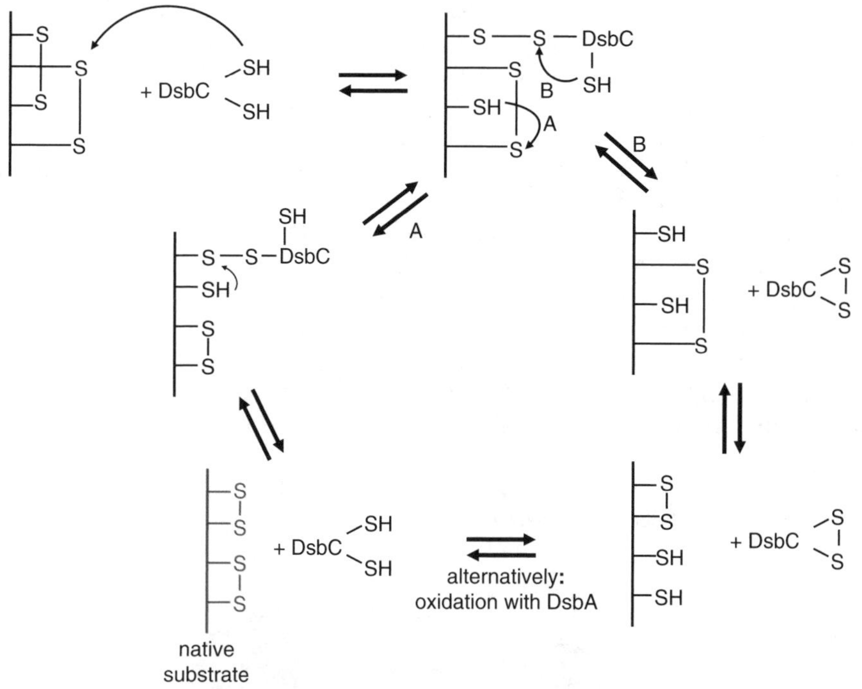

Figure 1.2.7 Models of the catalytic mechanism of DsbC. The first reaction step involves the formation of a mixed disulfide intermediate between Cys109 of DsbC and a cysteine from the polypeptide substrate. Subsequently, pathways **(A)** and **(B)** are possible mechanisms of DsbC catalyzed disulfide-bond isomerization. **(A)** Disulfide rearrangement in the substrate occurs while the substrate stays covalently bound to DsbC. **(B)** Complete reduction of one of the non-native disulfide bonds in the substrate and formation of oxidized DsbC, followed by intramolecular disulfide rearrangements in the released substrate and reoxidation of the substrate by either oxidized DsbC or oxidized DsbA.

1.2.3.2 Reactivation of DsbC: The Inner Membrane Electron Transporter DsbD

The oxidizing environment of the bacterial periplasm requires a specific, reductive pathway that maintains DsbC in its catalytically active, reduced state. This pathway is dependent on the electron transfer catalyst DsbD from the inner bacterial membrane,[59,65] which shuttles reducing equivalents from the cytoplasm directly to periplasmic DsbC. The reducing equivalents originate from cytoplasmic NADPH, which reduces cytoplasmic thioredoxin *via* thioredoxin reductase. Reduced thioredoxin then interacts with the cytoplasmic side of DsbD. In accordance with this mechanism, deletion of either DsbD, cytoplasmic thioredoxin or cytoplasmic thioredoxin reductase renders DsbC completely oxidized *in vivo*.[59,65–67]

DsbD is a 59-kDa inner membrane protein that consists of three domains: a periplasmic, N-terminal domain (nDsbD), a central transmembrane (TM) domain with eight predicted TM helices and a C-terminal domain (cDsbD) in the periplasm (Figure 1.2.8).[66,68,69] Mutational analyses showed that each of the domains contains one pair of conserved cysteine residues that is essential for DsbD activity.[66,68,69] An intriguing feature of DsbD is that it catalyzes electron transport across the inner membrane exclusively *via* inter- and intermolecular disulfide exchange reactions. DsbD has two additional periplasmic substrates besides DsbC which are also specifically reduced by the enzyme: the disulfide isomerase and DsbC homolog DsbG, and the periplasmic, thioredoxin domain of the cytochrome *c* maturation factor CcmG (see below).

An important step in establishing the electron flow in DsbD was the observation that co-expression of all individual DsbD domains restored DsbD

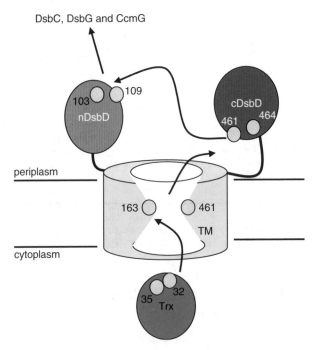

Figure 1.2.8   Proposed mechanism of DsbD-mediated electron transfer through the inner membrane *via* a single cysteine pair in the transmembrane (TM) domain. The model proposes an hourglass-like shape of the transmembrane domain that allows access to the catalytic Cys163/Cys461 pair of thioredoxin from the cytoplasmic side and of cDsbD from the periplasmic side. Cys 32 of thioredoxin (Trx) can form a mixed disulfide with Cys163 of the TM domain. It remains to be shown which of the two catalytic Cys residues of the TM domain interacts with cDsbD. Numbers refer to the catalytic cysteine residues in Trx and all DsbD domains, and the arrows indicate the direction of electron flow.

function *in vivo*.[67] These experiments, together with biochemical experiments and further *in vivo* data,[3,61,65–67,69–74] revealed that the electron transport mechanism mediated by DsbD is exclusively based on inter- and intermolecular disulfide exchange reactions and obviously independent of redox cofactors. Specifically, the cysteine pair Cys163/Cys285 in the TM domain accepts two electrons from cytoplasmic thioredoxin and passes them on to cDsbD in the periplasm. cDsbD then transfers the reducing equivalents to nDsbD, which then reduces the substrates DsbC, DsbG and CcmG.

There are two particularly fascinating aspects of the DsbD mechanism. The most surprising and unusual feature of DsbD is certainly its ability to shuffle electrons across the ~ 60 Å wide cytoplasmic membrane *via* a single cysteine pair. A detailed biochemical characterization of the TM domain, together with structure predictions, indicated that TM DsbD has an hourglass-like shape, where the catalytic Cys163/Cys285 pair is located close to the center of the hourglass and accessible for disulfide exchange with thioredoxin from the cytoplasmic side, and with cDsbD from the periplasmic side (Figure 1.2.8).[74] It has been established that Cys163 in the TM domain of DsbD is capable of forming a mixed disulfide with the nucleophilic Cys32 of thioredoxin,[67,75] but the cysteine residue interacting with cDsbD on the periplasmic side of the membrane is still unknown. The hourglass-like shape of the DsbD TM domain is also confirmed by a pseudosymmetry at the primary structure level around Cys163 and Cys285, which are located in the first and fourth predicted transmembrane segment. Two conserved proline residues located at positions -1 and $+2$ relative to each of the two cysteines are indicative of kinks in the transmembrane segments 1 and 4, in accordance with the hour-glass model.[74]

The second remarkable feature of DsbD is the ability of its N-terminal domain (nDsbD) to react specifically with four different proteins/domains *via* disulfide exchange, namely intramolecularly with cDsbD and intermolecularly with the three substrates DsbC, DsbG and CcmG. The structure of isolated nDsbD has been solved and revealed that the domain possesses an immu-noglobulin-like fold, with Cys109 being the nucleophilically active cysteine residue.[76] Moreover, the structures of both redox forms of isolated cDsbD have been determined and confirmed that cDsbD also has the thioredoxin fold with the active-site sequence CVAC.[3,72,77] Thus, all four reaction partners of nDsbD are thioredoxin-like proteins (see also below). Importantly, the structures of the kinetically trapped mixed disulfide species involving nDsbD have been also solved, except for the mixed disulfide between nDsbD and DsbG (Figure 1.2.9).[1,61,73] The structures of the mixed disulfide complexes between nDsbD and cDsbC, DsbC and CcmG revealed that nDsbD uses essentially the same interface to interact with its reaction partners,[73] with about 90% identical interface residues and very similar contact areas of about 1350 Å^2. The only exception is the mixed disulfide complex between nDsbC and DsbC, in which nDsbD is covalently bound to one of the active sites in the DsbC homodimer, and also interacts specifi-cally on the opposite side with the other DsbC subunit.[1] The nDsbD–CcmG

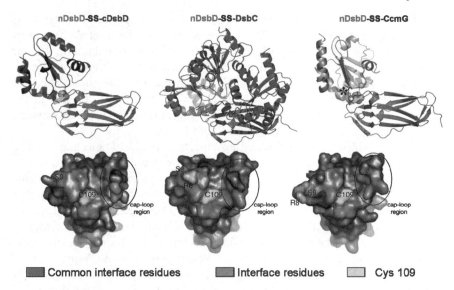

Figure 1.2.9 Structures of the kinetically stabilized mixed disulfide complexes between nDsbD and its reaction partners cDsbD,[61] DsbC[1] and CcmG.[73] **(Top panels)** Ribbon diagrams, with nDsbD indicated in red and cDsbD, DsbC and CcmG depicted in blue, green and grey, respectively. **(Bottom panels)** Space filling model of nDsbD in the respective structure, with all residues in the contact area colored in green (common interface in all three structures) and blue (specific interface).

mixed disulfide complex also forms additional, specific interactions with CcmG *via* its N-terminal segment. Overall, the ability of nDsbD to adapt to all four natural reaction partners is essentially conferred by steric compatibility.

Modeling of the non-natural interaction between nDsbD and DsbA (which would neutralize the oxidative DsbA/DsbB and the reductive DsbD/DsbC(DsbG,CcmG) pathways) shows that disulfide exchange with DsbA is sterically restricted (see also below). Overall, the almost identical contact areas of nDsbD with its partner proteins indicate that large movements of the periplasmic DsbD domains occur during the DsbD reaction cycle.[61,73,78]

With respect to the thermodynamics of the reductive pathway mediated by DsbD, the overall electron flow from cytoplasmic thioredoxin ($E^{\circ\prime} = -270$ mV) to the substrates DsbC, DsbG and CcmG ($E^{\circ\prime} = -140$, -129 mV and -203 mV, respectively) is energetically favorable.[61,62,73,79,81] The intrinsic redox potentials of nDsbD and cDsbD have values of -235 and -232 mV[73] and suggest that the electron flow within DsbD is energetically favorable as well. However, the picture of the energetics of the entire electron flow catalyzed by DsbD is still incomplete, as the redox potential of the DsbD TM domain has not yet been determined.

1.2.3.3 DsbG, a Structural Homolog of DsbC with Unknown Function

DsbG is the second disulfide isomerase in the *E. coli* periplasm,[66,79] and was discovered due to its ability to confer resistance to dithiothreitol when expressed *via* a multicopy plasmid.[80] DsbG is the least abundant of the periplasmic Dsb proteins.[79] Like DsbC, DsbG is a homodimeric protein with 28% sequence identity to DsbC.[64,80] The redox potential of DsbG (−129 mV) is oxidizing and similar to that of DsbA and DsbC (−122 and −140 mV, respectively).[60–61,79,81] DsbG is kept in the reduced state by DsbD and possesses disulfide isomerase activity in *E. coli*, as it improves the yield of heterologous proteins with multiple disulfide bonds in the periplasm.[79,82] The natural substrates of DsbG are, however, unknown, in particular because *E. coli dsbG* deletion mutants show no defect in periplasmic disulfide-bond formation.[2,79] In contrast to DsbC, DsbG does not exhibit significant disulfide isomerase activity *in vitro*.[79,81] In addition, DsbG shows an activity reminiscent of molecular chaperones in that it prevents unspecific aggregation of the classical chaperone substrates citrate synthase and luciferase.[83] In contrast to true chaperones, however, it does not improve the folding yield of these substrates but rather prevents folding by apparently irreversible binding to unfolded polypeptides. Together, the data indicate that DsbG is a protein disulfide isomerase with narrow substrate specificity.

The crystal structure of the DsbG homodimer[84] revealed a domain organization analogous to that of DsbC. The DsbG structure however differs from the DsbC structure[64] in that its dimensions are significantly larger due to the fact that the linker helix in DsbG that connects the dimerization domain with the thioredoxin-like domain is 2.5 turns longer compared to DsbC (Figure 1.2.6). This increases the size of the supposed polypeptide binding cleft relative to DsbC about two-fold. Another difference from DsbC is the fact that seven acidic residues from each subunit line the cleft between the thioredoxin domains of the DsbG dimer. These negatively charged surface patches in the presumed substrate binding site of DsbG support the view that DsbG is a specific isomerase for a hitherto unknown substrate, and not a general disulfide isomerase like DsbC.

1.2.3.4 The Cytochrome *c* Maturation Factor CcmG is a DsbD Substrate

The biogenesis of all *c*-type cytochromes requires the ligation of heme to apocytochrome *c*, where two thioether bonds are formed between the heme and the two cysteine residues in the active site of apocytochrome *c*, which has the sequence motif CXXCH. This ligation reaction can only occur when apocytochrome *c* is maintained in the reduced state in the periplasm. The disulfide reductase CcmG, also termed DsbE, is one of the eight proteins encoded by the *E. coli* Ccm gene cluster which is required for cytochrome *c* biogenesis, and the third known periplasmic substrate of DsbD. CcmG is a membrane-anchored, periplasmic thioredoxin-like protein of 20 kDa.[85–87] The thioredoxin-like domain of CcmG contains the active site sequence CPTC and has a redox potential of –203 mV.[73] In contrast to various other bacterial oxidoreductases, CcmG does not show reductase activity in the classical insulin reduction assay, and its reductase activity seems highly specialized and limited to the cytochrome *c* maturation pathway.[85,86,88] The crystal structure of CcmG from *Bradyrhizobium japonicum*[89] revealed a thioredoxin fold with several specific features that are required for cytochrome *c* maturation. These are a characteristic groove formed by two insertions in the fold, the N-terminal β-hairpin-like structure and a central insertion. The deletion of any of the two insertions suppresses cytochrome *c* formation.[89,90] Another unusual feature of CcmG compared to other thioredoxin-like proteins is an acidic region around the active site. Several conserved acidic residues contribute to the negative charge of the protein and are required for efficient cytochrome *c* biogenesis.[90]

The reducing equivalents required for cytochrome *c* maturation originate from cytoplasmic NADPH and flow to CcmG *via* thioredoxin and DsbD. Genetic studies showed that in the absence of DsbD, thioredoxin or thioredoxin reductase, cytochrome *c* maturation is suppressed and CcmG accumulates in the oxidized state.[66,87,91–93] The detection of the mixed-disulfide intermediate between *E. coli* DsbD and CcmG *in vivo*[67] and the direct observation of the reduction of a soluble CcmG variant with truncated membrane anchor by nDsbD *in vitro*[73] confirmed the direct interaction between CcmG and DsbD. However, the mechanism of the further electron transfer from CcmG to apocytochrome *c* remains to be established. The structure of the stabilized mixed disulfide complex between nDsbD and the soluble variant of *E. coli* CcmG (Figure 1.2.9) revealed that the N-terminal segment of nDsbD contributes to the specific recognition of CcmG, and that CcmG occupies essentially the same contact area as DsbC and cDsbD.[94]

1.2.4 Coexistence of the Oxidative Disulfide-bond Formation and the Reductive Disulfide Isomerization Pathways

As discussed above, Gram-negative bacteria contain catalysts performing both dithiol oxidation and disulfide reduction in the periplasm. It is obvious that both pathways, which have opposite directions of electron flow, must be kept separated in the periplasm to prevent their mutual neutralization. Figure 1.2.10 summarizes the measured rate constants of disulfide exchange between all periplasmic components of the dithiol oxidation and the reductive disulfide-bond isomerization pathway at pH 7.[48,61,73,78]

The data reveal that functional disulfide exchange reactions between components of the same pathway have rate constants between 10^5 and $10^6 \, M^{-1}s^{-1}$, while non-functional reactions occur with rates that are between three and six orders of magnitude slower. Thus, both pathways are separated by huge activation energy barriers that guarantee their co-existence. In structural terms,

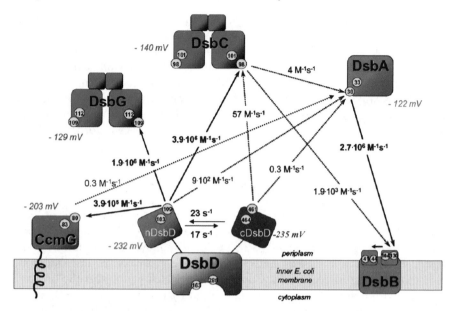

Figure 1.2.10   Complete kinetic picture of functional (solid lines) and non-functional (dotted lines) disulfide exchange reactions between components of the oxidative disulfide-bond formation pathway and the reductive disulfide isomerization pathway.[48,61,73,78] Arrows indicate the direction of electron flow.

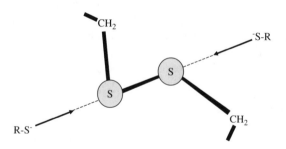

Scheme 1.2.2

these barriers can easily be rationalized based on the known structures of all redox-active Dsb proteins and their domains:[38,61,73,78] disulfide exchange reactions are S_N2 reactions with a transition state in which the negative charge of the attacking thiolate is distributed among all sulfur atoms involved. The attack by a thiolate anion on a disulfide bond, which has dihedral angles with respect to the sulfur–sulfur bond within 10° of +95° or −95°, can only occur along the axis of the sulfur–sulfur bond[95] (Scheme 1.2.2).

This defines rather stringent steric requirements for disulfide exchange reactions between proteins, because the tertiary structure context generally only allows attack of a disulfide bond from one side, and because steric clashes between the proteins can easily prevent the attack of a nucleophilic protein thiol from one protein on a catalytic disulfide bond of another protein. It appears that these steric restrictions are indeed underlying the kinetic barriers for non-functional disulfide exchange reactions between Dsb proteins in the periplasm. Modeling of potential mixed disulfide intermediates between "forbidden" Dsb pairs revealed steric clashes in all cases.[38,63,73,94]

1.2.5   Concluding Remarks

Oxidative protein folding in the *E. coli* periplasm is not a spontaneous process, but catalyzed by specialized Dsb proteins. A large body of *in vivo* and *in vitro* data supports the independent pathways of disulfide-bond formation (catalyzed by DsbA and DsbB) and the reductive pathway of disulfide-bond isomerization (catalyzed by DsbC and DsbD), which guarantees the "repair" proteins with non-native disulfides introduced by DsbA.

Despite considerable progress in the field, in both structural and mechanistic terms, there are major open questions on both dsb systems. The most intriguing open question is the mechanism underlying the electron transport of DsbD across the inner membrane. Structural information on the transmembrane domain of DsbD is thus eagerly awaited. Other important questions involve the molecular details of the *de novo* disulfide-bond formation in DsbB and the *in vivo* function of DsbG.

Besides the mechanistic insights into catalyzed disulfide-bond formation of the bacterial periplasm, the recently obtained results on the structure and function of Dsb proteins have strongly contributed to the improvement of *E. coli* expression systems for heterologous proteins with multiple disulfide bonds that are based on the co-expression of DsbA and DsbC.[52,96–98]

Acknowledgements

This work was supported by the Schweizerische Nationalfonds and the ETH Zurich.

References

1. P. W. Haebel, D. Goldstone, F. Katzen, J. Beckwith and P. Metcalf, *EMBO J.*, 2002, **21**, 4774–4784.
2. A. Hiniker and J. C. Bardwell, *J. Biol. Chem.*, 2004, **279**, 12967–12973.
3. J. H. Kim, S. J. Kim, D. G. Jeong, J. H. Son and S. E. Ryu, *FEBS Lett.*, 2003, **543**, 164–169.
4. C. B. Anfinsen, *Science*, 1973, **181**, 223–230.
5. V. P. Saxena and D. B. Wetlaufer, *Biochemistry*, 1970, **9**, 5015–5023.

6. J. C. Bardwell, K. McGovern and J. Beckwith, *Cell*, 1991, **67**, 581–589.
7. S. Kamitani, Y. Akiyama and K. Ito, *EMBO J.*, 1992, **11**, 57–62.
8. D. Missiakas, C. Georgopoulos and S. Raina, *Proc. Natl. Acad. Sci. USA*, 1993, **90**, 7084–7088.
9. S. J. Stafford, D. P. Humphreys and P. A. Lund, *FEMS Microbiol. Lett.*, 1999, **174**, 179–184.
10. F. E. Dailey and H. C. Berg, *Proc. Natl. Acad. Sci. USA*, 1993, **90**, 1043–1047.
11. J. A. Peek and R. K. Taylor, *Proc. Natl. Acad. Sci. USA*, 1992, **89**, 6210–6214.
12. J. L. Martin, *Structure*, 1995, **3**, 245–250.
13. J. L. Martin, J. C. Bardwell and J. Kuriyan, *Nature*, 1993, **365**, 464–468.
14. S. Kishigami, Y. Akiyama and K. Ito, *FEBS Lett.*, 1995, **364**, 55–58.
15. S. Kishigami, E. Kanaya, M. Kikuchi and K. Ito, *J. Biol. Chem.*, 1995, **270**, 17072–17074.
16. N. J. Darby and T. E. Creighton, *Biochemistry*, 1995, **34**, 3576–3587.
17. M. Wunderlich and R. Glockshuber, *Protein Sci.*, 1993, **2**, 717–726.
18. A. Zapun, J. C. Bardwell and T. E. Creighton, *Biochemistry*, 1993, **32**, 5083–5092.
19. M. Sone, Y. Akiyama and K. Ito, *J. Biol. Chem.*, 1997, **272**, 10349–10352.
20. M. Wunderlich, R. Jaenicke and R. Glockshuber, *J. Mol. Biol.*, 1993, **233**, 559–566.
21. J. Hennecke, A. Sillen, M. Huber-Wunderlich, Y. Engelborghs and R. Glockshuber, *Biochemistry*, 1997, **36**, 6391–6400.
22. M. Wunderlich, A. Otto, R. Seckler and R. Glockshuber, *Biochemistry*, 1993, **32**, 12251–12256.
23. Z. Shaked, R. P. Szajewski and G. M. Whitesides, *Biochemistry*, 1980, **19**, 4156–4166.
24. U. Grauschopf, J. R. Winther, P. Korber, T. Zander, P. Dallinger and J. C. Bardwell, *Cell*, 1995, **83**, 947–955.
25. M. Huber-Wunderlich and R. Glockshuber, *Fold Des.*, 1998, **3**, 161–171.
26. E. Mossner, M. Huber-Wunderlich and R. Glockshuber, *Protein Sci.*, 1998, **7**, 1233–1244.
27. E. Mossner, H. Iwai and R. Glockshuber, *FEBS Lett.*, 2000, **477**, 21–26.
28. J. W. Nelson and T. E. Creighton, *Biochemistry*, 1994, **33**, 5974–5983.
29. C. S. Sevier and C. A. Kaiser, *Nat. Rev. Mol. Cell Biol.*, 2002, **3**, 836–847.
30. L. W. Guddat, J. C. Bardwell, R. Glockshuber, M. Huber-Wunderlich, T. Zander and J. L. Martin, *Protein Sci.*, 1997, **6**, 1893–1900.
31. L. W. Guddat, J. C. Bardwell, T. Zander and J. L. Martin, *Protein Sci.*, 1997, **6**, 1148–1156.
32. J. C. Bardwell, J. O. Lee, G. Jander, N. Martin, D. Belin and J. Beckwith, *Proc. Natl. Acad. Sci. USA*, 1993, **90**, 1038–1042.
33. C. Guilhot, G. Jander, N. L. Martin and J. Beckwith, *Proc. Natl. Acad. Sci. USA*, 1995, **92**, 9895–9899.
34. M. Bader, W. Muse, D. P. Ballou, C. Gassner and J. C. Bardwell, *Cell*, 1999, **98**, 217–227.

35. M. W. Bader, T. Xie, C. A. Yu and J. C. Bardwell, *J. Biol. Chem.*, 2000, **275**, 26082–26088.
36. T. Kobayashi and K. Ito, *EMBO J.*, 1999, **18**, 1192–1198.
37. T. Kobayashi, S. Kishigami, M. Sone, H. Inokuchi, T. Mogi and K. Ito, *Proc. Natl. Acad. Sci. USA*, 1997, **94**, 11857–11862.
38. K. Inaba, S. Murakami, M. Suzuki, A. Nakagawa, E. Yamashita, K. Okada and K. Ito, *Cell*, 2006, **127**, 789–801.
39. G. Jander, N. L. Martin and J. Beckwith, *EMBO J.*, 1994, **13**, 5121–5127.
40. K. Inaba, Y. H. Takahashi, N. Fujieda, K. Kano, H. Miyoshi and K. Ito, *J. Biol. Chem.*, 2004, **279**, 6761–6768.
41. K. Inaba, Y. H. Takahashi, K. Ito and S. Hayashi, *Proc. Natl. Acad. Sci. USA*, 2006, **103**, 287–292.
42. H. Kadokura, M. Bader, H. Tian, J. C. Bardwell and J. Beckwith, *Proc. Natl. Acad. Sci. USA*, 2000, **97**, 10884–10889.
43. H. Kadokura and J. Beckwith, *EMBO J.*, 2002, **21**, 2354–2363.
44. H. Kadokura, F. Katzen and J. Beckwith, *Annu. Rev. Biochem.*, 2003, **72**, 111–135.
45. J. Regeimbal, S. Gleiter, B. L. Trumpower, C. A. Yu, M. Diwakar, D. P. Ballou and J. C. Bardwell, *Proc. Natl. Acad. Sci. USA*, 2003, **100**, 13779–13784.
46. Y. H. Takahashi, K. Inaba and K. Ito, *J. Biol. Chem.*, 2004, **279**, 47057–47065.
47. R. L. Searls, J. M. Peters and D. R. Sanadi, *J. Biol. Chem.*, 1961, **236**, 2317–2322.
48. U. Grauschopf, A. Fritz and R. Glockshuber, *EMBO J.*, 2003, **22**, 3503–3513.
49. K. Inaba, Y. H. Takahashi and K. Ito, *J. Biol. Chem.*, 2005, **280**, 33035–33044.
50. M. Berkmen, D. Boyd and J. Beckwith, *J. Biol. Chem.*, 2005, **280**, 11387–11394.
51. A. Hiniker and J. C. Bardwell, *Trends Biochem. Sci.*, 2004, **29**, 516–519.
52. K. Maskos, M. Huber-Wunderlich and R. Glockshuber, *J. Mol. Biol.*, 2003, **325**, 495–513.
53. M. Wunderlich and R. Glockshuber, *J. Biol. Chem.*, 1993, **268**, 24547–24550.
54. D. Missiakas, C. Georgopoulos and S. Raina, *EMBO J.*, 1994, **13**, 2013–2020.
55. V. E. Shevchik, G. Condemine and J. Robert-Baudouy, *EMBO J.*, 1994, **13**, 2007–2012.
56. A. Rietsch, D. Belin, N. Martin and J. Beckwith, *Proc. Natl. Acad. Sci. USA*, 1996, **93**, 13048–13053.
57. N. J. Darby, S. Raina and T. E. Creighton, *Biochemistry*, 1998, **37**, 783–791.
58. J. C. Joly and J. R. Swartz, *Biochemistry*, 1997, **36**, 10067–10072.
59. A. Rietsch, P. Bessette, G. Georgiou and J. Beckwith, *J. Bacteriol.*, 1997, **179**, 6602–6608.
60. A. Zapun, D. Missiakas, S. Raina and T. E. Creighton, *Biochemistry*, 1995, **34**, 5075–5089.

61. A. Rozhkova, C. U. Stirnimann, P. Frei, U. Grauschopf, R. Brunisholz, M. G. Grutter, G. Capitani and R. Glockshuber, *EMBO J.*, 2004, **23**, 1709–1719.
62. X. X. Sun and C. C. Wang, *J. Biol. Chem.*, 2000, **275**, 22743–22749.
63. M. W. Bader, A. Hiniker, J. Regeimbal, D. Goldstone, P. W. Haebel, J. Riemer, P. Metcalf and J. C. Bardwell, *EMBO J.*, 2001, **20**, 1555–1562.
64. A. A. McCarthy, P. W. Haebel, A. Torronen, V. Rybin, E. N. Baker and P. Metcalf, *Nat. Struct. Biol.*, 2000, **7**, 196–199.
65. D. Missiakas, F. Schwager and S. Raina, *EMBO J.*, 1995, **14**, 3415–3424.
66. J. Chung, T. Chen and D. Missiakas, *Mol. Microbiol.*, 2000, **35**, 1099–1109.
67. F. Katzen and J. Beckwith, *Cell*, 2000, **103**, 769–779.
68. E. H. Gordon, M. D. Page, A. C. Willis and S. J. Ferguson, *Mol. Microbiol.*, 2000, **35**, 1360–1374.
69. E. J. Stewart, F. Katzen and J. Beckwith, *EMBO J.*, 1999, **18**, 5963–5971.
70. D. Goldstone, P. W. Haebel, F. Katzen, M. W. Bader, J. C. Bardwell, J. Beckwith and P. Metcalf, *Proc. Natl. Acad. Sci. USA*, 2001, **98**, 9551–9556.
71. A. Hiniker, D. Vertommen, J. C. Bardwell and J. F. Collet, *J. Bacteriol.*, 2006, **188**, 7317–7320.
72. C. U. Stirnimann, A. Rozhkova, U. Grauschopf, R. A. Bockmann, R. Glockshuber, G. Capitani and M. G. Grutter, *J. Mol. Biol.*, 2006, **358**, 829–845.
73. C. U. Stirnimann, A. Rozhkova, U. Grauschopf, M. G. Grutter, R. Glockshuber and G. Capitani, *Structure*, 2005, **13**, 985–993.
74. S. H. Cho, A. Porat, J. Ye and J. Beckwith, *EMBO J.*, 2007, **26**, 3509–3520.
75. R. Krupp, C. Chan and D. Missiakas, *J. Biol. Chem.*, 2001, **276**, 3696–3701.
76. C. W. Goulding, M. R. Sawaya, A. Parseghian, V. Lim, D. Eisenberg and D. Missiakas, *Biochemistry*, 2002, **41**, 6920–6927.
77. D. A. Mavridou, J. M. Stevens, S. J. Ferguson and C. Redfield, *J. Mol. Biol.*, 2007, **370**, 643–658.
78. A. Rozhkova and R. Glockshuber, *J. Mol. Biol.*, 2007, **367**, 1162–1170.
79. P. H. Bessette, J. J. Cotto, H. F. Gilbert and G. Georgiou, *J. Biol. Chem.*, 1999, **274**, 7784–7792.
80. C. L. Andersen, A. Matthey-Dupraz, D. Missiakas and S. Raina, *Mol. Microbiol.*, 1997, **26**, 121–132.
81. M. van Straaten, D. Missiakas, S. Raina and N. J. Darby, *FEBS Lett.*, 1998, **428**, 255–258.
82. Z. Zhang, Z. H. Li, F. Wang, M. Fang, C. C. Yin, Z. Y. Zhou, Q. Lin and H. L. Huang, *Protein Expr. Purif.*, 2002, **26**, 218–228.
83. F. Shao, M. W. Bader, U. Jakob and J. C. Bardwell, *J. Biol. Chem.*, 2000, **275**, 13349–13352.
84. B. Heras, M. A. Edeling, H. J. Schirra, S. Raina and J. L. Martin, *Proc. Natl. Acad. Sci. USA*, 2004, **101**, 8876–8881.
85. R. A. Fabianek, H. Hennecke and L. Thony-Meyer, *J. Bacteriol.*, 1998, **180**, 1947–1950.

86. M. D. Page and S. J. Ferguson, *Mol. Microbiol.*, 1997, **24**, 977–990.
87. E. Reid, D. J. Eaves and J. A. Cole, *FEMS Microbiol. Lett.*, 1998, **166**, 369–375.
88. E. M. Monika, B. S. Goldman, D. L. Beckman and R. G. Kranz, *J. Mol. Biol.*, 1997, **271**, 679–692.
89. M. A. Edeling, L. W. Guddat, R. A. Fabianek, L. Thony-Meyer and J. L. Martin, *Structure*, 2002, **10**, 973–979.
90. M. A. Edeling, U. Ahuja, B. Heras, L. Thony-Meyer and J. L. Martin, *J. Bacteriol.*, 2004, **186**, 4030–4033.
91. H. Crooke and J. Cole, *Mol. Microbiol.*, 1995, **15**, 1139–1150.
92. R. Metheringham, K. L. Tyson, H. Crooke, D. Missiakas, S. Raina and J. A. Cole, *Mol. Gen. Genet.*, 1996, **253**, 95–102.
93. Y. Sambongi and S. J. Ferguson, *FEBS Lett.*, 1994, **353**, 235–238.
94. C. U. Stirnimann, M. G. Grutter, R. Glockshuber and G. Capitani, *Cell Mol. Life Sci.*, 2006, **63**, 1642–1648.
95. J. S. Richardson, *Adv. Protein Chem.*, 1981, **34**, 167–339.
96. Y. Kurokawa, H. Yanagi and T. Yura, *J. Biol. Chem.*, 2001, **276**, 14393–14399.
97. R. Levy, R. Weiss, G. Chen, B. L. Iverson and G. Georgiou, *Protein Expr. Purif.*, 2001, **23**, 338–347.
98. M. Schlapschy, S. Grimm and A. Skerra, *Protein Eng. Des. Sel.*, 2006, **19**, 385–390.

CHAPTER 1.3

The Periplasm of E. coli – Oxidative Folding of Recombinant Proteins

KATHARINA M. GEBENDORFER
AND JEANNETTE WINTER

Department Chemie, Lehrstuhl Biotechnologie, Technische Universität München, Lichtenbergstraße 4, 85747 Garching, Germany

1.3.1 *Escherichia coli* as Host for the Production of Recombinant Proteins – Benefits and Drawbacks

The enteric bacterium *Escherichia coli* has been extensively applied for the production of heterologous proteins over the past decades.[1] The rationales for its widespread use lay in a distinguished cost–value ratio. Not only is *E. coli* simple and inexpensive to cultivate, it rapidly generates biomass and is accessible to high-cell density fermentation. Bacterial expression systems have a relatively short process time due to an easier scale and shorter fermentation run times compared to eukaryotic systems. *E. coli*'s physiology and genetics are well studied and fairly well understood. Based on this knowledge and with the genome sequence being available, *E. coli* is easily accessible to genetic manipulation, therefore allowing for a fast development of an *E. coli*-based expression system. Further, a variety of expression vectors and modified host strains is available, which simplifies the customization of the expression system.[2] In addition, the target protein can be directed to the cytoplasm, periplasm

RSC Biomolecular Sciences
Oxidative Folding of Peptides and Proteins
Edited by Johannes Buchner and Luis Moroder
© Royal Society of Chemistry 2009
Published by the Royal Society of Chemistry, www.rsc.org

or cultivation media depending on its requirements for folding and accumulation. These facts make *E. coli* a versatile and valuable host for the tailor-made production of recombinant proteins.

Despite the advantages of an *E. coli*-based expression system, there are certain limitations in its application. Mammalian proteins in particular, which have a tremendous pharmaceutical potential, are often insufficiently produced in *E. coli*. Differences in mRNA stability, codon usage and translation efficiency are one reason (reviewed in ref. 3). Another concerns protein folding of the eukaryotic protein within the bacterial host. Eukaryotic proteins are typically larger than bacterial ones and they may be incompatible with the bacterial folding machinery.[4] Further, *E. coli* is unable to perform most of the post-translational modifications observed in eukaryotic proteins. As a consequence, eukaryotic proteins may be toxic to the host, unstable and tend to aggregate or become substrates of host proteases. Yet, *E. coli* may be improved to meet this demand (see also Section 1.3.5).[5–7] Further, many eukaryotic proteins retain their biological activity even if they are not glycosylated[8,9] and may be highly suitable for structure determination.[10]

1.3.2   Cytoplasm, Periplasm or Cultivation Media – Where to Direct the Target Protein?

The two *E. coli* compartments cytoplasm and periplasm (Figure 1.3.1) each exhibit unique characteristics that can be exploited compatibly to the nature of the target protein. The cytoplasm is the commonly used compartment for over-expression because it allows for high yields of the recombinant protein. It contains an extensive set of chaperones that assist the co- and post-translational folding of

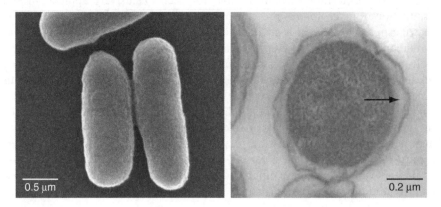

Figure 1.3.1   Electron micrographs of *E. coli* cells. Left, scanning electron micrograph at 35 000-fold magnification. Typical dimensions of *E. coli* cells are 0.6 μm width and 1.5 μm length. Right, transmission electron micrograph of a thin section of cells in stationary growth phase. Please note the expanded periplasm (indicated by an arrow) that is typical for stationary cells.

polypeptides. Frequently, protein aggregates, so-called inclusion bodies (see Chapter 2.2), are formed. While this is a drawback if a correctly folded, soluble protein was aimed for, it may also be advantageous. Inclusion bodies protect the included recombinant protein from proteolytic degradation, mainly consist of the recombinant protein and are fairly easy to isolate thereby avoiding extensive purification steps. Providing that the target protein readily reactivates under refolding conditions, the production of recombinant proteins in inclusion bodies remains the prior choice (reviewed in ref. 11 and 12). In most cases however, refolding is inadequate. Then, it may be favorable to direct the recombinant protein to the periplasm or the extracellular cultivation media.

Particularly if the function of the recombinant protein relies on correct disulfide-bond formation, the periplasm or the cultivation media are the destinations of choice. This is mainly due to their oxidizing redox potential that favors oxidative folding. The reducing redox potential of the cytoplasm (-0.24 V to -0.27 V) is driven by thioredoxin and the ratio of reduced/oxidized glutathione while the oxidizing environment of the periplasm (-0.165 mV) depends on the redox properties of the Dsb proteins and glutathione (see Chapter 1.2).[13] It should be noted that also the cytoplasm can be engineered to allow for oxidative protein folding. Stable disulfides can form in strains lacking thioredoxin reductase ($\Delta trxB$)[14] and even more rapidly in strains that are defective in the thioredoxin and glutathione pathway ($\Delta trxB$ *gor*, $\Delta trxB$ *gshA*, $\Delta trxB$ *gor ahpC*) (see Chapter 1.1).[15,16] Given that isomerization of wrong disulfide bonds does not occur, the utilization of such strains for biotechnological purposes may be limited.

Besides the favorable conditions for oxidative protein folding, the periplasm provides many more advantages over the cytoplasm. Considering the low overall amount of periplasmic proteins and the small volume of the periplasmic space, the secreted target protein is thus more concentrated and at the same time sufficiently enriched relative to other periplasmic proteins making its purification less cumbersome. Periplasmic inclusion bodies may form; the major difference is, however, that the included proteins may be easier to extract and may be fully active.[17–20] It should also be considered that the periplasm lacks ATP. Consequently, the folding of secreted proteins has to proceed without the assistance of ATP-dependent foldases. On the other hand, also ATP-dependent proteases are missing thereby limiting proteolytic degradation in the periplasm and procuring high product quality.[21] Advantageous for any protein, secretion is mediated by a signal sequence that is cleaved by the signal peptidase (see Section 1.3.4); therefore, the secreted protein contains its authentic N-terminus without the methionine extension. This also holds true if the recombinant protein is targeted to the cultivation media. However, secretion to the media is not very effective and suffers low productivity. While the secreted protein is indefinitely diluted in the media thereby certainly disfavoring aggregation and favoring folding as well as easing purification, it also has to be able to fold independently of any chaperones. Being aware of this deficiency, additives known to improve folding are frequently added to the media. They may also improve the folding of periplasmic recombinant proteins owing to the fact that porins in the outer membrane of *E. coli* allow small molecules to be exchanged between periplasm and cultivation media.[22]

1.3.3 Physiology and Properties of the Periplasm

The periplasm of *E. coli* is located between the inner and outer membrane (Figure 1.3.1). Under physiological conditions, the periplasm of a wild type strain is roughly 10 nm thick and its volume is about one tenth that of the cytoplasm (Figure 1.3.1, CyberCell database CCDB). It carries only about 80 000 proteins, which account for about 2% of the total cellular protein (CyberCell database CCDB). If cells are to live and grow they require a constant turgor pressure and concentration of constituents such as sugars, amino acids and nucleotides in the cytoplasm. Stock and co-workers showed that the cytoplasm and periplasm have similar osmolarity, which is higher than that of the external media.[23] To maintain a high osmolarity in the periplasm a Donnan potential is generated across the outer membrane,[23] which is mediated by a high concentration of periplasmic fixed anions. Membrane-derived oligosaccharides that contain eight to ten highly branched glucose molecules function as fixed anions.[24] With a molecular weight of 2200–2600 Da they are impermeable to the outer membrane.[24] This is especially important given that channels in the outer membrane of *E. coli*, so-called porins, allow molecules smaller than 600 Da to diffuse between the periplasm and the external media.[22,25,26] The rate of synthesis of membrane-derived oligosaccharides is controlled by the external osmotic pressure enabling *E. coli* to regulate its osmolarity. In combination with the strength of the outer membrane and peptidoglycan layer, the composition of the periplasm is therefore essential to facilitate a constant cellular turgor as well as shape of the bacteria.

The periplasm is devoid of ATP;[27] therefore, all enzymes including those relevant to protein folding and turnover have to function ATP independently. Nevertheless, few chaperones and proteases exist with DegP being the most prominent example. DegP switches between protease and chaperone activity in a temperature-dependent manner.[28] At physiological temperatures DegP primarily functions as general chaperone. At elevated temperatures, however, the protease activity of DegP becomes more prominent. In this case, DegP degrades abnormal, misfolded and oxidatively damaged proteins.[29,30] The increase in DegP concentration[31,32] and protease activity upon elevated temperatures ensure efficient protein quality control in the periplasm. Three more periplasmic proteases are known: OmpT, protease III (PtrA) and Prc (Tsp).[33–35] None of the periplasmic proteases are essential under non-stress conditions, and even their simultaneous deletion is not lethal to *E. coli*.[36]

Few other periplasmic folding catalysts have been identified. FkpA and SurA possess chaperone and peptidyl-prolyl isomerase activity,[37–39] while Skp fulfills only the former[40] and PpiA and PpiD only the latter activity. Skp is a general chaperone; it captures unfolded proteins, soluble and membrane proteins alike, as they emerge from the cytoplasm *via* the Sec translocation machinery.[41]

The concerted chaperone function of Skp, SurA and DegP is required for outer membrane protein assembly. Although all three function in parallel, SurA is the primary chaperone for outer membrane protein biogenesis while Skp and DegP rather constitute a backup pathway in this respect, whose importance increases under stress conditions.[42]

Most important for the oxidative folding of recombinant proteins are the periplasmic oxidoreductases (see Chapter 1.2). The Dsb proteins together with glutathione maintain the redox balance in the periplasm[13] with reduced but not oxidized glutathione being transported across the cytoplasm membrane *via* a specific transporter.[43] It seems, however, that glutathione functions as redox buffer and does not directly act on a protein disulfide.[13] DsbA is a strong oxidase and introduces disulfide bonds the very moment a protein enters the periplasm. Owing to its fast action, it is thought that DsbA oxidizes cysteines adjacent in the primary protein sequence, so-called consecutive cysteines. Nevertheless, DsbA was recently shown to correctly oxidize a protein with a non-consecutive disulfide.[13] DsbC and DsbG function as disulfide isomerases; their active site cysteines are maintained in the reduced state, which is a prerequisite for isomerase activity.[44,45] While no substrates are known for DsbG, DsbC has been shown to act on substrate proteins with multiple, non-consecutive disulfides.[46,47] Since many therapeutically relevant proteins contain multiple disulfide bonds, the joint action of DsbA and DsbC may enable their proper oxidative folding in the periplasm (see Section 1.3.5).

1.3.4   The Periplasm – How to Get There?

Compartmentalization of the *E. coli* cell is mediated by sealed membrane systems. As a consequence, a large subset of newly synthesized proteins must be inserted into membranes or transported across membranes to reach their desired destinations. Translocation to the periplasm occurs mainly *via* two pathways: the general secretion (Sec) pathway and the twin arginine translocation (Tat) pathway (reviewed in ref. 48). Both pathways exist in bacteria, archaea and eukaryotes. In eukaryotes, the Sec pathway is present in the endoplasmic reticulum and thylakoid membranes of chloroplasts; the latter also contains Tat translocons. The primary sequence of secreted proteins (preproteins) is distinctly different from that of proteins whose final destination remains the cytoplasm. Secreted proteins require an extension at their N-terminus that is recognized by the respective secretion apparatus. Upon completion of secretion, this sequence is removed. Therefore, it is not part of the structure and not required for the folding of the mature part.

1.3.4.1   Signal Sequences

The N-terminal amino acid sequence that designates a protein for secretion into the periplasm is the signal sequence or leader peptide (Figure 1.3.2). Signal sequences have a tripartite structure and are typically 15 to 30 amino acids long.

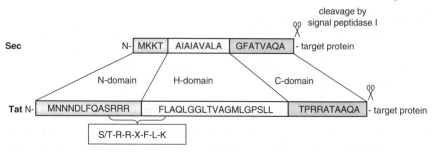

Figure 1.3.2   Schematic representation of signal sequences that direct preproteins to the Sec pathway or the Tat pathway. As an example, the amino acid sequences of the OmpA leader peptide (Sec pathway) and the TorA leader peptide (Tat pathway) are given (see ref. 50 for a detailed overview on sequences of Sec leader peptides). In general, signal sequences contain three domains: the polar N-region with a positive net charge, the hydrophobic uncharged H-region and the polar C-region that contains the cleavage site for signal peptidase I. Within the cleavage site small amino acids with neutral side-chains are found. Since this is usually alanine, a so-called Ala-X-Ala box is formed. Tat signal sequences are typically longer than those for the Sec pathway. They almost always contain a Ser/Thr-Arg-Arg-X-Phe-Leu-Lys motif, where X is any polar amino acid. The Arg-Arg motif is conserved and gives the pathway its name although a few exceptions are known where the first arginine is substituted by lysine.

The properties of the signal sequence depend on the translocation system they are targeting, either the Sec system or the Tat pathway. Signal sequences that target proteins to the Sec pathway are rich in hydrophobic amino acids like alanine, valine and leucine (Figure 1.3.2). The positively charged N-domain, located at the very N-terminus, is about two to ten amino acids long. It is followed by a hydrophobic H-domain of about ten to twenty amino acids.[49] The polar C-domain is typically less hydrophobic and contains the cleavage site for the signal peptidase. At this exact position the signal sequence is cleaved from the protein after it is transported across the cytoplasm membrane and enters the periplasm. Within the cleavage site the amino acid residues at positions −1 and −3 are required to have a small neutral side-chain, like alanine, glycine or serine, according to the −3, 1 rule.[49,50] In the majority of cases alanine is found at these positions and a so-called Ala-X-Ala box is formed. The cleavage site is not only determined by the primary sequence but also by the secondary structure at the cleavage junction of the preprotein.[51] Signal sequences targeting proteins to the Tat pathway are usually about 14 amino acids longer than those for the Sec pathway (Figure 1.3.2). These leader peptides almost always contain a Ser/Thr-Arg-Arg-X-Phe-Leu-Lys motif (X is a polar amino acid) at the interface of the N-domain and the H-domain. Their H-domain is typically less hydrophobic because it contains more glycine and threonine residues.[52,53] The Arg–Arg sequence between the N- and H-domain is fairly conserved and gives the twin arginine pathway its name. However, exceptions are known in nature, in which the first arginine is exchanged by lysine.[54–56]

1.3.4.2 Secretion of Unfolded Proteins *via* the Sec Pathway

Translocation *via* the Sec pathway requires certain components: an N-terminal signal sequence in the substrate protein, the ATP-dependent motor protein SecA and the membrane proteins SecYEG (Figure 1.3.3). Newly synthesized proteins are guided to the Sec translocase by two means, either *via* the signal recognition particle (SRP) or *via* the chaperone SecB. In either case the substrate proteins are maintained in an unfolded state, which is a prerequisite for Sec-dependent translocation. Transport *via* the ribonucleoprotein SRP occurs co-translationally; SRP binds to the signal sequence as soon as the secretory protein emerges from the ribosome. Subsequently, it mediates targeting of the ternary complex consisting of SRP, nascent polypeptide and ribosome to the Sec translocase.[57,58] Upon arrival, the nascent polypeptide–ribosome-complex is transferred to the Sec translocase *via* the aid of an SRP-receptor. By this pathway, most integral membrane proteins are inserted into the cytoplasm membrane. Importantly, the interaction of the ribosome and SecYEG[57,59] seems to be the driving force for membrane protein insertion. A subset of integral membrane proteins is inserted into the cytoplasmic membrane *via* the aid of YidC, which may act on its own or cooperate with the Sec system.[60] The majority of secretory proteins, however, are translocated by post-translational targeting *via* SecB. It appears that the decision on SRP- or SecB-mediated translocation is determined by a competition between SRP and trigger factor for binding to the nascent chain. Strongly hydrophobic N-termini and hydrophobic transmembrane stretches are preferentially recognized and bound by SRP. A trigger factor may, however, block SRP-nascent chain interaction thereby directing nascent preproteins to SecB.[61] SecB is a secretion specific chaperone, although it seems also to have general chaperone activity.[62,63] In contrast to SRP, which specifically binds to the signal sequence, SecB rather binds to different parts of long nascent secretory polypeptide chains thereby maintaining them unfolded.[64] The SecB-nascent chain complex specifically interacts with SecA upon which the preprotein is transferred to SecA and SecB is released.[65,66] SecA is peripherally associated with the membrane-spanning SecYEG complex, which forms the protein-conducting channel. SecA converts chemical energy (ATP) into conformational changes that together with the proton motive force mediate protein translocation.[67,68] Preprotein and ATP binding to SecA initiates translocation. Upon ATP binding, SecA inserts the N-terminal part of the preprotein onto the translocation pore. Upon ATP hydrolysis, the polypeptide is released from SecA but stays trapped in the SecYEG pore. SecA either dissociates or re-binds the trapped protein resulting in the translocation of a polypeptide stretch of 2 to 2.5 kDa. Subsequent ATP binding then causes membrane insertion of SecA and translocation of another segment of 2 to 2.5 kDa.[69–71] Repeated cycles of nucleotide-dependent SecA membrane insertion and de-insertion mediate the stepwise movement of the whole polypeptide chain through the translocation complex, with about 5 kDa translocated upon each complete cycle.[68,69] As the last step of protein translocation, the membrane bound signal peptidase I removes the signal sequence from the emerging polypeptide and thereby releases the mature protein.[72]

1.3.4.3   Secretion of Folded Proteins *via* the Tat Pathway

In contrast to the Sec pathway, which translocates the majority of secreted proteins, the Tat pathway transports selected substrates. The most remarkable feature of the Tat pathway is that the substrate proteins are transported in a folded state and may even contain disulfide bonds or cofactors.[73–75] To achieve that, the Tat transporter has selectively to move structured macromolecules across the membrane. At the same time, the permeability barrier of the membrane to ions and small molecules must not be compromised. This is possible because the membrane spanning Tat complex, consisting of the proteins TatA, TatB and TatC, forms a large pore with variable diameter. Translocation *via* the Tat pathway is independent of ATP but seems to depend on the proton motif force.[76,77] However, it remains unclear how the proton motif force provides energy for translocation and at which step of translocation it is required. Although translocation *via* the TatABC complex takes longer than translocation *via* the Sec pathway and the energetic cost may be higher,[78,79] Tat translocation offers an alternative for proteins with special folding requirements. Natural substrates of the Tat pathway are redox enzymes, which require cofactor insertion within the cytoplasm,[80] multimeric proteins, which assemble in the cytoplasm,[81] some membrane proteins[82] and proteins whose folding is not compatible with Sec export like the periplasmic amidases AmiA and

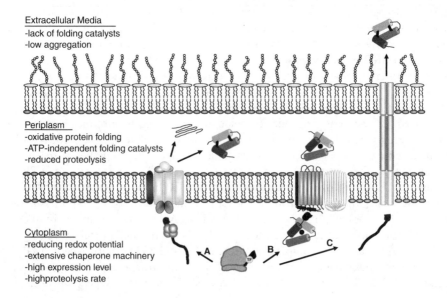

AmiC.[83] How proteins are targeted to the Tat complex is not entirely understood. It has been shown that the Tat signal sequence binds to the TatBC complex. TatC consists of six transmembrane helices[84] and initially binds the twin-arginine motif.[85] After recognition by TatC, TatB contacts the signal sequence and parts of the mature protein. TatB is anchored to the membrane *via* its N-terminal hydrophobic segment; the following cytoplasmic amphipathic helix and highly charged C-terminal region are critical for substrate recognition and translocation.[86–88] TatB supposedly participates in the transfer

---

Figure 1.3.3   Model of preprotein targeting and translocation in *E. coli*. Preproteins destined for secretion to the periplasm can be translocated *via* two different pathways (A and B). **(A)** The majority of preproteins undergo Sec-dependent export. The trigger factor (TF) associates with the nascent polypeptide chain as soon as it emerges from the ribosome. Upon further growth of the nascent chain, the TF dissociates and the polypeptide is transferred to SecB (depicted as homotetramer). The SecB-nascent chain complex is delivered to SecA and the membrane-spanning SecYEG complex. ATP-dependent conformational changes of SecA mediate translocation of the preprotein through the SecYEG pore. Once emerged into the periplasm, the signal sequence is cleaved off from the preprotein. The mature protein can either fold to its native state *via* the aid of periplasmic chaperones and Dsb proteins or may fail to fold and become degraded by periplasmic proteases. The SRP pathway, which mainly serves to insert integral membrane proteins into the cytoplasm membrane, is not depicted. Briefly, nascent preproteins with highly hydrophobic N-termini are captured by SRP before they can associate with TF. The ternary SRP-nascent chain–ribosome complex is delivered to the SRP receptor and SecYEG and continues as described for SecYEG. **(B)** Upon emerging from the ribosome the nascent preprotein folds into its native state and may acquire cofactors (depicted as a circle). Disulfide-bond formation requires an oxidizing cytoplasm. The folded protein is targeted to the TatBC receptor complex. TatC (depicted in grey) contains six transmembrane helices and first contacts the Arg–Arg motif of the Tat signal sequence. Afterwards, the cytoplasmic region of TatB (in dark grey) interacts with the N-terminus of the preprotein. Substrate proteins are transferred to TatA, which forms membrane-spanning oligomers of various sizes (in white) and supposedly constitutes the translocation pore (adapted from ref. 73). The pore size is variable and matches the requirement of the transported substrate protein. After translocation is completed and the signal sequence has been removed by signal peptidase I, TatA dissociates from TatBC and the mature protein is released into the periplasm. It should be noted that Tat-dependent translocation is mediated by the proton motif force but not ATP. **(C)** Secretion of preproteins to the external media can be achieved by active transport of preproteins *via* type 1 secretion/ABC transport and β-autotransporter. It is schematically depicted for type 1 secretion. Furthermore, passive transport from the cytoplasm or the periplasm to the media can be achieved by external or internal permeabilization or mutations that result in the lack of structural components (lipoproteins, outer membrane proteins) and defective membrane structure (for a review see ref. 166). Some hallmarks of the different compartments for recombinant protein production are listed (for details see Sections 1.3.2 and 1.3.3).

of the substrate to TatA before translocation.[85] The structure of TatA is similar to that of TatB but TatA has a shorter C-terminal region.[89] TatA is about 20 times more abundant than TatB and TatC and likely forms the Tat pore of variable diameter upon substrate binding to the TatBC complex.[90] Yet, three different complexes have been isolated: a TatB/TatC complex in 1:1 stoichiometry, a TatA/TatB complex with variable stoichiometry and a complex containing all three components with TatA in a large molar excess, which resembles the cellular ratio.[91–93] It is assumed that complexes of different composition may form pores of different sizes that exactly match the requirement of the folded Tat substrate protein. Restricting the pore size minimizes the space that is not occupied by the substrate thereby preventing leakage of ions and water. After translocation across the Tat pore, TatA dissociates from the TatBC complex, the signal sequence is removed by signal peptidase I and the mature protein is released. This last step is identical for proteins translocated *via* the Sec and the Tat pathway and leaves the secreted protein with its authentic N-terminus.

1.3.5 Biotechnological Application – the Periplasm as Production Compartment for Recombinant Proteins

The periplasm is the compartment of choice if high product quality is desired. It provides a reduced set of proteases and favorable redox conditions, thereby avoiding misfolding and aggregation of the recombinant protein. Commonly, recombinant proteins are secreted to the periplasm *via* the Sec pathway, which allows higher translocation efficiencies compared to the Tat pathway. However, some proteins may be determined to be secreted *via* Tat. These include proteins that may not be compatible with the Sec pathway and fold prior to translocation or proteins that rely on the assistance of ATP-dependent chaperones in the cytoplasm and fold only slowly in the periplasm meanwhile being degraded by DegP before reaching their native state.[94] As an example, green fluorescent protein (GFP) can be translocated into the periplasm *via* the Sec and the Tat pathway. However, GFP is only fluorescent if secreted *via* Tat, suggesting that GFP needs to fold in the cytoplasm prior to export in order to gain biological activity.[73,95,96] This should be taken into account when choosing the secretion pathway.

There are countless examples for the secretory expression of recombinant proteins in *E. coli*; various signal peptides have been tested with various client proteins. Obtained yields may be as high as 8.5 g per litre for insulin-like growth factor I where the majority accumulates in inclusion bodies[18] or 4.7 g per litre for alkaline phosphatase, which is mainly soluble in the periplasm.[97] Yet, in many cases, yields are low if not negligible, which is most likely due to the intrinsic folding requirements of the recombinant protein. It becomes obvious that there is not one signal peptide that is generally applicable to all recombinant proteins but that for each target protein the suitable signal peptide has to be identified experimentally. The same holds true for the production conditions; consideration of fusion partners, supplementation of the cultivation media with certain additives, genetic engineering of the target protein and fermentation process design are starting points for the optimization of secretory expression. Further, engineered *E. coli* strains that either lack certain proteases or co-secrete *E. coli* or foreign chaperones may be useful. We will focus here on a few examples of therapeutic proteins because of their high pharmaceutical value and their increasing use in clinics. The most prominent therapeutic proteins are antibodies and proinsulin, but growth hormones and tissue plasminogen activator will also be discussed.

1.3.5.1 Production of Antibodies and Antibody Fragments

Antibodies, or immunoglobulins, are Y-shaped molecules with the arms of the "Y" called Fab (fragment antigen binding) region and the base called Fc (fragment constant) region. The Fab region contains the light and heavy chains comprising the variable domains that specifically bind antigen while the Fc region is involved in modulating immune cell activity. scFv (single chain fragment variable) fragments contain only the variable domains joined by a flexible polypeptide linker. Fab and scFv fragments retain the antigen-binding specificity of the whole molecule but their specificity, stability and half-life can be engineered to meet the desired criteria of application.

A large number of antibody-based drugs are used in clinics or currently tested in clinical trials. Many forms of cancer including breast and colorectal cancer and non-Hodgkin lymphoma, as well as diseases such as rheumatoid arthritis, multiple sclerosis, psoriasis and some immune deficiencies, are currently treated by antibody therapy (for a review see ref. 98 and 99). Whole antibodies provide an excellent target-binding specificity in therapy; approved therapeutic antibodies are currently produced in transgenic mice and mammalian cells. Then again, antibody fragments are increasingly used in clinical imaging applications and *in vitro* diagnostics. As such they represent a large market share in the diagnostic industry.[100] The production of antibodies and antibody fragments for preclinical and clinical trials and diagnostics has been evaluated in many expression systems, including bacteria, yeast, plant, insect and mammalian cells (see also Chapter 1.7). Bacteria are even favored for expression of small Fab and scFv fragments.[9,99] One has to keep in mind, though, that *E. coli* lacks the ability to glycosylate proteins; aglycosylated antibodies can, however, comprise tight antigen binding and a long half-life.[9] Additionally, *E. coli* produced antibodies seem to lack some effector functions.[9] This property may even be desirable in certain therapeutic applications where the antibody is required to block certain targets (*e.g.*, ref. 101 and 102).

Most of the numerous examples for the secretory production of antibodies and antibody fragments in *E. coli* are proof-of-principle studies. The obtained yield would need broad optimization in order to meet the biotechnological production demand. However, there are some remarkable examples for the efficient production of full-length antibodies,[9] Fab fragments,[103,104] scFv fragments[105,106] and bi-specific antibodies (Table 1.3.1).[107–109] It should be noted that these systems cannot be directly compared given the diversity of antibodies produced and the non-uniformly indicated yield (volumetric or specific yield). Simmons and co-workers demonstrated that aglycosylated full-length IgGs can be successfully expressed and directed to the periplasm of *E. coli*.[9] They used a two-cistron system for the translation of heavy and light

Table 1.3.1   Secretory production of antibodies and antibody fragments.

Protein	Secretion via[a]	Obtained yield[b]	Process characteristics[c]	Ref.
IgG	PelB	$1 \, mg \, l^{-1}$	Combinatorial library[d]	161
	StII	$156 \, mg \, l^{-1}$	Fermentation	9
Fab	yBGL2	$560 \, mg \, l^{-1 \, e}$	Fermentation	103,104
	StII	$2.45 \, g \, l^{-1 \, f}$	Fermentation, protease-deficient strains	107–109
	n.a.	$0.5 \, mg \, l^{-1}$	Co-secretion of human PDI	162
	TorA	n.a.	Requires oxidizing cytoplasm	75
scFv	TorA	$0.24 \, mg \, OD^{-1}$	Requires oxidizing cytoplasm	75,119
	OmpA	$0.8 \, mg \, OD^{-1 \, g}$	Protease-deficient strains	163
	PelB	$0.025 \, mg \, OD^{-1}$	Arginine supplementation	134
		$50 \, mg \, g^{-1}$	Sorbitol, betain supplementation, co-secretion of Skp	105
		$3.3 \, g \, l^{-1 \, h}$	Fermentation	106
		$0.095 \, mg \, OD^{-1}$	Diabodies, triabodies	117
	PelB[i]	$1 \, mg \, l^{-1}$	Produced as diabodies	116
	Phage	n.a.	Co-secretion of Skp or FkpA	112,164

[a]Secretion was accomplished *via* the signal sequences PelB from *Erwinia carotovora*, StII from *E. coli* heat-stable enterotoxin, yBGL2 from yeast, *E. coli* TorA mediating secretion *via* the Tat pathway and *E. coli* OmpA.
[b]The given yield represents the maximum amount obtained. "$mg \, l^{-1}$" corresponds to mg native protein per litre broth; "$mg \, OD^{-1}$" relates to mg per litre and OD unit; "$mg \, g^{-1}$" corresponds to mg native protein per gram dry biomass; "n.a." means that no information was given in the respective reference.
[c]Cultivations were performed in shake flasks if not otherwise indicated; "fermentation" corresponds to high cell density fermentation.
[d]Full-length IgGs were isolated from combinatorial libraries expressed in *E. coli*.
[e]Corresponds to the amount of purified protein. Optical densities of 80 to 100 were obtained.
[f]Corresponds to the amount of functional $F(ab')_2$ heterodimers in the periplasm. Based on an OD 250,[107] the specific yield is $50 \, mg \, F(ab')_2$ per gram dry biomass. Carter and co-workers cultivated *E. coli* to OD 150.[109]
[g]Calculated based on the report that scFv accumulates to 30% of the periplasmic proteins.
[h]Corresponds to $25 \, mg \, g^{-1}$. *E. coli* cells were cultivated to OD 400 with almost maximum growth rate throughout the fermentation and without any substrate limitations.
[i]Diabodies were produced *via* two different signal sequences, PelB and the signal sequence from the phage fd gene 3.

chains. By optimizing their translation level and ratio, high expression levels were obtained and full-length antibodies could be assembled in the periplasm. This approach is applicable for several full-length antibodies; with up to 150 mg full-length IgG per litre from high cell density fermentations this process can likely compete with industrial production schemes. Further, the assembled antibodies show all properties expected of an aglycosylated antibody, including tight binding to antigen and the lack of certain effector functions. At the same time the full-length IgGs retain the long circulating half-life that is observed with antibodies derived from mammalian cells.[9]

Humphreys and co-workers succeeded in producing large quantities of a Fab fragment in the periplasm using the yeast BGL2 signal sequence. By optimizing the codon usage of the signal sequence and applying fed-batch fermentation strategies they obtained 260 mg purified Fab per litre.[104] In a subsequent study, they used the codon usage optimized signal peptide of the major coat protein of the bacteriophage M13.[103] The light and heavy chain sequences were expressed from a dual plasmid; upon mutating codons at the 5' of these genes and varying their expression ratio, functional Fab fragments were efficiently accumulated in the periplasm. Using such optimized constructs in high cell density fermentations they could purify up to 560 mg Fab per litre.[103] Many clinical applications of antibodies, however, require large quantities of bivalent and humanized fragments. Bispecific Fab fragments, F(ab')$_2$, have been produced in the periplasm with remarkable yields that make their technical secretory production in *E. coli* practical. Carter and co-workers[109] produced Fab' fragments that dimerize to F(ab')$_2$ fragments. Fab' fragments contain additional amino acids at the C-terminus of the heavy chain including one cysteine. This cysteine forms a disulfide bridge to a second Fab' therefore giving rise to bispecific F(ab')$_2$. They obtained secretion levels as high as 2 gram functional Fab' per litre upon high cell density fermentation. However, unlike anticipated, the Fab' fragments did not dimerize in the periplasm *via* the cysteine. Functional F(ab')$_2$ were instead obtained *in vitro* after affinity purification and chemical coupling with a maximum yield of approximately 750 mg per litre. The obtained fragments are as functional as F(ab')$_2$ obtained by proteolysis of intact antibodies.[109] To circumvent *in vitro* coupling, diabodies were created by genetically fusing the genes for the Fab' fragments.[108] By this approach, functional F(ab')$_2$ heterodimers (700 mg per litre) could be obtained directly from the periplasm after high cell density fermentation. The most impressive example is the production of functional F(ab')$_2$ heterodimers where the Fab' fragments are joined by a leucine zipper.[107] Proteolytic degradation limited the accumulation of antibody fragments. However, in protease deficient strains ($\Delta degP$ Δprc) additionally carrying a suppressor (Δspr), a remarkable 2.5 gram functional F(ab')$_2$ heterodimers per litre could be obtained (corresponds to 50 mg F(ab')$_2$ per gram dry biomass). Importantly, these *E. coli* strains could be cultivated in high cell density fermentations despite their mutations.

A similarly high yield of an antibody fragment was obtained by fusing the PelB signal sequence to the sequence of a miniantibody, an scFv fragment with a C-terminal linker that promotes homodimerization.[105,106] Horn and co-workers used an expression vector with high plasmid stability and a very strong Shine–Dalgarno sequence.[106] They produced functional bivalent miniantibodies based on chimeric scFv fragments. Upon high cell density fermentation they could obtain a maximum volumetric yield of 3.3 gram per litre, which corresponds to a maximum specific yield of 25 mg per gram dry biomass.[106] In a later study, Kraft and co-workers doubled the specific yield of a different miniantibody. By supplementing the cultivation media with sorbitol and betain as well as co-secreting the periplasmic chaperone Skp they obtained 50 mg functional miniantibody per gram dry biomass.[105] Other studies also demonstrated that co-secreted Skp increases the amount of active scFv in the periplasm.[110–112] Sorbitol and betain are osmolytes known to promote protein stability;[113,114] sorbitol is also a common buffer component in antibody formulations.[115] Similar to miniantibodies, multivalent scFv-based antibodies can be produced (reviewed in ref. 100). By using a linker sequence that is too short to allow pairing between both domains from one polypeptide, the domains are forced to pair with the complementary domains of a different polypeptide chain. Thereby, diabodies comprising two antigen-binding sites are created. When the heavy and light chains of two different antibodies are fused and paired, bispecific antibody fragments can be produced.[116] Even triabodies and tetrabodies can be created by modifying the length of the linker sequence between heavy and light chain.[117] However, the obtained yields are significantly lower than described for scFv fragments, miniantibodies and Fab fragments.

Research also aims at *E. coli* secretory production of engineered antibodies with improved or altered specificity, an increased expressibility and higher stability regarding production and half-life in applications. Random mutagenesis and phage display[118] are two auspicious methods to select for improved properties of antibodies and antibody fragments. With both methods, scFv fragments with improved folding and stability have been selected.[112,119] All of these parameters are critical for practical applications and may drive the production rate to high-level expression (for a review see ref. 120). Another example is the fusion of antibodies to toxins (*e.g.*, truncated versions of diphtheria toxin or *Pseudomonas* exotoxin), thereby generating immunotoxins. Immunotoxins combine the cell type specific targeting of antibodies and killing of the targeted cell by the internalized toxin moiety. Such chimeric molecules are routinely produced in *E. coli* and applied in the therapy of various forms of cancer.[121–123]

1.3.5.2 Secretory Production of Human Proinsulin

Proinsulin is the pre-form of insulin, which is applied to treat diabetes. Diabetes is the most common chronic metabolic disease in children and adolescents with considerable negative effects to health including damage to the heart, blood vessels, eyes, kidneys and nerves. According to the World Health Organization about 180 million people worldwide suffer from diabetes. Even though only a subset of diabetics require daily insulin application, the market value of insulin is tremendous. Proinsulin is a single polypeptide chain consisting of the A-chain and B-chain linked by the C-peptide. Upon cleavage of the C-peptide from the molecule, active insulin is released. Both proinsulin and insulin contain three disulfide bonds, two of which link the A- and B-chain. Commercial (pro)insulin production typically relies on *E. coli* that produces the recombinant human protein in inclusion bodies; either the A- and B-chains are separately produced or proinsulin is produced as a single polypeptide chain. In both cases, the protein has to be refolded *in vitro* under conditions favoring disulfide-bond formation.[124–126] The latter reference was the basis for Genentech's recombinant insulin production; Genentech's insulin was the first biotechnological product ever to be approved as a drug. Besides these established recombinant processes there are a few examples for the production of native proinsulin in *E. coli*. Although the productivity is significantly lower, some of the subsequently described examples have the potential to compete with established technical processes.

The first reports on the recombinant production of proinsulin in the *E. coli* periplasm were published almost three decades ago.[127–130,21] Even though they described extremely low amounts of native proinsulin, these reports were groundbreaking in terms of feasibility (see also ref. 131). More recently the secretory production of human proinsulin was massively improved by engineering the C-peptide of proinsulin, using fusion partners as well as media additives and co-secreting chaperones (see Table 1.3.2). One simple approach was the comparison of different *E. coli* strains and the optimization of the cultivation conditions; expression in LB media at 22 °C increased the yield of native proinsulin to 2.4 mg per gram dry biomass.[132] Upon supplementing the cultivation media with micromolar concentrations of Vectrase the yield was further improved to 3.6 mg per gram dry biomass.[132] Vectrase is a synthetic dithiol mimicking the active site of PDI (for a review on PDI see Chapter 1.7).[133] It is

Table 1.3.2 Secretory production of human proinsulin.

Secretion via[a]	Fusion partner	Amount of proinsulin[b]	Process characteristics	Ref.
PelB	–	0.074 mg l^{-1}	Co-secretion of chaperones	134
		3.6 mg g^{-1}	Supplementation of Vectrase	132
DsbA	DsbA	9.2 mg g$^{-1 c}$	Supplementation of arginine, ethanol	135
SpA	ZZ domain	3.5 mg g$^{-1 d}$	Cultivation in M9 media	142–144
		15 mg l$^{-1 e}$	Varied length of C-peptide	145
Ecotin	Ecotin	51.1 mg l$^{-1 f}$	Fermentation, supplementation of peptone	165

[a]Secretion *via* the signal sequence PelB from *Erwinia carotovora* and the signal sequences from *E. coli* DsbA, *Staphylococcus* protein A and *E. coli* ecotin.
[b]For all examples, including the fusion proteins, the yield is given as maximum amount of native proinsulin.
[c]Corresponds to 30 mg DsbA-proinsulin per gram.
[d]Corresponds to 6.4 mg Z-proinsulin per gram or 7.2 mg ZZ-proinsulin per gram. Up to 2.6 mg ZZ-proinsulin per gram is secreted to the media.
[e]Corresponds to 58 mg ZZ-proinsulin per litre broth. The length of the C-peptide was varied.
[f]Corresponds to 153 mg ecotin-proinsulin per litre; assuming 0.2 g dry biomass per litre and OD unit and an OD of 170, this corresponds to 1.5 mg native proinsulin per gram dry biomass.

small enough to enter the periplasm; where it seems to improve the correct disulfide-bond formation in proinsulin in analogy to its *in vitro* refolding.[132] The thus obtained yield is several times higher compared to a previous report that used the same expression construct.[134] Another approach by the same authors was the fusion of DsbA to the N-terminus of proinsulin. DsbA significantly increases the solubility and oxidative folding of its fusion partner, which subsequently increases the stability and therefore the amount of active proinsulin.[135] DsbA had been described before as an N-terminal fusion partner that allows periplasmic production of active enterokinase.[136] The overall maximum yield, 9.2 mg correctly folded proinsulin (without DsbA) per gram dry biomass, was obtained by using the media additives arginine and ethanol.[135] The arginine effect seems to mimic *in vitro* conditions where arginine is widely applied as a folding enhancer in protein refolding (*e.g.*, ref. 137 and 138). It suppresses protein interactions and increases the solubility of the unfolded species whereby irreversible aggregation is reduced.[139,140] Ethanol, on the other hand, is a powerful elicitor of the heat shock response,[141] which may increase the cell's capacity for folding. These properties seem beneficial for protein solubility and stability and eventually cause an increased accumulation of native proinsulin.

In a different fusion system, one or two Z domains of *Staphylococcus* protein A were fused to the N-terminus of proinsulin.[142–145] Mergulhao and co-workers studied the secretory expression of Z- and ZZ-proinsulin. They used plasmids with different copy numbers and promoters, different *E. coli* strains, and compared the secretion efficiency of cells grown in minimal or LB media. It seems that the use of medium copy number plasmids and M9 minimal media favor secretion of ZZ-proinsulin and that secretion across the inner membrane is limiting for production. Eventually, between 1.9 and 3.5 mg native proinsulin per gram dry biomass could be obtained,[142–144] which is comparable to the amount obtained from DsbA-proinsulin. It should be noted that some part of ZZ-proinsulin was secreted to the external media (up to 2.6 mg per gram dry biomass). Surprisingly, fairly similar yields were obtained when proinsulin was fused to either one or two Z domains, suggesting that the overall length of the fusion construct does not affect the yield.[146] In contrast, only insignificant amounts of proinsulin were obtained if it was produced without a Z domain.[142,145]

The authors concluded that proinsulin is protected from proteolytic degradation by the Z fusion partner.[142] Kang and Yoon varied the length of the C-peptide in a ZZ-proinsulin fusion and tested their secretion efficiency in rich media. The C-peptide of proinsulin consists of 35 amino acids; truncating it to 13 amino acids increased the amount of ZZ-proinsulin to 56 mg per litre.[145] This corresponds to *ca.* 12 mg proinsulin per litre, which is about a five times higher volumetric yield compared to DsbA-proinsulin.[135] When the complete C-peptide was deleted a slightly higher yield (58 mg per litre) was obtained. It should be noted, however, that this construct cannot be converted to mature insulin since it lacks the trypsin recognition sites that are usually located at the borders of the C-peptide. The 25-fold higher yield compared to the full-length construct suggested a higher secretion efficiency of the truncated fusion protein across the cytoplasmic membrane.[145]

A novel fusion system is ecotin, a 16 kDa periplasmic *E. coli* protein.[147] Malik and co-workers succeeded in producing 153 mg ecotin-proinsulin per litre culture, which corresponds to *ca.* 1.5 mg native proinsulin per gram dry biomass. They produced the fusion protein by high cell density fermentation in M9 minimal media that was supplemented with peptone. Even though the cultivations were performed at 25 °C, proteolytic degradation of periplasmic ecotin-proinsulin was a significant problem. This became obvious by comparing the yield and degradation products in wild type *E. coli* and a strain deficient in protease III, which is most important for proinsulin degradation in the periplasm. Using the protease deficient strain, a four-fold higher specific yield was obtained in shake flask cultivations. The volumetric yield, however, was much lower since this strain reaches only low final optical densities and did not produce ecotin-proinsulin under fermentation conditions. Taken together, the presented examples show that by careful optimization of the cultivation conditions and the consideration of fusion partners, reasonably high amounts of native proinsulin can be produced in the periplasm of *E. coli*.

1.3.5.3 Production of Other Therapeutic Proteins

Besides antibodies and insulin, other therapeutic proteins such as growth hormones, tissue plasminogen activator, nerve growth factor, granulocyte stimulating factor and prolactin are increasingly administered in clinics. Such recombinant proteins are typically produced in *E. coli* as inclusion bodies. Their native production in *E. coli* has been attempted with differing success. Some promising examples exist but in most cases secretory expression is hampered by insufficient expression levels. This may be a reason why they seem to be less in the focus of biotechnological research. A few representative examples will be discussed in this chapter.

Recombinant human growth hormone (hGH) was approved as a drug to treat hGH deficiency in 1985. Several diseases can now be treated with hGH; hGH stimulates growth and cell reproduction, and the most widely known function in the body is the increase of height throughout childhood. Using the OmpA and StII signal sequences, reasonable 15–25 mg native hGH per litre and OD unit could be obtained.[148,149] In contrast, only very low amounts of recombinant tissue plasminogen activator (tPA) can be produced in its native state in the periplasm of *E. coli*. In 1987, tPA was approved as a drug to treat non-hemorrhagic stroke and to dissolve blood clots in patients with acute myocardial infarction. Its secretory expression in *E. coli* seems insignificant. The co-secretion of Dsb proteins, yeast PDI or chaperones and the supplementation of the cultivation broth with arginine significantly improved the yield of folded tPA. Yet, a yield of up to 4 µg per litre and OD unit certainly cannot compete with *in vitro* refolding processes.[134,150,151] Also the amount of human nerve growth factor, which induces the differentiation and survival of particular target neurons, can be improved upon co-secretion of Dsb proteins. Nonetheless, the yield of native protein reaches only 85 µg per litre and OD unit.[152] The very low yields of tPA and NGF can probably be attributed to their intrinsic folding requirements. Full length tPA contains 17 disulfide bonds and the typically produced deletion variant, which can be administered in clinics, still contains nine disulfide bonds.[153] NGF contains three disulfide bonds forming a cysteine-knot motif and its pro-sequence seems to assist in its folding.[154,155] Such complicated formation of native structure is, to a low extent, possible in the *E. coli* periplasm, yet much more productive *in vitro* where folding conditions can be explored more elaborately.

Significantly higher yields of native protein could be obtained with human epidermal growth factor (hEGF), human leptin and granulocyte stimulating factor (GCSF). hEGF, a highly mitogenic protein that is for example applied in wound healing, can be efficiently secreted in *E. coli via* the Caf1 signal sequence from *Yersinia pestis*.[156] Its yield of 0.75 mg per litre is, however, still too low compared to its spontaneous *in vitro* folding.[157] Leptin, on the other hand, is efficiently produced in its active form *in vivo* as well as refolded from inclusion bodies *in vitro*.[158] It is a hormone with important effects in regulating body weight, metabolism and reproductive function. Upon co-secretion of DsbA the amount of soluble periplasmic leptin increased drastically and 160 mg protein per gram dry biomass (corresponding to 26% of total cellular protein) could be accumulated.[159] Similarly high secretion levels (22% of total protein) are obtained for GCSF, which is important for hematopoietic cell proliferation and was approved as a drug to treat chemotherapy-induced neutropenia in 1991.[160] In both cases, the efficiency of secretion of the recombinant proteins was dependent on the cultivation temperature and *E. coli* strain used.

1.3.6 Conclusions and Future Directions

Many promising and applicable secretory expression processes have been developed over the past 20 to 30 years. Folding enhancers and other beneficial factors were discovered and proteins as large as IgGs can now be natively produced in the *E. coli* periplasm. A timely and fruitful secretory production system should provide strong but tunable expression, efficient secretion, high stability of the recombinant protein and simple downstream processing. The periplasm of *E. coli* offers these properties and has thus been employed for many proteins, model proteins and therapeutic proteins alike. While *E. coli* has the intrinsic ability for efficient secretory expression, its application may be limited by the folding requirements of the target protein. Some proteins exhibit fast folding characteristics or can be engineered in this respect. They are exceptionally well secreted and natively accumulated in the periplasm (*e.g.*, scFv and Fab fragments, leptin, GCSF). Others are aggregation-prone and susceptible to proteases in their mature form but can be produced to high yields if their stability is increased by fusion partners (*e.g.*, proinsulin). And still other proteins comprise such complex folding that they are not accessible to native production in the *E. coli* periplasm (*e.g.*, tPA and NGF). Therefore, the expression and cultivation conditions for each individual recombinant protein have to be determined experimentally and known beneficial factors and production schemes can be used as guide but not as general rule.

In the future we will doubtlessly see incredible advances in the biotechnological application of the *E. coli* periplasm. A challenge to the more general applicability of *E. coli* and its periplasm will certainly be genetic engineering to enable post-translational modifications within the bacterial cell. Then proteins, which stringently require such modifications for stability in *E. coli* or functionality as therapeutic, may also be efficiently produced in the periplasm.

Acknowledgements

The authors would like to thank Petra Menhorn and Drs Robert Freedman and Stefan Gleiter for comments on the manuscript. Financial support by the Elitenetzwerk Bayern to K.G. as well as the Emmy-Noether program of the DFG and the Munich Center for Integrated Protein Science (CiPS^{M}) to J.W. is gratefully acknowledged.

References

1. O. Pines and M. Inouye, *Mol. Biotechnol.*, 1999, **12**, 25–34.
2. F. Baneyx, *Curr. Opin. Biotechnol.*, 1999, **10**, 411–421.
3. S. C. Makrides, *Microbiol. Rev.*, 1996, **60**, 512–538.
4. V. R. Agashe, S. Guha, H. C. Chang, P. Genevaux, M. Hayer-Hartl, M. Stemp, C. Georgopoulos, F. U. Hartl and J. M. Barral, *Cell*, 2004, **117**, 199–209.
5. S. I. van Kasteren, H. B. Kramer, D. P. Gamblin and B. G. Davis, *Nat. Protoc.*, 2007, **2**, 3185–3194.
6. M. Kowarik, S. Numao, M. F. Feldman, B. L. Schulz, N. Callewaert, E. Kiermaier, I. Catrein and M. Aebi, *Science*, 2006, **314**, 1148–1150.
7. L. C. Hsieh-Wilson, *Trends Biotechnol.*, 2004, **22**, 489–491.
8. D. Bottcher, E. Brusehaber, K. Doderer and U. T. Bornscheuer, *Appl. Microbiol. Biotechnol.*, 2007, **73**, 1282–1289.
9. L. C. Simmons, D. Reilly, L. Klimowski, T. S. Raju, G. Meng, P. Sims, K. Hong, R. L. Shields, L. A. Damico, P. Rancatore and D. G. Yansura, *J. Immunol. Methods*, 2002, **263**, 133–147.
10. A. Golden, S. S. Khandekar, M. S. Osburne, E. Kawasaki, E. L. Reinherz and T. H. Grossman, *J. Immunol. Methods*, 1997, **206**, 163–169.
11. B. Fahnert, H. Lilie and P. Neubauer, *Adv. Biochem. Eng. Biotechnol.*, 2004, **89**, 93–142.
12. R. Rudolph and H. Lilie, *FASEB J.*, 1996, **10**, 49–56.
13. J. Messens, J. F. Collet, K. Van Belle, E. Brosens, R. Loris and L. Wyns, *J. Biol. Chem.*, 2007, **282**, 31302–31307.
14. A. I. Derman, W. A. Prinz, D. Belin and J. Beckwith, *Science*, 1993, **262**, 1744–1747.

15. P. H. Bessette, F. Aslund, J. Beckwith and G. Georgiou, *Proc. Natl. Acad. Sci. USA*, 1999, **96**, 13703–13708.
16. M. Pohlschroder, W. A. Prinz, E. Hartmann and J. Beckwith, *Cell*, 1997, **91**, 563–566.
17. J. P. Arie, M. Miot, N. Sassoon and J. M. Betton, *Mol. Microbiol.*, 2006, **62**, 427–437.
18. J. C. Joly, W. S. Leung and J. R. Swartz, *Proc. Natl. Acad. Sci. USA*, 1998, **95**, 2773–2777.
19. G. A. Bowden, A. M. Paredes and G. Georgiou, *Biotechnology (NY)*, 1991, **9**, 725–730.
20. G. Georgiou, J. N. Telford, M. L. Shuler and D. B. Wilson, *Appl. Environ. Microbiol.*, 1986, **52**, 1157–1161.
21. K. Talmadge and W. Gilbert, *Proc. Natl. Acad. Sci. USA*, 1982, **79**, 1830–1833.
22. J. P. Rosenbusch, *Experientia*, 1990, **46**, 167–173.
23. J. B. Stock, B. Rauch and S. Roseman, *J. Biol. Chem.*, 1977, **252**, 7850–7861.
24. E. P. Kennedy, *Proc. Natl. Acad. Sci. USA*, 1982, **79**, 1092–1095.
25. H. Nikaido and E. Y. Rosenberg, *J. Gen. Physiol.*, 1981, **77**, 121–135.
26. G. M. Decad and H. Nikaido, *J. Bacteriol.*, 1976, **128**, 325–336.
27. J. E. Mogensen and D. E. Otzen, *Mol. Microbiol.*, 2005, **57**, 326–346.
28. C. Spiess, A. Beil and M. Ehrmann, *Cell*, 1999, **97**, 339–347.
29. J. Skorko-Glonek, D. Zurawa, E. Kuczwara, M. Wozniak, Z. Wypych and B. Lipinska, *Mol. Gen. Genet.*, 1999, **262**, 342–350.
30. K. L. Strauch, K. Johnson and J. Beckwith, *J. Bacteriol.*, 1989, **171**, 2689–2696.
31. J. Pogliano, A. S. Lynch, D. Belin, E. C. Lin and J. Beckwith, *Genes Dev.*, 1997, **11**, 1169–1182.
32. B. Lipinska, S. Sharma and C. Georgopoulos, *Nucleic Acids Res.*, 1988, **16**, 10053–10067.
33. K. C. Keiler, K. R. Silber, K. M. Downard, I. A. Papayannopoulos, K. Biemann and R. T. Sauer, *Protein Sci.*, 1995, **4**, 1507–1515.
34. F. Baneyx and G. Georgiou, *J. Bacteriol.*, 1991, **173**, 2696–2703.
35. F. Baneyx and G. Georgiou, *J. Bacteriol.*, 1990, **172**, 491–494.
36. H. J. Meerman and G. Georgiou, *Biotechnology (NY)*, 1994, **12**, 1107–1110.
37. S. Behrens, R. Maier, H. de Cock, F. X. Schmid and C. A. Gross, *EMBO J.*, 2001, **20**, 285–294.
38. J. P. Arie, N. Sassoon and J. M. Betton, *Mol. Microbiol.*, 2001, **39**, 199–210.
39. P. E. Rouviere and C. A. Gross, *Genes Dev.*, 1996, **10**, 3170–3182.
40. R. Chen and U. Henning, *Mol. Microbiol.*, 1996, **19**, 1287–1294.
41. T. A. Walton and M. C. Sousa, *Mol. Cell*, 2004, **15**, 367–374.
42. J. G. Sklar, T. Wu, D. Kahne and T. J. Silhavy, *Genes Dev.*, 2007, **21**, 2473–2484.
43. M. S. Pittman, H. C. Robinson and R. K. Poole, *J. Biol. Chem.*, 2005, **280**, 32254–32261.

44. P. H. Bessette, J. J. Cotto, H. F. Gilbert and G. Georgiou, *J. Biol. Chem.*, 1999, **274**, 7784–7792.
45. A. Rietsch, P. Bessette, G. Georgiou and J. Beckwith, *J. Bacteriol.*, 1997, **179**, 6602–6608.
46. M. Berkmen, D. Boyd and J. Beckwith, *J. Biol. Chem.*, 2005, **280**, 11387–11394.
47. A. Hiniker and J. C. Bardwell, *J. Biol. Chem.*, 2004, **279**, 12967–12973.
48. P. Natale, T. Brüser and A. J. Driessen, *Biochim. Biophys. Acta*, 2008, **1778**, 1735–1756.
49. A. P. Pugsley, *Microbiol. Rev.*, 1993, **57**, 50–108.
50. J. H. Choi and S. Y. Lee, *Appl. Microbiol. Biotechnol.*, 2004, **64**, 625–635.
51. J. Pratap and K. L. Dikshit, *Mol. Gen. Genet.*, 1998, **258**, 326–333.
52. B. C. Berks, F. Sargent and T. Palmer, *Mol. Microbiol.*, 2000, **35**, 260–274.
53. S. Cristobal, J. W. de Gier, H. Nielsen and G. von Heijne, *EMBO J.*, 1999, **18**, 2982–2990.
54. S. Molik, I. Karnauchov, C. Weidlich, R. G. Herrmann and R. B. Klosgen, *J. Biol. Chem.*, 2001, **276**, 42761–42766.
55. A. P. Hinsley, N. R. Stanley, T. Palmer and B. C. Berks, *FEBS Lett.*, 2001, **497**, 45–49.
56. Z. Ignatova, C. Hornle, A. Nurk and V. Kasche, *Biochem. Biophys. Res. Commun.*, 2002, **291**, 146–149.
57. K. Mitra, C. Schaffitzel, T. Shaikh, F. Tama, S. Jenni, C. L. Brooks III, N. Ban and J. Frank, *Nature*, 2005, **438**, 318–324.
58. Q. A. Valent, P. A. Scotti, S. High, J. W. de Gier, G. von Heijne, G. Lentzen, W. Wintermeyer, B. Oudega and J. Luirink, *EMBO J.*, 1998, **17**, 2504–2512.
59. A. Prinz, C. Behrens, T. A. Rapoport, E. Hartmann and K. U. Kalies, *EMBO J.*, 2000, **19**, 1900–1906.
60. M. van der Laan, N. P. Nouwen and A. J. Driessen, *Curr. Opin. Microbiol.*, 2005, **8**, 182–187.
61. K. Beck, L. F. Wu, J. Brunner and M. Muller, *EMBO J.*, 2000, **19**, 134–143.
62. R. S. Ullers, J. Luirink, N. Harms, F. Schwager, C. Georgopoulos and P. Genevaux, *Proc. Natl. Acad. Sci. USA*, 2004, **101**, 7583–7588.
63. C. A. Kumamoto, *Proc. Natl. Acad. Sci. USA*, 1989, **86**, 5320–5324.
64. L. L. Randall, S. J. Hardy, T. B. Topping, V. F. Smith, J. E. Bruce and R. D. Smith, *Protein Sci.*, 1998, **7**, 2384–2390.
65. P. Fekkes, J. G. de Wit, J. P. van der Wolk, H. H. Kimsey, C. A. Kumamoto and A. J. Driessen, *Mol. Microbiol.*, 1998, **29**, 1179–1190.
66. P. Fekkes, C. van der Does and A. J. Driessen, *EMBO J.*, 1997, **16**, 6105–6113.
67. N. Nouwen, B. de Kruijff and J. Tommassen, *Proc. Natl. Acad. Sci. USA*, 1996, **93**, 5953–5957.
68. A. Economou and W. Wickner, *Cell*, 1994, **78**, 835–843.
69. J. P. van der Wolk, J. G. de Wit and A. J. Driessen, *EMBO J.*, 1997, **16**, 7297–7304.

70. K. Uchida, H. Mori and S. Mizushima, *J. Biol. Chem.*, 1995, **270**, 30862–30868.
71. E. Schiebel, A. J. Driessen, F. U. Hartl and W. Wickner, *Cell*, 1991, **64**, 927–939.
72. R. Tuteja, *Arch. Biochem. Biophys.*, 2005, **441**, 107–111.
73. P. A. Lee, D. Tullman-Ercek and G. Georgiou, *Annu. Rev. Microbiol.*, 2006, **60**, 373–395.
74. L. Masip, J. L. Pan, S. Haldar, J. E. Penner-Hahn, M. P. DeLisa, G. Georgiou, J. C. Bardwell and J. F. Collet, *Science*, 2004, **303**, 1185–1189.
75. M. P. DeLisa, D. Tullman and G. Georgiou, *Proc. Natl. Acad. Sci. USA*, 2003, **100**, 6115–6120.
76. M. Alami, D. Trescher, L. F. Wu and M. Muller, *J. Biol. Chem.*, 2002, **277**, 20499–20503.
77. T. L. Yahr and W. T. Wickner, *EMBO J.*, 2001, **20**, 2472–2479.
78. N. N. Alder and S. M. Theg, *Cell*, 2003, **112**, 231–242.
79. N. N. Alder and S. M. Theg, *Trends Biochem. Sci.*, 2003, **28**, 442–451.
80. T. Palmer, F. Sargent and B. C. Berks, *Trends Microbiol.*, 2005, **13**, 175–180.
81. A. Rodrigue, A. Chanal, K. Beck, M. Muller and L. F. Wu, *J. Biol. Chem.*, 1999, **274**, 13223–13228.
82. K. Hatzixanthis, T. Palmer and F. Sargent, *Mol. Microbiol.*, 2003, **49**, 1377–1390.
83. T. G. Bernhardt and P. A. de Boer, *Mol. Microbiol.*, 2003, **48**, 1171–1182.
84. J. Behrendt, K. Standar, U. Lindenstrauss and T. Bruser, *FEMS Microbiol. Lett.*, 2004, **234**, 303–308.
85. M. Alami, I. Luke, S. Deitermann, G. Eisner, H. G. Koch, J. Brunner and M. Muller, *Mol. Cell*, 2003, **12**, 937–946.
86. M. G. Hicks, E. De Leeuw, I. Porcelli, G. Buchanan, B. C. Berks and T. Palmer, *FEBS Lett.*, 2003, **539**, 61–67.
87. P. A. Lee, G. Buchanan, N. R. Stanley, B. C. Berks and T. Palmer, *J. Bacteriol.*, 2002, **184**, 5871–5879.
88. E. De Leeuw, I. Porcelli, F. Sargent, T. Palmer and B. C. Berks, *FEBS Lett.*, 2001, **506**, 143–148.
89. M. R. Yen, Y. H. Tseng, E. H. Nguyen, L. F. Wu and M. H. Saier Jr., *Arch. Microbiol.*, 2002, **177**, 441–450.
90. U. Gohlke, L. Pullan, C. A. McDevitt, I. Porcelli, E. De Leeuw, T. Palmer, H. R. Saibil and B. C. Berks, *Proc. Natl. Acad. Sci. USA*, 2005, **102**, 10482–10486.
91. E. De Leeuw, T. Granjon, I. Porcelli, M. Alami, S. B. Carr, M. Muller, F. Sargent, T. Palmer and B. C. Berks, *J. Mol. Biol.*, 2002, **322**, 1135–1146.
92. A. Bolhuis, J. E. Mathers, J. D. Thomas, C. M. Barrett and C. Robinson, *J. Biol. Chem.*, 2001, **276**, 20213–20219.
93. F. Sargent, U. Gohlke, E. De Leeuw, N. R. Stanley, T. Palmer, H. R. Saibil and B. C. Berks, *Eur. J. Biochem.*, 2001, **268**, 3361–3367.
94. N. Ruiz and T. J. Silhavy, *Curr. Opin. Microbiol.*, 2005, **8**, 122–126.

95. J. D. Thomas, R. A. Daniel, J. Errington and C. Robinson, *Mol. Microbiol.*, 2001, **39**, 47–53.
96. B. J. Feilmeier, G. Iseminger, D. Schroeder, H. Webber and G. J. Phillips, *J. Bacteriol.*, 2000, **182**, 4068–4076.
97. J. H. Choi, K. J. Jeong, S. C. Kim and S. Y. Lee, *Appl. Microbiol. Biotechnol.*, 2000, **53**, 640–645.
98. T. A. Waldmann, *Nat. Med.*, 2003, **9**, 269–277.
99. P. J. Hudson and C. Souriau, *Nat. Med.*, 2003, **9**, 129–134.
100. A. Todorovska, R. C. Roovers, O. Dolezal, A. A. Kortt, H. R. Hoogenboom and P. J. Hudson, *J. Immunol. Methods*, 2001, **248**, 47–66.
101. L. Presta, P. Sims, Y. G. Meng, P. Moran, S. Bullens, S. Bunting, J. Schoenfeld, D. Lowe, J. Lai, P. Rancatore, M. Iverson, A. Lima, V. Chisolm, R.F. Kelley, M. Riederer and D. Kirchofer, *Thromb. Haemost.*, 2001, **85**, 379–389.
102. J. E. Thompson, T. J. Vaughan, A. J. Williams, J. Wilton, K. S. Johnson, L. Bacon, J. A. Green, R. Field, S. Ruddock, M. Martins, A. R. Pope, P. R. Tempest and R. H. Jackson, *J. Immunol. Methods*, 1999, **227**, 17–29.
103. D. P. Humphreys, B. Carrington, L. C. Bowering, R. Ganesh, M. Sehdev, B. J. Smith, L. M. King, D. G. Reeks, A. Lawson and A. G. Popplewell, *Protein Expr. Purif.*, 2002, **26**, 309–320.
104. D. P. Humphreys, M. Sehdev, A. P. Chapman, R. Ganesh, B. J. Smith, L. M. King, D. J. Glover, D. G. Reeks and P. E. Stephens, *Protein Expr. Purif.*, 2000, **20**, 252–264.
105. M. Kraft, U. Knupfer, R. Wenderoth, A. Kacholdt, P. Pietschmann, B. Hock and U. Horn, *Appl. Microbiol. Biotechnol.*, 2007, **76**, 1413–1422.
106. U. Horn, W. Strittmatter, A. Krebber, U. Knupfer, M. Kujau, R. Wenderoth, K. Muller, S. Matzku, A. Plückthun and D. Riesenberg, *Appl. Microbiol. Biotechnol.*, 1996, **46**, 524–532.
107. C. Chen, B. Snedecor, J. C. Nishihara, J. C. Joly, N. McFarland, D. C. Andersen, J. E. Battersby and K. M. Champion, *Biotechnol. Bioeng.*, 2004, **85**, 463–474.
108. Z. Zhu, G. Zapata, R. Shalaby, B. Snedecor, H. Chen and P. Carter, *Biotechnology (NY)*, 1996, **14**, 192–196.
109. P. Carter, R. F. Kelley, M. L. Rodrigues, B. Snedecor, M. Covarrubias, M. D. Velligan, W. L. Wong, A. M. Rowland, C. E. Kotts and M. E. Carver, *et al.*, *Biotechnology (NY)*, 1992, **10**, 163–167.
110. C. Mavrangelos, M. Thiel, P. J. Adamson, D. J. Millard, S. Nobbs, H. Zola and I. C. Nicholson, *Protein Expr. Purif.*, 2001, **23**, 289–295.
111. A. Hayhurst and W. J. Harris, *Protein Expr. Purif.*, 1999, **15**, 336–343.
112. H. Bothmann and A. Pluckthun, *Nat. Biotechnol.*, 1998, **16**, 376–380.
113. H. Wimmer, M. Olsson, M. T. Petersen, R. Hatti-Kaul, S. B. Peterson and N. Muller, *J. Biotechnol.*, 1997, **55**, 85–100.
114. T. Arakawa and S. N. Timasheff, *Biophys. J.*, 1985, **47**, 411–414.
115. Y. R. Gokarn, E. Kras, C. Nodgaard, V. Dharmavaram, R. M. Fesinmeyer, H. Hultgen, S. Brych, R. L. Remmele Jr., D. N. Brems and S. Hershenson, *J. Pharm. Sci.*, 2007, **97**, 3051–3066.

116. P. Holliger, T. Prospero and G. Winter, *Proc. Natl. Acad. Sci. USA*, 1993, **90**, 6444–6448.

117. A. M. Bayly, A. A. Kortt, P. J. Hudson and B. E. Power, *J. Immunol. Methods*, 2002, **262**, 217–227.

118. S. J. DeNardo, G. L. DeNardo, D. G. DeNardo, C. Y. Xiong, X. B. Shi, M. D. Winthrop, L. A. Kroger and P. Carter, *Clin. Cancer Res.*, 1999, **5**, 3213s–3218s.

119. B. Ribnicky, T. Van Blarcom and G. Georgiou, *J. Mol. Biol.*, 2007, **369**, 631–639.

120. A. Worn and A. Pluckthun, *J. Mol. Biol.*, 2001, **305**, 989–1010.

121. I. Pastan, R. Hassan, D. J. FitzGerald and R. J. Kreitman, *Annu. Rev. Med.*, 2007, **58**, 221–237.

122. D. J. FitzGerald, R. Kreitman, W. Wilson, D. Squires and I. Pastan, *Int. J. Med. Microbiol.*, 2004, **293**, 577–582.

123. B. Matthey, A. Engert and S. Barth, *Int. J. Mol. Med.*, 2000, **6**, 509–514.

124. M. Gorecki, J. R. Hartman and S. Mendelovitz, *Patent No. WO9620724*, 1996.

125. R. E. Chance and J. A. Hoffmann, *Patent No. US4421685*, 1983.

126. D. V. Goeddel, D. G. Kleid and K. Itakura, *Patent No. EP0055945*, 1982.

127. K. Talmadge, J. Brosius and W. Gilbert, *Nature*, 1981, **294**, 176–178.

128. K. Talmadge, J. Kaufman and W. Gilbert, *Proc. Natl. Acad. Sci. USA*, 1980, **77**, 3988–3992.

129. K. Talmadge, S. Stahl and W. Gilbert, *Proc. Natl. Acad. Sci. USA*, 1980, **77**, 3369–3373.

130. S. J. Chan, J. Weiss, M. Konrad, T. White, C. Bahl, S. D. Yu, D. Marks and D. F. Steiner, *Proc. Natl. Acad. Sci. USA*, 1981, **78**, 5401–5405.

131. S. J. Stahl and L. Christiansen, *Gene*, 1988, **71**, 147–156.

132. J. Winter, H. Lilie and R. Rudolph, *Anal. Biochem.*, 2002, **310**, 148–155.

133. K. J. Woycechowsky, K. D. Wittrup and R. T. Raines, *Chem. Biol.*, 1999, **6**, 871–879.

134. J. Schaffner, J. Winter, R. Rudolph and E. Schwarz, *Appl. Environ. Microbiol.*, 2001, **67**, 3994–4000.

135. J. Winter, P. Neubauer, R. Glockshuber and R. Rudolph, *J. Biotechnol.*, 2001, **84**, 175–185.

136. L. A. Collins-Racie, J. M. McColgan, K. L. Grant, E. A. DiBlasio-Smith, J. M. McCoy and E. R. LaVallie, *Biotechnology (NY)*, 1995, **13**, 982–987.

137. H. Lilie, E. Schwarz and R. Rudolph, *Curr. Opin. Biotechnol.*, 1998, **9**, 497–501.

138. J. Buchner and R. Rudolph, *Biotechnology (NY)*, 1991, **9**, 157–162.

139. T. Arakawa, D. Ejima, K. Tsumoto, N. Obeyama, Y. Tanaka, Y. Kita and S. N. Timasheff, *Biophys. Chem.*, 2007, **127**, 1–8.

140. K. R. Reddy, H. Lilie, R. Rudolph and C. Lange, *Protein Sci.*, 2005, **14**, 929–935.

141. T. K. Van Dyk, T. R. Reed, A. C. Vollmer and R. A. LaRossa, *J. Bacteriol.*, 1995, **177**, 6001–6004.

142. F. J. Mergulhao, M. A. Taipa, J. M. Cabral and G. A. Monteiro, *J. Biotechnol.*, 2004, **109**, 31–43.

143. F. J. Mergulhao, G. A. Monteiro, G. Larsson, A. M. Sanden, A. Farewell, T. Nystrom, J. M. Cabral and M. A. Taipa, *Appl. Microbiol. Biotechnol.*, 2003, **61**, 495–501.

144. F. J. Mergulhao, G. A. Monteiro, G. Larsson, M. Bostrom, A. Farewell, T. Nystrom, J. M. Cabral and M. A. Taipa, *Biotechnol. Appl. Biochem.*, 2003, **38**, 87–93.

145. Y. Kang and J. W. Yoon, *J. Biotechnol.*, 1994, **36**, 45–54.

146. F. J. Mergulhao and G. A. Monteiro, *J. Microbiol. Biotechnol.*, 2007, **17**, 1236–1241.

147. A. Malik, R. Rudolph and B. Sohling, *Protein Expr. Purif.*, 2006, **47**, 662–671.

148. K. C. Cheah, S. Harrison, R. King, L. Crocker, J. R. Wells and A. Robins, *Gene*, 1994, **138**, 9–15.

149. C. N. Chang, M. Rey, B. Bochner, H. Heyneker and G. Gray, *Gene*, 1987, **55**, 189–196.

150. X. Zhan, M. Schwaller, H. F. Gilbert and G. Georgiou, *Biotechnol. Prog.*, 1999, **15**, 1033–1038.

151. J. Qiu, J. R. Swartz and G. Georgiou, *Appl. Environ. Microbiol.*, 1998, **64**, 4891–4896.

152. Y. Kurokawa, H. Yanagi and T. Yura, *J. Biol. Chem.*, 2001, **276**, 14393–14399.

153. U. Kohnert, R. Rudolph, J. H. Verheijen, E. J. Weening-Verhoeff, A. Stern, U. Opitz, U. Martin, H. Lill, H. Prinz, M. Lechner, G.-B. Kresse, P. Bucket and S. Fisher, *Protein Engineer.*, 1992, **5**, 93–100.

154. M. Kliemannel, R. Golbik, R. Rudolph, E. Schwarz and H. Lilie, *Protein Sci.*, 2007, **16**, 411–419.

155. A. Rattenholl, M. Ruoppolo, A. Flagiello, M. Monti, F. Vinci, G. Marino, H. Lilie, E. Schwarz and R. Rudolph, *J. Mol. Biol.*, 2001, **305**, 523–533.

156. Y. L. Liu, L. M. Huang, W. P. Lin, C. C. Tsai, T. S. Lin, Y. H. Hu, H. S. Chen, J. M. Han, H. J. Wang and Y. T. Liu, *J. Microbiol. Immunol. Infect.*, 2006, **39**, 366–371.

157. J. Y. Chang, P. Schindler, U. Ramseier and P. H. Lai, *J. Biol. Chem.*, 1995, **270**, 9207–9216.

158. J. P. Varnerin, T. Smith, C. I. Rosenblum, A. Vongs, B. A. Murphy, C. Nunes, T. N. Mellin, J. J. King, B. W. Burgess, B. Junker, M. Chou, P. Hey, E. Frazier, D. E. MacIntyre, L. H. T. van der Ploeg and M. R. Tota, *Protein Expr. Purif.*, 1998, **14**, 335–342.

159. K. J. Jeong and S. Y. Lee, *Biotechnol. Bioeng.*, 2000, **67**, 398–407.

160. K. J. Jeong and S. Y. Lee, *Protein Expr. Purif.*, 2001, **23**, 311–318.

161. Y. Mazor, T. Van Blarcom, R. Mabry, B. L. Iverson and G. Georgiou, *Nat. Biotechnol.*, 2007, **25**, 563–565.

162. D. P. Humphreys, N. Weir, A. Lawson, A. Mountain and P. A. Lund, *FEBS Lett.*, 1996, **380**, 194–197.
163. M. Duenas, J. Vazquez, M. Ayala, E. Soderlind, M. Ohlin, L. Perez, C. A. Borrebaeck and J. V. Gavilondo, *Biotechniques*, 1994, **16**, 476–483.
164. H. Bothmann and A. Pluckthun, *J. Biol. Chem.*, 2000, **275**, 17100–17105.
165. A. Malik, M. Jenzsch, A. Lubbert, R. Rudolph and B. Sohling, *Protein Expr. Purif.*, 2007, **55**, 100–111.
166. A. Shokri, A. M. Sanden and G. Larsson, *Appl. Microbiol. Biotechnol.*, 2003, **60**, 654–664.

CHAPTER 1.4

Oxidative Protein Folding in Mitochondria

KAI HELL AND WALTER NEUPERT

Adolf-Butenandt-Institut für Physiologische Chemie, Ludwig-Maximilians-Universität München, Butenandtstrasse 5, 81377 München, Germany

1.4.1 Introduction

Mitochondria are organelles of eukaryotic cells fulfilling a variety of important functions. Their most prominent function is the generation of ATP by the process of oxidative phosphorylation, thereby providing the cell with a source of energy. In addition, many other metabolic processes take place in these organelles, such as fatty acid oxidation, the Krebs cycle, parts of the urea cycle and the biosynthesis of cofactors and amino acids. Furthermore, they harbor the machinery for biogenesis of Fe/S clusters, a process essential for viability of the cell. Mitochondria also play an important role in programmed cell death releasing pro-apoptotic proteins upon apoptotic stimuli.

Mitochondria contain a specific set of proteins: about 900 in *Saccharomyces cerevisiae* and presumably about 1500 in mammals.[1–3] The mitochondrial DNA, however, encodes only a small subset of mitochondrial proteins, eight in *S. cerevisiae* and thirteen in humans. The vast majority of mitochondrial proteins is encoded by nuclear DNA, synthesized in the cytosol and then post-translationally transported into the organelle. These proteins not only have to be imported into, but also sorted within mitochondria, as these organelles are made up by four distinct subcompartments: the outer and the inner mitochondrial membranes, the matrix space and the intermembrane space (IMS)

RSC Biomolecular Sciences
Oxidative Folding of Peptides and Proteins
Edited by Johannes Buchner and Luis Moroder
Published by the Royal Society of Chemistry, www.rsc.org

between both membranes. Translocation of proteins across the mitochondrial membranes and sorting to the subcompartments are mediated by oligomeric protein complexes, termed translocases, such as the TOM (translocase of the outer membrane) and the TIM (translocase of the inner membrane) complexes.[4–9]

Mitochondria originated from prokaryotic ancestors by a process of endosymbiosis.[10] Although mitochondria adapted to the conditions of intracellular life during evolution, as documented for instance by the gene transfer to the nucleus and the development of a protein import machinery, many basic mitochondrial properties and processes still resemble those of bacteria. An example is the redox state of the bacterial cytoplasm and its corresponding mitochondrial compartment, the matrix space. Both are strongly reducing compartments with highly conserved reduction systems, the thioredoxin and the glutathione/glutaredoxin systems, to protect proteins against oxidative damage (Berndt and Holmgren, Chapter 1.1).[11–15] Disulfide bonds are formed in the redox-active proteins of these systems, the thioredoxins, the glutaredoxins and their reductases, only transiently during the course of the redox processes. There are only a few exceptions of proteins containing disulfide bonds in the bacterial cytoplasm, *e.g.* the stress response heat shock protein Hsp33 upon exposure to oxidative stress (Ilbest et al., Chapter 1.8.3).[16] To our knowledge, there are no permanently disulfide-bonded proteins in the mitochondrial matrix. In both compartments, protein folding is not dependent on the formation of disulfide bonds, but rather mediated by ATP-dependent chaperones, the DnaK/Hsp70 and the GroEL/Hsp60 chaperone systems.[17,18] In contrast, protein folding in the bacterial periplasm is driven by oxidation of cysteine residues. This process and the isomerization of disulfide bonds in this compartment is reviewed in detail by Malojčić and Glockshuber in Chapter 1.2. Due to its oxidation and isomerization capacity the periplasm is suited for the oxidative folding of recombinant proteins, as discussed by Gebendorfer and Winter in Chapter 1.3. In this ATP-free, oxidizing compartment the generation of disulfide bonds is catalyzed by thiol oxidases of the Dsb family, DsbA and DsbB.[19,20] Although the corresponding mitochondrial compartment, the IMS, contains ATP, ATP-dependent chaperones are not known, and folding of proteins by oxidation takes place. This was rather unexpected, since the IMS is connected with the cytosolic compartment of the eukaryotic cell by the pore-forming protein porin (VDAC) in the outer membrane. Thus, it presumably has a reducing environment similar to the cytosol due to free diffusion of small molecules, such as reduced glutathione, between both compartments. The oxidative folding of IMS proteins is important for their import into the IMS. Although the components of the oxidative folding pathways are not conserved between mitochondria and bacteria, mitochondria appear to have retained the basic principles of protein oxidation during evolution and have adapted them as a driving force for protein import into the IMS. We will focus here on the components and mechanisms of formation of disulfide bonds in the IMS and the role of oxidative protein folding in mitochondrial protein import.

Table 1.4.1 Proteins with disulfide bonds in the IMS of mitochondria.

Protein	Donor of disulfide bonds	Function	Ref.
Ccs1	unknown	Copper chaperone for Sod1	79,81
Cox11	unknown	Cytochrome *c* oxidase assembly factor	82
Cox12[a]	unknown	Cytochrome *c* oxidase subunit 12	23
Cox17	Mia40-Erv1 disulfide relay system	Copper chaperone	26,27
Erv1	FAD[c]	Sulfhydryl oxidase	24,60
Mia40[b]	Erv1	Protein import receptor	32,44,46
Rieske FeS[a]	unknown	Subunit of cytochrome bc_1 complex	22
Qcr6[a]	unknown	Subunit of cytochrome bc_1 complex	21,22
Small Tim proteins	Mia40-Erv1 disulfide relay system	Protein import	31,83,84
Sco1[b]	unknown	Cytochrome *c* oxidase assembly factor	85
Sod1	Ccs1	Superoxide dismutase	79,81

The names of the proteins are in accordance with the yeast nomenclature. Proteins with a twin Cx_9C motif are not listed here, if disulfide bonds have not been experimentally confirmed.
[a]These proteins are subunits of respiratory chain complexes of the inner membrane protruding into the IMS.
[b]Sco1 and Mia40 (in fungi) are anchored to the inner membrane with the disulfide motifs facing the IMS.
[c]Erv1 forms disulfide bonds by the transfer of electrons to bound FAD. FAD is reoxidized by transfer of electrons to cytochrome *c*.

1.4.2 Disulfide Bonds in the IMS of Mitochondria

In view of the access of reducing low molecular components of the cytosol to the IMS it was quite surprising when first studies reported proteins in the IMS that contain disulfide bonds.[21–23] In two subunits of the cytochrome bc_1 complex, the Rieske FeS protein and the acidic hinge protein, and in the subunit 12 of the cytochrome c oxidase (Cox12) alpha-helical hairpin structures or loops were found to be stabilized by disulfide bonds in biochemical and crystallographic studies. Later on, disulfide bonds were reported in Ccs1, Cox11, Cox17, Erv1, Mia40, Sco1, Sod1 and the small TIM proteins (Table 1.4.1). Moreover, they are very likely to be present in proteins which contain similar cysteine motifs as the proteins with disulfide bonds. Thus, formation of disulfide bonds appears to be a rather common property of IMS proteins. The cysteine residues forming disulfide bonds are highly conserved and their oxidized state is crucial for the function of the proteins. Like with proteins of other compartments, disulfide bonds are required for the folding and the stabilization of IMS proteins. They are also important for the activity, e.g. in redox proteins, such as in the sulfhydryl oxidase Erv1.[24,25] Moreover, redox-active cysteine residues undergo dynamic changes of their redox status and shuttle between the thiol and the disulfide state. Thus, they provide molecular redox switches to regulate the functions and/or properties of the proteins. Such a regulatory function of cysteine residues appears to occur in Cox17. Reduction of the disulfide bonds in Cox17 affects its capacity and affinity for copper ions suggesting a redox-regulation of copper binding and release.[26,27] Furthermore, disulfide bonds play a role in protein transport and assembly processes.[5,15,28–31] It has been reported that oxidative protein folding triggers the import of small proteins into the IMS and their subsequent assembly into oligomeric complexes.[31–33] The studies of this import process have provided the first and so far only mechanistical insights into how proteins form disulfide bonds in mitochondria.

1.4.3 Protein Import into the IMS by Oxidative Protein Folding

Proteins of the IMS which are imported by oxidative protein folding are of relatively low mass, typically comprise a simple folding unit and lack a clea-vable mitochondrial targeting signal.[5,15,28–30] However, all of them contain highly conserved cysteine residues. The best-studied examples are the small TIM proteins and the copper chaperone Cox17, in which the cysteine residues are organized in a twin Cx_3C and a twin Cx_9C motif, respectively.[34,35] The term twin $Cx_{3/9}C$ motif is used to indicate further on either a twin Cx_3C or a twin Cx_9C motif. The two $Cx_{3/9}C$ segments are juxtaposed in antiparallel α-helices and linked by two disulfide bonds to form a hairpin-like structure.[36,37] Like almost all mitochondrial proteins these small twin $Cx_{3/9}C$ motif-containing proteins are encoded in the nucleus and synthesized in the cytosol. They are then transported across the outer membrane through the pore of the TOM complex.[38] In order to pass the TOM complex the proteins have to be in an unfolded reduced state.[31,38] Subsequently, they form disulfide bonds, thereby triggering their folding within the IMS.[31,32,39] In the folded state, proteins are trapped in the IMS, as they cannot diffuse back through the TOM complex. Thus, this folding trap drives unidirectional net translocation of proteins into the IMS. Recent results indicated that the formation of disulfide bonds in the imported twin $Cx_{3/9}C$ motif-containing proteins requires a disulfide relay system which consists of two essential components, Mia40 and Erv1.[32,40,41]

1.4.4 The Redox-dependent Import Receptor Mia40

The import receptor Mia40/TIM40 (mitochondrial import and assembly) is present throughout the eukaryotic domain of life. In fungi, the Mia40 homo-logs consist of a cleavable N-terminal mitochondrial targeting signal followed by a hydrophobic segment and a C-terminal hydrophilic domain. In contrast, homologs in higher eukaryotes are much smaller and lack targeting signals and hydrophobic segments. While these proteins are soluble in the IMS, the Mia40 proteins in fungi are N-terminally anchored with the hydrophobic segments in the inner membrane.[42–44] This membrane tethering, however, is not crucial for

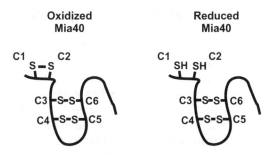

Figure 1.4.1 Schematic representation of the pattern of disulfide bonds in the oxidized and reduced states of Mia40. Presented is the highly conserved domain of Mia40. The cysteine residues are labeled according to their sequence position in Mia40. The residues C3, C4, C5 and C6 form the twin Cx_9C motif.

the function of Mia40.[42] The oligomeric state of native Mia40 and the significance of a possible oligomerization is an open question.[42,44,45]

In all species, the homologs of Mia40 have a highly conserved domain of about 60 amino acid residues with 6 invariant cysteine residues.[42,43,45] The cysteine residues are arranged in a CPC-Cx_9C-Cx_9C pattern. Some of these cysteine residues undergo redox switches because Mia40 can adopt two different redox states in mitochondria.[32,44] In its oxidized state, Mia40 forms three intramolecular disulfide bonds: the cysteine residues of the CPC motif are connected by one bond and the Cx_9C segments are linked by two disulfide bonds between the distal and proximal cysteine residues (Figure 1.4.1).[46] While the disulfide bonds in the twin Cx_9C motif are stable, the disulfide bond of the CPC motif is easily accessible to reducing agents.[46] This suggests a function of this cysteine pair in redox processes. The reduced form of Mia40 present in mitochondria still forms the two stable disulfide bonds, but the cysteine residues of the CPC segment are present in the thiol states (Figure 1.4.1). Besides their function in redox chemistry, the cysteine residues might play a role in metal binding and thereby in stabilization of the protein, since reduced Mia40 has the ability to bind zinc and copper ions *in vitro*.[43]

In *S. cerevisiae* and mammalian cells, Mia40 is essential for viability.[42–45,47] In the absence of functional Mia40, mitochondria contain reduced endogenous levels of the small twin $Cx_{3/9}C$ motif-containing proteins of the IMS, whereas proteins of other mitochondrial subcompartments and other IMS proteins are present in wild type amounts. The reduced levels are caused by defects in the import of these proteins.[42,43,45,48] On the other hand, the import rates of the small twin $Cx_{3/9}C$ motif-containing proteins are increased in mitochondria containing elevated levels of Mia40.[43] Mia40 is directly involved in their translocation, because it forms a disulfide intermediate with the incoming substrate protein during the import process.[32,45] Since the import of substrate proteins is inhibited in the presence of strong reducing agents, the formation of the disulfide intermediates between Mia40 and incoming substrate proteins plays an essential role for import.[32,45] Thus, Mia40 functions in the IMS as a receptor of small twin $Cx_{3/9}C$ motif-containing proteins which mediates their import into the IMS.

1.4.5   The FAD-dependent Sulfhydryl Oxidase Erv1

The sulfhydryl oxidase Erv1, located in the intermembrane space of mitochondria was identified as a protein essential for respiration and vegetative growth in yeast.[49,50] In mammalian cells, the Erv1 homolog is able to promote the liver regeneration by an unknown mechanism when added as purified protein.[51] Therefore, the human protein is termed ALR (augmenter of liver regeneration).

The members of the Erv1 family which are well conserved from fungi to plants and mammals constitute the class of Erv/ALR thiol oxidases, together with the Erv2 proteins which are present in the endoplasmic reticulum (ER) of fungi and Erv1-like sulfhydryl oxidases from viruses.[52–56] In order to generate disulfide bonds the Erv/ALR thiol oxidases shuttle electrons from thiol groups to the non-thiol electron acceptor flavine adenine dinucleotide (FAD). The proteins contain a highly conserved catalytic core domain of about 100 amino acid residues. In this core domain, FAD is non-covalently bound in a four-helix bundle structure next to a redox-active Cx_2C pair.[57–60] In addition, the core domains of Erv1 and Erv2 are stabilized by a structural disulfide-bonded cysteine pair. With the exception of the viral Erv1-like oxidases, the core domains of Erv/ALR thiol oxidases are fused to tail segments. These tail segments contain another pair of cysteine residues and appear to be rather unstructured and thus highly flexible.[24,58,59,61] Interestingly, in *S. cerevisiae* and in human Erv1, a Cx_2C motif is present in an N-terminal tail, whereas in Erv1 from *Arabidopsis thaliana* the cysteine residues are located as a Cx_4C motif in the C-terminal tail.[62,63] Based on the X-ray structures of Erv1 from *A. thaliana* and yeast Erv2 and on mutational analysis of proteins from this family, a model has been suggested in which a dithiol disulfide exchange reaction between reduced substrates and the oxidized shuttle disulfide pair in the flexible tail is used to introduce disulfide bonds into the substrates.[24,58,59,61] In this shuttle mechanism the position of the tail segments with respect to the core domains and the selective sequence context and spacing of the cysteine residues might allow the flexible tail to confer substrate specificity to distinct oxidases. The disulfide bond between the shuttle cysteine residues is regenerated by the transfer of electrons to the active site cysteine pair that is present in the core domain in the disulfide state. Subsequently, the active site cysteine pair passes the electrons on to the adjacent FAD which itself is recovered in its oxidized form by transfer of electrons to cytochrome *c* in the case of Erv1/ALR or to oxygen generating hydrogen peroxide.[25,56,64,65] According to crystallographic studies of Erv1 from *A. thaliana* and rat, as well as yeast Erv2, amino acid residues of the core domain mediate the formation of homodimers.[58–60] This allows cooperation of two subunits in the electron transfer process.[58,59,61] It has been suggested that the shuttle cysteine pair interacts in *trans* with the active site Cx_2C pair of the other subunit *via* an intermolecular disulfide bond to enable the transfer of disulfide bonds. Although the tail is located at the N terminus in Erv1 in yeast and mammals, a similar mechanism is likely to be present.[24,58]

Yeast Erv1 is essential for viability of cells.[49] Cells harboring a non-functional form of Erv1 have a variety of defects, such as loss of the mitochondrial genome, impaired respiration, altered mitochondrial morphology and distribution and defects in the biogenesis of cytosolic FeS cluster-containing proteins, as well as in the maturation and incorporation of heme into cytochrome c and cytochrome c peroxidase.[49–50,65–67] Similar to the phenotypes of Mia40 conditional mutants, cells lacking functional Erv1 have decreased amounts of small twin $Cx_{3/9}C$ motif-containing proteins in the IMS of mitochondria as a consequence of their impaired import and assembly.[32,39–41] Consistent with the observed import defects, the redox state of the import receptor of these proteins, Mia40, is affected in the absence of Erv1. Mia40 accumulates predominantly in the reduced non-functional form.[32] Erv1 transiently interacts with Mia40 *via* intermolecular disulfide bonds.[32,40] Thus, oxidation of Mia40 is driven by Erv1. Mia40 is the first identified physiological substrate of Erv1. Since a defect in Mia40 affects various other IMS proteins, the variety of defects observed in the absence of Erv1 might be largely explained by the effect of Erv1 on Mia40. It is also likely, however, that Erv1 directly mediates the formation of disulfide bonds in further substrate proteins in the IMS.

1.4.6 The Mia40-Erv1 Disulfide Relay System

The cooperation of Mia40 and Erv1 in the import of twin $Cx_{3/9}C$ motif-containing proteins into the IMS of mitochondria is described in the following model (Figure 1.4.2). After synthesis in the cytosol, small twin $Cx_{3/9}C$ motif-containing proteins are transported in an unfolded reduced state across the TOM complex into the IMS, where they interact with the oxidized form of Mia40 by generation of mixed disulfide intermediates. They are then released from Mia40 in the oxidized state, thereby triggering their folding. In the folded state, they cannot diffuse across the TOM complex and are trapped in the IMS. Thus, protein import into the IMS is driven by the Mia40-dependent oxidative "folding-trap"-mechanism. The transfer of disulfide bonds from Mia40 to imported proteins leaves Mia40 in a reduced non-functional state. To drive further rounds of import, Mia40 has to be reoxidized. The transfer of disulfide bonds from the thiol oxidase Erv1 to reduced non-functional Mia40 recovers Mia40 in its oxidized functional state. Oxidized Erv1 is regenerated by the transfer of electrons to the respiratory chain *via* cytochrome c as described below. In summary, Mia40 and Erv1 form a disulfide relay system which drives the import of proteins into the IMS by oxidative folding.

Biochemical and structural studies have begun to unravel the molecular mechanisms of this disulfide relay system in more detail. The disulfide relay system could be reconstituted with purified proteins.[46] Recombinant Mia40 is sufficient to selectively bind to substrate proteins and is able to oxidize a Mia40 variant which lacks the first two cysteine residues and thus resembles a typical substrate with a twin Cx_9C motif.[46,68,69] In the presence of catalytic amounts of Mia40, Erv1 is necessary and sufficient to oxidize substrate proteins. Erv1 on its own does not oxidize the substrate proteins.[46] This demonstrates transfer of disulfide bonds from Erv1 *via* Mia40 to substrate proteins without the need of

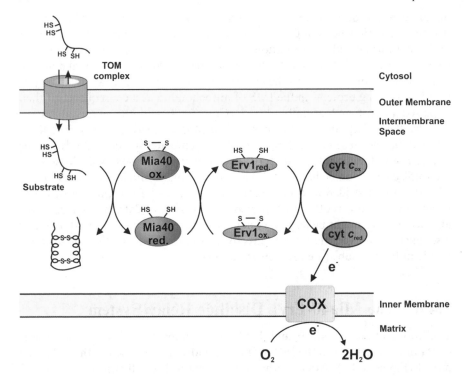

Figure 1.4.2   Model of oxidative protein import and folding by the disulfide relay
system of Mia40 and Erv1. Substrate proteins with twin Cx_3C or twin
Cx_9C motifs pass the TOM complex in the outer membrane in an
unfolded and reduced state. They interact with the import receptor
Mia40 *via* a transient intermolecular disulfide bond and are subsequently released in an oxidized folded form. In the folded state they are
unable to cross the outer membrane and are trapped in the IMS. The
oxidative folding of substrate proteins leads to a reduced form of Mia40.
To enable further rounds of import reduced Mia40 is reoxidized by the
sulfhydryl oxidase Erv1. Oxidized Erv1 is regenerated by transfer of
electrons to cytochrome *c* which shuttles the electrons *via* cytochrome *c*
oxidase (COX) to the final electron acceptor, oxygen. An alternative
pathway *via* molecular oxygen and cytochrome *c* peroxidase appears to
exist to reoxidize Erv1 and cytochrome *c* (see text).

any further components. The first two cysteine residues of Mia40 are essential
to mediate the oxidation of substrate proteins, as well as the interaction with
Erv1 and the reoxidation of Mia40. These residues appear to represent the
catalytically active disulfide bond, whereas the two disulfide bonds in the twin
Cx_9C motif seem to structurally stabilize the protein as is the case in the substrate proteins. Although Mia40 and Erv1 constitute a minimal disulfide relay
for protein oxidation in the IMS, the system in the cell may consist of more
components. A more complex system may increase the efficiency of oxidation
and/or have the ability to regulate its efficiency or its substrate specificity.

The mechanisms of disulfide transfer by Mia40 and the nature of its mixed
disulfide intermediates with substrate proteins and with Erv1 is still speculative.

For some small TIM proteins with twin Cx_3C motifs only the most amino-terminal cysteine residue is required for binding to Mia40.[68,69] A sequence or structural signal which allows selective recognition of this specific cysteine residue has, however, not been identified so far. In contrast, the other three cysteine residues of the twin Cx_3C motif are crucial for the assembly of the small TIM proteins into the native complexes and for their release from Mia40 in their oxidized state, rather than for binding to Mia40. Despite this selectivity of recognition, it is unknown whether Mia40 forms one or two intermolecular disulfide bonds with the substrates in the mixed disulfide intermediates. Since the substrates with twin $Cx_{3/9}C$ motifs are released fully oxidized, two disulfide bonds have to be generated in the substrates. So far, this process and the release are not understood at all. Interestingly, recent results support a role of functional Erv1 in mitochondria in the oxidation and the assembly of small TIM proteins at a step following the formation of the Mia40-substrate disulfide intermediate.[39,40] Moreover, the cysteine-rich protein Hot13 has been described to mediate the assembly of small TIM proteins into native oligomeric TIM complexes.[70] Further studies are required to determine whether Hot13 plays a role for the oxidative folding pathway of small TIM proteins.

In addition, the substrate spectrum of the disulfide relay system is not restricted to proteins with twin Cx_3C and Cx_9C motifs. The import of the sulfhydryl oxidase Erv1 depends on Mia40, although Erv1 does not contain a typical twin $Cx_{3/9}C$ motif.[48,71] Presumably, the mechanism of Erv1 import differs from those of substrates with twin Cx_3C motifs and Cx_9C motifs. Moreover, there are proteins with disulfide bonds in the IMS, as described, which do not belong to the class of proteins with twin $Cx_{3/9}C$ motifs. These proteins might still employ the disulfide relay system for protein oxidation, may be even uncoupled from protein import or they might use the sulfhydryl oxidase Erv1 directly for oxidation.

1.4.7 Cytochrome *c* Links the Disulfide Relay System to the Respiratory Chain of Mitochondria

Following the transfer of disulfide bonds to substrate proteins, Erv1 is reduced in the disulfide relay system. In order to keep Erv1 catalytically active and to enable further rounds of oxidation of Mia40, disulfide bonds have to be introduced into the reduced Erv1 by reoxidation. Molecular oxygen functions as an electron acceptor for flavin-dependent sulfhydryl oxidases producing hydrogen peroxide.[56] Although such an electron transfer from Erv1 to oxygen has been observed *in vitro*, cytochrome c has been proven to be a better electron acceptor of Erv1/ALR than oxygen.[25,64,65] These results suggest that, *in vivo*, Erv1 passes electrons to cytochrome *c*, thereby avoiding the production of deleterious hydrogen peroxide.[25,41] In consistence, Erv1 directly interacts with cytochrome *c in vitro* and in mitochondria.[65] Moreover, the lack of cytochrome *c* or the manipulation of the redox state of cytochrome *c* in mitochondria affects the reoxidation of Mia40 and thus the import of proteins into the IMS of mitochondria.[64] In summary, the disulfide relay system shuttles electrons from Erv1 to oxidized cytochrome *c*. Since cytochrome *c* feeds electrons into cytochrome *c* oxidase of the respiratory chain, thereby generating water from molecular oxygen, the reduction of cytochrome *c* by Erv1 links the disulfide relay system to the respiratory chain. In addition, cytochrome *c* is also able to interact with cytochrome *c* peroxidase, Ccp1, which functions as electron acceptor and reoxidizes cytochrome *c*.[65] To do so, Ccp1 has to be present in its oxidized form, which is generated by reduction of hydrogen peroxide to water.[72] Hydrogen peroxide, in turn, is produced upon reoxidation of Erv1 by molecular oxygen. Thus, Ccp1 has the capacity to link the two electron-accepting pathways of Erv1. Both pathways depend on the presence of oxygen. Presumably, another final electron acceptor for Erv1 or cytochrome *c* exists, at least in yeast, because these cells are viable under anaerobic conditions.[41,65] The use of cytochrome *c* as the preferred electron acceptor prevents the formation of deleterious hydrogen peroxide and may allow the oxidation of proteins even under low oxygen conditions. This might be important in particular in higher eukaryotes in which there are cells with limited oxygen supply under certain conditions.[64] On the other hand, the disulfide relay system may function as a sensor of the endogenous levels of molecular oxygen. Since oxygen is the final electron acceptor of the disulfide relay system, the activity of this system and its substrates may correlate with the concentration of oxygen in mitochondria.

1.4.8   Oxidative Protein Folding Drives Import of Sod1

In addition to the Mia40/Erv1 disulfide relay system, the oxidation folding of another component in the IMS, the copper/zinc superoxide dismutase Sod1, has been studied. It is a highly conserved antioxidant enzyme that catalyzes the disproportionation of superoxide to hydrogen peroxide and molecular oxygen.[73] The active homodimer consists of 16-kDa monomers which bind one copper and one zinc ion and are stabilized by an intramolecular disulfide bond. Stable folding of Sod1 is triggered by Ccs1, the copper chaperone for Sod1.[74–76] Ccs1 forms a transient intermolecular disulfide bond with Sod1.[77] This intermediate allows Ccs1 to introduce a disulfide bond and a copper ion into Sod1.

While the majority of Sod1 is present in the cytosol, about 1–5% of total Sod1 is located in the IMS of mitochondria in yeast.[78] Because of the very small volume of the IMS compared to the cytosol, the concentration of Sod1 in the IMS appears to be very high and presumably exceeds the cytosolic concentration. Its partner Ccs1 is also localized in the cytosol and in the IMS.[78] The import of Sod1 into the IMS follows an oxidative protein folding mechanism. In order to pass through the outer membrane Sod1 has to be present in an unfolded reduced form devoid of metal ions.[79] In the IMS, the unfolded Sod1 presumably binds to Ccs1 *via* a transient intermolecular disulfide bond. Subsequently, Ccs1 transfers the disulfide bond and the copper ion to Sod1, releasing the tightly folded form of Sod1, which cannot pass across the outer membrane and thus is trapped in the IMS. In consistence with such a mechanism, the amounts of Sod1 imported are correlated with the amount of Ccs1 present in the IMS.[79] Moreover, the cysteine residues involved in the formation of the disulfide crucial for the import of Sod1.[79] Thus, Ccs1 appears to act in the oxidative folding pathway of Sod1 as an import receptor similar to Mia40. Following release of Sod1, Ccs1 is also present in a reduced form and presumably needs to be reoxidized. Ccs1 may be oxidized by the Mia40/Erv1 disulfide relay system or by the sulfhydryl oxidase Erv1 directly. Both possibilities would explain why depletion of Erv1 affects the endogenous levels of Sod1 in the IMS.[32] Other mechanisms, however, such as direct oxidation by molecular oxygen, might exist as well, since oxidation of Sod1 by Ccs1 also takes place in the cytosol that is lacking the Mia40/Erv1 disulfide relay system.[80]

1.4.9   Conclusion and Perspectives

Oxidative protein folding in mitochondria has been discovered recently. The Mia40/Erv1 disulfide relay system catalyzes oxidative folding of proteins driving their import into the IMS and their assembly within. However, we are still beginning to unravel the pathways and mechanisms of this process. Future studies will elucidate whether this system and/or the thiol oxidase Erv1 alone mediate directly or indirectly the oxidation of all proteins with disulfide bonds in the IMS. Additional components might work as shuttle proteins in the oxidation of specific proteins, such as Ccs1 for Sod1. It cannot be excluded that, in addition, further protein oxidation systems exist in the IMS which have not been identified so far. The protein oxidation in the IMS might enable regulation of protein activities and control of mitochondrial processes by redox switches. It will be interesting to analyze how the activities of the disulfide relay system and their substrates are modulated. In addition, it will be revealing to compare the Mia40/Erv1 relay system with the disulfide relay systems in the bacteria and the endoplasmic reticulum.

Acknowledgements

We thank Nadia Terziyska and Soledad Funes for comments on the manuscript. This work was supported by grants from the Deutsche Forschungsgemeinschaft SFB594 B3 and B13 and He 2803/2-4.

References

1. J. Reinders, R. P. Zahedi, N. Pfanner, C. Meisinger and A. Sickmann, *J. Proteome Res.*, 2006, **5**, 1543–1554.
2. H. Prokisch, C. Andreoli, U. Ahting, K. Heiss, A. Ruepp, C. Scharfe and T. Meitinger, *Nucleic Acids Res.*, 2006, **34**, D705–D711.
3. S. W. Taylor, E. Fahy, B. Zhang, G. M. Glenn, D. E. Warnock, S. Wiley, A. N. Murphy, S. P. Gaucher, R. A. Capaldi, B. W. Gibson and S. S. Ghosh, *Nat. Biotechnol.*, 2003, **21**, 281–286.
4. W. Neupert and J. M. Herrmann, *Annu. Rev. Biochem.*, 2007, **76**, 723–749.
5. J. M. Herrmann and K. Hell, *Trends Biochem. Sci.*, 2005, **30**, 205–211.

6. C. de Marcos-Lousa, D. P. Sideris and K. Tokatlidis, *Trends Biochem. Sci.*, 2006, **31**, 259–267.
7. C. M. Koehler, *Annu. Rev. Cell Dev. Biol.*, 2004, **20**, 309–335.
8. T. Endo, H. Yamamoto and M. Esaki, *J. Cell Sci.*, 2003, **116**, 3259–3267.
9. P. Rehling, K. Brandner and N. Pfanner, *Nat. Rev. Mol. Cell. Biol.*, 2004, **5**, 519–530.
10. B. F. Lang, M. W. Gray and G. Burger, *Annu. Rev. Genet.*, 1999, **33**, 351–397.
11. A. Holmgren, *J. Biol. Chem.*, 1989, **264**, 13963–13966.
12. O. Carmel-Harel and G. Storz, *Annu. Rev. Microbiol.*, 2000, **54**, 439–461.
13. F. Aslund and J. Beckwith, *J. Bacteriol.*, 1999, **181**, 1375–1379.
14. G. T. Hanson, R. Aggeler, D. Oglesbee, M. Cannon, R. A. Capaldi, R. Y. Tsien and S. J. Remington, *J. Biol. Chem.*, 2004, **279**, 13044–13053.
15. C. M. Koehler, K. N. Beverly and E. P. Leverich, *Antioxid. Redox Signal.*, 2006, **8**, 813–822.
16. U. Jakob, W. Muse, M. Eser and J. C. Bardwell, *Cell*, 1999, **96**, 341–352.
17. F. U. Hartl and M. Hayer-Hartl, *Science*, 2002, **295**, 1852–1858.
18. B. Bukau and A. L. Horwich, *Cell*, 1998, **92**, 351–366.
19. H. Kadokura, F. Katzen and J. Beckwith, *Annu. Rev. Biochem.*, 2003, **72**, 111–135.
20. H. Nakamoto and J. C. Bardwell, *Biochim. Biophys. Acta*, 2004, **1694**, 111–119.
21. K. Mukai, T. Miyazaki, S. Wakabayashi, S. Kuramitsu and H. Matsubara, *J. Biochem.*, 1985, **98**, 1417–1425.
22. S. Iwata, J. W. Lee, K. Okada, J. K. Lee, M. Iwata, B. Rasmussen, T. A. Link, S. Ramaswamy and B. K. Jap, *Science*, 1998, **281**, 64–71.
23. T. Tsukihara, H. Aoyama, E. Yamashita, T. Tomizaki, H. Yamaguchi, K. Shinzawa-Itoh, R. Nakashima, R. Yaono and S. Yoshikawa, *Science*, 1995, **269**, 1069–1074.
24. G. Hofhaus, J. E. Lee, I. Tews, B. Rosenberg and T. Lisowsky, *Eur. J. Biochem.*, 2003, **270**, 1528–1535.
25. S. R. Farrell and C. Thorpe, *Biochemistry*, 2005, **44**, 1532–1541.
26. A. Voronova, W. Meyer-Klaucke, T. Meyer, A. Rompel, B. Krebs, J. Kazantseva, R. Sillard and P. Palumaa, *Biochem. J.*, 2007, **408**, 139–148.
27. F. Arnesano, E. Balatri, L. Banci, I. Bertini and D. R. Winge, *Structure*, 2005, **13**, 713–722.
28. K. Hell, *Biochim. Biophys. Acta*, 2008, **8**, 1783, 601–609.
29. J. M. Herrmann and R. Kohl, *J. Cell Biol.*, 2007, **176**, 559–563.
30. D. Stojanovski, J. M. Müller, D. Milenkovic, B. Guiard, N. Pfanner and A. Chacinska, *Biochim. Biophys. Acta*, 2008, **1783**, 610–617.
31. H. Lu, S. Allen, L. Wardleworth, P. Savory and K. Tokatlidis, *J. Biol. Chem.*, 2004, **279**, 18952–18958.
32. N. Mesecke, N. Terziyska, C. Kozany, F. Baumann, W. Neupert, K. Hell and J. M. Herrmann, *Cell*, 2005, **121**, 1059–1069.
33. K. Tokatlidis, *Cell*, 2005, **121**, 965–967.
34. C. M. Koehler, *Trends Biochem. Sci.*, 2004, **29**, 1–4.

35. D. M. Glerum, A. Shtanko and A. Tzagoloff, *J. Biol. Chem.*, 1996, **271**, 14504–14509.
36. S. Allen, H. Lu, D. Thornton and K. Tokatlidis, *J. Biol. Chem.*, 2003, **278**, 38505–38513.
37. C. T. Webb, M. A. Gorman, M. Lazarou, M. T. Ryan and J. M. Gulbis, *Mol. Cell*, 2006, **21**, 123–133.
38. T. Lutz, W. Neupert and J. M. Herrmann, *EMBO J.*, 2003, **22**, 4400–4408.
39. J. M. Muller, D. Milenkovic, B. Guiard, N. Pfanner and A. Chacinska, *Mol. Biol. Cell*, 2008, **19**, 226–236.
40. M. Rissler, N. Wiedemann, S. Pfannschmidt, K. Gabriel, B. Guiard, N. Pfanner and A. Chacinska, *J. Mol. Biol.*, 2005, **353**, 485–492.
41. S. Allen, V. Balabanidou, D. P. Sideris, T. Lisowsky and K. Tokatlidis, *J. Mol. Biol.*, 2005, **353**, 937–944.
42. M. Naoe, Y. Ohwa, D. Ishikawa, C. Ohshima, S. Nishikawa, H. Yamamoto and T. Endo, *J. Biol. Chem.*, 2004, **279**, 47815–47821.
43. N. Terziyska, T. Lutz, C. Kozany, D. Mokranjac, N. Mesecke, W. Neupert, J. M. Herrmann and K. Hell, *FEBS Lett.*, 2005, **579**, 179–184.
44. S. Hofmann, U. Rothbauer, N. Muhlenbein, K. Baiker, K. Hell and M. F. Bauer, *J. Mol. Biol.*, 2005, **353**, 517–928.
45. A. Chacinska, S. Pfannschmidt, N. Wiedemann, V. Kozjak, L. K. Sanjuan Szklarz, A. Schulze-Specking, K. N. Truscott, B. Guiard, C. Meisinger and N. Pfanner, *EMBO J.*, 2004, **23**, 3735–3746.
46. B. Grumbt, V. Stroobant, N. Terziyska, L. Israel and K. Hell, *J. Biol. Chem.*, 2007, **282**, 37461–37470.
47. E. A. Winzeler, *et al.*, *Science*, 1999, **285**, 901–906.
48. K. Gabriel, D. Milenkovic, A. Chacinska, J. Muller, B. Guiard, N. Pfanner and C. Meisinger, *J. Mol. Biol.*, 2007, **365**, 612–620.
49. T. Lisowsky, *Mol. Gen. Genet.*, 1992, **232**, 58–64.
50. H. Lange, T. Lisowsky, J. Gerber, U. Muhlenhoff, G. Kispal and R. Lill, *EMBO Rep.*, 2001, **2**, 715–720.
51. R. Pawlowski and J. Jura, *Mol. Cell Biochem.*, 2006, **288**, 159–169.
52. J. Gerber, U. Muhlenhoff, G. Hofhaus, R. Lill and T. Lisowsky, *J. Biol. Chem.*, 2001, **276**, 23486–23491.
53. C. S. Sevier, J. W. Cuozzo, A. Vala, F. Aslund and C. A. Kaiser, *Nat. Cell Biol.*, 2001, **3**, 874–882.
54. T. G. Senkevich, C. L. White, E. V. Koonin and B. Moss, *Proc. Natl. Acad. Sci. USA*, 2000, **97**, 12068–12073.
55. C. S. Sevier, H. Kadokura, V. C. Tam, J. Beckwith, D. Fass and C. A. Kaiser, *Protein Sci.*, 2005, **14**, 1630–1642.
56. D. L. Coppock and C. Thorpe, *Antioxid. Redox Signal.*, 2006, **8**, 300–311.
57. J. Lee, G. Hofhaus and T. Lisowsky, *FEBS Lett.*, 2000, **477**, 62–66.
58. E. Gross, C. S. Sevier, A. Vala, C. A. Kaiser and D. Fass, *Nat. Struct. Biol.*, 2002, **9**, 61–67.
59. E. Vitu, M. Bentzur, T. Lisowsky, C. A. Kaiser and D. Fass, *J. Mol. Biol.*, 2006, **362**, 89–101.

60. C. K. Wu, T. A. Dailey, H. A. Dailey, B. C. Wang and J. P. Rose, *Protein Sci.*, 2003, **12**, 1109–1118.
61. A. Vala, C. S. Sevier and C. A. Kaiser, *J. Mol. Biol.*, 2005, **354**, 952–966.
62. G. Hofhaus, G. Stein, L. Polimeno, A. Francavilla and T. Lisowsky, *Eur. J. Cell Biol.*, 1999, **78**, 349–356.
63. A. Levitan, A. Danon and T. Lisowsky, *J. Biol. Chem.*, 2004, **279**, 20002–20008.
64. K. Bihlmaier, N. Mesecke, N. Terziyska, M. Bien, K. Hell and J. M. Herrmann, *J. Cell Biol.*, 2007, **179**, 389–395.
65. D. V. Dabir, E. P. Leverich, S. K. Kim, F. D. Tsai, M. Hirasawa, D. B. Knaff and C. M. Koehler, *EMBO J.*, 2007, **26**, 4801–4811.
66. T. Lisowsky, *Curr. Genet.*, 1994, **26**, 15–20.
67. D. Becher, J. Kricke, G. Stein and T. Lisowsky, *Yeast*, 1999, **15**, 1171–1181.
68. D. Milenkovic, K. Gabriel, B. Guiard, A. Schulze-Specking, N. Pfanner and A. Chacinska, *J. Biol. Chem.*, 2007, **282**, 22472–22480.
69. D. P. Sideris and K. Tokatlidis, *Mol. Microbiol.*, 2007, **65**, 1360–1373.
70. S. P. Curran, D. Leuenberger, E. P. Leverich, D. K. Hwang, K. N. Beverly and C. M. Koehler, *J. Biol. Chem.*, 2004, **279**, 43744–43751.
71. N. Terziyska, B. Grumbt, M. Bien, W. Neupert, J. M. Herrmann and K. Hell, *FEBS Lett.*, 2007, **581**, 1098–1102.
72. A. Boveris, *Acta Physiol. Lat. Am.*, 1976, **26**, 303–309.
73. I. Fridovich, *Adv. Enzymol. Relat. Areas Mol. Biol.*, 1986, **58**, 61–97.
74. V. C. Culotta, L. W. Klomp, J. Strain, R. L. Casareno, B. Krems and J. D. Gitlin, *J. Biol. Chem.*, 1997, **272**, 23469–23472.
75. R. L. Casareno, D. Waggoner and J. D. Gitlin, *J. Biol. Chem.*, 1998, **273**, 23625–23628.
76. T. D. Rae, P. J. Schmidt, R. A. Pufahl, V. C. Culotta and T. V. O'Halloran, *Science*, 1999, **284**, 805–808.
77. A. L. Lamb, A. S. Torres, T. V. O'Halloran and A. C. Rosenzweig, *Nat. Struct. Biol.*, 2001, **8**, 751–755.
78. L. A. Sturtz, K. Diekert, L. T. Jensen, R. Lill and V. C. Culotta, *J. Biol. Chem.*, 2001, **276**, 38084–38089.
79. L. S. Field, Y. Furukawa, T. V. O'Halloran and V. C. Culotta, *J. Biol. Chem.*, 2003, **278**, 28052–28059.
80. Y. Furukawa and T. V. O'Halloran, *Antioxid. Redox Signal.*, 2006, **8**, 847–867.
81. A. L. Lamb, A. K. Wernimont, R. A. Pufahl, V. C. Culotta, T. V. O'Halloran and A. C. Rosenzweig, *Nat. Struct. Biol.*, 1999, **6**, 724–729.
82. L. Banci, I. Bertini, F. Cantini, S. Ciofi-Baffoni, L. Gonnelli and S. Mangani, *J. Biol. Chem.*, 2004, **279**, 34833–34839.
83. S. P. Curran, D. Leuenberger, W. Oppliger and C. M. Koehler, *EMBO J.*, 2002, **21**, 942–953.
84. S. P. Curran, D. Leuenberger, E. Schmidt and C. M. Koehler, *J. Cell Biol.*, 2002, **158**, 1017–1027.
85. C. Abajian and A. C. Rosenzweig, *J. Biol. Inorg. Chem.*, 2006, **11**, 459–566.

CHAPTER 1.5

Oxidative Folding
in the Endoplasmic Reticulum

SEEMA CHAKRAVARTHI, CATHERINE E. JESSOP
AND NEIL J. BULLEID

Faculty of Life Sciences, Michael Smith Building, University of Manchester, Manchester, M13 9PT, UK

1.5.1 Introduction

An essential step during the maturation of many membrane and secretory proteins in the endoplasmic reticulum (ER) is the formation of native disulfide bonds. Any two cysteines within a protein have the potential to form a disulfide bond. Hence, the formation of native disulfide bonds is a complex process and is the rate-limiting step in the biogenesis of many secreted or membrane proteins. These bonds are often crucial for the stability of the final protein structure, and the mispairing of cysteine residues can prevent proteins from attaining their native conformation and lead to misfolding.

Since disulfide bonds are covalent linkages they can dramatically increase the stability of the three-dimensional protein structure. This is particularly favorable for proteins that are exposed to the extracellular environment. Disulfide bonds maintain the integrity of many extracellular proteins and thus protect them from damage from factors such as oxidants and proteolytic enzymes. In addition, many receptors are exposed to a low pH environment when they are internalized and recycled *via* the endosome. Under these acidic conditions receptor integrity is maintained *via* disulfide bonding. Since disulfide bonds can have a large impact on the structure of proteins it is therefore unsurprising that

RSC Biomolecular Sciences
Oxidative Folding of Peptides and Proteins
Edited by Johannes Buchner and Luis Moroder
© Royal Society of Chemistry 2009
Published by the Royal Society of Chemistry, www.rsc.org

they are widely employed to regulate the activity of many proteins. Several transcription factors including Hsp33 are activated by the formation of disulfide bonds,[1] whereas a number of other proteins such as integrins are activated through the breaking of a disulfide bond.[2] In addition, the formation of a disulfide bond is essential for the catalytic activity of some metabolic enzymes such as ribonucleotide reductases, which are oxidized during their catalytic cycle and must be reduced again in order to restore activity.[3]

The formation of native disulfide bonds is also central to the folding process of many substrates: disulfide bonds may restrict energetically favorable changes that would otherwise result in a non-native conformation, ensure that the correct parts of a substrate are positioned to promote further folding or link subunits of a multimeric protein complex together. In addition, a disulfide-bonded intermediate has been observed in the folding pathway of a non-disulfide-bonded protein, suggesting that disulfide bonding could even play an important role in the folding of some non-disulfide-bonded proteins.[4] Thus, disulfide bonds greatly influence the structure and activity of many proteins and as such play an essential role in the regulation, catalysis and, most importantly, the folding of many proteins.

1.5.2   Biochemistry of Disulfide-bond Formation

Classic folding experiments by Anfinsen indicated that the primary sequence of a protein is sufficient to achieve the native conformation.[5] Although this is true for many proteins, the same cannot generally be said for proteins containing disulfide bonds. Formation of a disulfide bond between the thiol (-SH) groups of two cysteine residues generates two protons and two electrons. *In vitro*, disulfide bonds can be formed spontaneously by the loss of electrons from the two cysteine thiols, coupled with the gain of electrons by an available acceptor such as molecular oxygen. However, an intermediary, such as a transition metal or flavin, is required to overcome this kinetically sluggish reaction.

In vivo, the most common mechanism for the formation of protein disulfide bonds is a thiol–disulfide exchange reaction of free thiols with an already disulfide-bonded species. These reactions are catalyzed by cellular enzymes known as **thiol–disulfide oxidoreductases**. A thiol exchange reaction involves the spontaneous deprotonation of a free thiol to produce a thiolate anion (-S$^-$), which displaces one sulfur of the disulfide bond in the oxidized species. Thus, a transient covalent bond is formed between the proteins, then termed "mixed disulfides". A second exchange follows where the remaining thiolate anion attacks the mixed disulfide bond and resolves it. The net result is that one pair of cysteines is oxidized and the other reduced, and the intramolecular disulfide bond is effectively passed from one pair of cysteines to another (Figure 1.5.1). In eukaryotic cells, disulfide-bond formation proceeds predominantly in the lumen of the ER.

1.5.3   Folding Environment of the ER

The ER contains a rich cocktail of molecular chaperones and folding enzymes such as BiP, calnexin, calreticulin, peptidyl-proline *cis–trans* isomerase and

Figure 1.5.1   Thiol–disulfide exchange reaction between a CXXC-containing thio-redoxin-like oxidoreductase and a substrate. A thiolate anion ($-S^-$) is formed by the deprotonation of a free thiol. It displaces one sulfur of the disulfide bond in the oxidised species, resulting in a transient mixed disulfide bond between the two proteins. A second exchange reaction then follows where the remaining thiolate anion attacks the mixed di-sulfide bond and resolves it. The net result of thiol disulfide exchange is that the originally reduced protein is oxidised. This is due to the gain of electrons from the originally oxidised species, which itself is reduced. (Adapted from Sevier and Kaiser, 2002).

protein disulfide isomerase (PDI) and thus provides an environment highly optimized for efficient folding. Reflecting its role as the compartment for biogenesis of proteins destined for extracellular space, the physiochemical environment of the ER lumen differs in two main respects from that of the cytosolic compartment: the differences in calcium (Ca^{2+}) concentrations and redox conditions.

The ER luminal Ca^{2+} concentration is similar to extracellular concentra-tions.[6] Many of the ER resident proteins involved in protein folding such as BiP, PDI, calnexin and calreticulin have been shown to bind Ca^{2+}.[7,8] This elevated Ca^{2+} concentration has been shown to be essential for the correct folding of a number of proteins in the ER.[9,10] Ca^{2+} depletion within the ER inhibits protein folding and maturation[11,12] and facilitates protein degrada-tion.[13] Ca^{2+} can also regulate the formation of chaperone complexes in the ER.

The compartmentalization of the ER away from the cytosol allows the correct redox conditions for disulfide-bond formation to be established. The lumen of the ER is more oxidizing than the cytosol, which in turn enables a distinct set of folding catalysts to facilitate the formation of native disulfide bonds.[14] Glutathione is the major small molecule redox buffer in the ER.[14] Reduced glutathione (GSH) is a small tripeptide that can be easily oxidized to form a dimer linked by a disulfide bridge. This oxidized form is referred to as glutathione disulfide (GSSG). The ratio of reduced (GSH) to oxidized (GSSG) glutathione is $\sim 100:1$ in the cytosol.[14] This highly reducing environment dis-favors disulfide-bond formation. However, in the ER where disulfide-bond formation can occur, the ratio of GSH:GSSG is much more oxidizing at $\sim 3:1$.[14] This ratio of the concentration of GSH to GSSG is similar to that found in redox buffers effective in oxidative folding of proteins *in vitro*.[15] In order to allow efficient formation of disulfide bonds, the ER must closely regulate its redox potential. Under strongly reducing conditions, such as those prevalent in the cytosol, disulfide-bond formation is both kinetically and thermodynamically disfavored. Likewise, excessively oxidizing conditions result in misfolding due to incorrect intra- and intermolecular disulfide bonds. Thus there must exist a system that senses the redox state and alters the conditions accordingly.

In addition to maintaining optimum redox conditions, the ER is also abundant in enzymes necessary for efficient disulfide-bond formation. A growing family of ER oxidoreductases is thought to be responsible for catalyzing the formation, isomerization and reduction of disulfide bonds.[16] These oxidoreductases contain active sites homologous to the active site found in the cytosolic reductase thioredoxin, characterized by a pair of cysteine residues (CXXC) that shuttle between the disulfide and dithiol form.[17] The reactions that these enzymes catalyze require the individual active sites to be maintained in either the oxidized disulfide form for disulfide-bond formation, or the reduced dithiol form for isomerization or reduction of disulfide bonds[18] (Figure 1.5.2). How the active sites are maintained in either their reduced or oxidized state and how the ER maintains an environment conducive to disulfide-bond formation, isomerization and reduction has been the subject of intense speculation over the past 40 years.

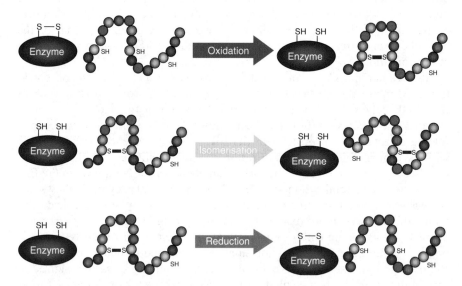

Figure 1.5.2   Oxidation, reduction and isomerisation reactions between an oxidoreductase and a substrate. In order for an oxidoreductase to form a disulfide bond or oxidise a substrate it must be oxidised. An oxidoreductase is only capable of accepting the electrons generated by the oxidation of a substrate when it is in an oxidised form. Conversely, for an oxidoreductase to reduce a substrate it must be in a reduced form so that it is able to donate electrons. The isomerisation of a disulfide bond requires that a bond is broken and then a new one spontaneously formed. To perform isomerisation or reduction reactions a disulfide bond in the substrate is broken, for which the oxidoreductase must be in a reduced form.

1.5.4 Thiol Disulfide Oxidoreductase Family

The eukaryotic ER oxidoreductase family is extensive with nearly 20 members being identified in humans alone[19] (Figure 1.5.3). In mammalian cells the key oxidoreductases described to date are protein disulfide isomerase (PDI), ER oxidase 1 (Ero1), ER protein 57 (ERp57), ERp72 and P5.

The CXXC motif allows the oxidoreductases to catalyze three essential reactions depending on the initial redox state of the enzyme. A disulfide bond may be formed through the oxidation of a substrate protein requiring the oxidoreductase to gain two electrons, non-native disulfide bonds may be shuffled or isomerized with no net exchange of electrons, or incorrect pairings may be broken or reduced, requiring the oxidoreductase to donate electrons to the substrate. The redox state of the oxidoreductase determines which reaction(s) it will perform. Therefore, for an oxidoreductase to be capable of forming a disulfide bond, it must itself be oxidized to be capable of accepting electrons. Conversely an oxidoreductase can only be functional as an isomerase or reductase when it is in a reduced form, to be capable of donating electrons. After each cycle of reduction

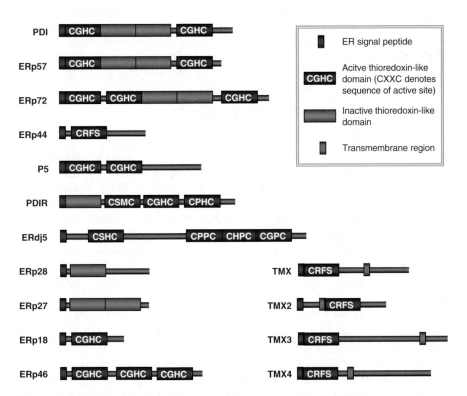

Figure 1.5.3 The domain structures of human protein disulfide isomerase homologues. The catalytic sites of PDI-like oxidoreductases are situated in domains homologous to thioredoxin. CXXC indicates the amino acid sequence of the active site. (Adapted from Ellgaard and Ruddock, 2005).

or oxidation, the oxidoreductase must be returned back to its original, active redox state before it can fulfill further catalytic rounds (Figure 1.5.2).

Although the requirement for an oxidase, which introduces disulfide bonds into newly synthesized proteins, is quite obvious, the function of disulfide-bond isomerization should not be underestimated. The bacterial, yeast and mammalian systems all have efficient oxidative pathways that introduce disulfide bonds into newly synthesized proteins which have arisen because the likelihood of bonds forming between incorrect cysteines is significant. In addition, the frequency of formation of non-native disulfides will increase dramatically as the number of cysteines within a given protein increases. Thus, there is a requirement for isomerases that rearrange incorrect disulfides during protein folding.[20] In addition, the role of a reductase is important in the reduction of disulfide bonds within terminally misfolded proteins prior to their retrotranslocation into the cytosol for degradation.[21,22]

1.5.5 Disulfide-bond Oxidation Pathway

Two pathways have been identified so far for the formation of disulfide bonds within proteins in the eukaryotic ER. The major pathway is composed of the membrane-associated flavoprotein Ero1 and the soluble thioredoxin-like protein PDI. More recently a second pathway involving another flavoenzyme, quiescin sulfhydryl oxidase (QSOX), has been reported (Figure 1.5.4).

1.5.5.1 Protein Disulfide Isomerase (PDI)

The search for enzymatic catalysts of oxidative refolding *in vitro* led to the isolation of protein disulfide isomerase (PDI).[5] PDI is a 55-kDa folding assistant and chaperone of the eukaryotic ER. It is a member of the thioredoxin superfamily and was one of the first identified thiol–disulfide oxidoreductases. PDI is a highly abundant ER luminal protein in yeast and mammalian cells, with a concentration approaching the millimolar range.[23] First isolated from the liver, it has since been discovered in a variety of tissues and organs, and is highly conserved between species. For detailed *in vivo* and *in vitro* roles of PDI refer to Chapter 1.7.

PDI of mammalian origin comprises four structural domains, **a, b, b′** and **a′** plus a 19-amino acid linker region between **b′** and **a′** domains[24] and a stretch of acidic residues at the C-terminus (Figure 1.5.5).[25] The **a** and the **a′** domains have high sequence similarity to thioredoxin, and each of these domains contains two catalytically active cysteine residues found in a CXXC motif. The **b** and **b′** domains are thioredoxin-like domains, but lack any reactive cysteines. PDI is a remarkably versatile enzyme. Depending on the redox environment and the characteristics of the substrate proteins, PDI can catalyze the formation, reduction or isomerization of disulfide bonds.[26] Detailed biochemical analysis has shown that the **a** and the **a′** domains of PDI are capable of catalyzing two kinds of disulfide reactions: (a) oxidation reactions in which

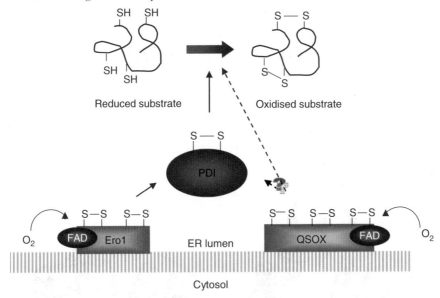

Figure 1.5.4 Disulfide bond formation in the eukaryotic ER. Oxidising equivalents can be introduced into the lumen of the ER by two parallel pathways. In the first, oxidising equivalents flow *via* Ero1 to PDI, which directly oxidises substrate proteins. In the second, QSOX might act similarly to Ero1 in the oxidation of PDI, although it can oxidise substrate proteins directly *in vitro*. Both Ero1 and QSOX can receive electrons ultimately from molecular oxygen in an FAD-dependent manner. In mammalian cells two isoforms of Ero1 are known to oxidise PDI.

intramolecular disulfide bonds of the CGHC motif are transferred to a pair of thiols in a substrate protein and (b) isomerization reactions in which disulfides are rearranged through the formation of mixed disulfides between the first cysteine residue of the CGHC motif and the substrate.[27] PDI is present in both the oxidized and reduced forms *in vivo*, but is more oxidized in yeast[28] compared to mammalian cells,[29] possibly because the yeast ER is more oxidizing. This mixture of redox states allows PDI to perform both oxidation and iso-merization/reduction reactions in cells.

A role for PDI in the catalysis of native disulfide-bond formation in the ER was first established by mutational analysis in yeast, where the PDI1 gene was shown to be essential for cell viability and oxidative protein folding.[30,31] It was first thought that it was the isomerase activity of PDI that made it essential, since yeast Pdi1p with active sites mutated to CXXS is functional as an isomerase but not an oxidase and is able to complement a PDI-deficient yeast strain.[30,32] However, this mutant also imparts increased sensitivity to the reducing agent dithiothreitol (DTT),[33] suggesting that the role of PDI could lie in the formation of disulfide bonds. In addition, expression of a single catalytic domain of PDI, which maintains oxidase activity but only minimal isomerase activity, is also able to sustain growth when the expression of endogenous PDI is repressed.[34]

A) PDI domain structure

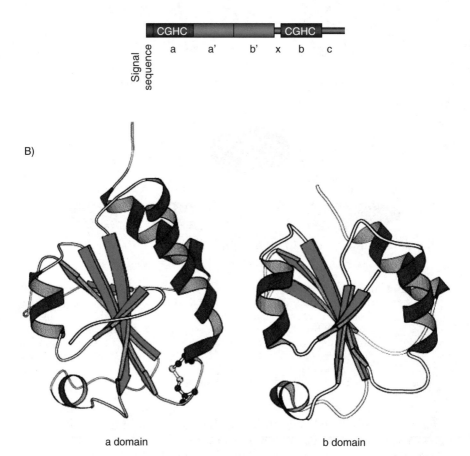

a domain   b domain

Figure 1.5.5   PDI Structure. A) Domain structure of PDI. The **a** and the **a'** domains
contain the CXXC active sites, while the **b** and the **b'** domains lack any
active-site cysteines. The highly acidic **c** region contains the -KDEL
motif for ER retention, while the **x** region is a linker of undefined
structure and function. B) Structures of the isolated **a** and **b** domains
of human PDI, based on a model suggested by NMR data (Kemmink
et al., 1997; Kemmink *et al.*, 1999). Both domains show characteristic
thioredoxin fold despite a low degree of sequence similarity. Active-site
cysteines of the **a** domain are shown at the lower right as a ball-and-stick
model.

The role of PDI as an oxidase is important since it is the only oxidoreductase
of which a significant proportion is oxidized *in vivo*,[28,29] where an oxidized
redox state is a prerequisite for oxidase activity. Evidence that Pdi1p engages
directly in thiol–disulfide exchange with ER proteins came from the detection
of disulfide-linked complexes between Pdi1p and newly synthesized secretory
proteins.[28] Oxidation of reduced cysteine pairs within substrate proteins by

PDI render the active site of PDI reduced. For further rounds of oxidation to proceed, the PDI active site needs to be reoxidized by an electron acceptor.

1.5.5.2 Oxidation by Ero1

Two independent experimental strategies led to the identification of ER oxidoreductin 1 (Ero1) as the enzyme responsible for oxidation of PDI (See Chapter 1.6).[35,36] Extensive biochemical and structural studies carried out on purified yeast Ero1p have elucidated a pathway for protein oxidation in the ER. Ero1p is a FAD-binding protein, which utilizes the oxidative power of molecular oxygen to drive disulfide-bond formation. Thus, components of the pathway include molecular oxygen as the source of oxidizing equivalents that are transferred directly by Ero1p *via* disulfide exchange to PDI, which then in turn oxidizes substrate proteins.

The Ero1 gene is highly conserved and present in a wide range of eukaryotes. Two functional homologs of yeast Ero1p, hEro1-Lα and hEro1-Lβ, have been identified in humans. Both of these can oxidize PDI and can complement the phenotypic defects associated with the yeast mutant *ero1-1* strain.[29,37,38] Overexpression of hEro1-Lα increases the proportion of PDI that is oxidized *in vivo*, and has been shown to increase secretion of various disulfide bonded substrate proteins such as J-chain subunits of immunoglobulins (JcM),[29] tissue plasminogen activator (tPA)[39] and adiponectin.[40] This indicates that the introduction of disulfide bonds is a rate-limiting step in the folding of substrate proteins.

In addition to an N-terminal signal sequence that targets the proteins to the ER, both hEros have two conserved redox active motifs, a CXXXXC motif near the N-terminus and a CXXCXXC motif towards the C-terminus. Studies in yeast have revealed that the N-terminal active site cysteines are involved in interactions with PDI and accept electrons from the PDI active sites. The electrons are then transferred to the C-terminal active sites which in turn pass on the electrons to molecular oxygen *via* the bound FAD.[41,42] The crystal structure of the yeast Ero1p core is consistent with the results obtained by site-directed mutagenesis of the Ero1p cysteines. As discussed in detail in the previous chapter, the crystal structure revealed that the first cysteine in the CXXCXXC motif is involved in a long-range intramolecular disulfide bond, while the second and third cysteines of the CXXCXXC motif are adjacent to the FAD cofactor, enabling them to be readily oxidized.[43] The long-range disulfide is important in sensing the cellular redox state and regulating the activity of Ero1p.[44] While such detailed analysis of the hEros is eagerly awaited, a comparison of the cysteine residues between the yeast and human Ero1 proteins indicate that workings of the hEros might be similar to the yeast protein.

An interesting difference between the yeast Ero1p and its two mammalian homologs is that the 127-residue tail essential for function and association to the ER membrane in yeast Ero1p[45] is absent from the hEros. Human Ero1s also lack any known ER retention motifs. Yet, the two oxidoreductases behave as peripheral membrane proteins when expressed in mammalian cells.[37,38,45]

This suggests that a stable interaction with another resident ER protein may be responsible for the luminal retention of Ero1s. hEro1-Lα and -β have been found to form mixed disulfides with ERp44, an ER protein with a C terminal RDEL motif.[46] It has since been shown that interactions with both PDI and ERp44 contribute to the retention of hEros in the ER.[47]

While there is a single Ero1 protein in yeast, the presence of two isoforms in humans can be explained to some extent by taking a closer look at their regulation and expression profiles. Ero1-Lα is strongly up-regulated by hypoxia.[49] Studies on the secretion of proangiogenesis factor VEGF (vascular endothelial growth factor) under hypoxic conditions suggests that the expression of Ero1-Lα is probably regulated *via* the HIF pathway and thus belongs to the classic family of oxygen regulated genes.[48] Ero1-Lβ, on the other hand, is up-regulated by ER stress that induces unfolded protein response (UPR).[38,49] The expression of the two Ero1-L isoforms therefore appears to be differently regulated; Ero1-Lα expression is mainly controlled by the cellular oxygen tension, whilst Ero1-Lβ is triggered mainly by the UPR.[50] Based on the expression patterns of these two genes, we can propose that Ero1-Lα is the major enzyme responsible for disulfide-bond formation in the ER lumen. If the synthetic load of the cell exceeds the oxidative capacity of Ero1-Lα, reduced cargo would accumulate in the ER thereby inducing a compensatory synthesis of Ero1-Lβ *via* the UPR pathway. The consequence of a prolonged ER stress is also an induction of Ero1α. CHOP is a transcription factor that is activated at multiple levels during ER stress and Ero1-Lα is a direct CHOP target gene.[51] It is also interesting to speculate that Ero1-Lα and -β might drive oxidative folding in substrate proteins *via* different PDI homologs present in the mammalian ER. However, to date Ero1-Lα and -β have only been shown to interact with PDI.[29]

Ero1-Lα and Ero1-Lβ transcripts also display differential tissue distribution.[38] Ero1-Lα is abundant in the esophagus, while Ero1-Lβ transcripts are abundant in the stomach, duodenum, pancreas, testis and pituitary glands. A study on Ero1-Lβ has shown that a notable proportion of this protein exists as a homodimeric pool at steady state.[52] In addition, Ero1-Lα and Ero1-Lβ heterodimers were also detected. The physiological role of such dimers is still unclear.

While the availability of purified Ero1p has allowed major advances to be made in the characterization of this enzyme, there are no structural data available for either of the human Ero1 proteins. Also, all *in vitro* enzyme assays carried out to date have used the yeast enzyme. Purification of the mammalian Ero1-L proteins will allow a systematic comparison of the FAD-binding properties and enzyme activities of these proteins with Ero1p.

Alternate electron acceptors for the Ero1 proteins also await identification. Ero1p is required for viability of yeast under anaerobic growth, indicating that its activity does not absolutely require oxygen.[53] However, Ero1p function *in vivo* is compromised under anaerobic conditions suggesting that the alternate electron acceptors are not as efficient as molecular oxygen in the regeneration of oxidized Ero1p.[41] Protein folding in mammalian cells has not been investigated under completely anaerobic conditions and hence it is unclear whether mammalian Ero1 can support disulfide-bond formation under these conditions.

1.5.5.3 Oxidation by QSOX

A second pathway for disulfide-bond formation in the yeast ER was identified by screening for proteins that could rescue an Ero1p mutant.[54] Over-expression of a luminal ER protein, Erv2p, was able to rescue *ero1Δ* yeast strains, leading to the discovery that Erv2p forms the basis of a disulfide-bond formation pathway that can function independently of Ero1p.

Erv2p is an FAD-binding thioloxidase and belongs to the ERV/ALR family of proteins.[55–58] Although Erv2p is not conserved in the strict sense between yeast and man, a family of proteins known as the thiol oxidase (QSOX) family, which contain an Erv2p homologous domain, have been detected in all multicellular plants and animals for which complete genome sequences exist.[59]

Two proteins belonging to the QSOX family exist in humans. The first to be discovered was a gene that was up-regulated when human fibroblasts reached quiescence with the elaboration of extracellular matrix,[59,60] hence called the Quiescin Q6 gene, now renamed as hQSOX1. An alternative splicing of the hQSOX1 gene generates a long (QSOX1a) and a short transcript (QSOX1b). The long form of the QSOX1 protein potentially retains a transmembrane segment that could allow the protein to be bound to the membrane. Indeed, when expressed in mammalian cells, the human QSOX1a protein is a transmembrane protein localized primarily in the Golgi apparatus.[61] However, in human-cultured fibroblasts, this long form has been detected in the culture medium.[60] A second gene was identified as sharing a high identity with hQSOX1 and encoding another member of the QSOX family. The corresponding protein, SOXN or hQSOX2, has been studied in human neuroblastoma cells.[62] The overall homology of the QSOX1 and QSOX2 proteins is $\sim 40\%$, the identity of functional regions such as the TRX-like or the ERV1 domain is as high as 68%.[62] Besides the sequence homology of QSOX1 and QSOX2, the relationship between the two genes is supported by their common structural features as both genes contain 12 exons of identical lengths and show a high GC content in the first exon. The differences in the *QSOX1* and *QSOX2* expression patterns with two transcripts of *QSOX1* in placenta, liver, lung and heart, but very weak expression levels in pancreas, brain, kidney and skeletal muscle,[63] indicate that the two genes have different functions or the same role in different tissues.

All members belonging to the QSOX family have fused their C-terminal Erv2p homologous FAD binding domain to a thioredoxin-like domain at their N terminal, which is in turn homologous to the **a**- and **a′**- domain of PDI (Figure 1.5.6).[59,64] Thus, while the ER oxidation system requires interaction between an Ero1/Erv2p and PDI, the QSOX enzymes have evolved by a fusion of these two proteins, providing them with the unique advantage of being able to introduce disulfide bonds into substrate proteins without the need to interact with additional proteins.

The FAD-dependent thioloxidases of the QSOX family catalyze the oxidation of thiols to disulfides by reducing oxygen to hydrogen peroxide following the reaction $2R\text{–}SH + O_2 \rightarrow R\text{–}S\text{–}S\text{–}R + H_2O_2$. Detailed enzymatic studies on avian

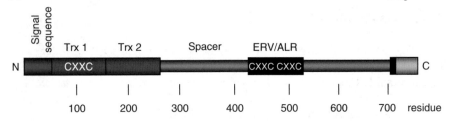

Figure 1.5.6 Schematic depiction of the domain structure of avian QSOX. The three
CXXC motifs have been indicated. Trx1 and Trx2 represent the two
thioredoxin-like domains and ERV/ALR represents the ERV1/ALR
FAD-binding domain. The solid box at the right is a single transmembrane span with the C-terminus facing the cytoplasm. The domains are
drawn approximately to scale.

QSOX indicate that the preferred substrates of this enzyme are peptides and
proteins, but not small monothiols such as glutathione.[65,66] The finding that,
in vitro, mixtures of avian QSOX1 and PDI can catalyze the rapid insertion
of the correct disulfide pairings in reduced RNases leads to the concept that a
major role of QSOXs is in the generation of disulfides *in vivo*. A recent finding
shows that unlike hEros, which can only complement a temperature-sensitive
yeast mutant *ero1-1* strain, hQSOX1a is able to rescue yeast completely depleted
of Ero1p. In this strain the over-expression of hQSOX1 is able to restore disulfide-
bond formation as evaluated by the folding and transport of carboxypeptidase
Y (CPY).[61]

While the intracellular localization and tissue distribution of the mammalian
QSOX enzymes have been extensively investigated,[67] the QSOX proteins have
not been found to be important in the secretion of any particular protein in any
given cell type. Also, no gene knock-down studies of the QSOX family members have been published. Whether this reflects that no effects have been
observed or that it is too difficult at present to evaluate the effects of such
experiments is not clear. On the basis of results obtained in yeast it can be
postulated that the Ero1-PDI pathway may be the major pathway for disulfide-
bond formation in the mammalian ER. Thus any effect that the QSOX enzymes
have on protein folding may only be visible in an Ero1 deletion background.
Study of oxidative protein folding in cells in which Ero1-Lα and Ero1-Lβ have
been knocked down will help in investigating the presence of Ero1 independent
pathways in mammalian cells.

1.5.6   Disulfide-bond Reduction Pathway

An investigation into the redox state of several ER oxidases within mammalian
cells has shown that PDI, ERp57, P5, PDIR and ERp72 are all in a predominantly reduced form at steady state.[29,68] After purification, these proteins

are spontaneously oxidized to form disulfide bonds within their active sites. This would suggest that a pathway exists to maintain these proteins in a reduced state within the cell. Hence, there is a requirement for both an oxidative pathway for disulfide-bond formation and a reductive pathway to allow reduction and iso-merization of non-native disulfides and for these pathways to co-exist in the same intracellular compartment.

While there is no net change in the redox state of the enzyme during iso-merization reactions, following each round of reduction an oxidoreductase becomes oxidized. In order for it to function in further rounds it must be returned to the reduced state. Recent work has demonstrated that reduced glutathione fulfills the role of a general reductant in the ER and is responsible for maintaining the oxidoreductases in a reduced state.

1.5.6.1 The Role of Glutathione in the ER

Glutathione (GSH) is a tripeptide (L-γ-glutamyl-L-cysteinyl-glycine) that is synthesized in the cytosol from the precursor amino acids glutamate, cysteine and glycine. The cell contains millimolar concentrations of GSH (up to 10 mM) that are maintained in this reduced form by a cytosolic NADPH-dependent reaction catalyzed by glutathione reductase. As stated earlier, glutathione is present in cells as a ratio of reduced (GSH) to oxidized (GSSG) glutathione. Evidence for the role of GSH as a reductant came from studies in which GSH synthesis was prevented. In mice, a null mutation in the gene encoding one subunit of γ-GCS (first enzyme involved in GSH synthesis) was found to be embryonic lethal. However, cell lines isolated from the mutants could grow indefinitely in a medium supplemented with the reducing agent N-acet-ylcysteine.[98] Similarly, a yeast strain lacking a functional copy of the GSH1 gene was unable to grow on minimal media unless supplemented with a reducing agent.[99] Since the function of glutathione can be complemented by the addition of reducing agents, this suggested that the main role for glutathione is to maintain reducing conditions within cells.

In order to determine whether GSH was directly responsible for maintaining the ER oxidoreductases in a reduced state *in vivo*, cells were subjected to strong oxidizing conditions and recovery of the ER oxidoreductases was studied.[68] Recovery of ERp57 from oxidation required the presence of cytosolic GSH. ERp57 was unable to recover in the presence of various small molecular reducing agents like NADH, NADPH, cysteamine or cysteine, indicating that, in mammalian cells at least, GSH specifically, rather than its reducing power, is responsible for maintaining the ER oxidoreductases in a reduced state. Furthermore, GSH was found to reduce ERp57 rapidly at physiological con-centrations *in vitro*, and biotinylated glutathione formed mixed disulfides with ERp57, confirming earlier work showing that glutathione directly interacts with ERp57 in intact cells *via* mixed disulfides.[69] Thus GSH is the main reductant responsible for maintaining the ER oxidoreductases in a reduced form, which are in turn necessary for the reduction or isomerization of non-native disulfide bonds.

Further evidence for the role of glutathione in controlling oxidative folding in mammalian cells came from two separate experiments. The effect of GSH depletion on the folding of two substrate proteins, tissue type plasminogen activator (tPA) and J chain of immunoglobulins (JcM), revealed that when the levels of total glutathione were diminished, the rate of formation of disulfide bonds increased.[39, 100] However, this increase in oxidation was accompanied by an increase in non-native disulfide-bond formation. In addition, Molteni *et al.* showed the absence or decrease in levels of cytosolic GSH resulted in PDI becoming more oxidized, limiting the ability of the enzyme to isomerize non-native disulfide bonds (Molteni *et al.*, 2004). Addition of GSH caused PDI to become reduced and restored normal disulfide formation rates.

The fact that glutathione can cross the ER membrane either through a specific transporter[70] or simply through pores in the membrane[71] suggests that this low molecular weight thiol could provide the necessary reducing equivalents to facilitate the reduction of folding proteins directly or *via* reduction of ER oxidoreductases such as ERp57. The results with ERp57 and PDI seem to indicate that GSH acts *via* the ER oxidoreductases. Depletion of GSH causes a change in the redox state of the ER oxidoreductases, which in turn makes them unavailable for reduction/isomerization reactions. However, the fact that GSH/GSSG can directly interact with substrate proteins cannot be completely ruled out. This is supported by the fact that a large proportion of glutathione in the ER lumen is found as mixed disulfides with proteins.[72] Such mixed disulfides could be formed either during the oxidation of proteins by GSSG or by the reduction of proteins by GSH. As stated earlier, the ratio of GSH:GSSG in the ER lumen is similar to those found in redox buffers. Many substrate proteins are able to fold spontaneously in a glutathione buffer in the absence of oxidoreductases. However, the reaction proceeds too slowly to be physiologically relevant and the rate of folding is increased dramatically in the presence of enzymes such as PDI[73] or ERp57.[74]

As the ER has been proposed to contain a relatively higher concentration of GSSG than the cytosol,[14] another theory put forward at this stage was that GSSG might be responsible for the oxidation of PDI which then in turn oxidizes the substrate proteins. However, the discovery of Ero1 as a provider of oxidizing equivalents for the formation of disulfide bonds[35,36,54] led researchers to question the hypothesis that GSSG is solely responsible for the oxidation of PDI in the ER. *In vitro* experiments showed that the Ero1-mediated oxidation of folding substrates is mainly independent from the bulk redox buffer. Ero1 can efficiently drive the oxidation of RNase A through PDI in the absence of GSH or GSSG, or even in the presence of excess GSH,[53] indicating that oxidative protein folding proceeds through the direct transfer of oxidizing equivalents between PDI and Ero1. However, these results do not rule out a role for GSSG as an alternative source of oxidizing equivalents for PDI.

While arguments against the ability of GSH to directly reduce non-native disulfides within substrate proteins are much weaker, oxidoreductases have a kinetic advantage over GSH in catalyzing reduction/isomerization reactions. Two molecules of GSH are required for reduction of each disulfide bond. The first molecule breaks the disulfide bond within substrate proteins and glutathionylates one of the cysteine residues; attack by a second molecule of GSH will result in release of GSSG and a reduced substrate protein. The reduction of disulfide bonds by a CXXC motif within the active sites of ER oxidoreductases appears to be a much more efficient and favorable reaction as the two cysteine residues within the active sites are perfectly positioned to carry out such reactions. Another advantage that the PDI family members have over GSH in interacting with substrate proteins is the presence of high-affinity substrate binding sites in the non-catalytic thioredoxin-like **b'** domains.[19]

1.5.7 Maintaining the Redox Balance of the ER

The ER is the site for oxidation, isomerization and reduction of disulfide bonds. Thus, there needs to exist a system within this organelle which keeps the oxidoreductases involved in disulfide-bond formation in an oxidized state, while those involved in reduction or isomerization are maintained in a reduced state. It is tempting to speculate that such segregation would rely upon limiting the interactions of the oxidoreductases with Ero1. While PDI in mammalian cells can be oxidized by Ero1 and catalyzes disulfide-bond formation,[29] ERp57 does not appear to be a substrate for Ero1 as judged by a lack of oxidation of ERp57 in cells over-expressing Ero1.[29] ERp57 has instead been shown to act as a reductase, at least *in vitro*, to allow the breaking of non-native disulfide bonds within MHC-class I heavy chain.[75] None of the other oxidoreductases discovered so far have been shown to interact with Ero1. Thus it is possible that PDI may be the only member of the oxidoreductase family capable of introducing disulfide bonds into proteins, due to its ability to cooperate with Ero1. All other members of the family may be involved in reduction or isomerization reactions.

In addition to restricting Ero1 interactions with oxidoreductases, Ero1 activity itself may be subject to feedback regulation as discovered in yeast.[44] The use of molecular oxygen as a terminal electron acceptor by Ero1, Erv2p and QSOX provides a robust driving force for disulfide formation. However, the complete four-electron reduction of molecular oxygen to water is kinetically sluggish, and the reduction of intermediates and by-products such as superoxide and hydrogen peroxide are highly reactive and damaging macromolecules. When the levels of such reactive oxygen species (ROS) exceed the cellular antioxidant capacity, a deleterious condition known as oxidative stress occurs (see Chapter 1.8). The idea that oxidative stress is associated with protein secretion has been suggested for Ero1p[76,77] and QSOX catalyzed oxidative processes.[59,78] The activity of these oxidases must therefore be controlled in order to prevent the ER from becoming over-oxidized.

The redox balance of the ER is crucial for optimal oxidative protein folding. A number of both internal and external factors can lead to disruption of the GSH/GSSG balance. Earlier it was described how the ER oxidoreductases are maintained in a reduced state by glutathione, which is necessary for them to facilitate disulfide rearrangement and reduction. Thus, to keep the oxidoreductases in an active, reduced state the ER must maintain its GSH:GSSG balance. Harmful reactive oxygen species produced during disulfide-bond formation can be neutralized by oxidizing GSH, thus leading to an increase in the levels of GSSG. In addition, the reduction of non-native disulfide bonds would also lead to an increase in the level of GSSG, particularly if the resulting cysteine residues remain reduced or reform native disulfide bonds through an Ero1-catalyzed oxidative pathway. Should Ero1 have a significant role in disulfide-bond formation then the net consequence would be an increase in the concentration of GSSG relative to GSH. However, if GSSG was also to oxidize PDI then the net result would be an increase in the concentration of GSH relative to GSSG. Therefore, balancing Ero1 and GSSG oxidation of PDI could theoretically regulate the [GSH]:[GSSG] ratio.

Another mechanism for preventing the redox environment in the ER from becoming unfavorably oxidizing is the selective transport of GSH but not GSSG from the cytosol into the ER lumen.[70,79] In contrast, conflicting studies prior to this suggest that GSSG might be preferentially transported into the ER lumen.[14] Such data must be viewed cautiously in the light that there are many technical difficulties to overcome when performing such experiments, such as contamination of ER measured glutathione concentrations with that from the cytosol and the oxidation or reduction of glutathione following cellular disruption. However, it is probable that a GSH, or indeed a GSSG transporter, exists to pump glutathione across the ER membrane in eukaryotes. In *Escherichia coli* GSH and cysteine are specifically transported to the periplasm by CydDC, an ATP-binding cassette (ABC)-type transporter.[80] ABC transporters might also function in the transport of glutathione in mammalian cells; it has already been shown that this bacterial transporter has similarities to the cystic fibrosis transmembrane conductance regulator (CFTR), which has been shown to be responsible for GSH flux from mammalian cells.[81] In addition, it is also possible that the ER membrane might be permeable to small molecules such as GSH, allowing transport to occur simply by diffusion.[71] This would provide an alternative mechanism for the introduction of reducing equivalents into the lumen of the ER.

It is important to maintain a suitable level of GSH in the lumen of the ER to buffer against the harmful effects of oxidizing equivalents and maintain a glutathione buffer that is optimal for protein folding. The fate of any GSSG that may be produced indirectly through Ero1 activity or any reduction of ER oxidoreductases by GSH, is unclear. It might be secreted from the cell, transported to the cytosol for reduction or reduced within the ER by an unknown reductase. Although it is clear that the redox environment in the ER is carefully maintained for optimal protein folding and protection of the cell against harmful oxidative species, there is currently still much that we do not know about how this delicate balance is maintained.

1.5.8 Substrate Recognition by PDI and its Homologs

The finding that there is a robust pathway for the reduction of oxidoreductases and the fact that the majority of the oxidoreductases are in a reduced state highlight the importance of the reduction and isomerization pathways in the cell. However, it also raises the question of why so many homologs are needed to fulfill these functions, a role that at first glance could potentially be fulfilled by PDI alone. To date, nearly twenty mammalian oxidoreductases have been identified and there are most likely more that have not yet been discovered. Many members of the PDI family may be present in any one cell at any time. Comparative studies of the *in vitro* activities of several of these enzymes demonstrate that they are capable of carrying out similar functions,[82] so the question as to why there are so many family members needs to be addressed. The PDI homologs differ significantly with respect to a number of features, such as size, active site sequences, levels of expression and by being membrane bound or soluble. Therefore functional differences between the homologs can be anticipated.

It is most likely that different oxidoreductases catalyze disulfide-bond formation and rearrangement in different subsets of substrates. Such a picture is certainly emerging for some of the oxidoreductases such as ERp44, ERp57 and PDIp. Anelli and colleagues initially identified ERp44 as a resident ER protein which is a binding partner for hEro1s.[83] ERp44 interacts with both Ero1-Lα and -β and is responsible for the retention of hEro1s in the ER. ERp44 has also been implicated in thiol-mediated retention of IgM molecules[84] and adiponectin.[85] ERp57 is distinct from PDI as it does not appear to be a substrate for Ero1,[29] at least *in vivo*. It has been shown to act as a reductase *in vitro* as it specifically reduces partially folded MHC class I molecules.[75] ERp57 binds to lectins calnexin and calreticulin and hence is thought to be specifically involved in disulfide-bond rearrangements within glycoprotein substrates.[86] More recently, ERp57 has been implicated in playing an important role in the uncoating of Simian Virus 40 (SV40) capsid on entry into cells.[87] Upon entering the ER, ERp57 specifically reduces the viral capsid protein VP1, which is linked to a network of neighbours, which helps to uncoat the virus. This isomerization reaction seems to be specific to ERp57 as depletion of PDI had no effect on the uncoupling of VP1 pentamers in cells. In addition to acting on different sets of substrates in the same cell, there are also examples of different tissue distributions for enzymes such as PDIp. This PDI-like oxidoreductase is expressed most highly in secretory cells such as the pancreas, where it may be required to assist in the oxidative folding of the exceptionally high substrate load passing through the secretory pathway.[88]

In order for an oxidoreductase to catalyze native disulfide-bond formation it must first bind to the substrate and specificity can be achieved by the oxidoreductases recognizing and binding their substrates in different manners. PDI binds its substrates primarily through a peptide binding site in the **b**′ domain, which is in a position homologous to that of the CXXC active site in the catalytic domains.[24] The precise peptide has been mapped to a small hydrophobic binding pocket. An equivalent binding site exists in PDIp which binds only a single tyrosine or tryptophan with no adjacent negative charge.[89] Although the **b**′ domain peptide binding site is the primary site for substrate recognition by PDI, contributions from other domains are also necessary for efficient binding.[90] Isolated **a** and **a**′ domains can catalyze thiol exchange reactions alone but the addition of the **b**′ domain is required for isomerization reactions. Only the full length molecule is fully catalytically active and is capable of performing disulfide arrangements involving more substantial changes in structure. Since the **b** domain of PDI has not been implicated in substrate binding, nor does it possess catalytic activity, it may have a structural role.

Whereas PDI interacts with its substrates directly through the **b**′ domain, ERp57 binds the lectins calnexin and calreticulin which in turn recruit newly synthesized glycoproteins. ERp57 binding to calnexin or calreticulin occurs mainly through the **b**′ domain and, to a lesser extent, the C-terminal positively charged region.[91] The residues in the **b**′ domain of ERp57 that are required for this interaction have been mapped to those required for peptide binding in the **b**′ domain of PDI. These bind to the acidic tip on an elongated hairpin loop that forms the P-domain of calreticulin.

Not only can calnexin and calreticulin present ERp57 with a precise subset of substrates but they also recruit glycoproteins at a precise point in their folding pathway. Calnexin and calreticulin only bind monoglucosylated glycoproteins that are formed following the trimming of the furthest two glucose residues of the oligosaccharide side-chain by glucosidase I.[92] Substrates are released following the trimming of a further glucose residue by glucosidase II but should the glycoprotein fail to be folded correctly it is reglucosylated by UDP glucose:glycoprotein glucosyl transferase whereupon it may again bind to calnexin or calreticulin and ERp57.[93,94] While the exact nature of ERp57 binding to substrates *via* calnexin and calreticulin is slowly coming to light, recent studies have shown that ERp57 is also able to interact with substrates independently of calnexin/calreticulin. This has been shown for substrates such as clusterin[95] and SV40.[87,95]

An analysis of the domain structures of glycoproteins interacting with ERp57 revealed that the enzyme interacts with a subset of proteins displaying disulfide-rich, secondary structure-poor domains, for example, the epidermal growth factor-like (EGF) domain.[1,87,95] Where a number of disulfide bonds are found close together in the primary structure, with no secondary structure to position the correct disulfides in proximity to each other there exists an increased potential for the formation of non-native disulfides. Therefore, ERp57 may be directed to such domains in substrates that are prone to forming non-native disulfide bonds in order to rearrange them. While the *in vivo* activities of some of the PDI homologs such as ERp57,[86] PDIp,[96] ERp44[83] and ERp72[97] are now coming to light (Table 1.5.1), the functions of many others (for example, TMX, ERp28 and PDIR) are poorly described. Understanding

Table 1.5.1 Eukaryotic homologs of PDI.

Protein	Length (amino acids)	ER-localization motif	Active site sequence	Specificity
PDI	508	KDEL	CGHC, CGHC	General peptide binding domain in **b'** domain
ERp57	505	QEDL	CGHC, CGHC	**b'** domain binds calnexin and calreticulin
ERp72	645	KEEL	CGHC, CGHC, CGHC	Calcium binding
P5	440	KDEL	CGHC, CGHC	Calcium binding
PDIR	519	KEEL	CSMC, CGHC, CPHC	
PDIp	525	KEEL	CGHC, CTHC	Pancreas-specific expression
ERp46	432	KDEL	CGHC, CGHC, CGHC	
ERp44	406	RDEL	CRFS	ER retention of proteins
ERdj5	793	KDEL	CSHC, CPPC, CHPC, CGPC	Contains three CXPC domains indicative of reductase activity
ERp18	172	EDEL	CGAC	
TMX	280	Unknown	CPAC	May modify substrates required for developmental processes
TMX2	296	KKDK	SNDC	
TMX3	454	KKKD	CGHC	
TMX4	349	Unknown	CPSC	

the mechanisms of action of the PDI family is critical for our understanding of native disulfide-bond formation.

1.5.9   Conclusion

Disulfide-bond formation in cells is a highly regulated and carefully controlled process facilitated by a family of enzymes. These ensure that substrates form only the correct, native bonds and thus a high yield of correctly folded protein is achieved. Although many of the basic mechanisms of oxidative folding in mammalian cells have been elucidated, the individual roles of each of the components are still debated or unknown. This situation not only inhibits our understanding of the biogenesis of a range of important proteins and hence associated disease states, but also prevents the effective manipulation of the cellular environment by the biotechnology industry for the efficient production of therapeutic proteins.

References

1. U. Jakob, W. Muse, M. Eser and J. C. Bardwell, *Cell*, 1999, **96**, 341–352.
2. P. J. Hogg, *Trends Biochem. Sci.*, 2003, **28**, 210–214.
3. A. Jordan and P. Reichard, *Annu. Rev. Biochem.*, 1998, **67**, 71–98.
4. A. S. Robinson and J. King, *Nat. Struct. Biol.*, 1997, **4**, 450–455.
5. R. F. Goldberger, C. J. Epstein and C. B. Anfinsen, *J. Biol. Chem.*, 1963, **238**, 628–635.
6. A. P. Somlyo, M. Bond and A. V. Somlyo, *Nature*, 1985, **314**, 622–625.
7. D. R. Macer and G. L. Koch, *J. Cell Sci.*, 1988, **91**, 61–70.
8. S. K. Nigam, A. L. Goldberg, S. Ho, M. F. Rohde, K. T. Bush and M. Sherman, *J. Biol. Chem.*, 1994, **269**, 1744–1749.
9. G. Kuznetsov, M. A. Brostrom and C. O. Brostrom, *J. Biol. Chem.*, 1992, **267**, 3932–3939.
10. H. F. Lodish, N. Kong and L. Wikstrom, *J. Biol. Chem.*, 1992, **267**, 12753–12760.
11. D. R. Wetmore and K. D. Hardman, *Biochemistry*, 1996, **35**, 6549–6558.
12. G. Kuznetsov, L. B. Chen and S. K. Nigam, *J. Biol. Chem.*, 1997, **272**, 3057–3063.
13. A. MacManus, M. Ramsden, M. Murray, Z. Henderson, H. A. Pearson and V. A. Campbell, *J. Biol. Chem.*, 2000, **275**, 4713–4718.
14. C. Hwang, A. J. Sinskey and H. F. Lodish, *Science*, 1992, **257**, 1496–1502.
15. T. E. Creighton, C. J. Bagley, L. Cooper, N. J. Darby, R. B. Freedman, J. Kemmink and A. Sheikh, *J. Mol. Biol.*, 1993, **232**, 1176–1196.
16. C. S. Sevier and C. A. Kaiser, *Nat. Rev. Mol. Cell Biol.*, 2002, **3**, 836–847.
17. D. M. Ferrari and H. D. Soling, *Biochem. J.*, 1999, **339**, 1–10.
18. R. B. Freedman, *Curr. Opin. Struct. Biol.*, 1995, **5**, 85–91.
19. L. Ellgaard and L. W. Ruddock, *EMBO Rep.*, 2005, **6**, 28–32.
20. A. Jansens, E. van Duijn and I. Braakman, *Science*, 2002, **298**, 2401–2403.
21. C. Fagioli and R. Sitia, *J. Biol. Chem.*, 2001, **276**, 12885–12892.
22. D. Tortorella, C. M. Story, J. B. Huppa, E. J. Wiertz, T. R. Jones, I. Bacik, J. R. Bennink, J. W. Yewdell and H. L. Ploegh, *J. Cell Biol.*, 1998, **142**, 365–376.
23. M. M. Lyles and H. F. Gilbert, *Biochemistry*, 1991, **30**, 613–619.
24. A. Pirneskoski, P. Klappa, M. Lobell, R. A. Williamson, L. Byrne, H. I. Alanen, K. E. Salo, K. I. Kivirikko, R. B. Freedman and L. W. Ruddock, *J. Biol. Chem.*, 2004, **279**, 10374–10381.
25. J. Edman and L. Ellis, *Nature*, 1985, **317**, 267–270.

26. N. J. Darby, R. B. Freedman and T. E. Creighton, *Biochemistry*, 1994, **33**, 7937–7947.
27. T. E. Creighton, D. A. Hillson and R. B. Freedman, *J. Mol. Biol.*, 1980, **142**, 43–62.
28. A. R. Frand and C. A. Kaiser, *Mol. Cell*, 1999, **4**, 469–477.
29. A. Mezghrani, A. Fassio, A. Benham, T. Simmen, I. Braakman and R. Sitia, *Embo J.*, 2001, **20**, 6288–6296.
30. M. L. LaMantia and W. J. Lennarz, *Cell*, 1993, **74**, 899–908.
31. B. Scherens, E. Dubois and F. Messenguy, *Yeast*, 1991, **7**, 185–193.
32. M. C. Laboissiere, S. L. Sturley and R. T. Raines, *J. Biol. Chem.*, 1995, **270**, 28006–28009.
33. B. Holst, C. Tachibana and J. R. Winther, *J. Cell Biol.*, 1997, **138**, 1229–1238.
34. R. Xiao, B. Wilkinson, A. Solovyov, J. R. Winther, A. Holmgren, J. Lundstrom-Ljung and H. F. Gilbert, *J. Biol. Chem.*, 2004, **279**, 49780–49786.
35. A. R. Frand and C. A. Kaiser, *Mol. Cell*, 1998, **1**, 161–170.
36. M. G. Pollard, K. J. Travers and J. S. Weissman, *Mol. Cell*, 1998, **1**, 171–182.
37. A. Cabibbo, M. Pagani, M. Fabbri, M. Rocchi, M. R. Farmery, N. J. Bulleid and R. Sitia, *J. Biol. Chem.*, 2000, **275**, 4827–4833.
38. M. Pagani, M. Fabbri, C. Benedetti, A. Fassio, S. Pilati, N. J. Bulleid, A. Cabibbo and R. Sitia, *J. Biol. Chem.*, 2000, **275**, 23685–23692.
39. S. Chakravarthi and N. J. Bulleid, *J. Biol. Chem.*, 2004, **279**, 39872–39879.
40. L. Qiang, H. Wang and S. R. Farmer, *Mol. Cell Biol.*, 2007, **27**, 4698–4707.
41. B. P. Tu and J. S. Weissman, *Mol. Cell*, 2002, **10**, 983–994.
42. A. R. Frand and C. A. Kaiser, *Mol. Biol. Cell*, 2000, **11**, 2833–2843.
43. E. Gross, D. B. Kastner, C. A. Kaiser and D. Fass, *Cell*, 2004, **117**, 601–610.
44. C. S. Sevier, H. Qu, N. Heldman, E. Gross, D. Fass and C. A. Kaiser, *Cell*, 2007, **129**, 333–344.
45. M. Pagani, S. Pilati, G. Bertoli, B. Valsasina and R. Sitia, *FEBS Lett.*, 2001, **508**, 117–120.
46. T. Anelli, M. Alessio, A. Bachi, L. Bergamelli, G. Bertoli, S. Camerini, A. Mezghrani, E. Ruffato, T. Simmen and R. Sitia, *Embo J.*, 2003, **22**, 5015–5022.
47. M. Otsu, G. Bertoli, C. Fagioli, E. Guerini-Rocco, S. Nerini-Molteni, E. Ruffato and R. Sitia, *Antioxid. Redox Signal.*, 2006, **8**, 274–282.
48. D. May, A. Itin, O. Gal, H. Kalinski, E. Feinstein and E. Keshet, *Oncogene*, 2005, **24**, 1011–1020.
49. B. Gess, K. H. Hofbauer, R. H. Wenger, C. Lohaus, H. E. Meyer and A. Kurtz, *Eur. J. Biochem.*, 2003, **270**, 2228–2235.
50. C. L. Andersen, A. Matthey-Dupraz, D. Missiakas and S. Raina, *Mol. Microbiol.*, 1997, **26**, 121–132.

51. S. J. Marciniak, C. Y. Yun, S. Oyadomari, I. Novoa, Y. Zhang, R. Jungreis, K. Nagata, H. P. Harding and D. Ron, *Genes Dev.*, 2004, **18**, 3066–3077.
52. S. Dias-Gunasekara, J. Gubbens, M. van Lith, C. Dunne, J. A. Williams, R. Kataky, D. Scoones, A. Lapthorn, N. J. Bulleid and A. M. Benham, *J. Biol. Chem.*, 2005, **280**, 33066–33075.
53. B. P. Tu, S. C. Ho-Schleyer, K. J. Travers and J. S. Weissman, *Science*, 2000, **290**, 1571–1574.
54. C. S. Sevier, J. W. Cuozzo, A. Vala, F. Aslund and C. A. Kaiser, *Nat. Cell Biol.*, 2001, **3**, 874–882.
55. J. Gerber, U. Muhlenhoff, G. Hofhaus, R. Lill and T. Lisowsky, *J. Biol. Chem.*, 2001, **276**, 23486–23491.
56. J. Lee, G. Hofhaus and T. Lisowsky, *FEBS Lett.*, 2000, **477**, 62–66.
57. T. Lisowsky, J. E. Lee, L. Polimeno, A. Francavilla and G. Hofhaus, *Dig. Liver Dis.*, 2001, **33**, 173–180.
58. T. G. Senkevich, C. L. White, E. V. Koonin and B. Moss, *Proc. Natl. Acad. Sci. USA*, 2000, **97**, 12068–12073.
59. C. Thorpe, K. L. Hoober, S. Raje, N. M. Glynn, J. Burnside, G. K. Turi and D. L. Coppock, *Arch. Biochem. Biophys.*, 2002, **405**, 1–12.
60. D. L. Coppock, C. Kopman, S. Scandalis and S. Gilleran, *Cell Growth Differ.*, 1993, **4**, 483–493.
61. S. Chakravarthi, C. E. Jessop, M. Willer, C. J. Stirling and N. J. Bulleid, *Biochem. J.*, 2007, **404**, 403–411.
62. I. Wittke, R. Wiedemeyer, A. Pillmann, L. Savelyeva, F. Westermann and M. Schwab, *Cancer Res.*, 2003, **63**, 7742–7752.
63. D. Coppock, C. Kopman, J. Gudas and D. A. Cina-Poppe, *Biochem. Biophys. Res. Commun.*, 2000, **269**, 604–610.
64. D. L. Coppock, D. Cina-Poppe and S. Gilleran, *Genomics*, 1998, **54**, 460–468.
65. K. L. Hoober, S. L. Sheasley, H. F. Gilbert and C. Thorpe, *J. Biol. Chem.*, 1999, **274**, 22147–22150.
66. K. L. Hoober and C. Thorpe, *Biochemistry*, 1999, **38**, 3211–3217.
67. E. J. Heckler, P. C. Rancy, V. K. Kodali and C. Thorpe, *Biochim. Biophys. Acta*, 2008, **1783**, 567–77.
68. C. E. Jessop and N. J. Bulleid, *J. Biol. Chem.*, 2004, **279**, 55341–55347.
69. M. Fratelli, H. Demol, M. Puype, S. Casagrande, I. Eberini, M. Salmona, V. Bonetto, M. Mengozzi, F. Duffieux, E. Miclet, A. Bachi, J. Vandekerckhove, E. Gianazza and P. Ghezzi, *Proc. Natl. Acad. Sci. USA*, 2002, **99**, 3505–3510.
70. G. Banhegyi, L. Lusini, F. Puskas, R. Rossi, R. Fulceri, L. Braun, V. Mile, P. di Simplicio, J. Mandl and A. Benedetti, *J. Biol. Chem.*, 1999, **274**, 12213–12216.
71. S. Le Gall, A. Neuhof and T. Rapoport, *Mol. Biol. Cell*, 2004, **15**, 447–455.

72. R. Bass, L. W. Ruddock, P. Klappa and R. B. Freedman, *J. Biol. Chem.*, 2004, **279**, 5257–5262.
73. J. S. Weissman and P. S. Kim, *Nature*, 1993, **365**, 185–188.
74. A. Zapun, N. J. Darby, D. C. Tessier, M. Michalak, J. J. Bergeron and D. Y. Thomas, *J. Biol. Chem.*, 1998, **273**, 6009–6012.
75. A. N. Antoniou, S. Ford, M. Alphey, A. Osborne, T. Elliott and S. J. Powis, *EMBO J.*, 2002, **21**, 2655–2663.
76. H. P. Harding, Y. Zhang, H. Zeng, I. Novoa, P. D. Lu, M. Calfon, N. Sadri, C. Yun, B. Popko, R. Paules, D. F. Stojdl, J. C. Bell, T. Hettmann, J. M. Leiden and D. Ron, *Mol. Cell*, 2003, **11**, 619–633.
77. B. P. Tu and J. S. Weissman, *J. Cell Biol.*, 2004, **164**, 341–346.
78. G. Mairet-Coello, A. Tury, A. Esnard-Feve, D. Fellmann, P. Y. Risold and B. Griffond, *J. Comp. Neurol.*, 2004, **473**, 334–363.
79. G. Banhegyi, M. Csala, G. Nagy, V. Sorrentino, R. Fulceri and A. Benedetti, *Biochem. J.*, 2003, **376**, 807–812.
80. M. S. Pittman, H. C. Robinson and R. K. Poole, *J. Biol. Chem.*, 2005, **280**, 32254–32261.
81. I. Kogan, M. Ramjeesingh, C. Li, J. F. Kidd, Y. Wang, E. M. Leslie, S. P. Cole and C. E. Bear, *EMBO J.*, 2003, **22**, 1981–1989.
82. H. I. Alanen, K. E. Salo, A. Pirneskoski and L. W. Ruddock, *Antioxid. Redox Signal.*, 2006, **8**, 283–291.
83. T. Anelli, M. Alessio, A. Mezghrani, T. Simmen, F. Talamo, A. Bachi and R. Sitia, *EMBO J.*, 2002, **21**, 835–844.
84. T. Anelli, S. Ceppi, L. Bergamelli, M. Cortini, S. Masciarelli, C. Valetti and R. Sitia, *EMBO J.*, 2007, **26**, 4177–4188.
85. Z. V. Wang, T. D. Schraw, J. Y. Kim, T. Khan, M. W. Rajala, A. Follenzi and P. E. Scherer, *Mol. Cell Biol.*, 2007, **27**, 3716–3731.
86. J. D. Oliver, H. L. Roderick, D. H. Llewellyn and S. High, *Mol. Biol. Cell*, 1999, **10**, 2573–2582.
87. M. Schelhaas, J. Malmstrom, L. Pelkmans, J. Haugstetter, L. Ellgaard, K. Grunewald and A. Helenius, *Cell*, 2007, **131**, 516–529.
88. M. G. Desilva, A. L. Notkins and M. S. Lan, *DNA Cell Biol.*, 1997, **16**, 269–274.
89. L. W. Ruddock, R. B. Freedman and P. Klappa, *Protein Sci.*, 2000, **9**, 758–764.
90. P. Klappa, L. W. Ruddock, N. J. Darby and R. B. Freedman, *EMBO J.*, 1998, **17**, 927–935.
91. S. J. Russell, L. W. Ruddock, K. E. Salo, J. D. Oliver, Q. P. Roebuck, D. H. Llewellyn, H. L. Roderick, P. Koivunen, J. Myllyharju and S. High, *J. Biol. Chem.*, 2004, **279**, 18861–18869.
92. D. N. Hebert, B. Foellmer and A. Helenius, *Cell*, 1995, **81**, 425–433.
93. J. D. Oliver, F. J. van der Wal, N. J. Bulleid and S. High, *Science*, 1997, **275**, 86–88.
94. F. J. Van der Wal, J. D. Oliver and S. High, *Eur. J. Biochem.*, 1998, **256**, 51–59.

95. C. E. Jessop, S. Chakravarthi, N. Garbi, G. J. Hammerling, S. Lovell and N. J. Bulleid, *EMBO J.*, 2007, **26**, 28–40.
96. R. B. Freedman, P. Klappa and L. W. Ruddock, *EMBO Rep.*, 2002, **3**, 136–140.
97. R. A. Mazzarella, M. Srinivasan, S. M. Haugejorden and M. Green, *J. Biol. Chem.*, 1990, **265**, 1094–1101.
98. Z. Z. Shi, J. Osel-Frimpong, G. Kala, S. V. Kala, R. J. Barrios, G. M. Habib, D. J. Lukin, C. M. Danney, M. M. Matzuk and M. W. Liebermann, *Proc. Natl. Acad. Sci. USA*, 2000, **97**, 5101–5106.
99. J. W. Cuozzo and C. A. Kaiser, *Nat. Cell. Biol.*, 1999, **1**, 130–135.
100. S. N. Molteni, A. Fassio, M. R. Crinolo, G. Filomeni, E. Pasqualetto, C. Fagioli and R. Sita, *J. Biol. Chem*, 2004, **279**, 32667–32673.

CHAPTER 1.6

The Ero1 Sulfhydryl Oxidase and the Oxidizing Potential of the Endoplasmic Reticulum

DEBORAH FASS[a] AND CAROLYN S. SEVIER[b]

[a] Department of Structural Biology, Weizmann Institute of Science, Rehovot 76100, Israel; [b] Department of Biology, Massachussetts Institute of Technology, Cambridge, MA 02139, USA

1.6.1 Introduction

As proteins pass through the organelles of the secretory pathway, they undergo a range of post-translational modifications including glycosylation and disulfide-bond formation.[1] The nature and purpose of carbohydrate additions are varied,[2] but disulfide bonding between cysteine side-chains is chemically homogenous and primarily serves to stabilize protein structures. Stabilization can occur on various levels. Classically, disulfide bonding is considered to increase the global thermodynamic stability of proteins relative to their denatured states by entropically restricting, and thereby raising the free energy of, the denatured forms.[3,4] Disulfide cross-linking also can improve the resistance of proteins to proteolytic degradation by limiting the flexibility of folded protein structures[5,6] and can covalently link the subunits of some multi-protein assemblies, such as antibodies, to prevent subunit dissociation when protein complexes are diluted upon secretion from the cell. Occasionally disulfides perform regulatory roles,[7] allowing for conformational and corresponding functional changes in proteins in response to alterations in the concentrations of oxidants or reductants in the cell or environment.

RSC Biomolecular Sciences
Oxidative Folding of Peptides and Proteins
Edited by Johannes Buchner and Luis Moroder
© Royal Society of Chemistry 2009
Published by the Royal Society of Chemistry, www.rsc.org

A range of small molecule oxidants can facilitate formation of disulfide bonds between cysteines in the test tube, but proteins acquire disulfides *via* a dedicated enzymatic cascade *in vivo*. In eukaryotes, biosynthetic disulfide-bond formation occurs within the lumen of the endoplasmic reticulum (ER). Within the ER, two opposing redox cascades ensure the net formation of disulfide bonds in secretory proteins while preventing the irreversible trapping of mis-paired cysteines. The presence of the ER-resident sulfhydryl oxidase Ero1[8,9] distinguishes the ER as an environment suitable for the formation of disulfide bonds. Specifically, Ero1 oxidizes an active-site di-cysteine motif of Protein Disulfide Isomerase (PDI),[11] which in turn oxidizes secretory protein substrates. The activity of Ero1 drives the thiol/disulfide equilibrium toward disulfides by catalyzing the overall transfer of electrons from reduced cysteine thiols to molecular oxygen.[10] Meanwhile, the import of reduced glutathione from the cytosol into the ER lumen, as well as the import of cysteines in nascent protein chains themselves, provides a steady supply of thiols to participate in reduction and isomerization of disulfides.[12] Glutathione reduces (mis)oxidized substrates directly and also reduces oxidized versions of PDI and other PDI-like proteins, making them available to catalyze reduction and rearrangement of disulfides in secretory proteins. The increased ratio of oxidized glutathione (GSSG) to reduced glutathione (GSH) observed in the ER relative to the cytosol is thus an outcome, rather than the origin, of the ER oxidizing potential.[13] In addition to accomplishing oxidation of cysteines in newly synthesized proteins in the ER, the ER oxidation cascade provides resistance against exogenous reducing agents,[8,9] whereas components of the reducing pathway buffer against exogenous and cellular oxidants such as reactive oxygen species.[13,14]

The coupling of genetic and biochemical studies with the high-resolution X-ray structures of yeast Ero1[15] and PDI[16] yields a general mechanism for how disulfide bonds are generated and transferred in the ER. Ero1 has a unique fold, but it shares certain structural features in common with other sulfhydryl oxidoreductases identified in bacteria and eukaryotes. A comparison of Ero1 with these non-homologous enzymes catalyzing the same reaction lends insight into the chemistry of Ero1. However, the finding that Ero1 is exceptional in being regulated on the protein level[17] adds a degree of sophistication to the basic mechanism. The structural and mechanistic studies on components of the yeast ER disulfide-bond formation pathway can be mined for their implications for oxidative protein folding in higher organisms, including humans.

1.6.2   Mechanism for Generation and Transfer of Disulfides by Ero1

1.6.2.1   A Route for Intramolecular Electron Transfer Supported by the Ero1 Structure

Ero1 comprises a large, ten-helix domain.[15] In addition to the helices, the domain also contains approximately 120 amino acid residues of polypeptide poor in secondary structure draped over what is arbitrarily designated the top of

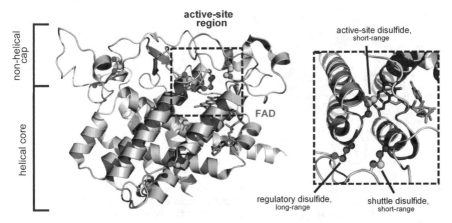

Figure 1.6.1 Structure of Ero1. On the left is a ribbon diagram of yeast Ero1 oriented with the helical core region below and the non-helical cap above. Cysteine side-chains are shown in ball-and-stick representation, with yellow sulfur atoms and green Cβ carbons. The bound FAD molecule is orange. The vicinity of the active site is boxed and shown in more detail, and as viewed from the mouth of the four-helix bundle, in the right panel. Disulfides in the active-site region are labeled according to function. The active-site disulfide is a Cys–X–X–Cys motif, the shuttle disulfide is a Cys–X_4–Cys motif and the nearby regulatory disulfide is formed from cysteines separated by ∼260 amino acid residues.

the domain (Figure 1.6.1). Yeast Ero1 has 14 cysteines in total, ten of which are paired in disulfide bonds in the non-helical cap or at the junction of this region and the helical core. Some of these disulfides are "short-range", linking cysteines close in primary structure, whereas others are "long-range", between cysteines distant from one another in sequence. Ero1 is a flavoenzyme,[18] with one molecule of flavin adenine dinucleotide (FAD) associated non-covalently. The isoalloxazine ring system of the FAD, the region of the cofactor active in redox chemistry, is buried at the mouth of a four-helix bundle within the helical core of the enzyme (Figure 1.6.1). Three of the disulfide bonds cluster within 12 Å of the FAD isoalloxazine. Two others are ∼25 and ∼35 Å away from the FAD.

Formation of a disulfide bond from two free cysteine thiol groups requires removal of two electrons, which Ero1 accomplishes by coupling disulfide formation to the reduction of non-thiol species *via* the bound FAD cofactor. One of the Ero1 cysteine pairs, present within a Cys–X–X–Cys sequence motif common to dithiol/disulfide oxidoreductases, is adjacent to the FAD. When this active-site cysteine motif becomes reduced, a disulfide can be regenerated by transfer of electrons to the FAD, and in turn to molecular oxygen or to another electron acceptor.[19] This seemingly simple process is actually very complex chemically, and many questions remain regarding the mechanism of oxygen reduction by this and other flavin-dependent oxidases.[20] It is clear, however, that the flavin-catalyzed reduction of oxygen to hydrogen peroxide is sufficiently favorable thermodynamically to make disulfide formation at the Ero1 active center essentially irreversible. Kinetically, re-oxidation of reduced

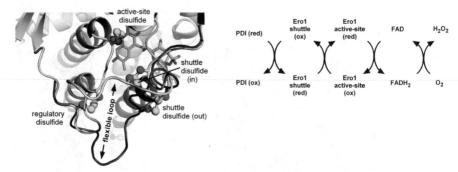

Figure 1.6.2   Ero1 shuttle disulfide transfers electrons from substrate to the catalytic center. The loop containing the shuttle disulfide was observed in different conformations in two crystal forms of Ero1, shown here colored tan and maroon. The series of two-electron transfer events that result in generation of a disulfide bond in a substrate protein and production of a molecule of hydrogen peroxide are summarized on the right. "Red" refers to a reduced di-cysteine motif, whereas "ox" indicates that the motif is in the oxidized, disulfide form.

Ero1-bound flavin is rapid compared to the rate of the reductive half-reaction, except at extremely low oxygen tension.[19]

The active-site disulfide of Ero1 is on a rigid part of the protein and relatively solvent inaccessible, minimizing direct access by reduced protein substrates. Instead, substrates (*e.g.*, PDI) undergo dithiol/disulfide exchange with another Ero1 disulfide, known as the "shuttle disulfide", present on a flexible and solvent exposed loop in the general vicinity of the active site (Figure 1.6.2). The reduced Ero1 shuttle disulfide is proposed to reduce the active-site disulfide by dithiol/disulfide exchange, in effect shuttling electrons from substrate to enzyme active site.[21,22] Indeed, mixed disulfides between shuttle and active-site cysteines have been detected.[22] Both the active-site and shuttle cysteines are typically required for enzyme activity *in vivo*, although mutations in Ero1 that allow the enzyme to bypass the requirement for the shuttle cysteines have been identified.[22] Use of two di-cysteine redox centers by Ero1 to sequentially relay electrons from substrate to its aromatic cofactor (*e.g.*, FAD) is a mechanism also observed in other disulfide-forming oxidoreductases, including the prokaryotic enzyme DsbB and the Erv flavoprotein family (discussed below).

1.6.2.2   Oxidation of PDI by Ero1

As noted above, Ero1 works in partnership with PDI to oxidize a range of substrates folding oxidatively in the ER. PDI consists of four domains, each with the thioredoxin fold, called the **a**, **b**, **b′** and **a′** domains in order from amino to carboxy terminus. Two of these domains, **a** and **a′**, contain the redox-active Cys–X–X–Cys motif. PDI is the most abundant protein in the ER lumen.[23] It participates in a number of processes required for oxidative protein folding, including cysteine oxidation, disulfide reduction and isomerization, and general chaperoning of unfolded or partially folded proteins.[24] Individual domains or sites within PDI may play particular roles in each of these activities, but there is some indication that the whole is more than the sum of its parts.[25]

Focusing on the role of PDI in the oxidation pathway headed by Ero1, we can appreciate how the kinetics of dithiol/disulfide exchange between Ero1 and PDI evolved to promote disulfide-bond formation in downstream substrate proteins. Ero1 has a kinetic preference for oxidation of the themodynamically less stable **a′** domain disulfide of PDI.[25,26] According to redox potential alone, the **a** domain, at −188 mV, should be more readily oxidized than the **a′** domain, at −152 mV[27] (Figure 1.6.3). The kinetic preference for Ero1 oxidation of the **a′** domain confers upon PDI the properties of a strong oxidant, creating an unstable **a′** disulfide that is prone to reduction. The **a** domain di-cysteine motif, disfavored for oxidation by Ero1, is biased toward the free thiol form. The presence of unfolded polypeptides may, by a mechanism not yet understood, further decrease the rate of PDI **a** domain oxidation by Ero1.[25] With its comparatively low redox potential, the PDI **a** domain can readily act as a nucleophile for reduction or isomerization of relatively unstable, mis-paired disulfides in substrate proteins.

It is not known how the kinetic preference for oxidation of the thermodynamically less stable PDI disulfide is established, but some insights may be obtained from the X-ray crystallographic structure.[16] In the PDI structure, the amino terminus of the protein packs against the helix bearing the Cys–X–X–Cys motif in the **a** domain, whereas in the **a′** domain the helix is exposed from the comparable direction. Packing of polypeptide against this helix was proposed to restrict access to the second cysteine of the Cys–X–X–Cys motif,[16] which may

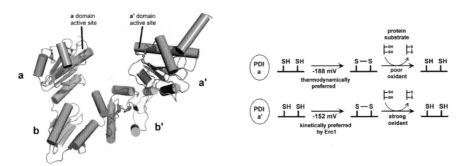

Figure 1.6.3 The preference of Ero1 for oxidation of the PDI **a′** domain creates a strong oxidant to further the disulfide-formation cascade. A cartoon of the PDI structure is shown with cylinders denoting helices and cysteines in ball-and-stick representation. On the right is a diagram illustrating the properties of the **a** and **a′** redox-active domains of PDI in relation to Ero1 and downstream protein substrates. Di-cysteine motifs that are oxidized readily by Ero1 because they have low redox potentials would make poor oxidants in the next step of the disulfide-bond formation cascade. Good oxidants must have high redox potential but nevertheless be able to transfer electrons to Ero1. A mechanism that minimizes electron transfer from di-cysteine motifs with low redox potential to Ero1 enables these motifs to participate in reduction/isomerization of downstream proteins.

hinder either formation or resolution of a mixed disulfide between Ero1 and PDI. Presumably, the packing of the amino-terminal region against the active-site helix in the PDI **a** domain plays less of a role in the kinetics of dithiol/disulfide exchange with small disulfide-containing oxidants such as oxidized glutathione, allowing for the low redox potential measured against glutathione redox buffers. The PDI structure also suggests why the **a** domain has the lower redox potential. An acidic residue in the vicinity of the second cysteine of the PDI **a** domain Cys–X–X–Cys motif is proposed to facilitate deprotonation of this cysteine and disulfide-bond formation,[16] as does a comparable acidic residue in thioredoxin.[28] In the PDI **a′** domain, however, this acidic residue is absent, and a leucine occupies its place, consistent with the less negative redox potential of the active site of this domain.

The yeast ER contains four other thioredoxin fold proteins that are potentially active in disulfide-bond formation and rearrangement in this compartment.[29–32] The extent to which some of these proteins function as substrates for Ero1 *in vivo* has been studied by trapping transient enzyme-substrate complexes. Trapping is facilitated by mutating the second cysteine in the Cys–X–X–Cys motif of putative Ero1 targets, preventing resolution of the mixed disulfides in the Ero1-substrate dithiol/disulfide exchange process. These covalent complexes can be recovered from whole cells by rapid acidification to protonate free thiols and prevent disulfide formation or rearrangement during cell lysis. Solubilization of cellular proteins at neutral pH in the presence of an alkylating agent then irreversibly blocks the free thiols. Using this procedure, PDI and Mpd2 were found in mixed-disulfide complexes with Ero1, but Mpd1 was not.[11] The Mpd1 structure determined by X-ray crystallography reveals a stretch of polypeptide packed against the active-site helix (unpublished data) just as in the **a** domain of PDI. The presence of this feature, which putatively diminishes reactivity with Ero1, may explain why Mpd1 (unpublished data), like the **a** domain of PDI, is not a preferred Ero1 substrate. Interestingly, the oxidation rates of both PDI and Mpd1 *in vitro* are dramatically accelerated by removal of a putative regulatory disulfide within Ero1, as described below, suggesting that under certain conditions even kinetically disfavored substrates become relevant Ero1 targets.

1.6.2.3   Comparison of Ero1 with the DsbB Intramembrane Sulfhydryl Oxidoreductase of Bacteria

The membrane-embedded sulfhydryl oxidoreductase from bacteria, DsbB, catalyzes disulfide-bond formation in the periplasmic space.[33] Despite a lack of sequence conservation between DsbB and Ero1 and the lipid *vs.* aqueous environments in which they function, these two proteins share remarkable mechanistic and structural similarities. Like Ero1, DsbB has a Cys–X–X–Cys disulfide bond adjacent to a non-covalently bound cofactor, in this case quinone rather than FAD. DsbB also has on a periplasmic loop a second disulfide bond, with a function analogous to that of the Ero1 shuttle disulfide. DsbB catalyzes oxidation of DsbA, an oxidoreductase functionally comparable to PDI and a powerful oxidant of proteins that acquire disulfide bonds while folding in the periplasm. Interestingly, the redox potentials of both DsbB disulfides, determined using small molecule thiol buffers, are lower than that of DsbA, suggesting that oxidation of DsbA by DsbB is thermodynamically disfavored.[35]

DsbB has evolved a remarkable mechanistic strategy to perform the seemingly thermodynamically "uphill" oxidation of DsbA, and it will be interesting to explore whether this mechanism has any parallels in Ero1. The structure of a DsbB/DsbA complex (Figure 1.6.4) suggests that a conformational change occurs after reduced DsbA attacks the DsbB shuttle disulfide, which distances the second shuttle cysteine from the intermolecular disulfide[35] and prevents the remaining DsbB shuttle cysteine from attacking the mixed disulfide and regenerating reduced DsbA and oxidized DsbB. Instead, the mixed disulfide is

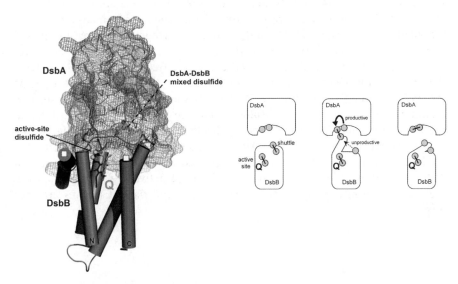

Figure 1.6.4   Structure of the DsbA–DsbB complex and mechanism to favor the productive reaction. DsbA is shown in green with a mesh surface. DsbB is blue with helices indicated by cylinders. The bound quinone is orange and labeled "Q". Disulfides are shown in ball-and-stick representation. The two endpoints of a periplasmic loop that could not be modeled into the electron-density map obtained by X-ray crystallography are shown as white squares. This loop contains the partner cysteine to that participating in the DsbA–DsbB mixed disulfide. Although the exact structure of this loop is unknown due to poor electron density, the cysteine residing on it cannot readily approach the intermolecular disulfide, which is secluded in a groove in the DsbA structure. The mechanism by which a conformational change in DsbB minimizes the unproductive back reaction is shown schematically on the right. Cysteines are shown as yellow balls; disulfides are indicated by a black line connecting two yellow balls. The conformational change that brings the DsbB shuttle cysteine participating in the mixed disulfide into the groove of DsbA distances it from the second DsbB shuttle cysteine, favoring the productive, forward reaction that generates a disulfide in DsbA.

resolved by attack of the second DsbA cysteine, releasing oxidized DsbA and reduced DsbB (Figure 1.6.4). DsbB restores its own disulfides by passing electrons to the respiratory chain *via* the bound quinone. The relative redox potentials of the active-site and shuttle disulfides of Ero1 are not known, so the need for such a mechanism in Ero1 is not established. More experiments, including structural studies of an Ero1/PDI complex, are necessary to address whether conformational changes upon mixed disulfide formation promote disulfide transfer from Ero1 to PDI.

Another mechanistic question raised in the DsbB system that may be relevant to Ero1 is whether oxidation of DsbA occurs as a sequential or concerted electron transfer between the cysteines in DsbA and the two cysteine pairs in DsbB (Figure 1.6.5). The potential for simultaneous formation of mixed disulfides between substrate and one of the shuttle cysteines and between the second shuttle cysteine and the active site was raised by a study of two fragments of DsbB that together reconstitute DsbB function when co-expressed.[36] Covalent assemblies of DsbA and both DsbB fragments were interpreted to reflect a coordinated process in which DsbA attacks the DsbB shuttle disulfide to form a mixed disulfide, and the single freed DsbB shuttle cysteine then attacks the DsbB active-site disulfide before the second DsbA cysteine resolves the complex (Figure 1.6.5). The alternative to this mechanism is a separate and complete dithiol/disulfide exchange reaction between DsbA and the shuttle disulfide, followed by dithiol/disulfide exchange between the shuttle and the active site (Figure 1.6.5). Although there is no evidence for formation of simultaneous disulfides between 1) PDI and an Ero1 shuttle cysteine and 2) the second Ero1 shuttle cysteine and the Ero1 active site, intermediates in the Ero1 reaction cycle have not been analyzed and their structures are thus unknown.

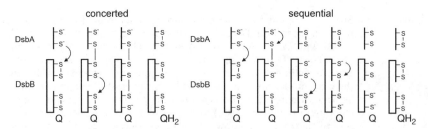

Figure 1.6.5 Two proposed mechanistic models for oxidation of DsbA by DsbB. The upper of the DsbB disulfides represents the shuttle disulfide, and the lower represents the active-site disulfide adjacent to the bound quinone (Q). The mechanism for resolving the double mixed disulfide in the concerted pathway is not shown. The concerted model is supported by the observation that both fragments of "split" DsbB were found in covalent association with DsbA.[36] The sequential model is proposed to be the rapid pathway for DsbA oxidation by DsbB.[34,35]

1.6.2.4 Comparison of Ero1 to Erv Sulfhydryl Oxidases

To our knowledge, Ero1 has not diversified evolutionarily to catalyze disulfide-bond formation in any context outside the early secretory pathway. Cysteine oxidation in compartments of the cell aside from the ER (*i.e.*, mitochondrial intermembrane space, Golgi, cytoplasm) is accomplished by members of a second family of eukaryotic sulfhydryl oxidases, which will be called the "Erv" family after a member identified and characterized early on, the protein "Essential for respiration and viability 1" or "Erv1".[37]

Erv sulfhydryl oxidases are flavoenzymes that share much in common with Ero1, although a lack of sequence homology suggests these two families may have evolved convergently.[15] A Cys–X–X–Cys motif is adjacent to the FAD in Erv enzymes as in Ero1.[38] The conformations of the bound FAD molecules are strikingly similar, especially considering that it is a non-standard configuration for this cofactor.[39] Erv enzymes also have a shuttle disulfide on a flexible segment of polypeptide that can potentially interact alternately with external substrates and the enzyme active site. However, the Erv sulfhydryl oxidases are typically dimeric, and the shuttle is positioned to undergo dithiol/disulfide exchange with the active site of the opposite subunit in the dimer. Erv enzymes have a compact, 5-helix fold in contrast to the large domain of Ero1. Interestingly, Ero1 is described in the Structural Classification of Proteins (SCOP) database[40] as comprising two four-helix bundles within its ten helices, whereas the Erv sulfhydryl oxidases have two four-helix bundles and ten helices per dimer.

Due to the inherent flexibility of the polypeptide segments containing the shuttle disulfides in Erv and Ero1 enzymes, structures determined by X-ray crystallography represent just some of a range of possible conformations for these regions. For the Erv family enzyme Erv2, one of the two conformations observed crystallographically for the shuttle disulfide region is within disulfide bonding distance of an active-site cysteine.[38] In contrast, both positions for the shuttle disulfide loop observed in the two Ero1 structures available place the shuttle cysteines too far from the active site to form the mixed-disulfide intermediate required for direct electron transfer from shuttle to active-site disulfide. However, a superposition of the Erv2 and Ero1 structures reveals that the space analogous to that occupied by the Erv2 shuttle disulfide as it approaches the active site is vacant in Ero1, indicating that the Ero1 shuttle could potentially occupy this position as well (Figure 1.6.6). The feasibility of the Ero1 shuttle reaching the active site is consistent with the isolation of a mixed-disulfide intermediate between the shuttle and active-site cysteines from living cells.[22]

1.6.3 Destination of Reducing Equivalents Derived from Cysteine Thiol Oxidation by Ero1

Characterization of an enzyme that catalyzes disulfide-bond formation is incomplete without an understanding of the physiologically relevant electron acceptors in the reaction. The product of a direct two-electron reduction of

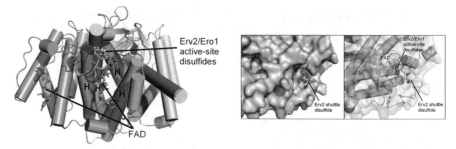

Figure 1.6.6   Superposition of Ero1 and Erv2. In the left panel, Ero1 (tan) and the Erv2
dimer (cyan) are superposed according to their active-site disulfides and
bound FAD. "H" indicates superposed helices of the four-helix bundle in
each structure. In the right panels, opaque and semi-transparent surfaces
of Ero1 are shown together with the isolated shuttle disulfide of Erv2 as it
appears in the superposition. The Ero1 shuttle cysteines may occupy a
similar position upon electron transfer to the active-site disulfide.

oxygen by sulfhydryl oxidases is hydrogen peroxide, and this reaction readily
occurs with recombinant Ero1 *in vitro*.[19] Correspondingly, *in vivo* experiments
have implicated Ero1 activity in the production of reactive oxygen species in the
ER of both yeast and worms.[41,42] It has been demonstrated also that chro-
mophores present in solution or proteins containing metal centers are reduced
by Ero1 supplied with thiol substrates.[19] This observation may explain how
Ero1 is required for protein secretion and cell survival under oxygen-depleted
conditions,[18] although it does not provide specific information on the relevant
in vivo electron acceptors. The bacterial enzyme DsbB and at least some of the
Erv family proteins do not appear to use oxygen as their direct electron
acceptors *in vivo*. DsbB transfers electrons derived from cysteine oxidation to
the electron-transport chain,[43,44] where they are used to reduce molecular
oxygen completely to water, whereas the mitochondrial Erv-type sulfhydryl
oxidases transfer electrons to cytochrome c, from which they may similarly
enter the electron transport chain.[45,46]

1.6.4   Regulation of Ero1 and the Maintenance of Redox Homeostasis in the ER

Initially, mutants of Ero1 were made to determine which of the conserved
cysteines are essential for activity.[5] Ero1 activity was found to depend on the
presence of the shuttle and FAD-proximal cysteines, whereas the third conserved disulfide in the vicinity of the active site (C90–C349) was not required.
Subsequently, mutants were made in cysteines for which the conservation was
less evident, including C150–C295, the disulfide most distant from the active
site. Like C90–C349, the C150–C295 disulfide was found not to be essential for
Ero1 function *in vivo*. Remarkably, when tested *in vitro*, the C150A/C295A
mutant of Ero1 oxidized various reduced substrates much faster than did the

wild-type enzyme.[17] This result implies that not only is the C150–C295 disulfide dispensable for Ero1 activity, it actually inhibits the enzyme. An *in vitro* analysis of the C90A/C349A Ero1 mutant revealed that loss of the C90–C349 disulfide also has subtle but reproducible effects on enzyme kinetics.

The potential for inhibitory disulfide bonds in Ero1 led to the realization that wild-type Ero1 must undergo a reductive activation process to obtain maximal activity. Ero1 activation by reducing substrates is evident from the unusual shape of its enzyme progress curves,[19] which show a slow initial enzyme turnover prior to reaching the maximal rate for that reaction. Monitoring by denaturing gel electrophoresis Ero1 blocked with a thiol-alkylating agent at various stages in the reaction shows that its mobility decreases in two discrete steps shortly after it encounters the reducing substrate.[17] This decrease in mobility reflects successive reduction of the long-range inhibitory C150–C295 and C90–C349 disulfides and the resulting change in hydrodynamic radius of the denatured enzyme. The timing of these disulfide reductions correlates with exit from the kinetic lag phase. Once reduced substrate has been consumed, these Ero1 disulfides become re-oxidized, and re-oxidation correlates with a return to the inhibited state. Thus, reduction/oxidation of long-range disulfide bonds in Ero1 and corresponding changes in Ero1 activity appear to occur in response to the concentration of thiols in the environment.

The importance of the capacity to regulate Ero1 activity *in vivo* has been suggested by the negative impact of a de-regulated mutant (C150A/C295A) on yeast physiology.[17] Over-expression of C150A/C295A causes hyper-oxidation of yeast cells, which is detrimental to cell growth. This growth inhibition is alleviated by addition of a counteracting reducing agent in the growth medium. The negative effects of the de-regulated Ero1 mutant in yeast imply that Ero1 activity is largely dampened by internal disulfide formation under normal growth conditions.

Disulfide bonds are used as redox sensors or to stabilize specific functional states of proteins in a variety of biological contexts, but in Ero1 the feedback loop is particularly tight and allows for rapid changes in enzyme activity. Ero1 appears to directly sense the product of its own activity and respond on the protein level. Explicitly, disulfide reduction by excess substrate activates Ero1 disulfide generating activity. As the thiol:disulfide ratio decreases, disulfides re-form in the enzyme, and catalysis of disulfide formation slows. This homeostatic mechanism complements regulation of Ero1 on the transcriptional level, which occurs as part of the unfolded protein response. In contrast, most other cases of redox control involve formation of a disulfide that activates the protein to perform a function not directly related to thiol chemistry, such as transcriptional activation or chaperoning. For example, a complicated series of disulfide-bond formation events activates the eukaryotic peroxide sensor and transcription factor Yap1.[47] The sugar sensor BglF in bacteria assumes a disulfide-bonded state upon stimulation with sugar,[48] which in turn activates the transcription antiterminator of the bgl operon. Such sensors transmit information about their redox states back to genes, and adjustments are made in the proteome *via* transcription and translation.

1.6.5 Ero1 Orthologs

Ero1 is a highly conserved eukaryotic protein, and Ero1 coding sequences can be found in almost all eukaryotic genomes with public sequence data, ranging from the minimalist genome of the fungi microsporidia to the expanded genome of the human cell. Notable exceptions are the sequenced genomes of several anaerobic parasitic protozoans (*Enatmoeba histolytica, Trichomonas vaginalis* and *Giardia lamblia*), which lack an apparent *ERO1* (or *ERV* family) gene. Intriguingly, *Giardia* encodes five PDI homologs[49,50] and secretes cysteine-rich disulfide cross-linked cyst wall proteins,[51] suggesting that a functional homolog of Ero1 likely exists. Perhaps these divergent protozoans, which share the ability to colonize human mucosa and persist in low-oxygen conditions, contain an alternative disulfide-generating enzyme that does not favor oxygen as an electron acceptor. Functional studies of Ero1 have focused on *Saccharomyces cerevisiae* and *Homo sapiens*; characterization of these Ero1 proteins suggests that the mechanism and pathway used by Ero1 to facilitate disulfide-bond formation have been preserved throughout eukaryota. Human and yeast Ero1 both show an activity dependent on two cysteine pairs[11,20] and facilitate disulfide-bond formation in substrate proteins using PDI as an intermediate.[52–54]

Despite the overall conservation of Ero1 function, organisms are divergent in the number of Ero1 proteins encoded within their genomes. For example, *H. sapiens* encode two functional Ero1 paralogs (Ero1-Lα and Ero1-Lβ) whereas *S. cerevisiae* contain a single Ero1. The significance of two Ero1 proteins encoded in human cells is not clear. No major differences in redox activities or substrate preference have been described between Ero1-Lα and Ero1-Lβ. Both human proteins are functional homologs of *S. cerevisiae* Ero1 and can complement the phenotypic defects of a temperature-sensitive *ERO1* mutant yeast strain.[55,56] However, neither human Ero1 can complement a yeast strain with a complete disruption of *ERO1*, which has been attributed to the absence of an extended carboxy-terminal region in both human Ero1s that is present in yeast Ero1 and may mediate ER retention in yeast cells.[57] Human Ero1s appear to be retained in the ER lumen by thiol-mediated retention, forming a mixed-disulfide bond with the PDI-like ER protein ERp44.[58] The human genes do possess unique transcriptional regulatory elements, and multiple *ERO1* genes may allow human cells to modulate the oxidizing potential of the ER under specific stress conditions.[56,59] Ero1 paralogs may also contribute to tissue specific oxidation pathways, as implied by the distinct tissue distribution of human Ero1-Lα and Ero1-Lβ.[56]

In addition to functional questions relating to the presence of Ero1 paralogs in some organisms, many interesting evolutionary questions remain. Two *ERO1* genes are observed amongst the majority of vertebrate genomes, while a single *ERO1* gene is found in most invertebrates and fungi. However, there are invertebrates, plants (*e.g.* reference 58) and fungi (*e.g.* reference 61) that contain two Ero1s. With the abundance of genome sequence data continually becoming available, it will be interesting to determine the stage in evolution when duplication of the *ERO1* gene occurred in different eukaryotes. Intriguingly, a comparison of the active-site motifs of Ero1 homologs has indicated that a "charged" β-like (referring to human Ero1-Lβ) active-site motif (Cys–Asp/Glu–Arg/Lys–Cys) is most common among Ero1 proteins. It has been proposed that this charged sequence may reflect an ancestral sequence from which the vertebrate α-like Ero1 proteins (Cys-Phe-Lys-Cys) diverged.[60]

1.6.6 Summary

Ero1 is the sulfhydryl oxidase responsible for setting the thiol/disulfide balance in the ER at a level appropriate for oxidative protein folding. The characteristically oxidizing conditions of the ER, as quantified by the ratio of oxidized to reduced glutathione, arise from the net flow of electrons from glutathione and cysteines in protein nascent chains to oxygen *via* PDI-like proteins and Ero1. Studies of yeast Ero1 suggest that Ero1 has the capacity to modulate its activity in response to changes in the rate of thiol import into the ER and to events or environmental conditions that alter the levels of available oxidants or reductants. Ero1 has evolved to directly sense the level of thiol substrates in the ER by requiring reductive activation of its sulfhydryl oxidase activity, a process that can be mimicked in part by the mutation of cysteines participating in regulatory disulfides distant from the enzyme active site. The essence of the catalytic mechanism, involving electron transfer from PDI to the exposed Ero1 shuttle disulfide, and then to the active-site disulfide and FAD at the catalytic center, appears to be shared by Ero1 enzymes from different organisms. It will be important to determine whether the phenomenon of reductive activation is also shared with orthologs in other species. Ero1 has features in common with the bacterial disulfide-bond generating enzyme DsbB and the eukaryotic Erv sulfhydryl oxidase family, such that insights into the structures and mechanisms of these other enzymes that catalyze disulfide-bond formation inform our understanding of Ero1.

References

1. E. Van Anken and I. Braakman, *Crit. Rev. Biochem. Mol. Biol.*, 2005, **40**, 191–228.
2. D. N. Hebert, S. C. Garman and M. Molinari, *Trends Cell Biol.*, 2005, **15**, 364–370.
3. P. J. Flory, *J. Am. Chem. Soc.*, 1956, **78**, 5222–5235.
4. C. N. Pace, G. R. Grimsley, J. A. Thomson and B. J. Barnett, *J. Biol. Chem.*, 1988, **263**, 11820–11825.
5. M. Sone, S. Kishigami, T. Yoshihisa and K. Ito, *J. Biol. Chem.*, 1997, **272**, 6174–6178.
6. D. LeBel and J. Paquette, *Biochem. Cell Biol.*, 1989, **67**, 281–287.
7. M. S. B. Paget and M. J. Buttner, *Ann. Rev. Genet.*, 2003, **37**, 91–121.

8. A. R. Frand and C. A. Kaiser, *Mol. Cell*, 1998, **1**, 161–170.
9. M. G. Pollard, K. J. Travers and J. S. Weissman, *Mol. Cell*, 1998, **1**, 171–182.
10. B. P. Tu and J. S. Weissman, *Mol. Cell*, 2002, **10**, 983–994.
11. A. R. Frand and C. A. Kaiser, *Mol. Cell*, 1999, **4**, 469–477.
12. S. Chakravarthi, C. E. Jessop and N. J. Bulleid, *EMBO Rep.*, 2006, **7**, 271–275.
13. J. W. Cuozzo and C. A. Kaiser, *Nat. Cell Biol.*, 1999, **1**, 130–135.
14. D. W. S. Stephen and D. J. Jamieson, *FEMS Micrbiol. Lett.*, 1996, **141**, 207–212.
15. E. Gross, D. B. Kastner, C. A. Kaiser and D. Fass, *Cell*, 2004, **117**, 601–610.
16. G. Tian, S. Xiang, R. Noiva, W. J. Lennarz and H. Schindelin, *Cell*, 2006, **124**, 61–73.
17. C. S. Sevier, H. Qu, N. Heldman, E. Gross, D. Fass and C. A. Kaiser, *Cell*, 2007, **129**, 333–344.
18. B. P. Tu, S. C. Ho-Schleyer, K. J. Travers and J. S. Weissman, *Science*, 2000, **290**, 1571–1574.
19. E. Gross, C. S. Sevier, N. Heldman, E. Vitu, M. Bentzur, C. A. Kaiser, C. Thrope and D. Fass, *Proc. Natl. Acad. Sci. USA*, 2006, **103**, 299–304.
20. A. Mattevi, *Trends Biochem. Sci.*, 2006, **31**, 276–283.
21. A. R. Frand and C. A. Kaiser, *Mol. Biol. Cell*, 2000, **11**, 2833–2843.
22. C. S. Sevier and C. A. Kaiser, *Mol. Biol. Cell*, 2006, **17**, 2256–2266.
23. T. Marquardt, D. N. Hebert and A. Helenius, *J. Biol. Chem.*, 1993, **268**, 19618–19625.
24. B. Wilkinson and H. G. Gilbert, *Biochim. Biophys. Acta*, 2004, **1699**, 35–44.
25. M. S. Kulp, E. M. Frickel, L. Ellgaard and J. S. Weissman, *J. Biol. Chem.*, 2006, **281**, 876–884.
26. B. Tsai and T. A. Rapoport, *J. Cell Biol.*, 2002, **159**, 207–215.
27. B. Wilkinson, R. Xiao and H. F. Gilbert, *J. Biol. Chem.*, 2005, **280**, 11483–11487.
28. D. M. LeMaster, P. A. Springer and C. J. Unkefer, *J. Biol. Chem.*, 1997, **272**, 29998–30001.
29. C. Tachibana and T. H. Stevens, *Mol. Cell Biol.*, 1992, **12**, 4601–4611.
30. H. Tachikawa, Y. Takeuchi, W. Funahashi, T. Miura, X. D. Gao, D. Fujimoto, T. Mizunaga and K. Onodera, *FEBS Lett.*, 1995, **369**, 212–216.
31. H. Tachikawa, W. Funahashi, Y. Takeuchi, H. Nakanishi, R. Nishihara, S. Katoh, X. D. Gao, T. Mizunaga and D. Fujimoto, *Biochem. Biophys. Res. Commun.*, 1997, **239**, 710–714.
32. Q. Wang and A. Chang, *EMBO J.*, 1999, **18**, 5972–5982.
33. J. C. Bardwell, J. O. Lee, G. Jander, N. Martin, D. Belin and J. Beckwith, *Proc. Natl. Acad. Sci. USA*, 1993, **90**, 1038–1042.

34. K. Inaba, Y. H. Takahashi and K. Ito, *J. Biol. Chem.*, 2005, **280**, 33035–33044.
35. K. Inaba, S. Murakami, M. Suzuki, A. Nakagawa, E. Yamashita, K. Okada and K. Ito, *Cell*, 2006, **127**, 789–801.
36. H. Kadokura and J. Beckwith, *EMBO J.*, 2002, **21**, 2354–2363.
37. T. Lisowsky, *Mol. Gen. Genet.*, 1992, **232**, 58–64.
38. E. Gross, C. S. Sevier, A. Vala, C. A. Kaiser and D. Fass, *Nat. Struct. Biol.*, 2001, **9**, 61–67.
39. O. Dym and D. Eisenberg, *Protein Sci.*, 2001, **10**, 1712–1728.
40. A. G. Murzin, S. E. Brenner, T. Hubbard and C. Chothia, *J. Mol. Biol.*, 1995, **247**, 536–540.
41. H. P. Harding, Y. Zhang, H. Zeng, I. Novoa, P. D. Lu, M. Calfon, N. Sadri, C. Yun, B. Popko, R. Paules, D. F. Stojdl, J. C. Bell, T. Hettmann, J. M. Leiden and D. Ron, *Mol. Cell*, 2003, **11**, 619–633.
42. C. M. Haynes, E. A. Titus and A. A. Cooper, *Mol. Cell*, 2004, **15**, 767–776.
43. T. Kobayashi, S. Kishigami, M. Sone, H. Inokuchi, T. Mogi and K. Ito, *Proc. Natl. Acad. Sci. USA*, 1997, **94**, 11857–11862.
44. M. Bader, W. Muse, D. P. Ballou, C. Gassner and J. C. Bardwell, *Cell*, 1999, **98**, 217–227.
45. S. R. Farrell and C. Thorpe, *Biochem.*, 2005, **44**, 1532–1541.
46. K. Bihlmaier, N. Mesecke, N. Terziyska, M. Bien, K. Hell and J. M. Herrmann, *J. Cell Biol.*, 2007, **179**, 389–395.
47. S. Okazaki, T. Tachibana, A. Naganuma, N. Mano and S. Kuge, *Mol. Cell*, 2007, **27**, 675–688.
48. Q. Chen, A. Nussbaum-Shochat and O. Amster-Choder, *J. Biol. Chem.*, 2001, **276**, 44751–44756.
49. A. G. McArthur, L. A. Knodler, J. D. Silberman, B. J. Davids, F. D. Gillin and M. L. Sogin, *Mol. Biol. Evol.*, 2001, **18**, 1455–1463.
50. L. A. Knodler, R. Noiva, K. Mehta, J. M. McCaffery, S. B. Aley, S. G. Svard, T. G. Nystul, D. S. Reiner, J. D. Silberman and F. D. Gillin, *J. Biol. Chem.*, 1999, **274**, 29805–29811.
51. F. D. Gillin, D. S. Reiner and J. M. McCaffery, *Annu. Rev. Microbiol.*, 1996, **50**, 679–705.
52. G. Bertoli, T. Simmen, T. Anelli, S. N. Molteni, R. Fesce and R. Sitia, *J. Biol. Chem.*, 2004, **279**, 30047–30052.
53. A. Mezghrani, A. Fassio, A. Benham, T. Simmen, I. Braakman and R. Sitia, *EMBO J.*, 2001, **20**, 6288–6296.
54. A. M. Benham, A. Cabibbo, A. Fassio, N. Bulleid, R. Sitia and I. Braakman, *EMBO J.*, 2000, **19**, 4493–502.
55. A. Cabibbo, M. Pagani, M. Fabbri, M. Rocchi, M. R. Farmery, N. J. Bulleid and R. Sitia, *J. Biol. Chem.*, 2000, **275**, 4827–4833.
56. M. Pagani, M. Fabbri, C. Benedetti, A. Fassio, S. Pilati, N. J. Bulleid, A. Cabibbo and R. Sitia, *J. Biol. Chem.*, 2000, **275**, 23685–23692.

57. M. Pagani, S. Pilati, G. Bertoli, B. Valsasina and R. Sitia, *FEBS Lett.*, 2001, **508**, 117–120.
58. T. Anelli, M. Alessio, A. Bachi, L. Bergamelli, G. Bertoli, S. Camerini, A. Mezghrani, E. Ruffato, T. Simmen and R. Sitia, *EMBO J.*, 2003, **22**, 5015–5022.
59. B. Gess, K. H. Hofbauer, R. H. Wenger, C. Lohaus, H. E. Meyer and A. Kurtz, *Eur. J. Biochem.*, 2003, **270**, 2228–2235.
60. D. P. Dixon, M. Van Lith, R. Edwards and A. Benham, *Antioxid. Redox Signal.*, 2003, **5**, 389–396.
61. K. Kettner, A. Blomberg and G. Rodel, *Yeast*, 2004, **21**, 1035–1044.

CHAPTER 1.7

Eukaryotic Protein Disulfide-isomerases and their Potential in the Production of Disulfide-bonded Protein Products: What We Need to Know but Do Not!

ROBERT B. FREEDMAN

Department of Biological Sciences, University of Warwick, Coventry CV4 7AL, UK

1.7.1 Introduction

This review will focus on what is NOT known about protein disulfide-isomerases, rather than on what IS known.

This choice is made for two reasons. The first is because a number of excellent recent reviews have appeared on different aspects of the topic, reflecting recent progress in many areas. Some address general issues (*e.g.* reference 1) but many have a more discrete focus *e.g.* the role of PDI and its homologs in the context of protein folding and quality control in the endoplasmic reticulum (ER),[2,3] the role of PDIs in the ER in relation to that organelle's wider functions in calcium homeostasis and signalling,[4] potential roles of PDIs in other cellular locations,[5] the functions of individual PDI domains and the synergies between them,[6] post-genomic elucidation of all the members of the human PDI family and analysis of

RSC Biomolecular Sciences
Oxidative Folding of Peptides and Proteins
Edited by Johannes Buchner and Luis Moroder
© Royal Society of Chemistry 2009
Published by the Royal Society of Chemistry, www.rsc.org

their potential roles,[7] structures of both eukaryotic and prokaryotic PDIs and elucidation of common features and themes,[8] analysis of the specificity of PDIs and their mechanisms of substrate recognition[9] and a very recent and challenging review of the relationship of biological role to structural features across the human PDI family.[10]

But the second – and more significant – reason for this approach is that a review of what is not known can highlight unsolved problems and disquieting gaps in knowledge and hence can provide an explicit agenda for future research. In this review, I will focus on several frustrating areas of ignorance which limit our ability to engineer PDI in cells and subcellular systems in ways that reliably enhance the capability of those systems to produce native disulfide-bonded proteins.

Human and mammalian extracellular proteins – including antibodies, cytokines and other serum factors – now constitute a major and rapidly growing sector in the pharmaceuticals industry. Much effort is engaged in attempts to produce such proteins industrially, in high yields and in an "authentic" state, where "authenticity" refers to conformation, disulfide bonding and other posttranslational modifications, most notably glycosylation. A full knowledge of the action of PDI in a cellular context would provide a rational basis for cell engineering to maximize product protein yield; this can be achieved in some cases, but not reliably or predictably. It should surely be possible to achieve that objective on a timescale of no more than a decade. This review highlights those areas of ignorance that are currently obstacles on the route to that objective.

In a much earlier review,[11] I identified a number of reasons why it was not generally accepted – at that time – that PDI played a role *in vivo* in the formation of native disulfide bonds in secretory proteins. These reasons included:

   i) no evidence of changes in the concentration or activity of the enzyme being correlated with changes in the pattern or amount of protein synthesis in a tissue, nothing that suggests a definite association with the synthesis of disulfide-bonded proteins
  ii) absence of clear evidence on the subcellular location of the enzyme (other than its being concentrated in microsomal fractions)
 iii) the fact that purified PDI catalyzed multiple related activities (oxidoreductions, disulfide isomerization) and that cells contained many enzymes which catalyzed at least some of these partial reactions
  iv) inadequate quantitative information on the enzyme's catalytic capability and on its substrate range and specificity
   v) no definite evidence for a requirement for the enzyme *in vivo*.

Over the subsequent few years, a considerable body of work was undertaken that addressed issues i)–iii) with the result that it became clear that PDI is found in the appropriate cells, at the appropriate time and in the appropriate subcellular location (ER lumen) to play a key role in protein folding and posttranslational modification in the secretory pathway.[12,13]

This circumstantial evidence for the cellular role of PDI accumulated at the precise period (the latter half of the 1980s) when the concept of "molecular chaperones" was gaining acceptance and so it promoted the view that PDI should be considered as part of the cellular folding machinery alongside other "molecular chaperones". However there are stringent criteria for confirming that putative "molecular chaperones" actually play a role in protein folding and/or assembly in the cell (see reference 14). This required that issue v) above be addressed by both genetic and cell biological approaches. Convincing evidence to confirm this role was provided over the subsequent few years with the findings a) that PDI can be cross-linked to nascent secretory proteins both in intact cells and in *in vitro* translation/translocation systems, b) that resolution and reconstitution of *in vitro* translation/translocation systems demonstrate that PDI is required for effective disulfide-bond formation in such systems and finally c) that the *PDI1* gene is required for the *in vivo* production of disulfide-bonded proteins in *S. cerevisiae* (reviewed in references 14 and 15).

Hence over the past 30 years, most of the gaps in knowledge of PDI that were originally identified have been filled in. (Our knowledge of issue iv) – the full catalytic capability of PDI – is still inadequate, but that is related to wider issues of mechanism of action that are addressed in Section 1.7.3.4 below). Over the same period, of course, our knowledge in other areas has advanced to such an extent that we now ask new and more demanding questions. And these questions expose just how limited our current knowledge of PDI is in a number of areas that are crucial to its potential exploitation.

1.7.2 Evidence that PDI is Rate or Yield Limiting in the Production of High-value Proteins

1.7.2.1 Oxidative Folding *in vitro*

Although the cellular role of PDI was formally proved by the early 1990s, some of the most dramatic demonstrations of the significance of PDI have been provided more recently in studies where supplementation of systems with PDI has profoundly increased their capability for producing folded proteins with native disulfide bonds.

1.7.2.1.1 Refolding and Post-translational Folding of Full-length Proteins

Catalysis of disulfide-linked folding by PDI was first demonstrated in simple model systems *in vitro* but has since been confirmed in more complex *in vitro* systems or with more demanding substrates. Two examples illustrate this.

Bovine pancreatic trypsin inhibitor (BPTI) is one of the classic small model proteins whose disulfide-linked folding pathway has been studied extensively.[16] At an early stage, it was demonstrated that all the disulfide oxido-reductions and isomerizations in the folding pathway were catalyzed by PDI[17] and subsequently the catalytic effects of PDI on the rate-determining isomerization steps were quantified.[18–20] When this reaction was studied in presence of a complex extract representing the total luminal content of the ER, catalysis of BPTI folding was observed but quantitatively this could be fully accounted for by the presence of PDI in the ER extract, suggesting that PDI was an effective catalyst in these conditions but there was no synergistic effect of other ER luminal proteins.[21] This issue was also explored in studies using a cell-free translation/translocation system;[22] pre-pro-BPTI was translated in a wheat germ lysate and co-translationally translocated into dog pancreas microsomes in reducing conditions to generate reduced pro-BPTI within the luminal compartment of the microsomes. On addition of oxidized glutathione, this product folded within the luminal environment to give oxidized pro-BPTI, with a half-time of approximately 2 minutes. Similar kinetics of oxidative folding were observed when recombinant reduced pro-BPTI was oxidized in the test-tube in presence of catalytic quantities of purified PDI. So, in this system, disulfide-linked folding catalyzed by PDI alone mimics the rate observed in the "quasi-physiological" environment of the microsomal lumen.

Antibodies are far more complex proteins than BPTI and their oxidative folding requires the formation of intramolecular disulfides within each domain, and also the specific formation of a number of inter-molecular disulfides to specify the familiar 4-chain structure. In contrast to BPTI, antibodies and their fragments such as Fabs are classic examples of "difficult" proteins for *in vitro* refolding, giving low yields and poor rates of formation of correct product (see also Chapter 2.2 for reoxidation of proteins *in vitro*). Nevertheless, even in this difficult case there is clear evidence of a positive effect of the addition of PDI.

PDI profoundly increased the yield of active Fab molecules after refolding of recombinant light and heavy chains from the reduced denatured state, generating recoveries of up to 50% compared to <10% in absence of PDI; work with truncated Fabs indicated that PDI was involved in formation of both intramolecular and intermolecular disulfides in this system.[23] This activity appeared to combine impact on thiol:disulfide chemistry with a "chaperone" effect; it required unmodified active-site Cys residues in the PDI but also required stoichiometric quantities of PDI. The presence of PDI generated high yields of refolded Fabs over a wide range of Fab concentration and over a wide range of redox conditions, whereas refolding yield in absence of PDI was strongly dependent on these two variables. Specifically, in presence of PDI, optimal recoveries of Fab activity were observed over a range of redox conditions from $[\text{GSH}]^2/[\text{GSSG}] = 0.1\,\text{mM}$ to $10\,\text{mM}$ whereas in absence of PDI there was negligible recovery of activity in relatively oxidizing conditions where $[\text{GSH}]^2/[\text{GSSG}] < 2\,\text{mM}$. The presence of PDI was required from the beginning of the refolding process suggesting that, in its absence, most of the unfolded Fab chains were rapidly converted to a misfolded state from which they could not be recovered by subsequent addition of PDI *i.e.*, that there was competition between misfolding and PDI-dependent refolding.

This competition was analyzed further in work exploring potential synergies between PDI and the chaperone BiP, a member of the hsp70 chaperone family which is known to interact with immunoglobulin heavy chains in the ER lumen of antibody-synthesizing cells. This work[24] showed that there was clear synergy between PDI and BiP in the ability to enhance the refolding of Fabs. Furthermore, the presence of BiP extended the period over which PDI could exert its enhancement effect; the "refoldability" of Fabs by PDI decayed with a half-time of 10–30 s whereas in the presence of BiP this decay had a half-time of *ca.* 300 s. This finding confirms a model in which BiP binding to Fab polypeptides maintains them in an unfolded but soluble state for a prolonged period during which PDI is able to facilitate the formation of disulfide bonds and associated refolding.

1.7.2.1.2 Co-translational Folding

Such work is an example of the "classical" approach to protein folding which focuses on the refolding of full-length proteins from the unfolded state. It is instructive, but in many ways it is a poor model for protein folding in the cell, particularly for multi-domain proteins, where it is likely that completed N-terminal domains can fold while subsequent domains are still being synthesized. A more realistic model is provided by coupled cell-free transcription/translation systems. These systems are based on *E. coli* S30 cell lysates and conventionally employ strongly reducing conditions, but they have now been modified and developed into efficient systems in which one can study co-translational protein folding and disulfide-bond formation; they have been shown to be capable of producing multi-domain and multi-disulfide-bonded proteins such as antibody fragments and serum proteases.[25–27]

Early studies using such a system gave the first indications that they required modification for production of disulfide-bonded proteins. It was observed that functional scFv products with antigen-binding activity were obtained only if disulfide formation and re-arrangement could take place during the translation process.[28] Addition of PDI led to a 3-fold increase in yield over that obtained in the presence of simple glutathione-based redox system; by contrast addition of DsbA had no such effect, indicating that isomerization rather than simple oxidation activity was required. In this system, the amount of functional scFv produced appeared to be limited entirely by correct disulfide-bond formation. A more stringent test of the capability of these systems to produce active folded disulfide-bonded molecules was the recent demonstration of their ability to generate a functional intact IgG molecule comprising full-length heavy and light chains.[29] This ability was not found in unsupplemented bacterial lysates; it was absolutely dependent on the addition of protein disulfide-isomerase activity, either in the form of human PDI or the bacterial periplasmic protein DsbC, which has protein disulfide-isomerase activity. In presence of either of these additional components, yields of up to 500 ng ml^{-1} were achieved. It was interesting that the addition of BiP together with PDI in this system showed no enhancement of activity over that found with PDI alone (*cf.* the result quoted above). However the *E. coli* lysate used for transcription/translation contains high levels of the BiP homolog DnaK, so it is probable that this functionally fulfills the role that BiP plays in the mammalian cell and in the *in vitro* refolding study reported above.[24]

All this work on the folding of functional antibodies or antibody fragments *in vitro* suggests that the formation of correct disulfide bonds is critical to their productive folding and that the presence of PDI is essential for the efficiency of this process. Very similar conclusions arise from studies on the production of disulfide-bonded proteins in intact cells (See Table 1.7.1).

1.7.2.2   Optimizing Production of Disulfide-bonded Proteins in *Escherichia coli*

Disulfide-bonded mammalian proteins characteristically form inclusion bodies when expressed in the bacterial cytoplasm.[30-32] While active protein can be recovered from such inclusions by solubilization and unfolding followed by controlled refolding, it would be attractive to make such proteins in the native state in a bacterial host, exploiting the simplicity and convenience of genetic manipulation and large-scale culture of bacteria. As discussed elsewhere in this volume (Chapters 1.2 and 1.3), bacteria contain machinery for catalyzing protein disulfide-bond formation and isomerization located in the periplasmic compartment. Hence considerable research has examined the capability and capacity of the disulfide formation machinery of the *E. coli* periplasm to permit folding of high-value mammalian proteins.

Few bacterial proteins contain multiple disulfide bonds, but it is nevertheless clear that bacteria have evolved the ability to perform the functions necessary to ensure formation of correct disulfides in such proteins. The key bacterial isomerase which functionally resembles eukaryotic PDI is DsbC; its role and properties were confirmed by the observations that isolated DsbC is highly active in catalyzing *in vitro* the rate-determining isomerizations in the BPTI oxidative folding pathway[33] and that mutants in the *dsbC* gene display a specific defect *in vivo* in the production of proteins containing multiple disulfide bonds.[34]

Many attempts have been made to exploit export to the bacterial periplasm as a strategy for producing high-value disulfide-bonded mammalian proteins and these have included studies on the ability of mammalian PDI to function effectively in the environment of the periplasm. When provided with a classical bacterial signal sequence directing export to the periplasm, human or rat PDI can be exported to the periplasm and can function there to complement mutants defective in the bacterial oxidoreductase DsbA.[35,36] Mammalian PDI can significantly increase the yields of active disulfide-bonded proteins exported to the periplasm, whether these are bacterial proteins – such as alkaline phosphatase or the degradative enzyme pectate lyase C[35] – or mammalian proteins such as BPTI.[36]

However, it is not clear that co-expression in the periplasm together with mammalian PDI is a universally successful strategy for optimizing production of complex mammalian disulfide-bonded proteins. In work on periplasmic expression of human tPA (tissue plasminogen activator), no enhancement of yield was observed on co-expression with either rat PDI or yeast (*S. cerevisiae*)

Table 1.7.1 Improvements in the yields of recombinant disulfide-bonded proteins as a result of expression in the presence of enhanced levels of PDI.

Expression system	Added* or co-over-expressed PDI	Target protein[a]	Maximum yield (various units)	Maximum yield enhancement (fold)	Ref.
E. coli cell-free	Bovine*	ScFv	30–48% of total	*ca.* 4-fold	28
E. coli cell-free	Human*	IgG	500 ng ml^{-1}	>50-fold	29
E. coli periplasm	Human	*E. carotova* pectate lyase C	–	>5-fold	35
E. coli periplasm	Rat	Bovine pancreatic trypsin inhibitor	–	15-fold	36
S. cerevisiae	Human	Tick antistasin	3 mg l^{-1}	>4-fold	39
S. cerevisiae	*S. cerevisiae*	Tick antistasin	16 mg l^{-1}	>24-fold	39
S. cerevisiae	*S. cerevisiae*	PDGF	1.5 mg l^{-1}	10-fold	40
S. cerevisiae	*S. cerevisiae*	*S. pombe* acid phosphatase	–	4-fold	40
S. cerevisiae	Rat	scFv	–	2-fold	44
S. cerevisiae	Rat PDI+*S. cerevisiae* BiP	scFv	–	8-fold	44
S. cerevisiae	*S. cerevisiae*	Transferrin	3 g l^{-1}	15-fold	46
Kluyveromyces lactis	*K. lactis*	Serum albumin	–	15-fold	45
Mouse hybridoma	Rat	IgG	150 mg l^{-1}	<2-fold	51
Chinese Hamster Ovary cells	Mouse	IgG	23 pg cell^{-1} day^{-1}	+37%	53
Trichoplusia ni (insect) cells	Mouse	IgG	–	+80%	54

[a]Target proteins are human, unless stated otherwise.

PDI, but there was a striking enhancement of active tPA production by co-over-expression of DsbC.[37] Similarly, it has proved possible to assemble intact full-length IgG molecules by export of both light and heavy chains to the *E. coli* periplasm without any specific manipulation of the enzymes for protein disulfide formation in the periplasm[38] and titres of antibody produced by fermentation of *E. coli* can exceed $100 \, \text{mg} \, l^{-1}$.

1.7.2.3 Optimizing Production of Disulfide-bonded Proteins in *Saccharomyces cerevisiae*

The secretory system in prokaryotes is considerably different from that in eukaryotes and hence it is possibly not surprising that inconsistent results have been obtained in studies of the effect of over-expression of mammalian PDIs in the bacterial periplasm. By contrast, it might be a reasonable assumption that over-expression of a eukaryotic PDI within the ER of a eukaryotic cell would increase that host cell's capacity for production of correctly disulfide-bonded protein products. This assumption has been tested using mammalian and yeast PDIs in various yeast species; the results have been generally positive, but only a limited number of cases have been studied.

In the mid 1990s, two groups demonstrated that enhancing the level of PDI in the yeast secretory pathway could increase the yield of various target heterologous proteins. Schulz *et al.*[39] noted that endogenous yeast PDI levels are low in *Saccharomyces cerevisiae* under the conditions in which heterologous secretory protein production is normally induced, and they therefore tested the effects of over-expression of either human PDI or *S. cerevisiae* PDI on the production of the leech anticoagulant protein antistasin (an inhibitor of factor Xa). Antistasin is a small highly disulfide-bonded protein – it contains 20 Cys residues and 10 disulfide bonds in a protein comprising only 119 residues in total – and hence it provides a stringent test of a cell's capacity for formation of native disulfide bonds. The effects depended on a number of factors – whether the PDI gene was integrated or introduced in a second plasmid, the nature of the signal sequence used to transfer the PDI into the ER, growth temperature, culture time *etc.* – but in all cases the results clearly showed enhancement of antistasin production with increases in PDI expression. In the best cases, over-expression of human PDI enhanced antistasin production 3-fold, while over-expression of *S. cerevisiae* PDI enhanced it by approximately 20-fold.

In parallel with this work, Robinson *et al.*[40] also approached this issue, working with cells in which an additional copy of the yeast PDI gene was chromosomally integrated in tandem with the native gene, but with the additional copy of the gene under control of the constitutive GAPDH promoter. This resulted in a 16-fold increase in the level of yeast PDI protein in the cells and so these cells were used as hosts to test for enhancement of heterologous secretory protein production. Production of the human platelet-derived growth factor (PDGF) B dimer was 10-fold enhanced in these cells and production of a fungal protein (*Schizosaccharomyces pombe* acid phosphatase) was enhanced

4-fold. Interestingly, this group noted that attempts to increase further the levels of PDI through over-expression of PDI from multicopy plasmids could have detrimental effects on secretion of heterologous disulfide-bonded proteins, possibly due to saturation of ER membrane translocation sites.

These studies did not show a clear relationship between PDI level and enhancement of yield of heterologous disulfide-bonded proteins, and yet the enhancements of yield were striking enough for it to be suggested that this might be a general strategy.[41,42] However, surprisingly little further work has been reported in this field. Hayano *et al*.[43] showed only a 50% increase in secretion of human lysozyme from *S. cerevisiae* cells in which the lysozyme and either human or yeast PDI genes had been co-integrated from yeast integrating plasmid vectors with multi-cloning sites, compared to cells containing only wild-type levels of yeast PDI. Shusta *et al*.[44] took a broader approach and investigated the secretory capacity of *S. cerevisiae* for single-chain antibody fragments (scFv fragments) in the presence of over-expressed yeast BiP (Kar2p) or rat PDI. Each of these factors separately enhanced maximal production of scFvs and there was synergy between them so that the combined effect of the factors together was to produce an 8-fold enhancement in yield of product (up to $20\,mg\,ml^{-1}$). Significantly these authors studied the effects of BiP and PDI at a range of scFv expression levels ranging from low copy to ER-saturating over-expression and observed enhanced yields at every level.

More recently, work of this kind has been extended to another yeast species, *Kluyveromyces lactis*. Using strains with a single additional integrated copy of the host PDI gene, it was demonstrated[45] that the increased level of PDI resulted in a 9-fold enhancement in the secretion of recombinant human serum albumin (rHSA). Human serum albumin is a multi-domain protein containing 17 disulfide-bonds; it is both a challenging model protein and a high-value protein product.

Many of the papers above also report negative results, where in specific circumstances the introduction of one or more additional PDI genes does NOT significantly increase the production of a specific disulfide-bonded target protein. It is probable that other cases with negative outcomes have not been reported in the literature. Inevitably, negative results are rarely followed up and hence we do not have a full and systematic picture of the circumstances in which PDI over-expression can and cannot aid the production of heterologous disulfide-bonded proteins in yeast. Even positive results may not be published in the conventional journal literature. However recent patents indicate that productive work is continuing in this field.[46,47]

1.7.2.4 Optimizing Production of Disulfide-bonded Proteins in Mammalian and Insect Cells

Given the economic value of hybridoma-derived monoclonal antibodies (Mabs), it is not surprising that considerable effort has been devoted to identifying and manipulating factors that might determine the yield of active Mab production by hybridomas and related mammalian cells (see reference 48 for a recent review). Using a systems engineering model of the synthesis and secretion of antibodies in hybridoma cells, as used in Mab production systems, Bibila and Flickinger[49,50] analyzed the sensitivity of production rate to changes in a number of the model parameters and used the results to propose genetic manipulation approaches to maximize the specific antibody secretion rate and the volumetric productivity of large-scale antibody production systems. It was recognized that the rate- and yield-determining steps for antibody production might alter with immunoglobulin chain expression level and that at higher rates of Mab production, the limiting steps were related to assembly and/or transport of the multi-chain antibody and not dependent on the cellular levels of heavy and light chain proteins or heavy and light chain mRNAs. By modeling transients in the rates of key processes they showed that the step of antibody assembly in the ER was a very good candidate for a rate-limiting step in the antibody production pathway in fast-growing hybridoma cells. In a simulation of increased PDI levels in the cells, they found that this resulted in proportional increases in the rate constant for antibody assembly and an increase in the secretion rate and final yield of Mab by 3- to 5-fold.

Attempts to test this prediction experimentally have produced a puzzling range of outcomes. In one early study,[51] a stable murine hybridoma cell line which secreted high levels of an IgG was used as host for generating transfectoma clones with inducible over-expression of rat PDI. Induction of PDI expression led to higher yields of antibody production from these cells, both in small volume shake-flask cultures and in a 1-l bioreactor. However, these cells showed only a very small increase in PDI levels and there was no enhancement of the specific antibody secretion rate. Induced cells showed greater longevity and appeared to be productive over a longer period, suggesting that the greater antibody yield was not directly related to the induction of PDI in these cells.

Another study[52] produced surprisingly negative outcomes. This study employed CHO (Chinese hamster ovary) cells which are widely used in the industry because of their high production rates, human-like glycosylation machinery and limited requirement for added proteins or serum in the culture medium. CHO cells secreting an interleukin (IL15) and a TNF receptor:Fc fusion protein were transformed with a vector encoding human PDI cDNA and 3- to 4-fold over-expression of the PDI was confirmed. In the IL-15 secreting cells there was little change in the rate or level of production of IL-15, but in the cells expressing the highly disulfide-bonded TNF receptor fusion protein, its secretion was decreased. It was established that intracellular levels of this target protein were considerably increased in the PDI-over-expressing cells, and that the target protein was co-localized with PDI in the ER. This led to the conclusion that the over-expressed PDI was acting to retain the target protein within the ER, rather than facilitating its folding and exit from the ER.

More recently, evidence that PDI levels in CHO cells are much lower than those in hybridomas has led to a re-examination of the effect of over-expressing PDI on the production of human antibodies by CHO cells. Borth *et al.*[53] generated cell lines over-expressing human PDI and BiP (separately and together) and showed that increasing the level of BiP led to decreases in the specific production rate by more than 30%, independent of whether PDI levels were increased in parallel. However, the cells transformed with PDI cDNA only showed a clear increase of specific antibody production rate (by 37%) linked to a 58% increase in PDI content. In parallel with this, Borth et al.[53] studied sister clones of CHO cells producing a human monoclonal antibody which differed in antibody secretion rate despite deriving from the same transfection; low- and high-secreting cells were obtained by cell sorting and then amplified independently. It was found that the high-secreting cells (with a 3-fold higher secretion rate than the low-secreting cells) had a PDI content that was 60–70% greater than that in low-secreting cells.

These studies in mammalian cells differ in many respects (host cell, culture conditions, target protein, quantitative level of PDI over-expression) but the contrasting results obtained make it clear that over-production of PDI is not a simple and general fix that can engineer mammalian cells to increase their production capacity for disulfide-bonded proteins.

Studies on insect cells expressing high-value proteins through infection with the baculovirus expression vector system have been more limited, but they lead to similar conclusions.[54] When Ig light and heavy chains were expressed in insect cells using this system, a substantial fraction of these polypeptides were found in an insoluble fraction, but when heterologous PDI was co-expressed, this led to greater solubility of the chains and a higher level of secretion of assembled IgG. Pulse chase studies suggested that the Ig chains initially moved into an insoluble fraction but that they could be rescued and assembled into antibody in the presence of enhanced levels of PDI. This is a promising result but it should be remembered that baculovirus-infected insect cells are not stable; by several days after infection, when the effects of co-expression of PDI were most apparent, their internal architecture and host machinery are severely compromised and their capacity for post-translational processing has deteriorated significantly.[55]

What emerges from all this work is that a much more sophisticated and systems-based approach is required in order to produce a general increase in capacity of cultured cells to produce disulfide-bonded secretory proteins. No single factor can be identified as rate- or yield-limiting in every case and it is not helpful to think in terms of expanding capacity at a single "bottle-neck". We are led to consider a much broader approach by proteomic studies which compare related cells which differ in secretory capacity. This was first carried out convincingly in a study that followed B cells after stimulation by bacterial lipopolysaccharide (LPS) which induces them to differentiate and mature into antibody-secreting plasma cells.[56] Such cells show sequential waves of up-regulation of different, functionally related classes of proteins as they build the metabolic machinery, protein synthesis capacity, folding machinery and secretory capacity that are required in order to function as highly active factories for antibody production. Most ER folding catalysts and chaperones are up-regulated in parallel, showing a consistent day-by-day increase and an average 6-fold increase over 4 days of differentiation, whereas cytosolic and mitochondrial chaperones are induced significantly by day 1 but then show a relative decline. Cytoskeletal proteins show a relative decline throughout the period, so that by day 4 the relative abundance of ER resident proteins compared to cytoskeletal proteins has increased by 15-fold compared to day 0.

What is not yet clear are the regulatory mechanisms that underlie the coordinated increase in expression of PDI and other ER luminal folding factors during the differentiation of B lymphocytes to plasma cells. The fact that this up-regulation occurs earlier in time than the increase in production of the immunoglobulin chains on which these proteins will act strongly suggests that this regulation is not based on the well-known "unfolded protein response" (UPR, see references in 4) in which the presence of an excess of unfolded or misfolded proteins in the ER activates specific signaling pathways that act on transcription factors which modulate the levels of components of the folding and degradative pathways.

The coordination of changes in level of ER resident proteins similar to those observed in lymphocyte differentiation has also been observed by comparing different mouse NS0 cell lines which vary in the production rate for recombinant monoclonal antibodies.[57] Four homogeneous NSO cell lines differing in specific production rate were analyzed, a large number of cellular proteins were quantitated and it was found that ER resident molecular chaperones known to interact directly with immunoglobulins during production and assembly, such as PDI, BiP and endoplasmin all showed a significant increase in abundance that correlated with increased specific antibody production rate.

1.7.3 What Limits our Ability to Enhance the Usefulness of PDI in the Production of High-value Proteins?

This section highlights areas of ignorance that constrain our ability to engineer PDI itself and to engineer cells by manipulation of PDI content to make them more effective factories for the production of challenging disulfide-bonded proteins. These areas of ignorance relate to levels of organization that are difficult to probe by simple methods; the spatial and temporal organization of folding factors and other proteins that manipulate partially folded proteins in the ER, the extent to which the various members of the PDI family in a given eukaryote have distinct or overlapping activities and specificities, the complete description of the flow of redox equivalents within the ER (including all the low molecular weight oxidants and reductants, as well as all the protein redox carriers and catalysts of electron transfer), and finally the mechanism of PDI itself in terms of the complete cycle of its actions on a protein substrate.

1.7.3.1 Functional Organization of Chaperones and Folding Factors in the ER

A number of complex functions relating to protein folding and targeting take place within the ER lumen and these functions involve a large number of components, many of which may have multiple roles. Thus productive protein processing in the ER lumen includes enzymes and chaperones that constitute machinery for folding, modification (*e.g.* N-glycosylation, hydroxylation, signal sequence removal) and assembly. In addition to this "production" machinery, there is quality assurance (QA) machinery, which retains within the ER any proteins which are not fully folded and assembled, and there is quality control machinery which initiates the removal of such defective protein products for destruction (ER-associated degradation, ERAD). Any given newly synthesized protein functionally interacts with a subset of this total machinery in the interval between its co-translational insertion across the ER membrane and its exit from the ER compartment, either productively along the secretory pathway or *en route* to its degradation. All these functional components coexist together with newly synthesized proteins within an ER luminal environment that has a high total concentration of protein and hence is very crowded at the molecular level.[2,3]

What is not clear is a) the extent to which there is long-term structural organization underpinning any of these multi-component functional modules, and b) the extent to which different ER resident proteins can function in more than one of these functional modules (production/QA/ERAD). In principle, one could envisage two extreme cases. In one, the ER resident proteins are effectively immobilized into organized and spatially defined functional chains or networks to generate a matrix through which the "substrate" proteins pass encountering the relevant components of the machinery in turn. In the other, all the ER resident proteins are highly dynamic with no long-term structural interactions, such that they can each access any potential substrate quickly and efficiently. There is some evidence in support of both of these contradictory models.

1.7.3.1.1 Association of PDI with Other Luminal Proteins

Cross-linking of cells producing immunoglobulin heavy chains has shown that the unassembled and incompletely folded chains are associated with up to 10 different ER-resident proteins, all of which are known to function in protein folding, modification and assembly.[58] This set of proteins includes PDI and other members of the PDI family including ERp72 and P5 (= CaBP1). Other evidence suggests that the majority of the identified ER proteins already exist as part of a complex in the absence of newly synthesized Ig H chain, although the presence of PDI in this pre-existing complex is less certain.

Evidence in favor of a much more dynamic situation in the ER was obtained by fluorescence studies of the diffusion of GFP-probes within the ER.[59] ER-targeted GFP was freely mobile in the ER lumen of fibroblasts, with an effective diffusion coefficient of approximately $10\,\mathrm{micron^2\,s^{-1}}$ and this value was not significantly altered by changes in the functional state of the ER induced by addition of puromycin, which inhibits release of newly synthesized proteins, or castanospermine, which blocks processing of N-glycosylated groups. When GFP was attached to a soluble ER-located chaperone (calreticulin), this considerably larger protein was less mobile with an effective diffusion co-efficient of $1.3\,\mathrm{micron^2\,s^{-1}}$, but this value was increased by puromycin treatment and decreased by castano-spermine treatment, implying that the dynamics of the chaperone were altered by changes in its functional interactions. In all conditions, calreticulin-GFP diffused considerably more freely than OST, the oligosaccharyltransferase which transfers preformed oligosaccharide to asparagine residues on nascent proteins and is known to be associated firmly with the translocation machinery. It is not clear if these measurements of *in vivo* protein dynamics are inconsistent with the finding of a pre-existing complex of several other chaperones and there is not sufficient broad-based evidence on this question to resolve the issues fully.

In the case of PDI itself, we know that it can form long-term associations with other polypeptides because it is found as a permanent functional component in two ER-located proteins, the prolyl-4-hydroxylase and microsomal triglyceride transfer protein (see references 13 and 14). We also know that PDI and other chaperones such as BiP and a cyclophilin-type prolyl-peptidyl-*cis/trans*-isomerase (PPI) can be found simultaneously in association with some newly synthesized proteins, such as immunoglobulins, but we do not know whether these factors are separately bound to the "substrate" protein or if they bind as a complex. Indeed there is some evidence in yeast, that a pre-existing complex of PDI and BiP may be dissociated by the presence of newly trans-located nascent proteins.[60]

1.7.3.1.2   PDI and the Retro-translocation Machinery

There is also evidence that PDI is a member not just of the "production" machinery, but also of the "quality control" machinery which prepares mis-folded proteins for "dislocation" through the ER translocon prior to their degradation in the cytoplasm. A very careful analysis of the sequential events in the biosynthesis and degradation of BACE457, a pancreas isoform of human beta-secretase, in human embryonic kidney cells (HEK293 cells) showed that this protein folds very inefficiently and mapped the states and interactions it passed through prior to its "dislocation" from the ER.[61] Misfolded BACE457 polypeptides dissociate from calnexin and form extensively disulfide-bonded complexes with PDI and BiP before reduction and dislocation. The data suggest that PDI is involved both in attempts to fold and form native disulfides in the substrate protein, and also later during reduction of intra- and inter-molecular disulfides, to prepare misfolded BACE457 for dislocation. The fully reduced protein is associated with PDI at the final stage before it disappears from the ER and hence there may be direct transfer of PDI-bound unfolded protein to the translocation machinery for return to the cytosol both in mammalian and yeast cells.[60,61]

This ability of PDI to function in reduction and dislocation events in the ER is relevant to the mechanism of action of some protein toxins. Ricin is a plant-derived protein comprising disulfide-linked A and B chains that is highly toxic to mammalian cells. The B chain directs the holotoxin to cell-surface receptors and mediates its retrotransport along the secretory pathway to the ER lumen, from which the A chain is translocated *via* the ER translocon to the cytosol where it exerts its toxic effects on cytosolic ribosomes. In mammalian cells transfected to synthesize ricin B chain, the presence of an excess of free B chains in the ER protects these cells from ricin toxicity, implying that the excess B chains exchange with incoming holotoxin, preventing liberation of free A chains.[62] The newly synthesized B chains are found to be associated with PDI, and it was observed that *in vitro* PDI could catalyze exchange between intact toxin and free B chains. These findings imply that in normally infected cells, PDI plays the key role of reducing the interchain disulfide bond, liberating the A chain for subsequent translocation from the ER to the cytosol; this activity of PDI has been confirmed *in vitro*.[62]

Results leading to a similar conclusion were reached in work on the mechanism of cholera toxin translocation to the cytoplasm. Cholera toxin comprises an A monomer non-covalently associated with a B chain pentamer; the A monomer is cleaved after secretion to generate A1 and A2 chains linked by a disulfide bond. As for ricin, cholera toxin is retrotransported through the secretory pathway of the target cell to the ER, and the reduction of the inter-A-chain disulfide in the ER is required to liberate the toxic A1 chain which then translocates to the cytosol to deliver its toxic effect. The A1–A2 disulfide bond is relatively unstable so that it can be reduced in the relatively oxidizing redox environment in the ER and this process is catalyzed by PDI *in vitro*.[63] Furthermore, there is evidence that cholera toxin is a substrate for reduction by PDI in intact intestinal cells and that this process occurs in the ER.[64] In a recent study,[65] it was observed that in cells in which the level of PDI was down-regulated by RNAi methodology, there was a reduction in transport of A1 chain to the cytosol, implying a role for PDI in this transport in normal cells. Furthermore, it was shown that over-expression of PDI stimulated the degradation *via* ERAD of a misfolded form of thyroglobulin. Both these lines of evidence imply that PDI in the ER is involved in presenting misfolded and unfolded proteins to the retrotranslocation machinery. In the case of cholera toxin,[65] there is also evidence that PDI-mediated reduction and unfolding of the protein is a necessary preliminary to retrotranslocation to the cytosol.

Taken together, this evidence suggests that PDI can interact functionally with a large number of other elements of the ER machinery, both in the productive pathway and also in the quality control pathway. However there is no evidence for long-term structural association with other components of this machinery, except in the well-defined hetero-oligomeric proteins prolyl-4-hydroxylase and microsomal triglyceride transfer protein.

1.7.3.2 Functional Significance of the Existence of Multiple Members of the PDI Family

For nearly 20 years it has been apparent that mammalian cells contain not just a single PDI, but also other proteins closely homologous in sequence to the "classic" enzyme first characterized in the 1960s and 1970s. By 1994, mammalian genes and gene products had been identified that represented 4 different homologs (PDI, ERp72, ERp60 – now known as ERp57 – and P5; see references in ref. 15), whereas 6 different human PDI family members were identified in a 1999 review.[66] This was further expanded by the publication of complete genome sequences, so that by 2005 a total of 14 human homologs had been published and a further 3 identified from sequences in public databases.[7] The most recent review[10] has added a further two family members (Hag2 and Hag3) and discusses a family that is now thought to contain 19 distinct human genes.

1.7.3.2.1 Deductions from Protein Sequence

Of these 19, 15 appear to encode soluble proteins that contain ER-signal sequences and C-terminal sequences that are potential ER-retention motifs, although some are divergent from the classical –KDEL or –KEEL. The other 4 contain probable membrane-spanning sequences. All are thought to be located within the ER, although it is possible that some move into more distal compartments of the secretory pathway.

Of the total of 19 human proteins, 12 contain 1 or more domains homologous to thioredoxin and with a classic active site dithiol motif ..CxxC.. Within this subset of 12, 4 contain only 1 ..CxxC.. domain (ERp18, TMX, TMX3 and TMX4), a further 4 contain 2 ..CxxC.. domains (PDI, ERp57, PDIp and P5), while 3 contain 3 ..CxxC.. domains (ERp72, ERp46 and PDIr) and 1 contains 4 such domains (ERdj5). The remaining 7 family members that lack the ..CxxC.. motif include 3 which contain thioredoxin-like domains where the "active-site" sequence is of the form ..CxxS.. (Hag3, Hag2, ERp44), a further 2 in which the corresponding sequences are of the form ..SxxC.. (PDILT, TMX2) and a final 2 which lack any such motif and are defined as family members because of their overall homology to PDI b domains (ERp27, ERp28/29).

Although the cellular functions of many of these proteins have not yet been defined, and only a subset has been expressed, purified and characterized in terms of chemical and enzymological properties, their redox, catalytic and binding properties will necessarily be constrained by these features of their amino acid sequence. Thus domains with the classic ..CxxC.. motif can alternate between the dithiol state and the disulfide state and hence proteins with these domains can act directly as redox components in two-electron-transfer reactions *i.e.* they can be net oxidants and reductants as well as acting as catalysts of oxidoreductions and disulfide isomerizations. On the basis of our knowledge of the mechanism of action of thioredoxins and DsbAs, we know that steric and electronic factors influence the pK and reactivity of the more N-terminal Cys residue in the ..CxxC.. motif. Hence we can infer that in domains of PDIs with the active-site sequence ..CxxS.. the single Cys residue can act as a nucleophile to break a pre-existing disulfide in a protein or low molecular weight substrate. Such domains therefore have the potential to act as catalysts of oxidoreductions in presence of excess of another thiol, dithiol or disulfide, and also have the potential to act as catalysts of isomerizations in the absence of any additional component. Properties of active sites with the sequence ..SxxC.. are less readily predicted; such sites could form disulfide bonds with partner proteins, but it is likely that the thiol group in this active site is less reactive both as an attacking nucleophile and as a leaving group than that in an active site of the sequence ..CxxS.. Domains that lack any ..CxxC..-like motif will not be active in thiol–disulfide interchange chemistry and are likely to function only in non-covalent binding interactions.

Prediction of functional role also requires some understanding of the ligand binding properties of PDIs. It is known[67] that although other domains contribute to substrate binding, the principal binding site in PDI for peptides and unfolded proteins is provided by the **b′** domain, and that this domain is required for isomerization activity towards protein substrates.[68] Thus functions that require binding of unfolded or misfolded substrates, rather than simple oxidoreductions, will be dependent on the presence of such a domain, which is found in ERp27, ERp44, ERp57, PDI, PDIp, PDILT, ERp72 and TMX3.

On the basis of analyses of this kind plus detailed analysis of the structural determinants of active site reactivity (*e.g.* reference 69; See section 1.7.3.4), Ellgaard and Ruddock,[7] predicted potential activities of members of the human PDI family. Extending that analysis to include new information,[10] one would predict a) that PDI, PDIp, ERp57, ERp72 and TMX3, which have canonical active-sites plus a **b′**-type domain, will have activity as protein disulfide-isomerases, b) that P5, ERp46 and possibly TMX would function as effective oxidants, c) that ERdj5 would act primarily as a reductant and d) that ERp18, PDIr and TMX4 would be relatively inactive in thiol–disulfide oxidoreductions because of the lack of certain key residues essential for the catalytic mechanism.

1.7.3.2.2   Evidence for the in vivo Roles of Human PDIs

Very few of these *a priori* predictions have been tested at the level of identifying these activities in the intact cell. For a few members of the family, particularly ERp57 and ERp44, some well-defined interactions with cellular substrates have been identified. There is also limited information concerning ERp72 and ERp28/29, but little can be said about most of the other family members. In most cases it is possible to infer that one particular PDI family member is "acting on" a particular cellular substrate protein, but it is much harder to identify the nature of that action. The data have been comprehensively reviewed up to 2007 by Appenzeller-Herzog and Ellgaard.[10]

The interaction of ERp57 with the glycoprotein-specific chaperones calnexin and calreticulin is well established,[70] and has now been characterized structurally.[71–75] This interaction allows this member of the PDI family to bind to newly synthesized N-glycosylated proteins and this interaction has been clearly demonstrated in a number of cases. Appenzeller-Herzog and Ellgaard[10] list evidence that, within this complex, ERp57 can act in the oxidative folding of immunoglobulin domains of MHC class I heavy chains and their homologs, and of thyroglobulin, Semliki Forest virus glycoprotein p62 and melanocyte tyrosinase. It is also involved in the multi-component complex required for loading peptides onto MHC class I molecules for presentation at the cell surface, in which a stable disulfide bond is formed between the **a** domain active site of ERp57 and tapasin, another component of the peptide loading complex.[76] Within the complex, ERp57 and also PDI itself are involved in the reduction of a disulfide bond in one of the domains of the MHC heavy chain,[77–79] a process which is key to the selection of high-affinity peptides. A mutational trap approach to generate long-lived disulfide-bonded linkages between ERp57 and substrates, combined with proteomics, has identified a large number of other potential cellular substrates of Erp57.[80] In all, 26 different proteins were identified, many of which contained small disulfide-rich domains such as the EGF domain. Mutational trapping of this kind cannot identify proteins on the productive oxidative folding pathway, only non-productive kinetically trapped species or proteins that are being reduced by ERp57 or being isomerized by cycles of reduction and oxidation.[7] Nevertheless, this is a useful approach and generates specific candidate substrates for further study.

A few *in vivo* substrates for ERp72 have been proposed on the basis of cross-linking (see references in 10) but interpretation is complicated by the fact that ERp72 can be cross-linked into the large multi-component chaperone complex described above (Section 1.7.3.1) and hence specific substrate interactions with ERp72 cannot be proven. The **b'** domain of ERp72 resembles that of ERp57,[7] and it does not readily bind model peptides *in vitro* like PDI,[84] but there is no evidence that ERp72 can interact with calreticulin or calnexin as observed for ERp57.[85] There is some evidence that ERp72 prevents dislocation of misfolded proteins from the ER to the cytoplasm and that in this respect it has the opposite action to that of PDI.[65]

PDI, ERp57 and ERp72 are observed in many cell types, whereas PDIp has a very restricted tissue-specific expression in the pancreas and brain. Its specificity for binding peptides is very well defined (see references in ref. 6) and in an *in vitro* translation/translocation system containing dog pancreas microsomes, many nascent proteins can be readily cross-linked to the PDIp present in these microsomal vesicles.[86,87] It is assumed that this tissue-specific PDI plays a role in the biosynthesis of disulfide-bonded proteins specific to the pancreas such as digestive enzymes, but there is no direct evidence for this.

The well-defined cellular role of ERp44 is to form long-term disulfide bonds with nascent and incompletely folded proteins and hence to effect "thiol-mediated retention" of such proteins in the ER.[81–83] The proteins retained by this mechanism include proteins destined for secretion such as Ig chains including the J chain, and components of adiponectin; in both these cases, ERp44 appears to be mediating the assembly of multimeric disulfide-bonded proteins. But other proteins are retained by binding to ERp44 including Ero-1, the source of oxidizing equivalents in the ER (see Section 1.7.3.3.2) and this may be the mechanism by which Ero-1 and possibly other endogenous proteins are retained within the ER compartment.

ERp28/29 is a family member with no active site thiol or dithiol groups and has no well-defined function in mammals, but it is homologous to a *D. melanogaster* protein, Wind, which has a well-defined *in vivo* function in ensuring the export from the ER of a key Golgi-located enzyme Pipe.[88] Rat ERp29 has been purified and its structural and functional properties are beginning to be defined[89,90] but – as yet – no *in vitro* activity relevant to protein maturation in the ER has been identified. Both rat ERp29 and *D. melanogaster* Wind comprise a **b**-type PDI domain and a C-terminal all-helical domain that is termed a D domain, and both dimerize *via* interactions between **b** domains.[88–90] The structure of Wind has been determined by X-ray diffraction[88] and there is currently very active work defining potential peptide binding sites in both the **b** and D domains.[91,92]

Erp18 and ERp27 are both relatively recently defined members of the family for which no function has yet been ascribed, but it is known that ERp18 has weak isomerase activity,[93] while ERp27 has a functional binding site for unfolded proteins located primarily in its **b'** domain.[94]

1.7.3.2.3   The PDI Family in Other Species

The previous section focused on the roles of the 19 human PDI family members, but it is clear that expansion of the PDI family has taken place repeatedly through evolution. The genome of *Saccharomyces cerevisiae* contains the one *PDI1* gene that has been discussed previously (see references 39–43) which is essential for viability;[95] the gene product PDI1p is clearly a homolog of mammalian PDI in terms of the number and order of its domains although there are a number of significant differences at the sequence level. But *S. cerevisiae* also contains four other non-essential homologs, which are normally expressed to much lower levels than that of PDI1p, but whose over-expression can to some extent complement deletion or disruption of the *PDI1* gene.[96] Of these, the gene products Mpd1p and Mpd2p each contain a single putative thioredoxin domain with the active-site sequence ..CxxC.. but neither is obviously homologous in overall architecture to any member of the human PDI family. Over-expression of Mpd1p can functionally replace PDI1p and restores viability to a *pdi1*-deleted strain in the absence of any of the other members of the family. Eug1p protein has an overall domain architecture similar to PDI1p and mammalian PDI but the active-site residues in its **a**-type domains have the sequence ..CxHS.. Over-expression of Eug1p can rescue the viability of a PDI-deleted strain, but only in the presence of wild-type levels of Mpd1p and Mpd2p. Pdi1p, MPD1p, Mpd2p and Eug1p all contain the canonical yeast ER-retention signal, namely –HDEL, at the C-terminus. There is also a further homolog Eps1p; this is a 700-residue protein which contains a single **a**-type domain with the active-site sequence ..CPHC.. (identical to that in bacterial DsbA), and a C-terminal transmembrane sequence. The function of this protein is not well defined.

There has been some debate over the functions of PDI1p in yeast and the nature of the function that is essential for viability. This has been explored by exploiting recombinant PDI mutants and domains whose catalytic properties can be defined *in vitro* and testing their ability to complement yeast lacking the functional *pdi1* gene.[97–99] Mutants lacking the C-terminal Cys residue in an active-site motif are able to restore function implying that the principal role of yeast PDI is in isomerization reactions,[97] specifically the correction of non-native disulfide bonds which certainly form in yeast.[98] However, account needs to be taken of the capabilities of the other members of the family in yeast; taking account of this, it has been shown that high levels of oxidoreductase activity (possibly 50–60% of those in wild-type yeast) are essential for yeast survival and growth, whereas only 5–6% of wild-type isomerase activity is essential for growth, although higher levels of isomerase activity are essential for folding and secretion of some (non-essential) disulfide-bonded proteins.[99]

The overall evolution of the PDI family has been analyzed[100,101] and in addition to mammals and yeasts, the number of family members has been established in flowering plants.[102] A genome-wide search of the *Arabidopsis thaliana* genome identified 104 genes encoding proteins with thioredoxin-like domains which included 22 PDI-like proteins that are orthologs of known PDIs. A similar 19-member family was identified in rice (*Oryza sativa*) and a 22-member family in maize (*Zea mays*). Detailed analysis of the sequences of these PDI-like proteins allowed them to be classified into 10 broad groups of which 5 resemble the canonical mammalian PDI, with two thioredoxin-like domains, while two others resembled the simpler Mpd1p and Mpd2p proteins found in yeast.

1.7.3.3 Functional Organization of the Flow of Redox Equivalents to Newly Synthesized Proteins in the ER: Linear Electron Transfer Chain or Network?

We have inadequate information on whether there are functionally significant structural interactions between the multiple components of the protein folding machinery in the ER (Section 1.7.3.1) and also limited information on the specific cellular roles of many of the multiple members of the PDI family (Section 1.7.3.2). These gaps in knowledge present us with an acute problem when we attempt to give an account of electron transfer within the ER lumen and of the flow of oxidizing equivalents to form disulfide bonds in newly synthesized proteins.

The classical electron-transfer chain of the mitochondrial inner membrane provides an instructive contrast. In that case, a variety of valuable tools was employed to determine the sequence of electron carriers and their interactions: i) the electron carriers are mainly membrane-bound, so that the membrane can be specifically dissociated into functional modules each comprising a series of carriers corresponding to a sequential set, ii) most of the carriers have intrinsic redox-state-dependent spectroscopic signals which allow their redox states to be determined without "trapping" or other external perturbation, iii) a variety of inhibitors is available that block electron-transfer at specific steps, allowing the detection of the redox status of carriers upstream and downstream of the block and iv) a variety of artificial electron donors and acceptors is available that interact at specific and well-defined intermediate points in the electron-transfer chain.

No comparable tools are available currently in relation to redox transfer within the ER lumen. Furthermore, given the number of components involved, there are no grounds *a priori* for expecting the ER system to comprise a simple linear chain of redox carriers (in which each component interacts with a maximum of two others); rather, a network of interactions might be anticipated in which any one component may accept electrons from more than one source and/or donate them to more than one recipient. In such circumstances, genetic approaches provide some important information, but the redundancy of components and the presence of alternative pathways make it difficult to assess roles and fluxes in the wild-type cell since pathways that are insignificant in wild-type cells may become significant in the presence of mutations.

1.7.3.3.1 Sources of Reducing Power (Electron Donors) and Redox Buffering

In cells actively synthesizing and secreting disulfide-bonded proteins, a major source of reducing power in the ER lumen is the nascent chains themselves; they contain cysteine (CysSH) residues that are incorporated by the protein synthesis machinery in the cytoplasm and are translocated across the ER membrane in this form. While this has been clear for many years, it is now also apparent that there is a further major source of reducing power delivered to the ER lumen, namely reduced glutathione (GSH), which is transported across the ER membrane by a bi-directional saturable transporter that is inhibited by anion-transport blockers.[103] This system does not transport the oxidized form of glutathione (GSSG) and hence must be regarded as a specific source of reducing equivalents.

The molecular identity of the ER-located GSH-transporter has not been established; however, a bacterial glutathione transporter has recently been identified in *E. coli* (CydDC) which is a heterodimeric ATP-binding-cassette transporter similar to the cystic fibrosis transmembrane conductance regulator (CFTR) in mammalian cells.[104] This bacterial transporter was first identified as a potential transporter of cysteine, but it transports GSH several-fold more rapidly than cysteine; like the ER membrane system, CydDC does not transport GSSG. It is possible that the ER-located GSH-transporter is also a member of the ABC-transporter family homologous to CFTR. It also seems probable that there is a broad analogy between periplasm and ER lumen in that both the bacterial periplasm and the ER lumen experience a flux of reduced proteins and GSH from the cytoplasm into the more oxidizing environment of a secretory compartment.

There is no published information on the status and concentration of forms of glutathione in the periplasm. However, the nature of the glutathione pool within the ER was characterized in a classic study that attempted to determine the GSH:GSSG ratio in the secretory pathway of living cells. Hwang *et al.*[105] used a radiolabeled Cys-containing N-glycosylatable tetrapeptide as an indicator of redox status within the secretory pathway. Cultured cells were incubated with radiolabeled acetyl-NYTC-amide and were then lysed in conditions that preserved the cellular redox state. Only peptide that had been transported to the ER would have become N-glycosylated and so this glycosylated fraction was isolated and its redox status analyzed by using HPLC to quantitate the forms of the labeled glycopeptides. By this approach, Hwang *et al.*, established both that glutathione is the principal redox buffer in the secretory pathway and that the ratio of GSH:GSSG was in the range 1:1 to 3:1, a ratio many times more oxidizing than that of the cytoplasm and corresponding to a redox potential in the secretory pathway of approximately $-170 \, \text{mV}$.

This work only considered low-molecular-weight thiols and disulfides, did not measure the total glutathione content of the ER lumen and took no account of the significance of protein-SSG mixed disulfides as contributors to the pool of oxidized glutathione and to the overall redox state of the ER lumen. However, the glutathione pool in the ER is known to comprise GSH, GSSG and glutathionylated proteins (*i.e.* protein-SSG mixed disulfides). Early studies suggested the presence of high levels of glutathionylated protein within microsomal fractions from mammalian tissues[106,107] and this has been confirmed more recently in a study[108] which reported that the glutathione pool in the lumen of rat liver ER is at a total concentration in the range 6–10 mM and that the various forms are in the ratios 3 GSH:1 GSSG:5 GSS-protein. This result is consistent with – but extends – the results of Hwang *et al.*[105] and emphasizes that protein-bound glutathione must be considered an important contributor to the overall pool of glutathione in the ER, to its redox buffering capacity and to its overall oxidizing redox potential.

None of this work addressed the sources of oxidizing equivalents in the ER lumen nor the pathways by which oxidizing equivalents are delivered to establish the oxidized redox status of luminal glutathione. Hwang *et al.* speculated that there might be preferential transport of GSSG from the cytoplasm, but that work predated the characterization of the ER membrane GSH transporter discussed above. In view of the apparent preferential transport of GSH over GSSG, it has to be concluded that the oxidized thiol–disulfide redox state in the ER lumen is the consequence of an oxidative enzyme system operating in the ER.

1.7.3.3.2   Sources of Oxidizing Equivalents for Protein Disulfide Formation

It has been clear for many years that PDI (alone) cannot be this oxidative enzyme system. The basic action of PDI is as a catalyst of thiol–disulfide interchange; it is an oxidoreductase that effectively transfers redox equivalents between a disulfide species as donor and a reduced (thiol or dithiol) acceptor. Depending on the other species present and their concentrations, this oxidoreductase activity can lead to net protein oxidation (in presence of an oxidant and a reduced protein), a net protein reduction (in presence of a reductant and a protein with poorly stable disulfide bonds, such as insulin) or net protein disulfide isomerization (which may involve distinct oxidation and reduction steps but may alternatively be a direct isomerization) (Figure 1.7.1).

Ai HS-T-SH + PDI(SS) → HS-T-SS-PDI-SH → T(SS) + HS-PDI-SH

Aii E(SS) + HS-PDI-SH → HS-E-SS-PDI-SH → HS-E-SH + PDI(SS)

Aiii GSSG + HS-PDI-SH → GSH + GSS-PDI-SH → 2 GSH + PDI(SS)

Bi T (SS) + HS-PDI-SH → HS-T-SS-PDI-SH → HS-T-SH + PDI(SS)

Bii PDI(SS) + 2 GSH → GSS-PDI-SH + GSH → HS-PDI-SH + GSSG

C HS*-T(S^S) + HS-PDI-SH→
$\qquad$ HS*-T(S^H)SS-PDI-SH→
$\qquad\qquad$ HS^-T(S*S) + HS-PDI-SH

Figure 1.7.1 All PDI-catalyzed reactions are thiol–disulfide oxidoreductions.
A: Net oxidation of reduced target protein (T) as in biosynthesis of protein disulfides. Ai Oxidation of target protein dithiol by oxidized PDI; Aii re-oxidation of reduced PDI by a redox-active disulfide within an Ero or QSOX protein (E); Aiii re-oxidation of reduced PDI by glutathione disulfide.
B: Net reduction of oxidized target protein (*e.g.* as in reduction of retro-translocating toxin proteins, or prior to ERAD of mis-oxidized proteins, or in reduction step of reduction/oxidation mechanism for correction of mis-oxidized proteins). Bi Reduction of target protein disulfide by reduced PDI (reverse of step Ai); Bii reduction of oxidized PDI by glutathione (reverse of step Aiii).
C: Net disulfide isomerization of target protein; note that one Cys residue, marked as HS*, is converted from the thiol to disulfide and that another, marked as −S^, is converted from the disulfide to the thiol state; however the net redox state of the target protein is unchanged.
Note: thiols are shown as –SH throughout, although the reactive group in every case is the thiolate anion.

Although PDI is an oxidoreductase it is NOT an oxidase *i.e.* it does not operate effectively with molecular oxygen as the oxidant. This is a matter of definition rather than mechanism and the point is succinctly and authoritatively addressed elsewhere.[109] (note, however, that much of the recent literature uses the term "oxidase" incorrectly to refer either to reactions in which PDI functions as a catalytic oxidoreductase with an oxidant other than molecular oxygen, or to stoichiometric actions in which it functions as a net oxidant. PDI and members of the PDI family CAN act as stoichiometric oxidants to introduce disulfide bonds into substrate proteins, but continuous turnover would then require re-oxidation of the PDI by a distinct oxidant).

Since PDI is NOT an oxidase and cannot transfer electrons to molecular oxygen, there was much speculation in the early literature about alternative potential sources of oxidation for nascent proteins in the ER, mostly with the assumption that molecular oxygen would be the ultimate oxidant, but proposing various candidates (flavins, quinones, low-molecular-weight disulfides, *etc.*) as proximate oxidants that would either oxidize the proteins directly (with

PDI functioning purely as an isomerase) or re-oxidize PDI that had acted as immediate oxidant of the nascent protein (for references to earlier literature see references 105, 110 and 111).

These earlier speculations have been made redundant in the past ten years by the substantial and convincing bodies of work carried out on intracellular sulfhydryl oxidases, particularly the Ero family. These enzymes are dealt with authoritatively elsewhere in this volume (see Chapters 1.5 and 1.6) and so only key issues and results will be highlighted here.

Genetic and structural work on *ero1* from *S. cerevisiae* and its gene product Ero1p have established that it is an essential component of the machinery of protein disulfide formation in the yeast ER and that it can act directly as an oxidant of PDI. It is therefore proposed that the core system for protein disulfide formation in the yeast ER comprises transfer of oxidizing equivalents from molecular oxygen *via* Ero1p to PDI and thence to reduced proteins. Aspects of the biochemistry of the system can be inferred from the crystal structure of Ero1p (see Chapter 1.6) but it should be noted that many details of the mechanism have not been explored in detail because of the lack of a good *in vitro* system for the Ero1p-dependent and PDI-dependent oxidation of reduced proteins. It should also be noted that the action of Ero1p generates stoichiometric quantities of hydrogen peroxide,[112] which may itself have oxidative roles *via* other pathways which have not yet been defined.

Despite these limitations, the structure of Ero1p persuasively suggests that it functions like other flavin-dependent oxidases in delivering electrons to molecular oxygen *via* the flavin moiety; in this case, a redox-active dithiol/disulfide group on a flexible part of the Ero1p molecule acts as an electron acceptor from a protein dithiol of PDI (becoming itself reduced while oxidizing PDI to the disulfide state), and then further transfers these electrons to another dithiol/disulfide group located close to the flavin moiety, which passes the electrons on to the flavin and thence to molecular oxygen. All the steps in the pathway from nascent protein are represented here as protein-dithiol/protein-disulfide oxidoreductions until the electrons are transferred to flavin within Ero1p itself.

It has also been established genetically that the Ero1 system in yeast can effect net transfer of oxidizing equivalents to reduced glutathione[113] and a similar conclusion has been drawn from cell biological and biochemical work on semi-permeabilized Hela cells.[114] However the lack of good *in vitro* assay systems for reconstituting this pathway makes it difficult to establish at what points GSH/GSSG interacts with other components of the system. For the same reason, there is also limited information on the protein substrates that can be oxidized by Ero1p, although mutational trapping experiments implicate both Pdi1p and Mdp2 as substrates of Ero1p in *S. cerevisiae* (see Chapter 1.6). Similarly, there is no good information as to which members of the mammalian PDI family can be oxidized directly by the mammalian Eros Ero1a and Ero1b. If a PDI homolog cannot interact directly to be oxidized by an Ero protein, then it must either be oxidized by a different member of the PDI family, or conceivably by another redox component which is itself oxidized by an Ero or a PDI. Multiple system components imply multiple potential pathways.

At this point it is clear that in both yeast and mammalian cells, Ero proteins can provide oxidizing equivalents to the ER lumen ultimately leading to the formation of protein disulfide bonds, and that GSH is an important source of reducing equivalents to the ER lumen. But the full pathways involved have not been established and the complexity of possible redox pathways linking these net donors and acceptors has been emphasized by Chakravarthi *et al.*[115] and elsewhere in this volume (see Chapter 1.5).

Furthermore, Ero proteins are not the only plausible sources of oxidizing equivalents in the ER lumen. Considerable progress has been made in analyzing the action and role of a distinct family of flavin-dependent sulfhydryl oxidases, the Erv and QSOX families (see references 109 and 116), which are located in the secretory pathway and can use molecular oxygen to oxidize dithiol containing substrates including proteins. Erv proteins such as yeast Erv2p contain a redox-active FAD located close to protein disulfides in an arrangement similar to that found in Ero1p (although lack of sequence homology suggests that Erv and Ero proteins are the result of convergent evolution, see Chapter 1.6). Oxidizing equivalents from oxygen are transferred to the flavin of Erv proteins and then to two cysteines of the Erv protein that can form a disulfide which is presumed to transfer oxidizing equivalents to exogenous protein substrates. By contrast, QSOX sulfhydryl oxidases are more complex, containing a redox-active thioredoxin-like domain similar to those found in PDI, plus a linker domain and an Erv domain. This suggests that QSOX sulfhydryl oxidases operate analogously to Erv enzymes except that oxidizing equivalents are finally transferred within the QSOX to the trx domain, which is the ultimate donor of oxidizing equivalents to reduced protein substrates (see reference 117). *S. cerevisiae* and other fungi contain Erv family members but lack the more complex QSOX enzymes which are found in all metazoans.[116]

The only question as to whether Erv and QSOX sulfhydryl oxidases can play a role in the initial oxidative folding of proteins and formation of protein disulfides concerns their subcellular location. These enzymes are encoded with a secretory signal sequence but also contain a transmembrane domain with no recognizable ER-retention motif. The information on their subcellular distribution is complex and inconclusive; in addition to a putative ER location, it appears that they can be found extracellularly, at the plasma membrane and in the Golgi. In a direct approach to this question, Chakravarthi *et al.*[118] expressed the long form of human QSOX1 in semi-permeabilized HT1080 cells and concluded that the recombinant QSOX was located primarily in the Golgi compartment.

1.7.3.3.3 Chain or Network?

A number of additional points should be noted.

i) Although there are clear similarities and analogies between the situation in *S. cerevisiae* and in mammalian cells, there are also clear differences in the machinery involved in these systems *e.g.* in the number of Ero proteins, the number and nature of the PDI family members, and in the presence of QSOX proteins in mammals but not in yeast.

ii) The assumption has been made that the significant product of Ero and Erv and QSOX action is an oxidized protein disulfide within the enzyme which can transfer oxidizing equivalents to exogenous proteins. However, hydrogen peroxide is produced in stoichiometrically equal amounts[112] and the possible oxidative action of this product towards protein disulfides needs to be examined further.

iii) Members of the PDI family containing trx domains with the active-site sequence ..CxxS.. are generally considered to be incapable of acting as net oxidants since they cannot form an intramolecular disulfide and interconvert between the dithiol and disulfide state. However, given the high levels of GSSG and protein-SSG within the ER it is likely that these active sites are glutathionylated to some extent *in vivo*. In that case, the Cys-SSG group within such a site can act as a glutathionyl donor and effect a net oxidation of protein dithiol (HS-P-SH) to disulfide:

$$Donor\text{-}SSG + HS\text{-}P\text{-}SH \rightarrow Donor\text{-}SH + GSS\text{-}P\text{-}SH \rightarrow GSH + P(SS)$$

iv) The relative concentrations of different components need to be taken into account when considering possible fluxes. Glutathione is present in the ER lumen at a total concentration of the order of 10 mM and PDI is present at a concentration approaching 1 mM in cells that are highly active in oxidative protein folding. Other components are present in lower quantities. Glutathione is therefore the quantitatively most significant component of the luminal redox system and can act as a true buffer. Hence the pathways by which electrons will flow through the network will alter dependent on circumstances. When oxygen is present and reduced protein substrate is not available, net flow of oxidizing equivalents (presumably *via* PDI and possibly *via* other PDI homologs) will favor conversion of GSH to GSSG and protein-SSG; when reduced nascent proteins appear, there will then be potential for net flux of oxidizing equivalents from these species to nascent protein thiols (either directly or *via* PDI), in addition to direct flux of oxidizing equivalents from Ero proteins. We must assume that the system operates as a network rather than a linear electron-transfer chain, but at present we have few reliable parameters to describe how the network functions.

1.7.3.4   Dynamic Description of the Action of PDI on Protein Substrates

The final and most significant gap in our knowledge of PDI concerns its mechanism of action on protein substrates, which limits our ability to design "improved" PDIs directed towards certain reactions and substrate proteins.

1.7.3.4.1 Catalysis of Thiol–Disulfide Interchange

Basic aspects of the mechanism of thiol–disulfide oxidoreductions catalyzed by PDI have been inferred by analogy from studies on thioredoxin and DsbA (see *e.g.* reference 15), and these have been amplified by studies on PDI using peptides as dithiol/disulfide redox substrates[119,120] or as substrates for disulfide isomerization.[121] Across the thioredoxin superfamily, the more N-terminal Cys side-chain in the dithiol form of the conserved active site motif ..CxxC.. is ionized at around neutral pH and can act as an attacking nucleophile to form a mixed disulfide with a disulfide-bonded substrate. Involvement of the side-chain of the other Cys residue of the active site motif can displace the bound substrate, releasing it in the reduced state with the enzyme now containing an active-site disulfide. This net reductive action is the main physiological activity of thioredoxin (which is then recycled to the reduced state by NADPH and thioredoxin reductase), while the reverse process represents the major oxidative action of DsbA (which is then reoxidized by the bacterial respiratory chain *via* DsbB). PDI can catalyze both these reactions, but can also catalyze other processes in which disulfide bonds within the bound substrate are rearranged. (See Chapter 1.5 for illustrations of the basic mechanisms of thiol–disulfide oxidoreductions and see Figure 1 of references 121 and 122 for clarification of the distinction between isomerization pathways and reduction/reoxidation pathways).

Until recently, there was little clue as to what kinetic and structural features of PDI might distinguish it from other members of the thioredoxin superfamily and better equip it for isomerization activity. However, Lappi *et al.*[69] noted the presence of a highly conserved (>90%) Arg residue in a specific loop within the structure of the **a**-type domains of the PDI family, which was not rationalized by known structural or mechanistic requirements and is not present in thioredoxins or DsbAs. They showed that the charged side-chain of this residue could be either distant from or close to the active site (moving from a distance of >2.0 nm to <0.6 nm) and presented evidence that this movement strongly perturbed the electronic properties and pK values of the active site Cys residues. As a result they presented a plausible hypothesis for how the more C-terminal Cys in the active site could be made more reactive as a result of this movement of the Arg side-chain and how this could shift the balance between favoring isomerization processes within the enzyme-bound substrate and favoring release of the substrate from the enzyme (see reference 69, Figure 8).

Other studies with model substrates have shown up differences in kinetic properties between PDI domains and glutaredoxin, another member of the thioredoxin superfamily which specifically catalyzes thiol–disulfide interchanges which involve the transfer of a glutathione (GS-) group.[123] There has been limited work of this kind, but such studies do suggest that further understanding is possible of how the mechanism of thiol–disulfide interchange differs subtly between different members of the large thioredoxin superfamily and, in particular, of what are the mechanistically distinctive features of protein disulfide isomerases.

1.7.3.4.2   *Folding, Unfolding and Chaperone Activities*

However, what has been obvious for several years is that PDIs are distinctive in being able to catalyze disulfide-isomerizations within large and highly structured protein substrates.[18–20] This activity is not shown by individual PDI **a** domains, or by thioredoxins or DsbA;[68] the multidomain structure of PDI (illustrated in Chapter 1.5) is essential for this activity because, while the **a** and **a'** domains are capable of catalyzing simple thiol–disulfide interchange reactions, all the domains contribute to the non-covalent binding of large protein substrates with the **b'** domain comprising the principal binding site.[67,124] It is the synergy between the functions performed by the various domains which allows PDI to bind complex protein substrates and to perform disulfide isomerization reactions that require considerable conformational changes in the protein substrate.[6]

The nature of this action of PDI is well illustrated by a study on the oxidative folding of lysozyme from the reduced unfolded state, followed by a wide range of physical methods.[125] During this process, some spectroscopic properties change with a time constant of 200 s, but the acquisition of native far UV CD and NMR signals occurs with a time constant of approximately 1000 s. Some enzyme activity is regained with a similar time constant (1000 s) but the regain of enzyme activity is biphasic and some occurs at a much slower rate (time constant of 15 000 s) implying that some oxidative refolding is *via* intermediates that are inactive despite having near-native structure. The major species present during this late phase is des-[74-94]-lysozyme, which shows extensive native folded structure but is nevertheless inactive and lacks the 76-94 disulfide bond which is deeply buried in the hydrophobic core of native lysozyme. In presence of a conventional GSH/GSSG redox buffer, PDI very effectively catalyzes the conversion of this intermediate to active fully-oxidized lysozyme. As an alternative to PDI, this conversion can be facilitated by quite a high concentration of urea (up to 4M)! This demonstrates directly that the des-[74-94]-lysozyme needs to be unfolded significantly in order for the formation of the missing disulfide bond to take place, and confirms that the action of PDI is both the chemical formation of the disulfide bond and the promotion of local unfolding in the substrate protein.

This work, and the classic work on BPTI reviewed earlier, provides us with a clear sense of how PDI acts. 1) It is able to bind to highly structured and extensively disulfide-bonded proteins which are not absolutely native in that they lack one or more of the native disulfides; we do not know what the precise structural feature is that PDI recognizes in such proteins, but they may contain limited local regions of non-native tertiary structure, or (more likely?) they may contain regions which are more dynamic and show greater flexibility than do native proteins. 2) By interacting initially with such regions, PDI is able to bind such proteins and facilitate more extensive unfolding which then allows access of the PDI active site (and possibly low molecular weight reagents) to what were previously buried cysteine side-chains (in either the thiol or disulfide state). 3) In this complex, thiol:disulfide interchanges can occur to put in place the fully correct set of disulfide bonds so that the protein can fold to its complete native structure and finally dissociate from PDI.

This is a fine and plausible picture but it is one painted with a very broad brush. We lack the atomic-level structural information on PDI–substrate complexes that would enable us to paint it more precisely. In particular, this picture of the action of PDI suggests that its ability to drive unfolding of extensively folded substrates must involve significant movements within the PDI–substrate complex. To characterize these will require structural and dynamic studies not simply on PDI itself but also on complexes between PDI and substrates that mimic intermediates in the PDI-facilitated unfolding/folding process.

The fact that non-covalent interactions – and the ability to unfold partially folded proteins – are key components of the normal action of PDI as a protein folding catalyst provides important insight to those situations where the action of PDI is described as that of a chaperone rather than an enzyme. Wang and colleagues[126,127] demonstrated that PDI can act as a classic chaperone by studying its action *in vitro* on non-physiological substrates that lack disulfide bonds and are found in cellular compartments other than the ER (*e.g.* rhodanese and glyceraldehyde-3-phosphate dehydrogenase). They demonstrated that the dithiol/disulfide active sites of PDI were not required for this "chaperone" activity[128] and this opened the way to distinguish between "enzymic" and "chaperone" activities of PDI on physiological substrates whose refolding does involve disulfide-bond formation. Such studies indicate that PDI assists the refolding of proinsulin[129] and phospholipase A2[130] both as a catalyst (in sub-stoichiometric quantities) and as a chaperone (when present in stoichiometric quantities).

1.7.3.4.3 Lessons from PDI Structures

Mammalian PDIs have been abundantly available as homogeneous proteins since the early 1980s, but no good crystals that diffract to high resolution have been reported. The most obvious explanation is that the molecule has some intrinsic flexibility or conformational heterogeneity that prevents the formation of a well-ordered crystal lattice. As a result, the high-resolution structural information available on mammalian PDIs derives from structural studies on individual domains, rather than full-length molecules, and the majority of these studies have been by multidimensional NMR. Following the pioneering studies of Creighton and Darby on the **a** and **b** domains of human PDI,[131,132] it was confirmed that the **a** domain had the classic thioredoxin structure and that the **b** domain, surprisingly, also had a thioredoxin-type fold. As a result, bearing in mind the internal sequence homologies, it was inferred that PDI comprised 4 thioredoxin-type structural domains, **a-b-b′-a′**. The high-resolution structures of several such domains have now been solved (see reference 10, Table 1, for pdb codes).

An X-ray structure has now been obtained for the combination of **b-b′** domains of human Erp57, which shows the two thioredoxin-fold domains linked by a short- 3-residue linker and with an extensive inter-domain interface (reference 74; pdb 2h81). It is very similar in overall structure to two of the domains of calsequestrin, the abundant calcium-binding protein found in the sarcoplasmic reticulum of muscle cells.[133] Calsequestrin shows limited sequence homology to PDIs but its structure (pdb 1A8Y) shows that it comprises three thioredoxin-fold domains that resemble the **b** and **b′** domains of PDI in lacking the dithiol active site motif. Although calsequestrin has not been regarded as a member of the PDI family, it is clear that in structural terms it should be considered in that light.

In the case of yeast PDI (PDI1p), a high-resolution structure of the full-length protein is available[134] and this has allowed interesting comparisons to be made at a structural level between eukaryotic PDIs and their bacterial homologs.[8,135] The structure (pdb: 2B5E) confirms the presence of four thioredoxin-fold domains and shows that they are organized in an annular or horseshoe shape. It also demonstrates features previously inferred from studies on mammalian PDIs, namely i) that the active site motifs in **a** and **a′** domains are close to each other in space[111] despite being at opposite ends of the polypeptide sequence, ii) that there is a distinct linker region (x) between the **b′** and **a′** domains,[136,137] which is in an extended conformation in the yeast PDI1p structure, and iii) that there is a hydrophobic site on the **b′** domain which could function as a protein substrate binding site and forms the core of an extended binding site with contributions from adjacent domains.[6,67,137]

In the yeast PDI1p structure, this extended binding site is seen as a well-defined cleft and a remarkable feature of the crystal structure is that part of a symmetry-related PDI molecule occupies the binding cleft making extensive interactions between the two PDI molecules. The angle between the **b** and **b′** domains in the yeast PDI1p structure differs by 15 degrees from that observed in the mammalian ERp57 **bb′** structure and it could be thought that the insertion of a symmetry-related molecule into the binding cleft of the yeast PDI1p reflects how a protein substrate might bind, forcing apart the **b** and **b′** domains. It is at least a reasonable working hypothesis at this point that the yeast structure defines a conformation in which PDI has a protein substrate bound in its binding cleft. This interaction may have diminished the internal motion and flexibility that has prevented crystallization of PDIs from other species, allowing the formation of diffracting crystals in this case.

The overall "horseshoe" shape observed for yeast PDI1p is almost certainly a general feature of PDIs with a similar domain composition. Small angle X-ray diffraction studies have been carried out on full-length human PDI[138] and human ERp57;[74] the models of PDI generated by this lower resolution method show an annular arrangement of domains and a very good fit with the yeast PDI1p structure. At low resolution it is not possible to observe details of asymmetry, but the X-ray structure shows that the yeast PDI1p "horseshoe" is clearly asymmetric with the **a** and **a′** domains having different relationships with the hydrophobic substrate-binding cleft. Indeed, there is evidence that the active sites of the yeast **a** and **a′** domains have different standard redox potentials[139] and potentially different roles. Studies with a reconstituted electron-transfer system comprising Ero1p, PDI1p and reduced ribonuclease support the hypothesis that the two domains have different functions[140] with the **a'** domain acting as an oxidant, transferring a disulfide bond into a bound protein substrate, while the **a** domain is in the reduced state capable of supporting isomerase activity.

This is an interesting hypothesis, but there is no current evidence supporting this model for mammalian PDI. The **a** and **a'** domains of human PDI appear to have very similar standard redox potentials[141] and there appears to be a good structural rationalization for this: in the **a** and **a'** domains of human PDI and in the **a** domain of yeast PDI1p there is a buried ion-pair involving a glutamate side-chain that is very close to the dithiol/disulfide active site; this residue is known to be critical in determining the redox properties of the dithiol/disulfide couple. This ion-pair is absent from the **a'** domain of yeast PDI1p where the glutamate is replaced by a leucine. In this respect the **a'** domain of yeast PDI1p is highly unusual and quite different from the **a**-type domains of the majority of PDIs.

The yeast PDI1p structure is a very important advance and a crucial step towards understanding how PDI operates, but it is a static representation of one functional state. Structural and dynamic studies of several states in the PDI functional cycle, including those of PDI–substrate complexes, will be required to give us a full understanding of how PDI operates.

Acknowledgements

I thank the BBSRC for support and I thank Nader Amin, Neil Bulleid, Peter Klappa, David James, Lloyd Ruddock, Ateesh Sidhu, Katrine Wallis, Chih-chen Wang and Richard Williamson for valuable discussions over several years.

References

1. B. Wilkinson and H. F. Gilbert, *Biochim. Biophys. Acta*, 2004, **1699**, 35–44.
2. R. Sitia and I. Braakman, *Nature*, 2003, **426**, 891–894.
3. E. van Anken and I. Braakman, *Crit. Revs. Biochem. Mol. Biol.*, 2005, **40**, 191–228.
4. A. Goerlach, P. Klappa and T. Kietzmann, *Antiox. Redox Signal.*, 2006, **8**, 1391–1418.
5. C. Turano, S. Coppari, F. Altieri and A. Ferraro, *J. Cell. Physiol.*, 2002, **193**, 154–163.
6. R. B. Freedman, P. Klappa and L. W. Ruddock, *EMBO Rep.*, 2002, **3**, 136–140.
7. L. Ellgaard and L. W. Ruddock, *EMBO Rep.*, 2005, **6**, 28–32.
8. C. W. Gruber, M. Čemažar, B. Heras, J. L. Martin and D. J. Craik, *TIBS*, 2006, **31**, 455–463.
9. F. Hatahet and L. W. Ruddock, *FEBS J.*, 2007, **274**, 5223–5234.
10. C. Appenzeller-Herzog and L. Ellgaard, *Biochim. Biophys. Acta*, 2008, **1783**, 535–548.

11. R. B. Freedman and H. C. Hawkins, *Biochem. Soc. Trans.*, 1977, **5**, 348–357.

12. R. B. Freedman, *TIBS*, 1984, **9**, 438–441.

13. R. B. Freedman, *Cell*, 1989, **57**, 1069–1072.

14. R. B. Freedman, in *Protein Folding*, T. E. Creighton, ed., W. H. Freeman and Co., 1992, pp. 455–539.

15. R. B. Freedman, T. R. Hirst and M. F. Tuite, *TIBS*, 1994, **19**, 331–336.

16. D. P. Goldenberg, *TIBS*, 1992, **17**, 257–261.

17. T. E. Creighton, D. A. Hillson and R. B. Freedman, *J. Mol. Biol.*, 1980, **142**, 43–62.

18. J. S. Weissmann and P. S. Kim, *Nature*, 1993, **365**, 185–188.

19. N. J. Darby, P. E. Morin, G. Talbo and T. E. Creighton, *J. Mol. Biol.*, 1995, **249**, 463–477.

20. T. E. Creighton, N. J. Darby and J. Kemmink, *FASEB J.*, 1996, **10**, 110–118.

21. A. Zapun, T. E. Creighton, P. J. E. Rowling and R. B. Freedman, *Proteins*, 1992, **14**, 10–15.

22. T. E. Creighton, C. J. Bagley, L. Cooper, N. J. Darby, R. B. Freedman, J. Kemmink and A. Sheikh, *J. Mol. Biol.*, 1993, **232**, 1176–1196.

23. H. Lilie, S. McLaughlin, R. B. Freedman and J. Buchner, *J. Biol. Chem.*, 1994, **269**, 14290–14296.

24. M. Mayer, U. Kies, R. Kammermeier and J. Buchner, *J. Biol. Chem.*, 2000, **275**, 29421–29425.

25. M. Hoffmann, C. Nemetz, K. Madin and B. Buchberger, *Biotechnol. Annu. Rev.*, 2004, **10**, 1–30.

26. D.-M. Kim and J. R. Swartz, *Biotechnol. Bioeng.*, 2004, **85**, 122–129.

27. J. R. Swartz, *J. Ind. Microbiol. Biotechnol.*, 2006, **33**, 476–485.

28. L. Ryabova, D. Deplancq, A. S. Spirin and A. Plückthun, *Nat. Biotechnol.*, 1997, **15**, 79–84.

29. S. Frey, M. Haslbeck, O. Hainzl and J. Buchner, *Biol. Chem.*, 2008, **389**, 37–45.

30. F. Baneyx and M. Mujacic, *Nat. Biotechnol.*, 2004, **22**, 1399–1408.

31. F. R. Schmidt, *Appl. Microbiol. Biotechnol.*, 2004, **65**, 363–372.

32. K. Graumann and A. Premstaller, *Biotechnol. J.*, 2006, **1**, 164–186.

33. A. Zapun, D. Missiakis, S. Raina and T. E. Creighton, *Biochemistry*, 1995, **34**, 5075–5089.

34. A. Rietsch. D. Belin, N. Martin and J. Beckwith, *Proc. Nat. Acad. Sci. USA*, 1996, **93**, 13048–13053.

35. D. P. Humphreys, N. Weir, A. Mountain and P. A. Lund, *J. Biol. Chem.*, 1995, **270**, 28210–28215.

36. M. Ostermeier, K. de Sutter and G. Georgiou, *J. Biol. Chem.*, 1996, **271**, 10616–10622.

37. J. Qiu, J. R. Swartz and G. Georgiou, *Appl. Env. Microbiol.*, 1998, **64**, 4891–4896.

38. L. C. Simmons, D. Reilly, L. Klimowski, T. S. Raju, G. Meng, P. Sims, K. Hong, R. L. Shields, L. A. Damico, P. Rancatore and D. G. Yansura, *J. Immunol. Meth.*, 2002, **263**, 133–147.

39. L. D. Schultz, H. Z. Markus, K. J. Hofmann, D. L. Montgomery, C. T. Dunwiddie, P. J. Kniskern, R. B. Freedman, R. W. Ellis and M. F. Tutie, *Ann. NY. Acad. Sci*, 1994, **721**, 148–157.

40. A. S. Robinson, V. Hines and K. D. Wittrup, *Biotechnology*, 1994, **12**, 381–384.

41. M. F. Tuite and R. B. Freedman, *TIBTech*, 1994, **12**, 432–434.

42. K. D. Wittrup, *Curr. Opin. Biotechnol.*, 1995, **6**, 203–208.

43. T. Hayano, M. Hirose and M. Kikuchi, *FEBS Lett.*, 1995, **377**, 505–511.

44. E. V. Shusta, R. T. Raines, A. Pluckthun and K. D. Wittrup, *Nat. Biotechnol.*, 1998, **16**, 773–777.

45. W.-G. Bao and H. Fukuhara, *Gene*, 2001, **272**, 103–110.

46. D. Sleep, G. Shuttleworth and C. J. A. Finnis, Gene expression technique, 2005, WO2005/061718.

47. C. J. A. Finnis, D. Sleep and G. Shuttleworth, Gene expression technique, 2006, WO2006/067511.

48. D. M. Dinnis and D. C. James, *Biotechnol. Bioeng.*, 2005, **91**, 180–189.

49. T. A. Bibila and M. C. Flickinger, *Biotechnol. Bioeng.*, 1992, **39**, 251–261.

50. T. A. Bibila and M. C. Flickinger, *Biotechnol. Bioeng.*, 1992, **39**, 262–272.

51. K. Kitchin and M. C. Flickinger, *Biotechnol. Prog.*, 1995, **11**, 565–574.

52. R. Davis, K. Schooley, B. Rasmussen, J. Thomas and P. Reddy, *Biotechnol. Prog.*, 2000, **16**, 736–743.

53. N. Borth, D. Mattanovich, R. Kunert and H. Katinger, *Biotechnol. Prog.*, 2005, **21**, 106–111.

54. T.-A. Hsu, S. Watson, J. J. Eiden and M. J. Betenbaugh, *Prot. Express. Purif.*, 1996, **7**, 281–288.

55. E. Ailor and M. J. Betenbaugh, *Curr. Opin. Biotechnol.*, 1999, **10**, 142–145.

56. E. van Anken, E. P. Romijn, C. Maggioni, A. Mezghrani, R. Sitia, I. Braakman and A. J. Heck, *Immunity*, 2003, **18**, 243–253.

57. C. M. Smales, D. M. Dinnis, S. H. Stansfield, D. Alete, E. A. Sage, J. R. Birch, A. J. Racher, C. T. Marshall and D. C. James, *Biotechnol. Bioeng.*, 2004, **88**, 474–488.

58. L. Meunier, Y.-K. Usherwood, K. T. Chung and L. M. Hendershot, *Mol. Biol. Cell*, 2002, **13**, 4456–4469.

59. E. L. Snapp, A. Sharma, J. Lippincott-Schwartz and R. S. Hegde, *Proc. Natl. Acad. Sci. USA*, 2006, **103**, 6536–6541.

60. P. Gillece, J. M. Luz, W. J. Lennarz, F. J. de la Cruz and K. Romisch, *J. Cell Biol.*, 1999, **147**, 1443–1456.

61. M. Molinari, C. Galli, V. Piccaluga, M. Pieren and P. Paganetti, *J. Cell Biol.*, 2002, **158**, 247–257.

62. R. A. Spooner, P. D. Watson, C. J. Marsden, D. C. Smith, K. A. H. Moore, J. P. Cook, J. M. Lord and L. M. Roberts, *Biochem. J.*, 2004, **383**, 285–293.

63. I. Majoul, D. Ferrari and H. D. Soling, *FEBS Lett.*, 1997, **401**, 104–108.

64. P. A. Orlandi, *J. Biol. Chem.*, 1997, **272**, 4591–4599.

65. M. L. Forster, K. Sivick, Y.-N. Park, P. Arvan, W. I. Lencer and B. Tsai, *J. Cell Biol.*, 2006, **173**, 853–859.

66. D. M. Ferrari and H.-D. Soling, *Biochem. J.*, 1999, **339**, 1–10.
67. P. Klappa, L. W. Ruddock, N. J. Darby and R. B. Freedman, *EMBO J.*, 1998, **17**, 927–935.
68. N. J. Darby, E. Penka and R. Vincentelli, *J. Mol. Biol.*, 1998, **276**, 239–247.
69. A. K. Lappi, M. F. Lensink, A. I. Alanen, K. E. H. Salo, M. Lobell, A. H. Juffer and L. W. Ruddock, *J. Mol. Biol.*, 2004, **335**, 283–295.
70. L. Ellgaard and E. M. Frickel, *Cell Biochem. Biophys.*, 2003, **39**, 223–247.
71. M. R. Leach, M. F. Cohen-Doyle, D. Y. Thomas and D. B. Williams, *J. Biol. Chem.*, 2002, **277**, 29686–29697.
72. E. M. Frickel, R. Riek, I. Jelesarov, A. Helenius and K. Wuthrich, *Proc. Natl. Acad. Sci. USA*, 2002, **99**, 1954–1959.
73. S. Pollock, G. Kozlov, M.-F. Pelletier, J. F. Trempe, J. Jansen, D. Sitnikov, J. J. M. Bergeron, K. Gehring, I. Ekiel and D. Y. Thomas, *EMBO J.*, 2004, **23**, 1020–1029.
74. G. Kozlov, P. Maattanen, J. D. Schrag, S. Pollock, M. Cygler, B. Nagar, D. Y. Thomas and K. Gehring, *Structure*, 2006, **14**, 1331–1339.
75. L. W. Ruddock, *Structure*, 2006, **14**, 1209–1210.
76. D. R. Peaper, P. A. Wearsch and P. Creswell, *EMBO J.*, 2005, **24**, 3613–3623.
77. B. Park, S. Lee, E. Kim, K. Cho, S. R. Riddell, S. Cho and K. Ahn, *Cell*, 2006, **127**, 369–382.
78. A. Kienast, M. Preuss, M. Winkler and T. P. Dick, *Nat. Immunol.*, 2007, **8**, 864–872.
79. A. N. Antoniou, S. Ford, M. Alphey, A. Osborne, T. Elliott and S. J. Powis, *EMBO J.*, 2002, **21**, 2655–2663.
80. C. E. Jessop, S. Chakravarthi, N. Garbi, G. J. Hammerling, S. Lovell and N. J. Bulleid, *EMBO J.*, 2007, **26**, 28–40.
81. B. Kramer, D. M. Ferrari, P. Klappa, N. Pohlmann and H.-D. Soling, *Biochem. J.*, 2001, **357**, 83–95.
82. T. Solda, N. Garbi, G. J. Hammerling and M. Molinari, *J. Biol. Chem.*, 2005, **281**, 6219–6226.
83. J. Volkmer, S. Guth, W. Nastainczyk, P. Knippel, P. Klappa, V. Gnau and R. Zimmermann, *FEBS Lett.*, 1997, **406**, 291–295.
84. J. G. Elliott, J. D. Oliver, J. Volkmer, R. Zimmermann and S. High, *Eur. J. Biochem.*, 1998, **252**, 372–377.
85. T. Anelli, M. Alessio, A. Mezghrani, T. Simmen, F. Talamo, A. Bachi and R. Sitia, *EMBO J.*, 2002, **21**, 836–844.
86. T. Anelli, M. Alessio, A. Bachi, L. Bergamelli, G. Bertoli, S. Camerini, A. Mezghrani, E. Ruffato, T. Simmen and R. Sitia, *EMBO J.*, 2003, **22**, 5015–5022.
87. Z. V. Wang, T. D. Schraw, J. Y. Kim, T. Khan, M. W. Rajala, A. Follenzi and P. E. Scherer, *Mol. Cell. Biol.*, 2007, **27**, 3716–3731.
88. Q. Ma, C. Guo, K. Barnewitz, G. Sheldrick, H.-D. Soling, I. Uson and D. M. Ferrari, *J. Biol. Chem.*, 2003, **278**, 44600–44607.

89. E. Liepinsh, M. Baryshev, A. Shapiro, M. Ingelman-Sundberg, G. Otting and S. Mkrtchian, *Structure*, 2001, **9**, 457–471.
90. M. J. Hubbard, J. E. Mangum and N. J. McHugh, *Biochem. J.*, 2004, **383**, 589–597.
91. K. Barnewitz, C. Guo, M. Sevvana, Q. Ma, G. Sheldrick, H.-D. Soling and D. M. Ferrari, *J. Biol. Chem.*, 2004, **279**, 39829–39837.
92. U. Lippert, D. Diao, N. N. Barak and D. M. Ferrari, *J. Biol. Chem.*, 2007, **282**, 11213–11220.
93. H. I. Alanen, R. A. Williamson, M. J. Howard, A. K. Lappi, H. P. Jantti, S. M. Rautio, S. Kellokumpu and L. W. Ruddock, *J. Biol. Chem.*, 2003, **278**, 28912–28920.
94. H. I. Alanen, R. A. Williamson, M. J. Howard, F. S. Hatahet, K. E. H. Salo, A. Kauppila, S. Kellokumpu and L. W. Ruddock, *J. Biol. Chem.*, 2006, **281**, 33727–33738.
95. R. Farquhar, N. Honey, S. J. Murant, P. Bossier, L. Shculz, D. Montgomery, R. W. Ellis, R. B. Freedman and M. F. Tuite, *Gene*, 1991, **108**, 81–89.
96. P. Norgaard, V. Westphal, C. Tachibana, L. Also, B. Holst and J. R. Winther, *J. Cell Biol.*, 2001, **152**, 553–562.
97. M. C. A. Laboisierre, S. L. Sturley and R. T. Raines, *J. Biol. Chem.*, 1995, **270**, 28006–28009.
98. P. T. Chivers, M. C. A. Laboissiere and R. T. Raines, *EMBO J.*, 1996, **15**, 2659–2667.
99. R. Xiao, B. Wilkinson, A. Solovyov, J. R. Winther, A. Holmgren, J. Lundstrom-Ljung and H. F. Gilbert, *J. Biol. Chem.*, 2004, **279**, 49780–49786.
100. S. Kanai, H. Toh, T. Hayano and M. Kikuchi, *J. Mol. Evol.*, 1998, **47**, 200–210.
101. A. G. McArthur, L. A. Knodler, J. D. Silberman, B. J. Davids, F. D. Gillin and M. L. Sogin, *Mol. Biol. Evol.*, 2001, **18**, 1455–1463.
102. N. L. Houston, C. Fan, Q.-Y. Xiang, J.-M. Schulze, R. Jung and R. S. Boston, *Plant Physiol.*, 2005, **137**, 762–778.
103. G. Banhegyi, L. Lusini, F. Puskas, R. Rossi, R. Fulceri, L. Braun, V. Mile, P. di Simplicio, J. Mandl and A. Benedetti, *J. Biol. Chem.*, 1999, **274**, 12213–12216.
104. M. S. Pittman, H. C. Robinson and R. K. Poole, *J. Biol. Chem.*, 2005, **280**, 32254–32261.
105. C. Hwang, A. J. Sinskey and H. F. Lodish, *Science*, 1992, **257**, 1496–1502.
106. K. R. Harrap, R. C. Jackson, P. G. Riches, C. A. Smith and B. T. Hill, *Biochim. Biophys. Acta*, 1973, **310**, 104–110.
107. J. Isaacs and F. Binkley, *Biochim. Biophys. Acta*, 1977, **497**, 192–204.
108. R. Bass, L. W. Ruddock, P. Klappa and R. B. Freedman, *J. Biol. Chem.*, 2004, **279**, 5257–5262.
109. C. Thorpe and D. L. Coppock, *J. Biol. Chem.*, 2007, **282**, 13929–13933.
110. R. B. Freedman, in *Glutathione: Metabolism and Physiological Functions*, J. Vina, ed., CRC Press, 1990, pp. 125–134.

111. H. C. Hawkins, M. de Nardi and R. B. Freedman, *Biochem. J.*, 1991, **275**, 341–348.
112. E. Gross, C. S. Sevier, N. Heldman, E. Vitu, M. Bentzur, C. A. Kaiser, C. Thorpe and D. Fass, *Proc. Nat. Acad. Sci. USA*, 2006, **103**, 299–304.
113. J. W. Cuozzo and C. A. Kaiser, *Nature. Cell. Biol.*, 1999, **1**, 130–135.
114. S. N. Molteni, A. Fassio, M. R. Ciriolo, G. Filomeni, E. Pasqualetto, C. Fagioli and R. Sitia, *J. Biol. Chem.*, 2004, **279**, 32667–32673.
115. S. Chakravarthi, C. E. Jessop and N. J. Bulleid, *EMBO Reports*, 2006, **7**, 271–275.
116. D. L. Coppock and C. Thorpe, *Antioxid. Redox. Signal.*, 2006, **8**, 300–311.
117. S. Raje and C. Thorpe, *Biochemistry*, 2003, **42**, 4560–4568.
118. S. Chakravarthi, C. E. Jessop, M. Willer, C. J. Stirling and N. J. Bulleid, *Biochem. J.*, 2007, **404**, 403–411.
119. N. J. Darby, R. B. Freedman and T. E. Creighton, *Biochemistry*, 1994, **33**, 7937–7947.
120. L. W. Ruddock, T. R. Hirst and R. B. Freedman, *Biochem. J.*, 1996, **315**, 1001–1005.
121. E. A. Kersteen, S. R. Barrows and R. T. Raines, *Biochemistry*, 2005, **44**, 12168–12178.
122. M. Schwaller, B. Wilkinson and H. F. Gilbert, *J. Biol. Chem.*, 2003, **278**, 7154–7159.
123. M. J. Peltoniemi, A.-R. Karala, J. K. Jurvansuu, V. L. Kinnula and L. W. Ruddock, *J. Biol. Chem.*, 2006, **281**, 33107–33114.
124. P. Klappa, H. C. Hawkins and R. B. Freedman, *Eur. J. Biochem.*, 1997, **248**, 37–42.
125. B. Van den Berg, E. W. Chung, C. V. Robinson, P. L. Mateo and C. M. Dobson, *EMBO J.*, 1999, **18**, 4794–4803.
126. H. Cai, C.-C. Wang and C. L. Tsou, *J. Biol. Chem.*, 1994, **269**, 24550–24552.
127. J.-L. Song and C.-C. Wang, *Eur. J. Biochem.*, 1995, **231**, 312–316.
128. H. Quan, G. Fan and C.-C. Wang, *J. Biol. Chem.*, 1995, **270**, 17078–17080.
129. J. Winter, P. Klappa, R. B. Freedman, H. Lilie and R. Rudolph, *J. Biol. Chem.*, 2002, **277**, 310–317.
130. Y. Yao, Y.-C. Zhou and C.-C. Wang, *EMBO J.*, 1997, **16**, 651–658.
131. J. Kemmink, N. J. Darby, K. Dijkstra, M. Nilges and T. E. Creighton, *Biochemistry*, 1996, **35**, 7684–7691.
132. J. Kemmink, N. J. Darby, K. Dijkstra, M. Nilges and T. E. Creighton, *Curr. Biol.*, 1997, **7**, 239–245.
133. S. Wang, W. R. Trubble, H. Liao, C. R. Wesson, A. K. Dunker and C.-H. Kang, *Nat. Struct. Biol.*, 1998, **5**, 476–493.
134. G. Tian, S. Xiang, R. Noiva, W. J. Lennarz and H. Schindelin, *Cell*, 2006, **124**, 61–73.
135. B. Heras, M. Kurz, S. R. Shouldice and J. L. Martin, *Curr. Opin. Struct. Biol.*, 2007, **17**, 1–8.

136. R. B. Freedman, P. J. Gane, H. C. Hawkins, R. Hlodan, S. H. McLaughlin and J. W. Parry, *Biol. Chem.*, 1998, **379**, 321–328.
137. A. Pirneskoski, P. Klappa, M. Lobell, R. A. Williamson, L. J. Byrne, H. I. Alanen, K. E. Salo, K. I. Kikvirikko, R. B. Freedman and L. W. Ruddock, *J. Biol. Chem.*, 2004, **279**, 10374–10381.
138. S.-J. Li, X.-G. Hong, Y.-Y. Shi, H. Li and C.-C. Wang, *J. Biol. Chem.*, 2006, **281**, 6581–6588.
139. B. Wilkinson, R. Xiao and H. F. Gilbert, *J. Biol. Chem.*, 2005, **280**, 11483–11487.
140. M. S. Kulp, E.-M. Frickel, L. Ellgaard and J. S. Weissman, *J. Biol. Chem.*, 2006, **281**, 876–884.
141. N. J. Darby and T. E. Creighton, *Biochemistry*, 1995, **34**, 16770–16780.

Cellular Responses to Oxidative Stress

MARIANNE ILBERT, CAROLINE KUMSTA
AND URSULA JAKOB

Department of Molecular, Cellular and Developmental Biology, University of Michigan, Ann Arbor, MI, USA

1.8.1 Oxidative Stress: An Imbalance in Favor of Pro-oxidants

1.8.1.1 Reactive Oxygen Species

Oxygen, one of the most abundant chemical elements in the universe, presents an interesting paradox for all living organisms: while the vast majority of complex life requires oxygen for its existence, oxygen is also a highly reactive molecule that can damage living organisms by producing reactive oxygen species (ROS).[1,2] These dangerous by-products of oxygen are formed during normal metabolism, mainly due to incomplete electron transfer in the respiratory chain. ROS are generally very small highly reactive molecules that attack a wide variety of macromolecules. They include hydrogen peroxide (H_2O_2), hypochlorous acid (HOCl) and free radicals such as hydroxyl radicals ($^{\bullet}OH$) and superoxide anions (O_2^-).[3] To counteract these ROS, all aerobically growing organisms constitutively express proteins and synthesize small molecules that have antioxidant properties.[4] The glutaredoxin and thioredoxin systems are the primary redox balancing protein systems in bacteria that remove unwanted

RSC Biomolecular Sciences
Oxidative Folding of Peptides and Proteins
Edited by Johannes Buchner and Luis Moroder
Published by the Royal Society of Chemistry, www.rsc.org

oxidative thiol modifications in cytosolic proteins.[5-7] A redox buffer consisting of the reduced (*i.e.*, GSH) and oxidized (*i.e.*, GSSG) forms of the cysteine-containing tripeptide glutathione maintains an overall reducing environment in the cytosol, with a redox potential of $-260\,mV$ in many pro- and eukaryotes. In addition, cells possess multiple constitutively expressed antioxidant proteins that actively remove ROS. For instance, superoxide anions are destroyed by superoxide dismutases (SOD), which convert superoxide anions to hydrogen peroxide. Hydrogen peroxide, in turn, is broken down into non-reactive alcohols or water by catalases as well as various peroxidases.[2]

A variety of environmental stress conditions, including UV or metal stress, pathogen invasion, herbicide action and oxygen shortage, as well as many physiological or pathological conditions (see below) lead to the cellular accumulation of reactive oxygen species and generate a condition that is commonly known as oxidative stress. Oxidative stress is defined as "an imbalance between pro-oxidants and anti-oxidants in favor of the pro-oxidants".[8] It can lead to severe oxidative damage of cellular biomolecules and often leads to cell death. To counteract this hazard, organisms have developed very specific response mechanisms that allow them to quickly sense and respond to accumulating ROS. This chapter gives an overview of the current knowledge of the bacterial oxidative stress response.

1.8.1.2 The Deleterious Effects of Oxidative Stress

Over the past few years, studying and understanding the effects of oxidative stress as well as the mechanisms of cellular response and repair systems has become a major focus in the scientific community.[9-12] Oxidative stress has been implicated in the development of a wide variety of pathologies including chronic inflammatory processes, atherosclerosis, diabetes mellitus, Alzheimer's disease and Parkinson's disease.[13-16] Accumulation of ROS has also been postulated to be one of the major culprits of the eukaryotic aging process.[17-19] At this point, it is still unclear for most of these processes whether accumulating ROS is the cause or a consequence. It is undeniable, however, that increased concentrations of ROS, which have been found to accompany all these conditions, have deleterious effects on most cellular macromolecules. Excess levels of ROS have been shown to lead to DNA damage, oxidation of polydesaturated fatty acids in lipids, oxidation of amino acids in proteins, and inactivation of enzymes by oxidation of their cofactors.[20,21] Because individual ROS have distinct biological properties, they react with different biological targets. Hydroxyl radicals, for instance, are particularly unstable and react rapidly and non-specifically with most biomolecules. In contrast, the major targets of superoxide are proteins that contain iron-sulfur ([4Fe-4S]) clusters, and H_2O_2 is known to preferentially react with the thiol groups of cysteine residues in proteins.[16,22]

1.8.1.3 Cellular Responses to Oxidative Stress

Oxidative stress is a ubiquitous phenomenon encountered by every aerobically living organism. Pathogenic bacteria, for instance, are exposed to oxidative stress conditions during host defense, where activated phagocytes produce lethal concentrations of H_2O_2 and HOCl to kill invading pathogens.[23,24] More recently, evidence has been presented that suggests oxidative stress is not restricted to the innate immune response, but might be used as an effective strategy to limit bacterial colonization at mucosal barrier epithelia.[25]

To protect themselves against the harmful effects of ROS, cells have developed specialized stress response mechanisms in addition to the constitutive defense strategies described above. The basic principles of oxidative stress response are universally conserved from bacteria to eukaryotes, although the exact mechanisms by which they sense ROS vary depending on the type and severity of stress condition and the organism.

Much of our current knowledge about mechanisms of oxidative stress response comes from studies in bacteria, particularly *Escherichia coli*, where specific stress responses can be easily provoked by the external addition of chemical oxidants that elevate intracellular ROS levels.[26,27] Additionally, mutant strains are available that are constitutively exposed to oxidative stress conditions, either due to alterations in their response pathways (*i.e.*, *oxyR*, *soxR*) or in their ability to remove ROS properly (*e.g.*, *sod*, *katG*, *ahpC*). Exposure of *E. coli* to oxidative stressors like H_2O_2 or O_2^-, for instance, induces global regulatory responses that lead to the over-expression of proteins involved in detoxifying the oxidants (*e.g.*, catalase, SOD), restore the redox balance (*e.g.*, thioredoxin, glutaredoxin) and repair cellular damage (*e.g.*, Dps, YtfE, UvrA/B). Sensing of ROS occurs at the posttranslational level, where the respective transcriptional regulators undergo specific ROS-mediated protein modifications that alter their DNA binding activity and activate the proteins. Similar post-translational ROS-sensing mechanisms have also been identified for a number of non-transcription factor proteins (*e.g.*, Hsp33, RsrA, OhrR), which apparently serve as a first line of defense against oxidative stress conditions.[28,29]

To illustrate a cell's response against specific oxidative stress conditions, we will focus on two *E. coli* proteins that have been intensely researched over the past few years: OxyR, the oxidative stress transcription factor of *E. coli*, which senses micromolar levels of H_2O_2 *via* highly reactive cysteine residue(s) and induces the expression of a plethora of antioxidant and repair enzymes; and Hsp33, a highly specialized molecular chaperone, which utilizes two inter-dependent sensor regions to sense oxidative stress conditions that lead to protein unfolding. Both proteins are essential to protect *E. coli* against oxidative stress conditions. In this chapter, we will illustrate the capability of these proteins to rapidly and fully reversibly sense ROS and, in response, modulate their protein conformation and activity.

1.8.1.4 Cysteines: The Building Blocks of ROS-sensing Nano-switches

One central feature that provides OxyR, Hsp33 and a rapidly increasing number of redox sensing proteins (*e.g.*, RsrA, PDI)[28,30] with the ability to sense and respond to accumulating ROS is the presence of one or more unusually reactive cysteine residues.[28,31] These cysteine residues, which play either structural or functional roles in redox-regulated proteins, undergo reversible oxidative modifications. These, in turn, cause the dramatic changes in conformation and/ or activity that are necessary for the rapid oxidative stress response.[31,32]

Cysteine residues have been well documented to serve as preferred building blocks of ROS-sensing molecular nano-switches.[28,33] What makes the cysteine residue so well suited to act as ROS sensor are the unique chemical features of its thiol group. Thiol oxidation has been shown to lead to the formation of a variety of different oxidative modifications.[34] The sensitivity of a cysteine's thiol groups is influenced by neighboring charges, its localization within the protein and its pK_a.[32,35] The type of oxidative thiol modification that is formed depends on the individual protein as well as on the sort and extent of reactive oxygen species that the protein is exposed to. Sulfenic acid (-SOH), for instance, is thought to be the first reaction product of thiols reacting with H_2O_2.[36,37] It is a highly reactive intermediate, which is either stabilized by nearby charges (*e.g.*, AhpC, Prx-2)[35,38] or rapidly condenses with available proximal thiol groups to form a disulfide bond (*e.g.*, Hsp33, OxyR).[39] Alternatively, it can react with glutathione, leading to S-glutathionylation (-S-GSH)[40] or it can be further oxidized to sulfinic acid (-SO_2H) (*e.g.*, Prx-2).[41] Most importantly, all of these oxidative thiol modifications are fully reversible either by enzymes of the redox balancing systems (*i.e.*, thioredoxin, glutaredoxin)[42,43] or by specialized reductases such as sulfiredoxins.[44] This allows proteins to be rapidly switched off once reducing conditions are restored. It is this unique ability of cysteines to cycle between different stable redox states, each of which is associated with a distinct protein conformation and activity, that provides the molecular basis for redox-regulated proteins such as OxyR and Hsp33.

1.8.2 OxyR: A Redox-regulated Transcription Factor

1.8.2.1 Discovery of an H_2O_2-response Regulator in *E. coli*

OxyR, which was first discovered in *Salmonella enterica* and *E. coli*, is one of the main transcription factors responsible for the rapid cellular response of bacteria to H_2O_2 treatment.[45] The first indication that OxyR is involved in antioxidant defense came from mutations that rendered OxyR constitutively active. Expression of these *oxyR* variants in *E. coli* substantially increased resistance against H_2O_2 stress.[46,47] In contrast, strains carrying the *oxyR* deletion were found to be hypersensitive to H_2O_2 and failed to activate the expression of genes involved in antioxidant defense.[46,48]

OxyR is a member of the LysR family, which comprises the largest number of transcriptional regulators in bacterial cells.[47,49] LysR family members vary in size from 30 to 35 kDa and, apart from a few exceptions, function as homo-tetramers *in vitro* and *in vivo*. The conserved N-terminal domain contains a helix-turn-helix DNA binding motif, whereas the C-terminal part of the protein is characterized as the regulatory domain.[49] LysR transcription factors usually

bind to DNA regions adjacent to or overlapping with the promoter to positively regulate the expression of target genes and to negatively regulate their
own expression.

OxyR is a typical LysR family member. It is 34 kDa in size and is composed
of the N-terminal DNA-binding motif and the C-terminal regulatory domain,
which also contains residues important for its tetramerization.[47,50,51] Most of
OxyR's target genes contain the OxyR binding site upstream of the –35 region,
which is the optimal position for recruiting RNA polymerase to the promoter
region.[52,53]

1.8.2.2   The OxyR Regulon

Once activated by peroxide treatment, OxyR regulates a plethora of genes that
are involved in detoxification of the oxidants, protection against further oxidative damage and repair.[53] Specifically, the OxyR regulon comprises genes
encoding (i) enzymes such as catalase (*katG*) and alkyl hydroperoxide-NADPH
oxido-reductase (*ahpCF*), which are involved in the direct degradation of
hydrogen peroxide; (ii) proteins including GSH reductase (*gorA*), glutaredoxin
(*grxA*), and thioredoxin reductase (*trxC*), which restore the intracellular thiol
disulfide balance; (iii) a repressor of iron transport (*fur*), which apparently
serves to prevent Fe^{2+}-mediated Fenton reactions that would lead to the
additional production of highly reactive and toxic hydroxyl radicals; (iv) proteins involved in DNA protection (*dps*); (v) the small regulatory RNA *oxyS*;
and (vi) several other proteins such as biofilm promoting proteins (*agn43*, *fhuF*)
that enhance bacterial survival.[53,54]

Whereas all of these genes are under direct OxyR control, a number of
additional genes are only indirectly controlled by OxyR *via* its induction of the
small regulatory RNA *oxyS*. OxyS regulates expression of at least 20 downstream targets by affecting their mRNA stability or translation efficiency.[55,56]
A number of these downstream targets have not yet been characterized. DNA
microarrays and two-dimensional gel experiments performed in *E. coli* not only
revealed the presence of overlapping regulons, but also demonstrated the
existence of OxyR-independent, alternative stress response pathways that are
activated by peroxide.[54] A systematic identification of all genes regulated by
particular transcriptional factors will help us map the complex regulatory
networks involved in the cellular response that allows bacteria such as *E. coli* to
survive H_2O_2-mediated oxidative stress.

1.8.2.3   Redox Regulation of OxyR's Function

Exposure of bacteria to peroxide stress does not change the subcellular localization of OxyR or alter the levels of OxyR protein. These observations suggested early on that posttranslational modifications of OxyR are likely to be
responsible for the H_2O_2-mediated induction of the OxyR response.[57] Since
then, considerable effort has been applied toward understanding the exact
mechanism of OxyR's redox-regulation and its sophisticated mode of action.

1.8.2.3.1 Functional Mechanism of OxyR

OxyR is constitutively bound to DNA sequences that are close to the promoter region of its target genes. While OxyR functions under non-stress conditions as repressor of a subset of genes including its own, exposure of cells to micromolar concentrations of H_2O_2 rapidly activate OxyR, which in turn leads to the induction of the OxyR regulon.[53,54,58] Central to this rapid response is the presence of highly conserved cysteine residue(s) in OxyR, which are exquisitely sensitive to peroxide-mediated oxidation (see below). DNA foot-printing studies demonstrated that the reduced and oxidized forms of OxyR differ in their association with DNA. Whereas the reduced inactive form of OxyR binds two adjacent major grooves separated by one helical turn, the oxidized active form of OxyR binds four consecutive major grooves.[52,57] Oxidation-mediated conformational changes in OxyR trigger these profound changes in DNA recognition, which are accompanied by the ability of OxyR to recruit RNA polymerase to the promoter.[52,57,59] Thus far, no evidence for a direct interaction between OxyR and RNA polymerase has been presented. Recently, two residues at the surface of oxidized OxyR were proposed to serve as contact sites between oxidized OxyR and RNA polymerase; however, further experiments are required to test this attractive hypothesis.[60]

1.8.2.3.2 The Central Role of OxyR's Cys199

In contrast to most other members of the LysR family whose activity is directly controlled by the binding of effector molecules to the C-terminal regulatory domain,[49,61] OxyR's activity is primarily controlled by the redox status of one exquisitely H_2O_2-sensitive cysteine (Cys199) in its C-terminal regulatory domain. Substitution of Cys199, which is conserved among all members of the OxyR family, completely abolishes OxyR's activity both *in vitro* and *in vivo*.[51,62,63] Cys208, an equally conserved cysteine, appears to play a redox-regulatory role in OxyR as well. This conclusion is also based on mutant studies, which revealed that OxyR variants harboring Cys208 mutations show low constitutive levels of transcriptional activity.[64] Absence of all four non-conserved cysteine residues, on the other hand, does not influence the activity of OxyR.

Early mutational studies suggested that Cys199 plays the major role in the H_2O_2 sensing of OxyR presumably by forming a highly reactive sulfenic acid.[51] Subsequent *in vivo* studies led to the proposal by Storz and co-workers that rapid condensation of the sulfenic acid intermediate with nearby Cys208 forms a transiently stable intramolecular disulfide bond.[63] Mass spectrometry and *in vivo* labeling assays agree with this conclusion and also suggest the presence of disulfide-bond formation as a direct consequence of H_2O_2 sensing.[65] If this is the case, why do Cys199 and Cys208 mutants differ in their phenotype and activity? While Cys199 mutants are completely inactive, Cys208 mutants show low constitutive activity. This result would clearly indicate that the presence of Cys199 is sufficient for redox sensing and OxyR activity. Interestingly, however, double mutants lacking both Cys199 and Cys208 are not inactive, as expected, but behave very similarly to the single Cys208 mutant protein.[65] This result suggests that the introduced mutations in Cys208 alter the conformation of the reduced form of OxyR, thereby rendering the protein constitutively active. This would explain why Cys208 mutants are sensor blind and show constitutively low activity, without excluding that both cysteines play important roles in the redox regulation of OxyR.[65]

1.8.2.3.3 The Structure of OxyR's Redox-switch

The crystal structure of OxyR's regulatory domain, which has been solved in both the reduced and oxidized forms, provides excellent insight into the architecture of OxyR's redox switch.[66] In the reduced form, the two critical cysteine residues are separated by 17 Å. In the oxidized form, on the other hand, these two cysteines form the expected intramolecular disulfide bond,[63] thereby not only bringing the two cysteines into close proximity, but also introducing substantial overall conformational changes. The observed structural rearrangements include the conversion of a β-strand into an α-helix, as well as the formation of a new β-strand. These structural changes affect the oligomeric interfaces of OxyR and result in a slight rotation of the monomers relative to each other. These studies agree well with results obtained from a recently conducted computational modeling study of full-length OxyR.[67] Kona *et al.* expanded the modeling to include OxyR's N-terminal DNA-binding domain (whose structure has not yet been solved) to explain the different DNA binding patterns observed with the reduced and the oxidized form.[67]

What is the basis for the high reactivity of Cys199, which is the main player of OxyR's redox switch? Analysis of the crystal structure of OxyR in its reduced form provides insight into the environment of Cys199, an important aspect in understanding the exquisite sensitivity of these cysteines towards H_2O_2. Cys199 is flanked by two basic residues H198 and R201. In addition, a third, albeit less conserved, positively charged amino acid, Arg266, is also in close proximity to Cys199. This provides a highly positive electrostatic potential, which will likely lower the pK_a of Cys199 and lead to the deprotonation of the cysteine's sulfhydryl group at physiological pH. The resulting thiolate anion is much more reactive than its uncharged counterpart,[35] which would explain its high sensitivity towards H_2O_2 and other oxidants. Note that it is this preference for positive or aromatic amino acids in the vicinity that characterizes most redox-sensitive cysteines characterized to date.[28]

The crystal structure in combination with modeling studies and recent kinetic experiments led to a new model, which assumes that sulfenic acid formation rather than disulfide-bond formation might be the driving force for OxyR's conformational changes and activation[65–67] (Figure 1.8.1). In this model, sulfenic acid formation at position Cys199 increases the size and charge of this residue. These changes are predicted to destabilize the Cys199 side-chain, which discharges from the inter-domain pocket and results in the formation of a flexible loop around the cysteine. This initial conformational change and the

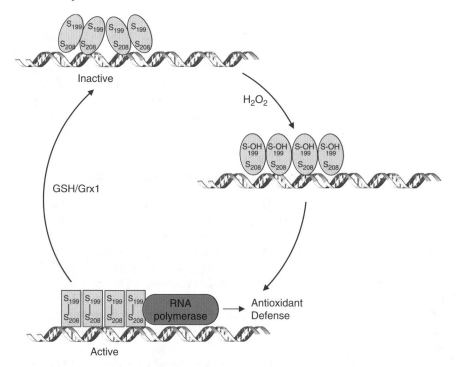

Figure 1.8.1 OxyR: Master regulator of the oxidative stress response. OxyR is a DNA-binding homotetrameric protein, which switches from the reduced "repressor" conformation to the oxidized "activator" conformation upon exposure to micromolar concentrations of H_2O_2. Two highly conserved cysteines Cys199 and Cys208 appear to play the central role in OxyR's functional regulation. Presence of H_2O_2 is first sensed by Cys199, which is rapidly oxidized to sulfenic acid (SOH). Sulfenic acid formation is proposed to induce a partial conformational change in OxyR, which brings Cys199 into close proximity to Cys208. Subsequent intramolecular disulfide-bond formation between Cys199 and Cys208 causes large structural rearrangements in OxyR, which change OxyR's interaction with DNA and are necessary for the recruitment of RNA polymerase. This leads to the expression of numerous antioxidant genes. Antioxidant gene expression is turned off by the glutathione/glutaredoxin system (Grx1), which reduces OxyR's oxidative thiol modifications and returns OxyR into its inactive state.

flexibility of Cys208 promote the disulfide-bond formation, which then triggers the reorganization and stabilization of the flexible regions.[67]

The activation of OxyR must be very fast and the conformational changes must be sufficiently stable to allow the efficient induction of the stress responsive genes. *In vivo* studies revealed that OxyR is indeed fully oxidized within 30 seconds after the start of the stress treatment. Because the kinetics of OxyR reduction are slow compared to the oxidation kinetics, OxyR remains oxidized and active for at least 5 to 10 min after the removal of H_2O_2 from the growth media.[68] Subsequent titration experiments revealed that submicromolar concentrations of H_2O_2 are sufficient to transiently activate OxyR even when present in a highly reducing redox environment, suggesting that OxyR's disulfide bond is meta-stable relative to the surrounding GSH/GSSG redox buffer.[68] *In vitro* and *in vivo* experiments showed that OxyR is directly reduced by glutaredoxin reductase 1 (Grx1), a component of the major disulfide-reductase systems in the cell.[63] Importantly, the gene encoding Grx1 belongs to the OxyR regulon, which nicely illustrates the auto-regulatory capacity of the OxyR dependent stress response.

1.8.2.3.4   The Capacity of OxyR to Micromanage Stress

Although hydrogen peroxide is clearly the best-characterized activator of OxyR, other oxidative stressors have also been shown to induce the OxyR regulon. Stamler and co-workers showed that reactive nitrogen species (*e.g.*, S-nitroso-glutathione) lead to the S-nitrosylation of Cys199, whereas treatment with GSSG causes glutathionylation of OxyR's active site cysteine.[62,69] These oxidation products are thought to generate conformational changes in OxyR that are distinct from the conformational changes caused by H_2O_2 stress. *In vitro* studies suggested that these alternative OxyR conformations might exert different DNA binding patterns, which could in turn lead to different stress responses.[62]

The concept that one single redox-active residue is capable of differentially responding to different oxidants is clearly a very intriguing one. However, *in vivo* studies are necessary to elucidate the physiological relevance of these different OxyR species. DNA microarray technology and complementation assays to characterize the overall gene expression in response to different stress conditions would demonstrate whether OxyR can indeed activate distinct stress responses depending on the detected signals. A recent study of Lu *et al.*[70] using various genetic "oxidative stress probes" to quantitatively and kinetically monitor the expression of eight oxidative stress genes at once revealed distinct response patterns and expression rates under different stress conditions. This study confirms Stamler and co-workers' idea of a more complicated gene regulation upon oxidative stress.

1.8.2.4   Biotechnological Application of OxyR

The high sensitivity of OxyR to very low concentrations of H_2O_2 makes OxyR's regulator domain well suited for biotechnological applications. Recently, Belousov and co-workers designed a genetically encoded highly sensitive H_2O_2 biosensor termed HyPer, which works by employing the major conformational changes that OxyR undergoes upon H_2O_2 treatment.[71] Ingeniously, the authors inserted non-fluorescent circular permuted yellow fluorescent protein (cpYFP) into OxyR's regulatory domain, positioning it between the two active-site cysteines Cys199 and Cys208. In the absence of H_2O_2, both cysteines are reduced and cpYFP is non-fluorescent. Upon disulfide-bond formation, however, cpYFP becomes properly folded and fluorescent. *In vitro* and *in vivo* assays demonstrated the high sensitivity and specificity of HyPer towards H_2O_2, making this an excellent probe to detect micromolar concentrations of H_2O_2 *in vivo*. It remains to be seen what other biotechnological applications could make use of the extensive conformational changes that accompany OxyR's disulfide-bond formation. Pomposiello and Temple, for instance, suggested using the redox-switch activity of OxyR as part of a "chemo-mechanical micromachine", which would allow one to apply the torsional work done by OxyR on a nano-scale level.[72]

1.8.3 Hsp33: A Chaperone Specialized for Oxidative Stress Protection

1.8.3.1 The Redox-regulated Chaperone Holdase Hsp33

Molecular chaperones are present in all living organisms. The main function of molecular chaperones is to assist folding of nascent polypeptides under non-stress conditions and to prevent non-specific protein aggregation during stress conditions.[73] It is therefore not surprising that many molecular chaperones belong to the family of heat shock proteins (Hsp), which are over-expressed under conditions that lead to protein unfolding and aggregation in the cell.[74]

Mechanistically, molecular chaperones can be separated into two distinct groups: chaperone foldases and chaperone holdases. Chaperone foldases, such as the well-characterized DnaK-(Hsp70) system or the GroEL/ES machinery (Hsp60/Hsp10),[75] use ATP binding and hydrolysis to actively support the folding of proteins. Chaperone holdases such as the small heat shock proteins (sHsps)[76] or Hsp33,[77] on the other hand, function independently of ATP. They form highly efficient binding platforms for protein folding intermediates and effectively prevent protein aggregation. Unable to use ATP binding and hydrolysis to regulate substrate binding and release, Hsps and Hsp33 developed alternative mechanisms to regulate their substrate affinity.[78] Small Hsps use temperature-induced conformational changes to switch from a low- to a high-affinity binding state under heat shock conditions.[79] Hsp33, on the other hand, uses a cysteine-containing thiol switch to rapidly sense and respond to oxidative stress conditions that lead to protein unfolding.[80]

Hsp33, which is encoded by the gene *hslO*, was first discovered by Chuang and Blattner as part of their global analysis of the *E. coli* heat shock regulon.[81] Like most other heat shock-regulated chaperones, Hsp33 (cellular concentration $\sim 1.5\,\mu M$) is constitutively expressed under non-stress conditions and up-regulated about two-fold under conditions that induce the heat shock response (*e.g.*, elevated temperatures). Sequence analysis of Hsp33 revealed that the 32.8 kDa cytoplasmic protein is highly conserved. Homologs of Hsp33 can be found in almost all prokaryotic species that have been sequenced so far. Moreover, Hsp33 appears to be present in the mitochondria of a few eukaryotic parasites including *Chlamydomonas reinhardtii*, *Dictyostelium discodeum* and members of the *Trypanosomatidae* family.[82] Sequence alignments revealed an absolutely conserved C-X-C$_{(27-32)}$C-X-Y-C motif in the C-terminal part of Hsp33. These four cysteines constitute the redox-regulated nano-switch in Hsp33, which is central to its very rapid and highly specific activation under oxidative stress conditions. Interestingly, two other proteins, protein disulfide isomerase (PDI) and peroxiredoxin (Prx-2), have recently been shown to function as redox-regulated chaperones as well.[83,84] This makes Hsp33 a member of a growing group of molecular chaperones that use the redox status of functionally or structurally important cysteine residues to regulate their binding affinity for unfolded substrate proteins.

1.8.3.2   Mechanism of Hsp33's Redox Regulation

Hsp33 is regulated on the transcriptional level by the heat shock response and on the post-translational level by the redox environment of the cytosol. Thus, Hsp33's cellular concentration as well as its chaperone activity are directly controlled by and tied to cellular stress conditions. This section summarizes recent advances in our understanding of how Hsp33 senses and responds to oxidative stress conditions and why cells require such a highly specialized molecular chaperone.

1.8.3.2.1   A Zinc Center as a Redox Switch

Central to the redox regulation of Hsp33 are the four absolutely conserved cysteines located within the last 50 amino acids of its C-terminus. That cysteines play a major part in Hsp33's regulation became evident very early on. Activity assays using purified Hsp33 revealed that incubation with oxidants quickly activated its chaperone function, while incubation with thiol-reducing agents inactivated the protein.[77] Subsequent mechanistic studies demonstrated that under reducing inactivating conditions, Hsp33's four conserved cysteines coordinate one zinc(II) ion with high affinity ($K_a > 10^{17} M^{-1}$, 25 °C at pH 7.5). Oxidizing, activating conditions, on the other hand, lead to the formation of two intramolecular disulfide bonds and the concomitant release of zinc.[39,85] This finding made Hsp33 one of the first proteins known to use a cysteine-coordinating zinc center (previously considered as highly stable and redox inert) as a redox-sensitive functional nano-switch.

The NMR structure of Hsp33's C-terminus[86] as well as the crystal structure of full-length Hsp33 of *Bacillus subtilis*[87] and *Thermotoga maritima*[88] revealed the presence of a tetrahedral zinc site in reduced Hsp33. All four cysteines were found to be the same distance from the zinc ion and from each other. This zinc coordination appears to play several important roles in the redox regulation of Hsp33. Zinc coordination protects Hsp33's redox-sensitive cysteines against non-specific air oxidation (M. Ilbert and U. Jakob, unpublished results). At the same time, zinc binding dramatically increases the reactivity of cysteines towards ROS such as H_2O_2. As in other known redox-sensitive proteins such as the antisigma factor RsrA[89] and protein kinase C,[90] zinc acts as a Lewis acid and accepts electron pairs from the cysteine.[91] This substantially lowers the pK_a of the thiol group and leads to deprotonation and the formation of the highly reactive thiolate anion at neutral pH. In addition, the high affinity binding of zinc provides enormous stabilization energy to Hsp33's C-terminal domain, which lacks hydrophobic amino acids and harbors several residues in highly unfavorable secondary structure motifs.[86] Under reducing conditions, zinc binding enables Hsp33's C-terminus to form a very compact and stable folding unit, which is critical for maintaining Hsp33 in its inactive conformation. In contrast, zinc release in response to oxidative stress-induced thiol modification, triggers extensive conformational changes and rapidly converts Hsp33's C-terminus into a natively unfolded polypeptide. This unfolding, in turn, is critical for the activation of Hsp33 as a molecular chaperone.[80]

1.8.3.2.2 Hsp33's Activation Requires Dual Stress Sensing

Early on, it became evident that Hsp33's activation is not simply triggered by the presence of ROS such as H_2O_2, but also requires elevated temperatures (*i.e.*, oxidative heat stress). Incubation in millimolar concentrations of H_2O_2 did not lead to significant activation unless temperatures were above 40 °C. *In vivo* experiments confirmed this stringent requirement and revealed that Hsp33 protects bacteria against oxidative heat stress but not against peroxide stress at 30 °C or heat stress alone (>49 °C).[80,92,93] Subsequent mechanistic studies showed that low concentrations of unfolding reagents (*e.g.*, 1M guanidinium-hydrochloride) can substitute for the elevated temperature requirement in Hsp33's activation process.[80] This result suggested that it is not heat shock temperature *per se* that is required for the rapid oxidative activation of Hsp33, but the presence of mildly unfolding conditions. This regulation apparently prevents Hsp33's activation under pure H_2O_2 stress conditions, which do not lead to protein unfolding and, therefore, do not require the presence of such a potent chaperone holdase.[80]

Detailed structural analysis revealed that Hsp33 utilizes two interdependent stress-sensor regions that must be simultaneously triggered to activate its chaperone function[80] (Figure 1.8.2). Whereas oxidants are, as previously described, sensed by the four redox-sensitive cysteines, unfolding conditions are sensed by a thermolabile linker region that connects the zinc center with the N-terminal substrate binding site. How does Hsp33 link its activation to the simultaneous presence of these two stress signals? In our current working model, oxidants such as H_2O_2 are first sensed by the two distal cysteines (Cys265 and Cys268) of Hsp33's zinc center. These cysteines are then rapidly oxidized to form an intramolecular disulfide bond, which is sufficient to trigger zinc release and cause unfolding of the zinc binding domain.[80,94] While this first redox event is necessary for the activation of Hsp33, it is not sufficient. Hsp33 that has been oxidized with H_2O_2 at 30 °C *in vitro* does not exhibit chaperone activity even though these cysteines are oxidized and the zinc center is unfolded. To fully activate Hsp33, the second disulfide bond between Cys232 and Cys234 must be formed.[39] This,

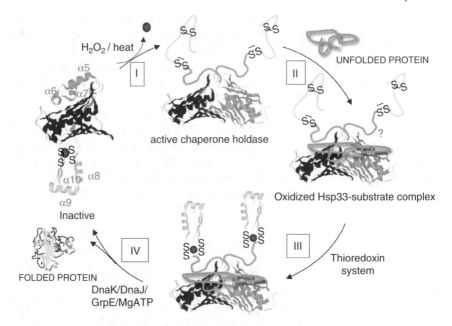

Figure 1.8.2   Hsp33: A chaperone specialized for oxidative stress protection. Under non-stress conditions, Hsp33 is monomeric and inactive. All four invariant cysteines are reduced and bind one zinc ion (red) forming a compactly folded zinc-center (yellow). Upon exposure to oxidative heat stress (step I), two intramolecular disulfide bonds form. This apparently triggers the unfolding of the zinc binding domain and linker region (green), exposes hydrophobic surfaces on the N-terminal substrate binding domain of Hsp33 (blue) and causes the formation of highly active Hsp33 dimers (step II). Cellular redox systems such as the thioredoxin system (Trx) convert oxidized Hsp33 dimers into active reduced Hsp33 dimers (step III). Inactivation of Hsp33 and substrate release requires non-stress conditions and the presence of the DnaK-system, which supports the refolding of the substrate proteins (step IV).

however, can only occur when mild unfolding conditions are present that trigger the unfolding of an adjacent "folding sensor" present in Hsp33's linker region. This region is thermostable in reduced Hsp33 ($T_m > 65\,°C$), but becomes thermolabile ($T_m \sim 40\,°C$) once the zinc center is oxidatively unfolded.[80] Unfolding of the folding sensor region either by elevated temperatures or low concentrations of denaturants is apparently critical to provide access to the two remaining cysteines. They are subsequently oxidized and form the second intramolecular disulfide bond in Hsp33. Formation of the second disulfide bond apparently induces Hsp33's dimerization and its activation as a molecular chaperone[93] (Figure 1.8.2). In addition, disulfide-bond formation locks both the linker region and the zinc-binding domain in a natively unfolded conformation and maintains Hsp33 in an active conformation until reducing non-stress conditions are restored.[80] This model of Hsp33's activation process, which involves the successive rather than the simultaneous formation of two disulfide bonds, explains how Hsp33 intimately links its activation to both oxidizing and protein unfolding conditions. It agrees with earlier observations showing that disulfide bonds in Hsp33 only occur between adjacent cysteines[39] even though all four cysteines are an equal distance from each other in the tetrahedral zinc-binding site.[86]

Oxidative unfolding of Hsp33's linker region and activation of Hsp33 coincide with the exposure of large hydrophobic patches on the surface of the molecule.[94,95] Because most other well-characterized molecular chaperones such as DnaK and GroEL have been shown to utilize similar hydrophobic surfaces to bind partially unfolded substrate proteins,[96,97] this result strongly suggested that linker unfolding might induce the exposure of Hsp33's substrate binding site. Although Hsp33's substrate binding site remains to be identified, mutational analysis suggests that it is most likely present in the structurally stable N-terminal domain.[98] A cysteine-free variant of Hsp33, for instance, which contains an unfolded zinc-binding domain, is inactive as a molecular chaperone at low temperatures. Temperature-induced unfolding of the linker region, however, activates this Hsp33 variant and exposes the very same hydrophobic surfaces that characterize activated Hsp33.[80] All these experiments indicate that the linker region masks the substrate binding site under non-stress conditions and exposes the hydrophobic surfaces upon its unfolding. Hsp33's crystal structure supports this conclusion and shows that the linker region is folded on top of the N-terminal domain, where it buries a largely hydrophobic surface of about $4900 \, \text{Å}^2$.[98,99] However, further experiments are required to ultimately define Hsp33's substrate-binding site.

1.8.3.2.3 The Hsp33–DnaK Connection

What triggers substrate release in chaperone holdases? In the case of Hsp33, an obvious requirement for substrate release is the return to reducing conditions.[100] *In vivo* thiol trapping analysis revealed that the two disulfide bonds in Hsp33 are rapidly reduced either by the thioredoxin or the glutaredoxin system[100] (Figure 1.8.2). This presumably leads to the re-incorporation of zinc and the refolding of Hsp33's zinc center. Surprisingly, however, this reduction of Hsp33's disulfide bonds is not sufficient for substrate protein release.[100] It produces the reduced Hsp33 dimer, a chaperone-active species, which remains very stably associated with the unfolded substrate proteins. Formation of this active reduced Hsp33 dimer seems to be necessary for Hsp33's interaction with the ATP-dependent chaperone machinery DnaK/DnaJ/GrpE.[100] Substrate proteins are rapidly released from the reduced Hsp33 dimers only in the presence of the DnaK-system; this in turn leads to the inactivation of Hsp33. The released substrate proteins are then refolded either by the DnaK-system alone or in concert with the GroEL/ES system[100] (Figure 1.8.2).

The maintenance of a stable Hsp33-substrate complex even under reducing conditions makes excellent physiological sense. It prevents Hsp33 from releasing its substrate proteins into a cellular condition in which the original redox status may have been restored, but is not yet permissive for protein folding. Strict dependence of Hsp33's substrate protein release on the presence of the active DnaK-system links the dissociation of the highly aggregation-sensitive substrate proteins to conditions where ATP-dependent chaperones are available to support their refolding. Interestingly, a similar cooperation between different chaperone systems has also been observed for sHsps. Similar to Hsp33, sHsps interact with the eukaryotic DnaK-homolog Hsp70 to induce substrate release and to support the refolding of their substrate proteins.[101,102] As in the case of Hsp33, however, the precise mechanism of communication between the various players requires further investigation.

1.8.3.3 Hsp33: Central Member of a Multi-chaperone Network

The chaperone function of Hsp33 is rapidly activated by oxidative stress conditions that lead to protein unfolding, while neither oxidative stress conditions alone (*e.g.*, H_2O_2) nor heat shock temperatures activate Hsp33.[80,92] This specific regulation of Hsp33's chaperone function properly reflects its *in vivo* requirement. Hydrogen peroxide alone does not cause protein unfolding and, therefore, does not require such a potent chaperone.[80] Elevated temperatures, on the other hand, induce protein unfolding; but protein folding intermediates are effectively protected against aggregation by the ATP-dependent DnaK-system,[75] again apparently precluding the need for Hsp33. In contrast, the chaperone activity of Hsp33 becomes essential for survival of bacteria when peroxide stress is combined with unfolding conditions or exposure to more potent oxidants that directly cause cellular protein unfolding (Winter, Graf and Jakob, unpublished observations). We have demonstrated that Hsp33 as an ATP-*in*dependent chaperone compensates for the transient loss of function of the ATP-dependent DnaK-system under oxidative heat stress conditions.[92] The inactivation of the DnaK-system is caused by the rapid decrease in cellular ATP levels, which is known to accompany oxidative stress conditions in both pro- and eukaryotic cells.[92,103] Activated by the very conditions that inactivate DnaK, Hsp33 now binds to about 80% of all aggregation-sensitive *E. coli* proteins.[92] Substrate proteins of Hsp33 include metabolic enzymes as well as proteins involved in transcription and cell division. Comparison of Hsp33's substrate specificity under oxidative heat stress conditions with substrate proteins that are protected by the DnaK-system against heat-induced unfolding revealed a substantial substrate overlap. This overlap in substrate specificity appears to become particularly important upon return to reducing non-stress conditions. Hsp33 can now present its substrate proteins to the re-activated DnaK-system, which in turn induces substrate release and supports the refolding of the proteins to their native state.[100] This makes Hsp33 the central player of a multi-chaperone network that is critical for protecting proteins against protein aggregation under oxidative stress conditions.

1.8.4   Oxidative Stress and Redox Regulation: Turning Lemons into Lemonade

For many years it has been known that increased concentrations of intracellular ROS cause a highly stressful and dangerous condition called oxidative stress. Oxidative stress has been shown to exert its deleterious effects on proteins, DNA and lipids. Only in the past ten years, however, has it become evident that a group of proteins are not damaged by ROS, but specifically use them to quickly regulate their protein activity. In this chapter, we reviewed current knowledge about two members of this rapidly growing class of redox-regulated proteins, OxyR and Hsp33. Both proteins use ROS-mediated changes in the oxidation status of critical cysteines to induce large conformational changes. These changes are essential for the proteins' activity and allow them to play active roles in protecting bacteria against the lethal consequences of oxidative stress. While oxidation of the chaperone Hsp33 unmasks a previously buried highly hydrophobic binding site for unfolded proteins, oxidation of OxyR's critical cysteines exposes an RNA polymerase binding site essential for inducing antioxidant gene transcription.

Although individual facets of Hsp33's and OxyR's regulation can be found in other redox-regulated proteins, no single common mechanism has been identified. This makes the characterization of redox-regulated proteins both challenging and fascinating. Many redox-regulated proteins, including the glycolytic enzyme glyceraldehyde-3-phosphate dehydrogenase (GapA), the oxidative transcription factor in yeast, Yap1p, as well as other transcription factors such as p53, are similar to OxyR in that they use one or more reactive cysteine residues as their redox sensor.[104–106] Antisigma factor RsrA and protein kinase C, on the other hand, are similar to Hsp33 in that they use a redox-sensitive zinc center that switches from a zinc-bound reduced state to a zinc-free disulfide-bonded state upon exposure to oxidative stress conditions.[90,107,108] A very different ROS-sensing mechanism is employed by the transcription factors SoxR and FNR. Both proteins use an iron-sulfur cluster as a redox-sensor. Although oxidation reactions occur on the metal rather than on the sulfur, they lead to equally dramatic conformational rearrangements and changes in protein function.[9,109]

Over the past few years, a number of global methods have been developed to identify proteins with redox-sensitive cysteines.[110] A common conclusion of many of these studies was that surprisingly few cytosolic proteins harbor redox-sensitive and potentially redox-regulated cysteines. A recently conducted quantitative redox proteomics approach confirmed these results. These studies revealed that only a very limited number of *E. coli* proteins have cysteine residues that become substantially (>40%) oxidatively modified during oxidative stress *in vivo*. Subsequent functional studies demonstrated that many of these proteins play important roles in the oxidative stress protection of *E. coli*.[111] This result confirmed earlier observations, which showed that redox-regulated proteins are generally involved in protecting cells and organisms against oxidative stress conditions. The goal will now be to identify and characterize redox-regulated proteins. These studies will provide us with the very real possibility of discovering eukaryotic proteins that protect cells against the toxic effects of ROS that accumulate during aging and many diseases; they will also facilitate the discovery of prokaryotic proteins that could serve as targets for more effective antimicrobial therapies.

References

1. K. J. Davies, *Biochem. Soc. Symp.*, 1995, **61**, 1–31.
2. I. Fridovich, *Ann. NY Acad. Sci.*, 1999, **893**, 13–8.
3. V. J. Thannickal and B. L. Fanburg, *Am. J. Physiol. Lung Cell Mol. Physiol.*, 2000, **279**, L1005–L1028.
4. F. Aslund and J. Beckwith, *Cell*, 1999, **96**, 751–753.
5. S. M. Deneke, *Curr. Top. Cell Regul.*, 2000, **36**, 151–180.
6. J. Moskovitz, *Curr. Pharm. Des.*, 2005, **11**, 1451–1457.
7. E. S. Arner and A. Holmgren, *Eur. J. Biochem.*, 2000, **267**, 6102–6109.
8. H. Sies, *Am. J. Med.*, 1991, **91**, 31S–38S.
9. D. Touati, *Redox Rep.*, 2000, **5**, 287–293.
10. G. Storz and J. A. Imlay, *Curr. Opin. Microbiol.*, 1999, **2**, 188–194.
11. E. Cabiscol, J. Tamarit and J. Ros, *Int. Microbiol.*, 2000, **3**, 3–8.
12. T. Rabilloud, M. Chevallet, S. Luche and E. Leize-Wagner, *Expert Rev. Proteomics*, 2005, **2**, 949–956.
13. F. Bonomini, S. Tengattini, A. Fabiano, R. Bianchi and R. Rezzani, *Histol. Histopathol.*, 2008, **23**, 381–390.
14. P. Kovacic and J. D. Jacintho, *Curr. Med. Chem.*, 2001, **8**, 773–796.
15. P. Kyselova, M. Zourek, Z. Rusavy, L. Trefil and J. Racek, *Physiol. Res.*, 2002, **51**, 591–595.
16. W. Droge, *Physiol. Rev.*, 2002, **82**, 47–95.
17. E. R. Stadtman, *Ann. NY Acad. Sci.*, 2001, **928**, 22–38.
18. W. Droge, *Adv. Exp. Med. Biol.*, 2003, **543**, 191–200.
19. T. Finkel and N. J. Holbrook, *Nature*, 2000, **408**, 239–247.
20. R. T. Dean, S. Fu, R. Stocker and M. J. Davies, *Biochem. J.*, 1997, **324** (Pt 1), 1–18.
21. W. C. Burhans and M. Weinberger, *Nucleic Acids Res.*, 2007, **35**, 7545–7556.
22. M. J. Davies, S. Fu, H. Wang and R. T. Dean, *Free Radic. Biol. Med.*, 1999, **27**, 1151–1163.
23. S. J. Klebanoff, *J. Leukoc. Biol.*, 2005, **77**, 598–625.
24. R. A. Miller and B. E. Britigan, *Clin. Microbiol. Rev.*, 1997, **10**, 1–18.
25. E. M. Ha, C. T. Oh, Y. S. Bae and W. J. Lee, *Science*, 2005, **310**, 847–850.
26. S. Dukan, S. Dadon, D. R. Smulski and S. Belkin, *Appl. Environ. Microbiol.*, 1996, **62**, 4003–4008.
27. J. A. Imlay and S. Linn, *J. Bacteriol.*, 1987, **169**, 2967–2976.
28. M. S. Paget and M. J. Buttner, *Annu. Rev. Genet.*, 2003, **37**, 91–121.

29. M. Ilbert, P. C. Graf and U. Jakob, *Antioxid. Redox Signal.*, 2006, **8**, 835–846.
30. P. Ghezzi, *Biochem. Soc. Trans.*, 2005, **33**, 1378–1381.
31. K. Linke and U. Jakob, *Antioxid. Redox Signal.*, 2003, **5**, 425–434.
32. D. Barford, *Curr. Opin. Struct. Biol.*, 2004, **14**, 679–686.
33. N. M. Giles, A. B. Watts, G. I. Giles, F. H. Fry, J. A. Littlechild and C. Jacob, *Chem. Biol.*, 2003, **10**, 677–693.
34. G. I. Giles and C. Jacob, *Biol. Chem.*, 2002, **383**, 375–388.
35. L. K. Moran, J. M. Gutteridge and G. J. Quinlan, *Curr. Med. Chem.*, 2001, **8**, 763–772.
36. A. Claiborne, J. I. Yeh, T. C. Mallett, J. Luba, E. J. Crane 3rd, V. Charrier and D. Parsonage, *Biochemistry*, 1999, **38**, 15407–15416.
37. L. B. Poole, P. A. Karplus and A. Claiborne, *Annu. Rev. Pharmacol. Toxicol.*, 2004, **44**, 325–347.
38. L. B. Poole, *Arch. Biochem. Biophys.*, 2005, **433**, 240–254.
39. S. Barbirz, U. Jakob and M. O. Glocker, *J. Biol. Chem.*, 2000, **275**, 18759–18766.
40. M. Fratelli, E. Gianazza and P. Ghezzi, *Expert Rev. Proteomics*, 2004, **1**, 365–376.
41. V. Prouzet-Mauleon, C. Monribot-Espagne, H. Boucherie, G. Lagniel, S. Lopez, J. Labarre, J. Garin and G. J. Lauquin, *J. Biol. Chem.*, 2002, **277**, 4823–4830.
42. A. P. Fernandes and A. Holmgren, *Antioxid. Redox Signal.*, 2004, **6**, 63–74.
43. H. Nakamura, *Antioxid. Redox Signal.*, 2005, **7**, 823–828.
44. H. A. Woo, H. Z. Chae, S. C. Hwang, K. S. Yang, S. W. Kang, K. Kim and S. G. Rhee, *Science*, 2003, **300**, 653–656.
45. G. Storz and L. A. Tartaglia, *J. Nutr.*, 1992, **122**, 627–630.
46. M. F. Christman, R. W. Morgan, F. S. Jacobson and B. N. Ames, *Cell*, 1985, **41**, 753–762.
47. M. F. Christman, G. Storz and B. N. Ames, *Proc. Natl. Acad. Sci. USA*, 1989, **86**, 3484–3488.
48. G. Storz, M. F. Christman, H. Sies and B. N. Ames, *Proc. Natl. Acad. Sci. USA*, 1987, **84**, 8917–8921.
49. M. A. Schell, *Annu. Rev. Microbiol.*, 1993, **47**, 597–626.
50. I. Kullik, J. Stevens, M. B. Toledano and G. Storz, *J. Bacteriol.*, 1995, **177**, 1285–1291.
51. I. Kullik, M. B. Toledano, L. A. Tartaglia and G. Storz, *J. Bacteriol.*, 1995, **177**, 1275–1284.
52. M. B. Toledano, I. Kullik, F. Trinh, P. T. Baird, T. D. Schneider and G. Storz, *Cell*, 1994, **78**, 897–909.
53. M. Zheng, X. Wang, B. Doan, K. A. Lewis, T. D. Schneider and G. Storz, *J. Bacteriol.*, 2001, **183**, 4571–4579.
54. M. Zheng, X. Wang, L. J. Templeton, D. R. Smulski, R. A. LaRossa and G. Storz, *J. Bacteriol.*, 2001, **183**, 4562–4570.
55. A. Zhang, S. Altuvia and G. Storz, *Nucleic Acids Symp. Ser.*, 1997, **36**, 27–28.

56. A. Zhang, S. Altuvia, A. Tiwari, L. Argaman, R. Hengge-Aronis and G. Storz, *EMBO. J.*, 1998, **17**, 6061–6068.
57. G. Storz, L. A. Tartaglia and B. N. Ames, *Science*, 1990, **248**, 189–194.
58. G. Storz and M. Zheng in *Bacterial Stess Responses*, G. Storz and R. Hengge-Aronis eds., ASM Press, 2000, pp. 47–59.
59. K. Tao, N. Fujita and A. Ishihama, *Mol. Microbiol.*, 1993, **7**, 859–864.
60. X. Wang, P. Mukhopadhyay, M. J. Wood, F. W. Outten, J. A. Opdyke and G. Storz, *J. Bacteriol.*, 2006, **188**, 8335–8342.
61. T. Clark, S. Haddad, E. Neidle and C. Momany, *Acta Crystallogr. D. Biol. Crystallogr.*, 2004, **60**, 105–108.
62. S. O. Kim, K. Merchant, R. Nudelman, W. F. Beyer Jr., T. Keng, J. DeAngelo, A. Hausladen and J. S. Stamler, *Cell*, 2002, **109**, 383–396.
63. M. Zheng, F. Aslund and G. Storz, *Science*, 1998, **279**, 1718–1721.
64. M. Zheng and G. Storz, *Biochem. Pharmacol.*, 2000, **59**, 1–6.
65. C. Lee, S. M. Lee, P. Mukhopadhyay, S. J. Kim, S. C. Lee, W. S. Ahn, M. H. Yu, G. Storz and S. E. Ryu, *Nat. Struct. Mol. Biol.*, 2004, **11**, 1179–1185.
66. H. Choi, S. Kim, P. Mukhopadhyay, S. Cho, J. Woo, G. Storz and S. Ryu, *Cell*, 2001, **105**, 103–113.
67. J. Kona and T. Brinck, *Org. Biomol. Chem.*, 2006, **4**, 3468–3478.
68. F. Aslund, M. Zheng, J. Beckwith and G. Storz, *Proc. Natl. Acad. Sci. USA*, 1999, **96**, 6161–6165.
69. A. Hausladen, C. T. Privalle, T. Keng, J. DeAngelo and J. S. Stamler, *Cell*, 1996, **86**, 719–729.
70. C. Lu, C. R. Albano, W. E. Bentley and G. Rao, *Biotechnol. Bioeng.*, 2005, **89**, 574–587.
71. V. V. Belousov, A. F. Fradkov, K. A. Lukyanov, D. B. Staroverov, K. S. Shakhbazov, A. V. Terskikh and S. Lukyanov, *Nat. Methods*, 2006, **3**, 281–286.
72. P. J. Pomposiello and B. Demple, *Trends Biotechnol.*, 2001, **19**, 109–114.
73. R. J. Ellis, S. M. van der Vies and S. M. Hemmingsen, *Biochem. Soc. Symp.*, 1989, **55**, 145–153.
74. S. Lindquist and E. A. Craig, *Annu. Rev. Genet.*, 1988, **22**, 631–677.
75. B. Bukau and A. L. Horwich, *Cell*, 1998, **92**, 351–366.
76. M. Haslbeck, T. Franzmann, D. Weinfurtner and J. Buchner, *Nat. Struct. Mol. Biol.*, 2005, **12**, 842–846.
77. U. Jakob, W. Muse, M. Eser and J. C. Bardwell, *Cell*, 1999, **96**, 341–352.
78. J. Winter and U. Jakob, *Crit. Rev. Biochem. Mol. Biol.*, 2004, **39**, 297–317.
79. T. M. Franzmann, P. Menhorn, S. Walter and J. Buchner, *Mol. Cell*, 2008, **29**, 207–216.
80. M. Ilbert, J. Horst, S. Ahrens, J. Winter, P. C. Graf, H. Lilie and U. Jakob, *Nat. Struct. Mol. Biol.*, 2007, **14**, 556–563.
81. S. E. Chuang and F. R. Blattner, *J. Bacteriol.*, 1993, **175**, 5242–5252.
82. P. C. Graf and U. Jakob, *Cell Mol. Life Sci.*, 2002, **59**, 1624–1631.
83. B. Tsai, C. Rodighiero, W. I. Lencer and T. A. Rapoport, *Cell*, 2001, **104**, 937–948.

84. H. H. Jang, K. O. Lee, Y. H. Chi, B. G. Jung, S. K. Park, J. H. Park, J. R. Lee, S. S. Lee, J. C. Moon, J. W. Yun, Y. O. Choi, W. Y. Kim, J. S. Kang, G. W. Cheong, D. J. Yun, S. G. Rhee, M. J. Cho and S. Y. Lee, *Cell*, 2004, **117**, 625–635.
85. U. Jakob, M. Eser and J. C. Bardwell, *J. Biol. Chem.*, 2000, **275**, 38302–38310.
86. H. S. Won, L. Y. Low, R. D. Guzman, M. Martines-Yamout, U. Jakob and H. J. Dyson, *J. Mol. Biol.*, 2004, **341**, 893–899.
87. I. Janda, Y. Devedjiev, U. Derewenda, Z. Dauter, J. Bielnicki, D. R. Cooper, P. C. Graf, A. Joachimiak, U. Jakob and Z. S. Derewenda, *Structure*, 2004, **12**, 1901–1907.
88. L. Jaroszewski, R. Schwarzenbacher, D. McMullan, P. Abdubek, S. Agarwalla, E. Ambing, H. Axelrod, T. Biorac, J. M. Canaves, H. J. Chiu, A. M. Deacon, M. DiDonato, M. A. Elsliger, A. Godzik, C. Grittini, S. K. Grzechnik, J. Hale, E. Hampton, G. W. Han, J. Haugen, M. Hornsby, H. E. Klock, E. Koesema, A. Kreusch, P. Kuhn, S. A. Lesley, M. D. Miller, K. Moy, E. Nigoghossian, J. Paulsen, K. Quijano, R. Reyes, C. Rife, G. Spraggon, R. C. Stevens, H. van den Bedem, J. Velasquez, J. Vincent, A. White, G. Wolf, Q. Xu, K. O. Hodgson, J. Wooley and I. A. Wilson, *Proteins*, 2005, **61**, 669–673.
89. W. Li, A. R. Bottrill, M. J. Bibb, M. J. Buttner, M. S. Paget and C. Kleanthous, *J. Mol. Biol.*, 2003, **333**, 461–472.
90. R. Gopalakrishna and S. Jaken, *Free Radical. Biol. Med.*, 2000, **28**, 1349–1361.
91. K. E. Hightower and C. A. Fierke, *Curr. Opin. Chem. Biol.*, 1999, **3**, 176–181.
92. J. Winter, K. Linke, A. Jatzek and U. Jakob, *Mol. Cell*, 2005, **17**, 381–392.
93. J. Graumann, H. Lilie, X. Tang, K. A. Tucker, J. H. Hoffmann, J. Vijayalakshmi, M. Saper, J. C. Bardwell and U. Jakob, *Structure*, 2001, **9**, 377–387.
94. P. C. Graf, M. Martinez-Yamout, S. VanHaerents, H. Lilie, H. J. Dyson and U. Jakob, *J. Biol. Chem.*, 2004, **279**, 20529–20538.
95. B. Raman, L. V. Siva Kumar, T. Ramakrishna and C. Mohan Rao, *FEBS Lett.*, 2001, **489**, 19–24.
96. S. Rudiger, A. Buchberger and B. Bukau, *Nat. Struct. Biol.*, 1997, **4**, 342–349.
97. W. A. Houry, D. Frishman, C. Eckerskorn, F. Lottspeich and F. U. Hartl, *Nature*, 1999, **402**, 147–154.
98. S. J. Kim, D. G. Jeong, S. W. Chi, J. S. Lee and S. E. Ryu, *Nat. Struct. Biol.*, 2001, **8**, 459–466.
99. J. Vijayalakshmi, M. K. Mukhergee, J. Graumann, U. Jakob and M. A. Saper, *Structure*, 2001, **9**, 367–375.
100. J. H. Hoffmann, K. Linke, P. C. Graf, H. Lilie and U. Jakob, *EMBO J.*, 2004, **23**, 160–168.
101. J. R. Glover and S. Lindquist, *Cell*, 1998, **94**, 73–82.
102. M. Haslbeck, *Cell Mol. Life Sci.*, 2002, **59**, 1649–1657.

103. H. Osorio, E. Carvalho, M. del Valle, M. A. Gunther Sillero, P. Moradas-Ferreira and A. Sillero, *Eur. J. Biochem.*, 2003, **270**, 1578–1589.
104. R. Rainwater, D. Parks, M. E. Anderson, P. Tegtmeyer and K. Mann, *Mol. Cell Biol.*, 1995, **15**, 3892–3903.
105. S. Kuge, M. Arita, A. Murayama, K. Maeta, S. Izawa, Y. Inoue and A. Nomoto, *Mol. Cell Biol.*, 2001, **21**, 6139–6150.
106. B. Brune and S. Mohr, *Curr. Protein Pept. Sci.*, 2001, **2**, 61–72.
107. J. G. Kang, M. S. Paget, Y. J. Seok, M. Y. Hahn, J. B. Bae, J. S. Hahn, C. Kleanthous, M. J. Buttner and J. H. Roe, *EMBO J.*, 1999, **18**, 4292–4298.
108. B. Hoyos, A. Imam, R. Chua, C. Swenson, G. X. Tong, E. Levi, N. Noy and U. Hammerling, *J. Exp. Med.*, 2000, **192**, 835–845.
109. P. J. Kiley and H. Beinert, *Curr. Opin. Microbiol.*, 2003, **6**, 181–185.
110. L. I. Leichert and U. Jakob, *Antioxid. Redox Signal.*, 2006, **8**, 763–772.
111. L. I. Leichert, F. Gehrke, H. V. Gudiseva, T. Blackwell, M. Ilbert, A. K. Walker, J. R. Strahler, P. C. Andrews and U. Jakob, *Proc. Natl. Acad. Sci. USA*, 2008, **105**, 8197–8202.

CHAPTER 2
Oxidative Folding of Proteins in vitro

CHAPTER 2.1
The Role of Disulfide Bonds in Protein Folding and Stability

MATTHIAS JOHANNES FEIGE AND
JOHANNES BUCHNER

Department Chemie, Technische Universität München, Lichtenbergstrasse 4,
85747 Garching, Germany

2.1.1 Introduction

Disulfide bonds add covalent cross-links to the linear polypeptide chain. It is therefore intuitive to assume that this posttranslational modification has a pronounced impact on the folding and stability of proteins. These covalent linkages are found mostly in extracellular proteins where they contribute significantly to the stability of the respective protein. An extension to the intrinsic stabilization of extracellular proteins by disulfide bridges is the stabilization of quaternary structure by intermolecular cystine links. This is an effective strategy to prevent dissociation in the extracellular environment which, due to the often low concentrations, would in many cases be irreversible. However, in the context of discussing the effects on protein structure and folding, these disulfides are set apart from the intra-chain disulfides.

RSC Biomolecular Sciences
Oxidative Folding of Peptides and Proteins
Edited by Johannes Buchner and Luis Moroder

Interestingly, disulfide bonds in proteins are highly conserved; only Trp residues are even more conserved.[1] The natural selection for disulfide bonds in secreted proteins seems to be due to the simple and reversible redox chemistry involved in their formation and breaking and their intrinsic stability. Only at extreme conditions such as 100 °C and alkaline pH values do they spontaneously break or rearrange.[2] Disulfide bridges come in different flavors, reflecting the different purposes they can fulfill in a protein. In addition to their effect on the global stability of a polypeptide chain, they may have important effects on the structure, stability and dynamics of local structural elements. This is especially evident for the so-called cysteine knots, in which disulfide bridges cross each other and thus determine the local topology[3] and for "allosteric" disulfide bonds which serve to regulate protein function.[4] In this review we will focus exclusively on the dissection of the effects of structural, in particular non-local, disulfide bonds on protein stability and folding.

The effect a disulfide bond will have on a given protein will of course depend on the position of the disulfide bond in the structure of the protein. The complexity of the influence of disulfide bonds is reflected by the finding that stabilizing disulfide bonds can be found either in the core of proteins as such or surface-localized. An important issue in this context is the distance between two Cys residues in the sequence of a polypeptide chain. As will be pointed out in this chapter, the stabilizing effect depends to a significant extent on the number of amino acids spanned between the two residues. In agreement with this notion the average distance of two Cys residues in proteins was found to be 15 amino acids.[1]

Disulfide-bonded proteins have been studied early on in the history of protein folding. Notably, one of the first proteins analyzed extensively, RNase A, contains several disulfide bridges. In seminal work, Christian Anfinsen showed that the completely reduced and denatured protein could spontaneously regain its native structure including the correct disulfide bonds.[5] Efforts to analyze the impact of disulfide bonds on protein folding and stability mechanistically have contributed enormously to the generation of concepts for these processes. Furthermore, engineered disulfide bonds have become valuable tools to explore them in detail. In this chapter we will discuss the basic models resulting from these studies and we will outline the thermodynamic and kinetic implications the presence of a disulfide bond has on unfolded, folding and folded proteins.

2.1.2 Stabilization of Proteins by Disulfide Bonds

In the early days of protein science, the native state of a protein was believed to be one, rather fixed, conformation and the unfolded state to be a random coil.[6] An ideal random coil is devoid of any long-range interactions except excluded volume effects. It behaves as a freely joined chain of segments of defined length.[7] In this description, the impact of a covalent cross-link between two defined residues of the polypeptide chain, such as a disulfide bridge, was thought to be exclusively on the unfolded state. The fixed geometry of the native state should be left essentially unaltered by a bond between two residues which are in proximity, yet the freedom of the random coil polypeptide should be significantly decreased. Restricting the conformational space of the unfolded state clearly reduces its entropy. Hence, the entropy change for the reaction to the ordered native state was thought to be less negative with a net stabilization of the folded protein as a result. A quantitative description of the phenomenon was developed by Schellman, Flory, Poland and Scheraga.[8–10] The decrease in entropy of the unfolded state is derived from the probability that two otherwise free elements of the chain are now found in a defined volume element (v). Mathematically and based on polymer theory, the problem is described in Equation (2.1.1) (where R is the gas constant, l the average length of a statistical segment of the chain composed of N segments; in proteins l is assumed to be 3.8 Å corresponding to one amino acid):

$$\Delta S = -R \ln[3/(2\pi l^2 N)^{3/2}]v \qquad (2.1.1)$$

A major point of discussion has been the adequate choice of v. A value of 57.9 Å^3 based on the closest possible approach of two thiols is still mostly in use.[11] Hence, Equation (2.1.1) can be simplified to Equation (2.1.2) where n is the number of amino acids bridged by the disulfide bond:

$$\Delta S = -2.1 - 3/2 \; R \ln(n) \qquad (2.1.2)$$

Based on a study of Ribonuclease T1 with zero, one and two intact disulfide bonds, Equation (2.1.2) was developed by Pace *et al.*[11] The authors did not only find a good correlation between n and $\Delta\Delta G$ upon removal of disulfide bonds in RNase T1, but additionally agreement between experimental data for lysozyme, RNaseA and the antibody C_L domain.[11]

The above equations have two consequences. Conceptually, the stabilization of a protein is thought to be an entirely entropy-driven process with an impact exclusively on the random coil unfolded state. Practically, the stabilization exerted by a disulfide bridge should increase with the number of amino acids between the two cysteines. This theory treats the unfolded polypeptide chain as a system devoid of any intra- or intermolecular interactions. However, the water surrounding a protein is an important factor shaping the free energy landscape of the polypeptide chain during folding and in the native state.[12] Sometimes it can be regarded as being a part of the native structure.[13] The impact of this scenario with respect to disulfide bonds has been addressed by Doig and Williams in a widely recognized publication in 1991.[14] The authors argued that disulfide bonds might lead to a significantly decreased solvent accessible surface in the unfolded state. Hence, hydrophobic residues as well as hydrogen-bond donors and acceptors might become buried. This is supposed to lead on the one hand to a larger entropy of the solvent in disulfide-containing proteins due to buried hydrophobic residues. Consequently, the hydrophobic effect should be less pronounced for these proteins. On the other hand, hydrogen bonding to water in the more compact unfolded state will be possible to a lesser extent, and hence folding to the native state will be enthalpically more favorable. According to the authors, the enthalpic contribution has to be considered as the major stabilizing factor of disulfide bridges. The model will therefore be called *solvent enthalpy model* in the subsequent paragraphs. As in the *chain entropy model*, eventually occurring differences in the native state such as induced strain or reduced dynamics are also neglected in this model. Both models are summarized in Figure 2.1.1 where the reduced conformational freedom of a disulfide-linked polypeptide chain is visualized as well as hydrophobic clustering in the unfolded state or reduced solvent-protein interactions which might be present for a disulfide-linked protein.

(A) chain entropy model

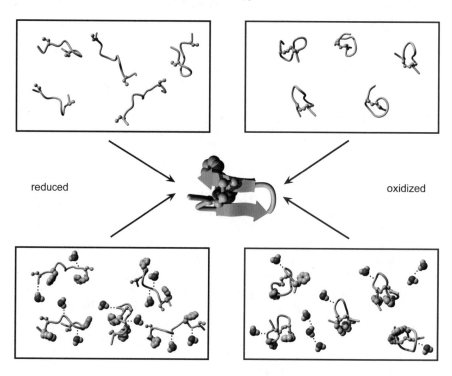

reduced   oxidized

(B) solvent enthalpy model

Figure 2.1.1   Graphical representation of the chain entropy and the solvent enthalpy model. In the chain entropy model, the major stabilizing factor exerted by a disulfide bridge is believed to be the conformational restriction of the unfolded state which renders the folding reaction entropically less unfavorable (A). In the solvent enthalpy model (B), two major effects are attributed to disulfide bridges. First, hydrophobic interactions are believed to be more pronounced in unfolded, disulfide-bridged proteins decreasing the hydrophobic effect for a protein folding reaction. Second, and numerically more important, disulfide bridges are assumed to inhibit hydrogen bonding to water in the unfolded state, making the folding reaction enthalpically more favorable. In both theories, the native state is believed to be unaffected. To illustrate both models, a hypothetical 12-residue β-hairpin structure with a single disulfide bridge was designed.

A variety of experimental evidence argues against both theories rendering the problem much more complex yet more insightful. Site-specific mutagenesis offered the possibility to introduce artificial disulfide bonds at defined positions within a protein and allowed to explore the effects of disulfide bonds rigorously.[15–18] Many results of these studies were not compatible with predictions based on the prevailing theories concerning the effects of disulfide bonds on protein stability. Most discrepancies arose from efforts to stabilize a variety of proteins by the introduction of new disulfide bridges. In contrast to expectations, destabilization was the result of a significant proportion of these attempts.[19–21] High-resolution structures of the engineered proteins helped to rationalize these unexpected effects. It turned out that strain was imposed on the native state by the disulfide cross-links in several cases.[22] Two particularly insightful examples are studies on barnase[19,23] and staphylococcal nuclease.[21] Clarke and Fersht introduced three artificial disulfide bonds into barnase. One was found to destabilize the native state, one stabilized it to an extent predicted by the chain entropy model and one to a much lesser extent.[19] Furthermore, the bridge connecting fewer residues was found to be more stabilizing than the one encompassing more residues. For staphylococcal nuclease, no stabilization was found for all disulfide constructs, but in this case a peptide bond *cis/trans* equilibrium was shifted in the native state and the catalytic activity of all mutants was reduced.[21] The NMR H/D exchange analysis of the barnase mutants revealed an altered dynamics of the native state imposed by the presence of the covalent cross-links.[23] In the study on staphylococcal nuclease, strain on the native state, as reflected by the alteration of the *cis/trans* equilibrium, was evoked to explain the observed unexpected alterations.[21] A more general comparison between experimental data for disulfide-bridged proteins with native or engineered disulfide bonds and the chain entropy model, which predicts that disulfide bonds stabilize proteins in a strictly loop length dependent manner, can be found in Figure 2.1.2. A clear deviation between experimental data and predictions is evident from Figure 2.1.2 for the major part of the proteins.

Protein	Cross-linked residues	Loop size	T (°C)	$\Delta\Delta G^{experimental}$ (kJ/mol)	$\Delta\Delta G^{theoretical}$ (kJ/mol)
human C_L^λ	26-87	60	25	16.7	17.8
human C_L^κ	26-86	61	25	19.7	17.9
tendamistat	11-27	17	25	24.3	13.2
	45-73	29	25	20.9	15.1
barnase	43-80	38	25	9.6	16.1
	70-92	23	25	-2.5	14.3
	85-102	18	25	18.8	13.4
CD2	13-66	54	25	29.3	17.5
	15-64	50	25	-6.3	17.2
	79-90	12	25	10.0	11.9

Figure 2.1.2 Comparison of experimental data and theory for the chain entropy model. Experimental data on the stability of proteins in the reduced and oxidized state are summarized in the table on the left.[11,20,70,75] A comparison to the predicted stabilization based on Equation (2.1.2) is shown on the right. The chain entropy model predicts a larger stabilization *via* disulfide bridges at higher temperatures and longer loop lengths.

One clear conclusion can be drawn from the two mentioned and a variety of other studies:[15,16,18,22,24–29] the native state is not left unaffected by a disulfide bridge. A major assumption underlying the chain entropy model as well as the solvent enthalpy model hence does clearly not hold in all cases. Strain is often found to be imposed on the native state by the presence of covalent cross-links. This can be reflected in structural changes[21,22] or more subtle changes, like alterations of the dynamics of the native state.[30] The important consequence is that the enthalpy as well as the entropy of the native state are not unlikely to be altered if two residues in the polypeptide are covalently cross-linked. Furthermore, the alterations in dynamics and structure are not always a global but sometimes a local context-specific effect. β-sheets and loops are thought to be more suited to dissipate induced strain than α-helical elements, and more dynamic parts of the structure are influenced to a greater extent. In addition, changes in the solvent-exposed hydrophobic surface which have been observed for disulfide mutants of interleukin-4 have a significant effect on the enthalpy of the native state.[26] Changes in the dynamics, *e.g.* the vibrational normal modes of a protein, will also influence the entropy of the native state as has been shown in a molecular dynamics study by Karplus and co-workers.[30] Consequently, a net destabilization due to a loss of native state entropy can be expected in some cases if disulfide bonds are introduced into a protein.[30] In summary, novel experimental as well as theoretical insights clearly argue against the simple chain entropy model, at least concerning one side of the equation, the native state. The same holds for the solvent enthalpy model by Doig and Williams, which is additionally at odds with thermodynamic parameters derived for some disulfide-bridged proteins which showed that disulfide bridges do not necessarily stabilize the native state enthalpically.[26] But what about the unfolded state? Do the assumptions underlying the theory hold? In other words, is it correct to assume a random coil almost devoid of any interactions except for hydrogen bonding to the solvent? Clearly not in all cases. Residual structure which is believed to be important for folding pathways has been detected in a variety of proteins.[31–35] In particular, residual hydrophobic interactions or fluctuating α-helical elements seem to be a rather general than an exceptional feature of proteins, in particular under mildly denaturing conditions.[32,34,36–38] Importantly, the structural features of the unfolded state like residual structure, which has been reported for the unfolded state of barnase,[39] or the tendency towards irreversible inactivation have in some cases been shown to be influenced by the presence or absence of disulfide bridges.[40,41] Hence, disulfide bonds very likely not only influence the conformational freedom of the unfolded chain but can also introduce structure which might also be protective against irreversible aggregation.[40] Even apparently minor structural changes like clustering of some hydrophobic residues will influence the enthalpy and entropy of the unfolded state. In addition, as outlined above, dynamic phenomena, strain induced on the native state and an impact on the native structure are completely omitted in these theories. Although some proteins could be adequately described by one of the theories, it comes as no surprise that the chain entropy model as well as the solvent enthalpy model fail to describe the major part of experimental data (Figure 2.1.2).

How can the different findings on the divergent effects disulfide bridges have on the stability of different proteins be summarized in a comprehensive way? Hardly by any theory describing the unfolded polypeptide chain as a construct composed of identical elements and devoid of structural features. The decreased entropy of an unfolded and cross-linked polypeptide chain has to be taken into account as developed in the chain entropy model. Furthermore, the solvent enthalpy model based on the decreased hydrophobic effect and hydrogen bonding in the unfolded state should be included. And to be added to these models are interactions eventually present in the unfolded state due to the disulfide bridge, which may not be localized directly around the bond but can be present as long-range residual structure.[39] This will clearly have an effect on the enthalpy and entropy balance for the reaction to the native state. The same holds for decreased dynamics of the native state, locally or globally, as well as enthalpically unfavorable strain or enthalpically favorable induced proximity of interacting residues. In summary, the effect of a disulfide bridge on the stability of a protein can be easily assessed experimentally, yet its molecular explanation is almost as diverse as the protein under investigation. The key factors giving rise to the net effect are most likely all known but, as for the prediction of the native state of a protein, their contribution to the overall effect are blurred in their sum as well as their mutual influence on each other. It seems therefore highly rewarding to use a combined empirical and theoretical approach. Exact stability data, if possible together with structural data on the native as well as the unfolded state of the protein under investigation, are a prerequisite for the detailed understanding of the effect of a disulfide bridge. They should be complemented by theoretical approaches like molecular dynamics simulations of the native and the unfolded state to obtain a more complete picture. This might be different in detail for different proteins but in summary these analyses will most likely reveal general principles.

2.1.3 Disulfide Bonds in Protein Folding Reactions

One of the major questions in biophysical chemistry is how a linear polypeptide chain specifically adopts its intricate three-dimensional structure in a reasonable amount of time. A variety of mutational approaches,[42–50] high-resolution structural techniques[51–54] and ultrafast perturbation methods[55–58] have recently provided deep insight into this phenomenon of biological self-organization. Nowadays, proteins are believed to fold with a certain heterogeneity *via* multiple individual pathways to the native state,[59] yet distinct features of the protein dictate the general trajectories of the folding process. In particular, residues whose interactions define the overall topology of a protein are assumed to be in contact in the transition state and are believed to be key elements in a protein folding reaction in general.[60–62] While there seems to be consensus concerning the general scheme of events, a variety of details are still under intense debate, like the role of residual structure in the unfolded state,[35] the role of folding inter-mediates on the way to the native state[63] and the heterogeneity of the transition states.[61] In all these questions, disulfide bridges have played a pronounced role as factors influencing these processes as well as tools to elucidate them.

Independent of its pathway, every protein begins its folding reaction in the unfolded state, be it at the ribosome or under denaturing conditions. There-fore, the nature of the unfolded state of proteins has gained much attention again recently, in part fuelled by the development of novel experimental techni-ques.[34,64] In this context, deviations from a random coil are regarded as important elements of protein folding in general. They will not only have an impact on the net stability balance of a protein, as outlined above, but further-more likely shape the energy landscape on the way to the native state. Preformed interactions in the unfolded state might therein have a rather controversial effect on a folding reaction. If native and not detrimentally influencing the ability of the remaining polypeptide chain to explore the necessary conformational space, they might lead to faster and more efficient folding to the native state. Yet if non-native or too stable, the opposite might hold. Examples for both cases have been reported for disulfide-bridged proteins, which *per se* possess preformed correct tertiary interactions. In the case of RNase T1 (Figure 2.1.3), deceleration of the folding kinetics was observed which has been attributed to decreased chain flexibility in the presence of a disulfide bridge.[65] In this context it is important to note than RNase T1 possesses two disulfide bridges, one connecting a small N-terminal β-turn and one connecting the C-terminus to this N-terminal β-turn (Figure 2.1.3). Accordingly, the protein will almost be a completely looped structure in its unfolded state, which will clearly have an effect on the dynamics of the whole polypeptide chain and might interfere with the establishment of a folding nucleus, if not in proximity of the disulfide bridges, which is unlikely due to the solvent exposure of the disulfide bridges in the native state. This has been confirmed for one of the disulfide bonds connecting residues 2 and 10.

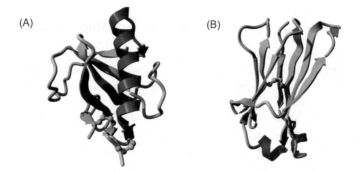

(A)   (B)

Figure 2.1.3   Structural comparison of RNase T1 and the murine C_L domain. RNaseT1 (A) possesses two solvent-exposed disulfide bridges. The murine immunoglobulin G C_L domain has one disulfide bridge buried in the hydrophobic core of the protein (B).

Its deletion led to a stability reduction of RNase T1 but had no impact on the folding mechanism of the protein.[66]

For C_L, the constant domain of the antibody light chain, on the other hand, its single intrinsic disulfide bridge accelerated folding up to ~100-fold (Figure 2.1.3).[67,68] In the case of this protein, the disulfide bond is found in the hydrophobic core of the protein and part of the folding nucleus.[68] Accelerated folding due to preformed interactions which facilitate the way to the native state were hence expected and experimentally observed. For many other proteins, either a deceleration or an acceleration of the respective folding reactions have been observed in the presence of their natural intrinsic disulfide bonds.[20,69–74] Hence, disulfide bonds are in general far from being inert in kinetic terms. Very often, different disulfide bridges within the same protein had different effects on the folding rates. A comprehensive study in this respect has been carried out for the all-β protein CD2, where thirteen different artificially introduced disulfide bridges showed a markedly different impact on the folding behaviour.[20] Similar results were obtained for unfolding reactions which were either left unaffected by disulfide bonds or their rates were reduced.[20,75] The experimental findings are in agreement with simple lattice-based simulations, where disulfide bonds inside the folding nucleus were found to be accelerating yet decelerating if outside.[76] In some simulations folding has been found to be influenced by a disulfide bond to a lesser extent than expected, revealing a larger dynamics of a real polypeptide chain in comparison to a chain moving on a lattice.[76] Despite the heterogeneous effects of a disulfide bridge on protein folding/unfolding kinetics at first glance, it can be more easily rationalized than the effect on native state stability. The key lies in the transition state for folding and unfolding. Acceleration of folding is expected when residues are cross-linked which have to come into contact early in a protein folding reaction, if not weighed out by entropy/enthalpy compensation in the transition state. Analogously, deceleration of unfolding is expected when two residues, whose interactions are broken in the transition state for unfolding are covalently linked. Both effects can provide structural information about the otherwise hardly accessible transition state as exemplified for the immunoglobulin proteins C_L and CD2 where the transition state for folding could be stabilized by disulfide bridges[20,68] or for barnase, where the transition state for unfolding could be destabilized by a disulfide bond.[75] Often, disulfide-bridged proteins are found to fold less cooperatively.[70] This might be caused by the population of disulfide-stabilized intermediates in agreement with the stabilization of the transition state and subsequent partially folded states. If stabilization of partially folded structures becomes too strong, either by native or non-native interactions, this can even result in a net deceleration of a protein folding reaction as has been observed for CD2.[71] Attempts to increase the folding rate by multiple disulfide bonds had the opposite effect: deceleration of the folding reaction by the over-stabilization of a partially folded state.[71] This highlights the role of cooperativity for efficient protein folding which can be beneficially or detrimentally influenced by preformed tertiary interactions like disulfide bridges.

2.1.4   Conclusions

Disulfide bonds are one of the most widespread structural elements stabilizing the native state of a protein. Their stabilizing role likely arises from a variety of effects imposed on the unfolded as well as the native state. A destabilization of the unfolded state due to the restriction of the conformational freedom as well as decreased protein-solvent interaction will favor folding to the native state which, additionally, can be stabilized by induced proximity of energetically favorable interactions. In most cases, in particular for naturally selected disulfide-bond positions, the stabilizing effects of this covalent bond are more pronounced than eventually occurring destabilizing effects. These can include a stabilization of the unfolded state due to residual or non-native structure. Additionally, strain as well as reduced dynamics imposed on the native state can significantly decrease the stability of the native state. The sum of all these factors will be the net stabilizing effect of a disulfide bridge.

Having a closer look at the kinetic impact of disulfide bridges is highly insightful. By accelerated folding or decelerated unfolding, entropic, enthalpic and structural conclusions about otherwise almost inaccessible transitions states can be drawn. Furthermore, how disulfide bonds are positioned in natural proteins might not only help in the design of proteins with improved characteristics but also hold a clue for the specific structural features a protein has been selected for. In this respect, it is particularly revealing to look at disulfide bridge positions in an evolutionary perspective. For many extracellular proteins which do not have to undergo a large variety of binding reactions requiring large scale dynamics, loop structures and flexible parts are often found to be disulfide-linked.[77] This, on the one hand, directly reduces susceptibility to protease digestion and, on the other hand, might generally reduce unfolding rates by linking parts which are likely to come apart early in an unfolding reaction. If large flexibility is needed, as *e.g.* in members of the immunoglobulin superfamily, which are optimized for molecular recognition processes, another evolutionary strategy has prevailed. Here, disulfide bridges are found in the folding nucleus,[77] where they seem to accelerate refolding if unfolding occurs. At the same time this strategy permits flexibility where needed and also results in a net stabilization of the native state. Positions of evolutionary selected disulfide bonds might hence not only provide insight into folding nuclei but additionally provide a basis for the rational stabilization of homologous proteins.

The elucidation of the versatile effects disulfide bonds have on protein structure, stability and folding has significantly extended our knowledge about how proteins fold and function and it paved the path to influence their properties in desirable ways.

References

1. J. M. Thornton, *J. Mol. Biol.*, 1981, **151**, 261–287.
2. T. J. Ahern and A. M. Klibanov, *Science*, 1985, **228**, 1280–1284.

3. N. W. Isaacs, *Curr. Opin. Struct. Biol.*, 1995, **5**, 391–395.

4. B. Schmidt and P. J. Hogg, *BMC Struct. Biol.*, 2007, **7**, 49.

5. C. B. Anfinsen and E. Haber, *J. Biol. Chem.*, 1961, **236**, 1361–1363.

6. C. Tanford, K. Kawahara and S. Lapanje, *J. Biol. Chem.*, 1966, **241**, 1921–1923.

7. L. J. Smith, K. M. Fiebig, H. Schwalbe and C. M. Dobson, *Fold. Des.*, 1996, **1**, R95–106.

8. J. A. Schellman, *C.R. Trav. Lab Carlsberg. (Chim.)*, 1955, **29**, 230–259.

9. P. J. Flory, *J. Am. Chem. Soc.*, 1956, **78**, 5222–5235.

10. D. C. Poland and H. A. Scheraga, *Biopolymers*, 1965, **3**, 379–399.

11. C. N. Pace, G. R. Grimsley, J. A. Thomson and B. J. Barnett, *J. Biol. Chem.*, 1988, **263**, 11820–11825.

12. R. Lumry and S. Rajender, *Biopolymers*, 1970, **9**, 1125.

13. B. P. Schoenborn, A. Garcia and R. Knott, *Prog. Biophys. Mol. Biol.*, 1995, **64**, 105–119.

14. A. J. Doig and D. H. Williams, *J. Mol. Biol.*, 1991, **217**, 389–398.

15. R. T. Sauer, K. Hehir, R. S. Stearman, M. A. Weiss, A. Jeitlernilsson, E. G. Suchanek and C. O. Pabo, *Biochemistry*, 1986, **25**, 5992–5998.

16. J. E. Villafranca, E. E. Howell, S. J. Oatley, N. H. Xuong and J. Kraut, *Biochemistry*, 1987, **26**, 2182–2189.

17. T. Vogl, R. Brengelmann, H. J. Hinz, M. Scharf, M. Lotzbeyer and J. W. Engels, *J. Mol. Biol.*, 1995, **254**, 481–496.

18. C. M. Johnson, M. Oliveberg, J. Clarke and A. R. Fersht, *J. Mol. Biol.*, 1997, **268**, 198–208.

19. J. Clarke, K. Henrick and A. R. Fersht, *J. Mol. Biol.*, 1995, **253**, 493–504.

20. J. M. Mason, N. Gibbs, R. B. Sessions and A. R. Clarke, *Biochemistry*, 2002, **41**, 12093–12099.

21. A. P. Hinck, D. M. Truckses and J. L. Markley, *Biochemistry*, 1996, **35**, 10328–10338.

22. B. A. Katz and A. Kossiakoff, *J. Biol. Chem.*, 1986, **261**, 5480–5485.

23. J. Clarke, A. M. Hounslow and A. R. Fersht, *J. Mol. Biol.*, 1995, **253**, 505–513.

24. H. Kikuchi, Y. Goto, K. Hamaguchi, S. Ito, M. Yamamoto and Y. Nishijima, *J. Biochem.*, 1987, **102**, 651–656.

25. R. Kuroki, K. Inaka, Y. Taniyama, S. Kidokoro, M. Matsushima, M. Kikuchi and K. Yutani, *Biochemistry*, 1992, **31**, 8323–8328.

26. D. C. Vaz, J. R. Rodrigues, W. Sebald, C. M. Dobson and R. M. M. Brito, *Protein Sci.*, 2006, **15**, 33–44.

27. B. Asgeirsson, B. V. Adalbjornsson and G. A. Gylfason, *Biochim. Biophys. Acta*, 2007, **1774**, 679–687.

28. A. Shimizu-Ibuka, H. Matsuzawa and H. Sakai, *Biochemistry*, 2004, **43**, 15737–15745.

29. A. McAuley, J. Jacob, C. G. Kolvenbach, K. Westland, H. J. Lee, S. R. Brych, D. Rehder, G. R. Kleemann, D. N. Brems and M. Matsumura, *Protein Sci.*, 2008, **17**, 95–106.

30. B. Tidor and M. Karplus, *Protein. Struct. Funct. Genet.*, 1993, **15**, 71–79.

31. A. Lapidot and C. S. Irving, *J. Am. Chem. Soc.*, 1977, **99**, 5489–5490.
32. D. Neri, M. Billeter, G. Wider and K. Wüthrich, *Science*, 1992, **257**, 1559–1563.
33. V. L. Arcus, S. Vuilleumier, S. M. Freund, M. Bycroft and A. R. Fersht, *Proc. Natl. Acad. Sci. USA*, 1994, **91**, 9412–9416.
34. K. H. Mok, L. T. Kuhn, M. Goez, I. J. Day, J. C. Lin, N. H. Andersen and P. J. Hore, *Nature*, 2007, **447**, 106–109.
35. E. R. McCarney, J. E. Kohn and K. W. Plaxco, *Crit. Rev. Biochem. Mol. Biol.*, 2005, **40**, 181–189.
36. D. Eliezer, J. Chung, H. J. Dyson and P. E. Wright, *Biochemistry*, 2000, **39**, 2894–2901.
37. J. Yao, J. Chung, D. Eliezer, P. E. Wright and H. J. Dyson, *Biochemistry*, 2001, **40**, 3561–3571.
38. C. S. Le Duff, S. B. Whittaker, S. E. Radford and G. R. Moore, *J. Mol. Biol.*, 2006, **364**, 824–835.
39. J. Clarke, A. M. Hounslow, C. J. Bond, A. R. Fersht and V. Daggett, *Protein Sci.*, 2000, **9**, 2394–2404.
40. L. J. Perry and R. Wetzel, *Science*, 1984, **226**, 555–557.
41. R. Wetzel, L. J. Perry, W. A. Baase and W. J. Becktel, *Proc. Natl. Acad. Sci. USA*, 1988, **85**, 401–405.
42. A. R. Fersht, A. Matouschek and L. Serrano, *J. Mol. Biol.*, 1992, **224**, 771–782.
43. A. Matouschek, L. Serrano, E. M. Meiering, M. Bycroft and A. R. Fersht, *J. Mol. Biol.*, 1992, **224**, 837–845.
44. A. Matouschek, L. Serrano and A. R. Fersht, *J. Mol. Biol.*, 1992, **224**, 819–835.
45. A. Matouschek, J. T. Kellis Jr., L. Serrano, M. Bycroft and A. R. Fersht, *Nature*, 1990, **346**, 440–445.
46. A. Matouschek, J. T. Kellis Jr., L. Serrano and A. R. Fersht, *Nature*, 1989, **340**, 122–126.
47. L. Serrano, J. T. Kellis Jr., P. Cann, A. Matouschek and A. R. Fersht, *J. Mol. Biol.*, 1992, **224**, 783–804.
48. L. Serrano, A. Matouschek and A. R. Fersht, *J. Mol. Biol.*, 1992, **224**, 847–859.
49. L. Serrano, A. Matouschek and A. R. Fersht, *J. Mol. Biol.*, 1992, **224**, 805–818.
50. D. P. Raleigh and K. W. Plaxco, *Protein Pept. Lett.*, 2005, **12**, 117–122.
51. D. M. Korzhnev, X. Salvatella, M. Vendruscolo, A. A. Di Nardo, A. R. Davidson, C. M. Dobson and L. E. Kay, *Nature*, 2004, **430**, 586–590.
52. T. L. Religa, J. S. Markson, U. Mayor, S. M. V. Freund and A. R. Fersht, *Nature*, 2005, **437**, 1053–1056.
53. U. Mayor, N. R. Guydosh, C. M. Johnson, J. G. Grossmann, S. Sato, G. S. Jas, S. M. V. Freund, D. O. V. Alonso, V. Daggett and A. R. Fersht, *Nature*, 2003, **421**, 863–867.
54. H. Q. Feng, Z. Zhou and Y. W. Bai, *Proc. Natl. Acad. Sci. USA*, 2005, **102**, 5026–5031.

55. T. L. Religa, C. M. Johnson, D. M. Vu, S. H. Brewer, R. B. Dyer and A. R. Fersht, *Proc. Natl. Acad. Sci. USA*, 2007, **104**, 9272–9277.
56. H. Ma, C. Wan and A. H. Zewail, *J. Am. Chem. Soc.*, 2006, **128**, 6338–6340.
57. W. A. Eaton, V. Munoz, S. J. Hagen, G. S. Jas, L. J. Lapidus, E. R. Henry and J. Hofrichter, *Annu. Rev. Biophys. Biomol. Struct.*, 2000, **29**, 327–359.
58. J. Kubelka, T. K. Chiu, D. R. Davies, W. A. Eaton and J. Hofrichter, *J. Mol. Biol.*, 2006, **359**, 546–553.
59. K. A. Dill and H. S. Chan, *Nat. Struct. Biol.*, 1997, **4**, 10–19.
60. K. Lindorff-Larsen, P. Rogen, E. Paci, M. Vendruscolo and C. M. Dobson, *Trends Biochem. Sci.*, 2005, **30**, 13–19.
61. K. Lindorff-Larsen, M. Vendruscolo, E. Paci and C. M. Dobson, *Nat. Struct. Mol. Biol.*, 2004, **11**, 443–449.
62. M. Vendruscolo, E. Paci, C. M. Dobson and M. Karplus, *Nature*, 2001, **409**, 641–645.
63. D. J. Brockwell and S. E. Radford, *Curr. Opin. Struct. Biol.*, 2007, **17**, 30–37.
64. K. Sugase, H. J. Dyson and P. E. Wright, *Nature*, 2007, **447**, 1021–1025.
65. M. Mucke and F. X. Schmid, *J. Mol. Biol.*, 1994, **239**, 713–725.
66. L. M. Mayr, D. Willbold, O. Landt and F. X. Schmid, *Protein Sci.*, 1994, **3**, 227–239.
67. Y. Goto and K. Hamaguchi, *J. Mol. Biol.*, 1982, **156**, 911–926.
68. M. J. Feige, F. Hagn, J. Esser, H. Kessler and J. Buchner, *J. Mol. Biol.*, 2007, **365**, 1232–1244.
69. S. R. K. Ainavarapu, J. Brujic and J. M. Fernandez, *Biophys. J.*, 2007, **92**, 225–233.
70. N. Schönbrunner, G. Pappenberger, M. Scharf, J. Engels and T. Kiefhaber, *Biochemistry*, 1997, **36**, 9057–9065.
71. J. M. Mason, M. J. Cliff, R. B. Sessions and A. R. Clarke, *J. Biol. Chem.*, 2005, **280**, 40494–40499.
72. M. E. Denton, D. M. Rothwarf and H. A. Scheraga, *Biochemistry*, 1994, **33**, 11225–11236.
73. S. H. Lin, Y. Konishi, B. T. Nall and H. A. Scheraga, *Biochemistry*, 1985, **24**, 2680–2686.
74. A. Yokota, K. Izutani, M. Takai, Y. Kubo, Y. Noda, Y. Koumoto, H. Tachibana and S. Segawa, *J. Mol. Biol.*, 2000, **295**, 1275–1288.
75. J. Clarke and A. R. Fersht, *Biochemistry*, 1993, **32**, 4322–4329.
76. V. I. Abkevich and E. I. Shakhnovich, *J. Mol. Biol.*, 2000, **300**, 975–985.
77. L. A. Mirny and E. I. Shakhnovich, *J. Mol. Biol.*, 1999, **291**, 177–196.

CHAPTER 2.2

Strategies for the Oxidative in vitro Refolding of Disulfide-bridge-containing Proteins

RAINER RUDOLPH AND CHRISTIAN LANGE

Institut für Biochemie and Biotechnologie, Martin-Luther-Universität Halle-Wittenberg, Kurt-Mothes-Str. 3, 06120 Halle (Saale), Germany

2.2.1 Introduction

The advent of recombinant DNA technology three decades ago opened the possibility for the heterologous expression of almost any protein in microbial host organisms or eukaryotic cell culture systems. This has enabled the production even of those proteins that had been almost impossible to obtain from natural sources, as well as the manipulation of protein sequences for specific purposes. The engineering, production, and purification of proteins as tools for research and development, as well as for therapeutic uses, today represent mature branches of applied technology.[1,2]

Some important practical issues have to be solved in the development of the production process for any given new protein. One of these problems, which requires special attention when working with microbial host systems, concerns the formation of intraprotein disulfide bonds. Most proteins that are of interest for medical applications are either natively secreted or exposed to the extracellular space as part of membrane-bound receptor complexes. Examples for marketable therapeutic products that are derived of extracellular proteins include antibodies and antibody fragments, peptide hormones like insulin and

RSC Biomolecular Sciences
Oxidative Folding of Peptides and Proteins
Edited by Johannes Buchner and Luis Moroder
© Royal Society of Chemistry 2009
Published by the Royal Society of Chemistry, www.rsc.org

human growth hormone, as well as thrombolytic proteins like tissue-type plasminogen activator and hirudin. As a consequence, many of these proteins contain specific disulfide bonds that provide structural stability and are in most cases a necessary prerequisite for function.[3,4] Throughout the present book, the current state of knowledge on the fascinating aspects of disulfide-bond formation *in vivo* and the roles that this modification may play for the structure and function of proteins are highlighted.

The correct formation and preservation of disulfide bonds in the final product has to be guaranteed by any protein production process. When proteins are expressed into the cytoplasmic space of microbial hosts, the formation of disulfide bonds is generally not possible due to the reducing environment of the cytosol.[4–6] The concomitantly reduced stability of the native conformation of naturally disulfide bond-containing proteins aggravates a general problem of high-yield heterologous expression, namely the formation of insoluble, biologically inactive aggregates of the produced protein that are deposited in the form of inclusion bodies[7–9] (Figure 2.2.1). The formation of inclusion bodies *in vivo* can be explained quite well by models taking into account the kinetic competition between folding of the over-expressed protein and its aggregation.[10] A folding reaction that is impaired by the inability to form stable native disulfide bonds will consequently direct the protein product into the solid state.

Several workarounds exist for this problem. They include production in different expression systems, *e.g.*, yeast,[1,11] animal cell culture[12] and plant tissue,[13] expression into the bacterial periplasm[14] or secretion into the culture medium,[15] as well as the engineering of variants of the desired protein product that are sufficiently stable without internal disulfide bonds.[16] Each of these possibilities is associated with its own specific advantages and drawbacks, as discussed in Chapters 1.3 and 1.7 of this book.

Inclusion bodies obtained by bacterial over-expression are significantly enriched in recombinant protein, and the sequestered insoluble product is well protected from the action of intracellular proteases and other degrading enzymes.[17] These facts combined make high-yield expression into inclusion bodies followed by solubilization and *in vitro* refolding a straightforward, and in most cases viable, option for the development of protein production processes.

The generalized protocol for the recovery of active protein from inclusion body material[8,18–21] requires, as a first step, the solubilization of the inclusion bodies in denaturing agents like urea or guanidinium chloride (GuHCl)[i] in the presence of a reducing thiol reagent like, *e.g.*, dithiothreitol (DTT) or dithioerythritol (DTE). The reducing agent has to be included in this step in order to break disulfide bonds that may have accidentally formed during cell breakup and sample handling on air. In general, the reducing agent then needs

---

[i] Abbreviations used: BiP, heavy chain-binding protein; BPTI, bovine pancreatic trypsin inhibitor; DTNB, 5,5′-dithio-bis(2-nitrobenzoic acid); DTE, dithioerythritol; DTT, dithiothreitol; E^0, biochemical standard redox potential; EDTA, ethylene diamine tetraacetic acid; GSH, reduced glutathione; GSSG, oxidized glutathione; GSeH, reduced selenoglutathione; GSeSeG, oxidized selenoglutathione; GuHCl, guanidinium chloride; PDI, protein disulfide isomerase (EC 5.3.4.1); RNase, endoribonuclease (EC 3.1.27); Sec, selenocysteine.

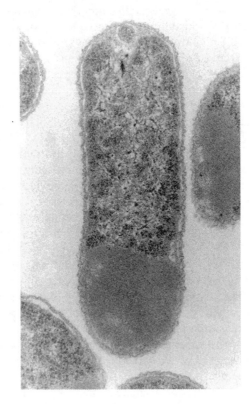

Figure 2.2.1 *Escherichia coli* cell filled with inclusion bodies (lower third of cell volume).

to be removed under conditions where the re-oxidation of the free cysteine thiol groups is prevented, and oxidative *in vitro* folding is initiated by transfer of the reduced-denatured protein into an oxidizing renaturation buffer. This very general process may be carried out according to a wide variety of strategies. The conditions for the refolding process have to be carefully chosen and controlled in order to obtain a good yield of correctly folded product. The desired reaction competes with a number of unproductive side reactions, mainly the aggregation of non-native species, and the formation of mis-folded species, that may contain non-native disulfide bonds.[10,22]

The choice of refolding conditions is dependent on the nature of each individual protein. No generally applicable protocols are available, and to date almost any newly developed *in vitro* refolding requires a largely empirical optimization. However, a series of guidelines does exist that aid the researcher in this task, and protocols are available that may serve as starting points for the development of an optimized process for any given refolding problem, with a reasonably high probability of success. One strategy that is commonly applied in this context involves the rapid dilution of reduced-denatured protein into a buffer system that contains, on the one hand, additives like, *e.g.*, L-arginine or glycerol that serve the purpose of preventing unspecific aggregation and enhancing the stability of the native state, respectively, and, on the other hand a thiol redox system, *e.g.*, a mixture of reduced and oxidized glutathione (GSH/GSSG), that allows for the controlled formation of disulfide bridges.[21,23] A major issue with this approach is the need to work at low protein concentrations, due to the already mentioned aggregation tendency of the unfolded and non-native protein species that are present after the dilution of the denatured inclusion body material into the refolding buffer. This might be remedied by applying the pulsed renaturation method, in which small aliquots of denatured protein are added to the refolding mixture with sufficient time in between to allow for the renaturation reaction to proceed beyond the early, aggregation-prone states.[23,24] Many more different ways of performing the task are conceivable, and have been applied.

The purpose of the present chapter is to give an overview of the methods that have been successfully used for the oxidative *in vitro* folding of disulfide-containing proteins. Emphasis will be put on reviewing the various strategies that have been explored for controlling and optimizing the involved thiol redox chemistry.

2.2.2 Chemical Systems for the *in vitro* Formation of Disulfide Bridges

The sulfur chemistry of proteins has been intensively studied since the early days of protein biochemistry.[25–27] The sulfur centers in cysteine and methionine residues, and in protein-bound iron–sulfur clusters play an essential role in the biochemistry of life. Sulfur in cysteines has been found in formal oxidation numbers between +6 and −2, although not all of them have been identified within a physiological context. Apart from being involved in the ligation of transition metal ions, redox catalysis and signaling, the cystine disulfide bridges in many proteins are critical for the formation and stabilization of protein tertiary structure. The *in vitro* refolding of a disulfide bridge-containing protein from its reduced denatured state requires the formal oxidation of the sulfur atoms in free cysteines (−2) to disulfides (−1). *In vivo*, during the processing of newly synthesized proteins in the endoplasmic reticulum and the periplasm, respectively, as well as in most of the redox systems discussed in this chapter, this change in oxidation number is effected by thiol-disulfide exchange reactions.

In principle, the number of possible combinations of free cysteines to form disulfide bridges in proteins may be huge. While for proteins that contain two disulfide bonds, like, *e.g.*, single-chain F_v fragments, three combinations are theoretically conceivable, the number of combinations for tissue-type plasminogen activator, which contains 17 disulfide bonds and an additional free cysteine, is 2.2×10^{20} (Table 2.2.1). If disulfide pairing were to occur randomly during oxidative protein folding, the number of possible combinations would drown out any chance of obtaining native protein. However, it has become clear that folding proteins do not randomly sample conformational space, but are directed by forces driving the collapse into the native state along more or

Table 2.2.1 Number of possible combinations of $2n$ cysteines to form n disulfide bonds.

n	Number of combinations[a]
1	1
2	3
3	15
4	105
5	945
6	10395
7	135135
8	2027025
9	34459425
10	654729075
11	$1.374931058 \times 10^{10}$
12	$3.162341432 \times 10^{11}$
13	$7.905853581 \times 10^{12}$
14	$2.134580467 \times 10^{14}$
15	$6.190283354 \times 10^{15}$
16	$1.918987840 \times 10^{17}$
17	$6.332659871 \times 10^{18}$
18	$2.216430955 \times 10^{20}$
. . .	. . .

[a]$(2n - 1) \times (2n - 3) \times (2n - 5) \times \ldots \times 3 \times 1$

less clearly defined pathways. While oxidative folding *in vivo* is supported by the cellular chaperone and foldase systems, oxidative refolding *in vitro* has to depend on the physicochemical driving forces alone. In any case, the native and correctly disulfide-bridged state is thermodynamically favored under physiological conditions, and if the refolding reaction occurs under thermodynamic control, the protein molecule should eventually be able to reach it.

The kinetic pathways for the oxidative refolding of proteins containing multiple native disulfide bonds are often complex and may involve multiple consecutive rearrangements. This has been studied extensively for bovine pancreatic trypsin inhibitor (BPTI),[28–30] while the elucidation of the pathways for the oxidative folding of the natively dimeric human macrophage colony-stimulating factor β[31] and hen egg-white lysozyme[32] represent more recent examples. The complexity of oxidative refolding pathways along with competing unproductive side reactions like, *e.g.*, aggregation of non-native protein species frequently leads to oxidative *in vitro* refolding reactions proceeding under kinetic control. The choice of the chemical system responsible for the formation of disulfide bridges may play a critical role in determining the yield of renaturation.

2.2.2.1 Transition Metal-catalyzed Air Oxidation

The most readily available electron sink for the oxidative *in vitro* refolding of proteins is atmospheric oxygen. The earliest refolding experiments naturally made use of air oxidation. In a series of ground-breaking works performed during the late 1950s by Anfinsen and others, the hypothesis was firmly established that for most proteins the three-dimensional structure necessary for activity is encoded in the primary sequence alone. This body of work relied on the capability of the molecular oxygen present in air to enable the re-formation of native disulfide bonds in reduced denatured RNase A.[33–35]

One of the major drawbacks of using air as the oxidating agent is the slow kinetics of direct sulfur–oxygen reactions under physiologic conditions. As a consequence, reaction systems for the oxidative refolding of proteins by air are dependent on the presence of catalytic amounts of transition metal ions that act as redox mediators. Cu^{2+} was shown to be the most effective catalytic metal ion in this respect.[36] This limitation had not been apparent in the first studies on air oxidation due to the presence of trace amounts of Cu^{2+} even in double-distilled water, but addition of EDTA to a refolding buffer was quickly found to abolish the oxidative renaturation of proteins by air.

One of the key parameters determining the efficiency of air oxidation is the transport of oxygen from the gas phase into the solution, which may present problems of being reproducibly controlled in laboratory scale renaturation. For large-scale refolding processes, vigorous stirring is required to bring about the necessary mass transfer of oxygen into the solution. In this case, the induced interfacial stress may lead to loss of protein to aggregation.[37]

Although a number of active industrial processes and patents make use of air as unbeatably inexpensive oxidizing agent, literature reports about improved methods for *in vitro* refolding of disulfide-containing proteins by air oxidation have become few and far between compared to the ever-growing body of literature on methods for the production of proteins from inclusion body material. In a recent example, the *in vitro* refolding of prochymosin from inclusion bodies by controlled air oxidation was described.[38] Under the reported conditions, renaturation yields by air oxidation were higher than when a standard thiol exchange redox system consisting of reduced and oxidized glutathione was used.

2.2.2.2 Thiol–Disulfide Exchange Systems

Redox systems for the oxidative refolding of proteins need to enable thermodynamic control of the refolding process, *i.e.*, the disulfide isomerization of protein species with non-native disulfide bridges, which might have formed early in the renaturation reaction, must be possible in order to allow the protein to relax into its thermodynamically preferred, native state. The chemical system that is used in order to effect the formation of disulfide bridges has to work in a way that the reaction is reversible. Rapid oxidation of immobilized reduced denatured trypsin with dehydroascorbate, *e.g.*, with an oxidizing biochemical standard redox potential, $E^{0\prime}$, of 0.06 V, led to an inactive species with non-native disulfide bonds, while oxidation by a redox system consisting of reduced and oxidized glutathione (GSH/GSSG), with a $E^{0\prime}$ of 0.23 V, resulted in a good recovery of active enzyme.[39] Glutathione is a naturally occurring tripeptide (γ-Glu-Cys-Gly)

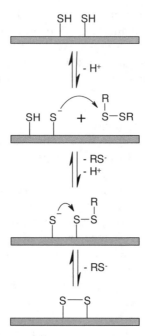

Scheme 2.2.1   Thiol–disulfide exchange.

that serves as a cellular redox buffer and was long thought to be essential for the formation of protein disulfide bridges *in vivo*. The kind of redox system represented by GSH/GSSG, consisting of a pair of reduced and oxidized thiol, is the most straightforward way to control the redox potential for the oxidative *in vitro* refolding of disulfide bridge-containing proteins, as the reaction between redox reagent and protein follows a very simple mechanism (Scheme 2.2.1). No further compounds acting as redox mediators are necessary, and no free-radical species are ever involved in the reactions, as electrons are only transferred pairwise.

Examples for the successful oxidative *in vitro* refolding of proteins by GSH/GSSG abound in the research literature. Since their effectiveness as reagents for the oxidative reactivation of reduced denatured lysozyme was first demonstrated,[40] the pair of reduced and oxidized glutathione have been the most commonly used oxido-shuffling reagents for protein *in vitro* refolding, although other low-molecular-weight thiols such as cystine/cysteine, cystamine/cysteamine or bis-β-hydroxyethyl disulfide/2-mercaptoethanol have also been found to be effective. The *in vitro* refolding of recombinant hirudin, *e.g.*, was reported to proceed approx. five times faster in the presence of cysteine/cystine than in the presence of GSH/GSSG.[41]

It is generally not possible to predict the optimal composition of the renaturation buffer from first principles, so the key parameters of the refolding reaction, like the concentrations and the ratio of GSSG to GSH, pH and temperature, have to be optimized on an empirical basis. GSSG and GSH are generally employed in molecular ratios between 1:1 and 1:10 in the sub-millimolar to millimolar concentration range. The redox potential in the renaturation buffer, which is proportional to $\log([GSH]^2/[GSSG])$, is kept slightly reducing in order to guarantee reversibility of the thiol–disulfide exchange. Under these conditions, the driving force for the formation of disulfide bridges is largely derived from the stabilization energy of the native protein,[42] which ensures that native disulfide bonds are favored over non-native random combinations of cysteines. The empirical optima for the GSH/GSSG ratio in oxidative *in vitro* refolding have been found to be quite broad in many cases.[43] Complex refolding problems, however, may be susceptible to small changes in the composition of the redox buffer and therefore may require careful fine-tuning of the refolding conditions as demonstrated, *e.g.*, for the case of murine αβ T-cell receptor.[44]

The overall process for protein *in vitro* refolding may be simplified by *not* removing the necessary reducing agent, *e.g.*, DTT or DTE, from the solubilized inclusion body material before renaturation is initiated. In this approach, the refolding buffer contains only oxidized dithiol, *e.g.*, GSSG, and the corresponding reduced free thiol is generated *in situ* by reaction with the carried-over reducing agent. This method was demonstrated by the successful oxidative *in vitro* refolding of a recombinant F_{ab} antibody fragment.[45] A process for the production of recombinant human growth differentiation factor-5 was described that made use of a similar strategy.[46] In this case, the inclusion body material was solubilized in the presence of 32 mM cysteine, followed by dilution into refolding buffer and oxidative renaturation on air. Although EDTA was present in the reaction mixture, apparently enough cysteine was gradually oxidized to cystine over the time-course of the reaction to form an effective redox environment for the formation of native disulfide bridges.

As an alternative to the rapid dilution of solubilized inclusion body material into an oxidizing refolding buffer, *in vitro* refolding may also be carried out through dialysis steps. Although dialysis presents technical problems when upscaling of protein production is desired, it may provide a valuable alternative option for protein renaturation on the laboratory scale. This strategy has the distinct advantage that the conditions during the oxidative refolding of a given protein may be changed in a stepwise, controlled manner. It was applied for developing a dialysis protocol for the *in vitro* refolding of antibody fragments from reduced solubilized inclusion body material, in which the concentration of the denaturing agent guanidinium chloride was successively reduced to produce compact folding intermediates, and the oxidative formation of disulfide bridges was only induced during the last dialysis steps by addition of GSSG.[47,48] The success of this approach may be due to a reduced probability for the formation of non-native disulfide bonds during the early stages of refolding.

In any case, the success of the oxidative *in vitro* refolding of proteins does not depend on the redox system for disulfide-bond formation alone. As already mentioned, the suppression of unproductive side reactions, namely unspecific aggregation of non-native protein species by additives like L-arginine plays a critical role. Additionally, for many recombinantly expressed proteins and peptide hormones that lack their native pro- or pre-pro-peptides, the naturally favored pathways of folding and even the thermodynamic driving force into the native conformation may be compromised. The pro-region of BPTI was shown to facilitate the correct disulfide-bond formation during oxidative refolding in a GSH/GSSG redox buffer.[49] A single cysteine residue in this pro-region was responsible for directing the protein into the native form along a different, faster sequence of disulfide exchange reactions with a different set of folding intermediates. Although the small peptide hormone guanylyl cyclase-activating peptide II (GCAP II) contains only four cysteine residues, the native, correctly disulfide-bridged form could only be obtained as a minor by-product of oxidative *in vitro* refolding of the synthetic peptide in a GSH/GSSG redox buffer, whereas the recombinant pro-protein quantitatively folded into a structure with the native disulfide bonds in place.[50] The peptide hormones of the neurotrophin family are homodimeric, and their monomer units contain three native disulfide bridges in a tightly constricted cystine knot topology. The oxidative *in vitro* refolding of recombinant human nerve growth factor (NGF), a member of the neurotrophin family, proceeded significantly faster, and with higher yield, when the pro-region was present in the recombinant protein.[51] This enhancement of renaturation kinetics and yield was also observed when a fusion protein construct of NGF and the pro-region of the related neurotrophin-3 was expressed into inclusion bodies and refolded *in vitro*.[52] In an extreme example of thermodynamically limited oxidative folding, insulin-like growth factor-I does not quantitatively form its native disulfide bridges under redox conditions that allow for reversible thiol disulfide exchange reactions, *i.e.*, the native form is not the most stable one.[53,54] Its oxidative folding *in vivo* is directed by specific binding proteins,[55] and its recombinant production for therapeutic purposes relies on fusion protein constructs. In the end, the success of the oxidative *in vitro* refolding of recombinant proteins critically depends on sequence-specific properties, and engineering the expressed protein construct may be required to solve challenging renaturation problems.

2.2.2.3   Mixed Disulfides

Instead of transferring the reduced denatured protein into an oxidizing redox buffer in order to effect oxidative refolding, another approach can be chosen, in which oxidation to a disulfide and isomerization to the native pattern of disulfide bridges do not occur simultaneously. Upon oxidation in the presence of denaturing agents or under kinetic control,[39] proteins may form fully oxidized, non-native forms with a scrambled disulfide bridge pattern. This has been mainly exploited for studying disulfide exchange reactions. A very-well-known example is disulfide-scrambled RNase A,[56] which has been serving as a model for non-native proteins and as substrate for systems that catalyze disulfide-exchange reactions for a long time.

From the point of view of *in vitro* refolding, however, systems in which the cysteine residues of reduced denatured proteins from inclusion body material are quantitatively and reversibly modified have proven more useful. The non-native protein species, in which the cysteine residues have been modified and thereby protected against, *e.g.*, further oxidation may be more easily handled than the reduced denatured protein, and may in some cases be more advantageous starting materials for the renaturation reaction.

It is possible to reversibly modify the free cysteines of a reduced protein by reaction with an excess of free low-molecular-weight dithiols. The resulting mixed disulfide may then be used as starting material for *in vitro* refolding. In some cases, this approach offers advantages over direct oxidation. Modification of the reduced denatured protein with cysteine or glutathione may be expected to reduce the hydrophobic character of the unfolded polypeptide chain and therefore contribute to a decreased tendency towards unspecific aggregation in the refolding mixture. As quantitatively modified mixed disulfides already represent fully oxidized protein states, dithiols do not necessarily have to be included in the refolding reaction mixture. In principle, small amounts of free thiols are sufficient to catalyze the elimination of the modifying compound under conditions where renaturation is driven by the thermodynamic stability of the native state. The feasibility of this method has been demonstrated with a series of model systems. The examples include the *in vitro* refolding of lysozyme from its mixed disulfides with cysteine[57] and of RNase T$_1$ from its glutathione-modified form.[58] Mixed disulfides with glutathione were also successfully used as starting materials for the *in vitro* refolding of serine proteases.[59,60] In this case, the elimination of GSSG from the mixed disulfides was catalyzed by a cysteine/cystine redox buffer. In the context of these studies, the renaturation of the two-chain form of neo-chymotrypsinogen was reported as well,[61] which requires the formation of an interchain disulfide bond to adopt its native state. Interestingly, in a case where the direct comparison was made, the pH optima for the *in vitro* refolding from mixed disulfides were shifted by approx. one log unit into the acidic range compared to direct oxidative renaturation from reduced denatured protein.[62]

Another method for reversible modification of cysteines is the S-sulfonation[63–65] of proteins with sodium sulfite. This reagent cleaves disulfide bonds to form one S-sulfonylcysteine residue and one free cysteine residue (Scheme 2.2.2). In the presence of an oxidizing agent, *e.g.*, air/Cu^{2+},[63] or sodium tetrathionate,[64] the oxidative sulfitolysis reaction proceeds to the complete modification of all accessible free cysteines. The reaction is readily reversible, and in the presence of free thiols, two S-sulfonated cysteines can form a disulfide bridge under elimination of sulfite from the protein.

The re-synthesis of active insulin from the separate A and B chains under denaturing conditions represents an early example for the use of S-sulfonated peptides as model systems for protein renaturation. Reaction of an excess of reduced A chain with the S-sulfonated B chain gave yields of up to 80% active renatured protein.[66] More recently, several protocols were published for the production of proinsulin from solubilized inclusion body material that had been subjected to oxidative sulfitolysis.[67–69] The elimination of sulfite from the

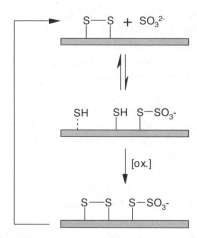

Scheme 2.2.2 Oxidative sulfitolysis.

modified denatured proinsulin and its concomitant *in vitro* refolding were brought about by partial reduction with β-mercaptoethanol. A rather complex process was devised for the large-scale production of human calcitonin from a multimeric fusion protein precursor over-expressed into *E. coli* inclusion bodies.[70] Efficient solubilization of the inclusion bodies required the modification of lysine residues with citraconic acid and oxidative sulfitolysis of the insoluble fraction. After cleavage of the fusion protein by trypsin, acid-catalyzed de-citraconilation and C-terminal clipping and amidation by carboxypeptidase Y, the mature peptide hormone was finally obtained by de-sulfonation with glutathione.

Larger disulfide-bridged proteins that have served as examples for successful *in vitro* refolding from S-sulfonated forms include isoforms of platelet-derived growth factor,[71,72] and prethrombin-2.[73] In these cases, the de-sulfonation of the refolded proteins was achieved by incubation in GSH/GSSG redox buffers.

2.2.2.4   Enzymatic Catalysis of Disulfide-bond Formation *in vitro*

In vivo, the reshuffling of disulfide bonds in proteins is catalyzed by protein disulfide isomerases (PDIs). PDI activity was discovered more than 40 years ago in microsomal preparations of animal tissue.[74] The principal form of PDI in eukaryotes is present in high concentrations in the endoplasmic reticulum. A large family of homologous PDI-like proteins exists in eukaryotes as well as in prokaryotes.[5,6,75,76] Many of these proteins have been shown to fulfill roles in maintaining the physiological redox balance and the processing of disulfide bridge-containing proteins in the endoplasmic reticulum and the periplasm, respectively, as reviewed elsewhere in this book. All family members contain one ore more domains that share the fold of the small cytoplasmic redox protein thioredoxin. The active centers of thioredoxins and PDI-like proteins contain a special pair of cysteines within a CXXC sequence. One of the cysteines shows an unusually low thiol pK_a value.[77] This cysteine alone is capable of catalyzing disulfide exchange reactions, and single-cysteine mutants were found to promote the renaturation of disulfide-scrambled RNase A, while oxidative *in vitro* refolding of the same protein required the full disulfide oxidoreductase activity and therefore the presence of both active cysteines in PDI.[78] The biochemical standard redox potential of the disulfide/CXXC couple in mammalian PDI, *e.g.*, is approx. $-0.18\,V$, with a pK_a value of 6.7.[79] Due to the low pK_a value of the active cysteine, PDI and its homologs retain to some degree the ability to act as catalysts of oxidative *in vitro* refolding at relatively low pH values,[80,81] in contrast to low-molecular-weight aliphatic thiols. Within the scope of this chapter, it was of interest to look for examples where PDI family members have been employed as active reagents for protein *in vitro* refolding.

The effectiveness of PDI as enhancer of the oxidative *in vitro* refolding for the production of complex proteins from inclusion bodies was demonstrated using an F_{ab} fragment as model.[82] The maximum yield was obtained in a more oxidizing redox buffer compared to the optimal conditions in the absence of PDI. The renaturation yield of a genetically engineered immunotoxin consisting of an $F(ab')_2$ antibody fragment and a tumor necrosis factor α domain was also significantly improved by the presence of PDI.[83] The oxidative folding of *Ragi* bifunctional inhibitor, a protein with a complex pattern of five overlapping disulfide bridges, was effectively promoted by the bacterial disulfide isomerase DsbC.[84] Curiously, any disulfide bond-containing protein may have some residual disulfide oxidoreductase activity. The oxidative disulfide bridge formation in biological peptides obtained by chemical synthesis, including several conotoxins and BPTI, was promoted by albumins that had been added in high concentrations as crowding agents.[85]

Although PDI from the endoplasmic reticulum acts as chaperone, *i.e.*, as suppressor of aggregation, along with its role as catalyst of disulfide exchange reactions, its effect as enhancer of oxidative *in vitro* refolding may be improved by the presence of additional folding helper proteins. The PDI-catalyzed oxidative renaturation of an F_{ab} antibody fragment was synergistically enhanced by the presence of heavy chain-binding protein (BiP), a chaperone protein from the endoplasmic reticulum.[86] The homologous oxidoreductase ERp57 facilitated the renaturation of the same F_{ab} fragment in a manner similar to PDI. However, in this case, no synergistic action with BiP was observed,[87] while the oxidative *in vitro* refolding of the model proteins α-lactalbumin and RNase B by yet another PDI-like protein, ER60, was shown to be significantly improved in the presence of BiP.[88]

In all examples mentioned so far, PDI or one of its homologs was employed in stoichiometric amounts. Sub-stoichiometric concentrations of PDI had been reported earlier to promote the aggregation of denatured protein *in vitro* by formation of covalently cross-linked products,[89] *i.e.*, adding PDI in insufficient amounts may be potentially counterproductive for *in vitro* refolding. A requirement for large amounts of active enzyme as an ingredient of the reaction mixture represents a severe practical limitation for the use of PDI-like proteins as catalysts of oxidative *in vitro* refolding, especially in production-scale processes. On the other hand, the yield of *in vitro* refolding of proinsulin in a GSH/GSSG redox system was found to be significantly enhanced by catalytic amounts of bovine liver PDI as well as by a single-site mutant of the enzyme,[90] while an increase in the rate of refolding could only be observed at molar PDI/proinsulin ratios approaching unity.

As an aside, it seems that high local concentrations of PDI-like proteins in the vicinity of a target protein may be most conveniently reached by co-expression as direct fusion proteins. Using this approach, it has been possible to avoid the formation of inclusion bodies altogether. *P. haemolytica* glycoprotease, *e.g.*, could only be obtained as a soluble product in the cytosol of *E. coli* when it was expressed as fusion protein with thioredoxin, from which the active enzyme was recovered by enterokinase cleavage.[91] Apparently the correct formation of native disulfide bonds in the glycoprotease had been enabled by the thioredoxin part of the fusion protein.

The *in vivo* formation of disulfide bonds in the endoplasmic reticulum and the bacterial periplasm, respectively, has been intensively studied for decades. Two flavoprotein sulfhydryl oxidases of the endoplasmic reticulum, Ero1p[92,93] and Erv2p,[94] were identified in yeast a few years ago, followed by the identification of their homologs in other eukaryotes, including humans. Both are dependent on oxidized flavin adenin dinucleotide (FAD) as cofactor and capable of using molecular oxygen as terminal electron sink. The functional homolog of Ero1/Erv2 in bacteria, DsbB, is coupled to the respiratory chain.[95,96] All three proteins contain an active cysteine pair that transfers oxidizing equivalents to PDI and DsbA, respectively, by thiol–disulfide exchange. These in turn are responsible for the formation of disulfide bridges in the newly expressed proteins being processed in the endoplasmic reticulum and periplasm, respectively. Along with these discoveries, the long-held belief that glutathione is essential for the formation of disulfide bonds *in vivo* was refuted.[92] The high concentrations of GSH/GSSG present in the endoplasmic reticulum now are ascribed the role of redox buffer and protectant against oxidative stress.[97]

A number of chemical species may act as electron sinks for the Ero1 and DsbB systems, respectively, and an even greater number may be envisaged to act as oxidizing agents for the *in vitro* refolding of reduced denatured proteins. When the full enzyme complement of the endoplasmic reticulum is present, *e.g.*, in liver microsome preparations, dehydroascorbate may act as an effective oxidizing agent for disulfide-bond formation.[98]

From the point of view of oxidative *in vitro* folding for protein production, however, atmospheric oxygen is the most attractive source of oxidizing power. The sulfhydryl oxidase activities that can be isolated from readily available sources like skim milk membrane[99,100] in the form of a metalloprotein, and from egg white[101] as a flavoprotein, were shown to be able to catalyze the *in vitro* formation of disulfide bonds in a number of protein substrates. However, this reaction is practically irreversible and randomly generates non-native disulfide bonds. Therefore, oxidative refolding of RNase A to its active, native form by hen egg white sulfhydryl oxidase on air required the presence of catalytic amounts of PDI and of reduced glutathione.[102] An additional concern is the generation of hydrogen peroxide as by-product of the sulfhydryl oxidase reaction, which might lead to oxidative damage of the refolding protein.

2.2.3 Alternative Approaches to Oxidative *in vitro* Folding

2.2.3.1 Dithiols

As mentioned above, the activity of PDI-like proteins is dependent on the presence of the CXXC sequence motif, and these two cysteines present in the enzyme's active center cycle between the reduced thiol and disulfide forms during the catalyzed thiol exchange reactions. It would obviously be an advantage for the development of oxidative *in vitro* refolding methods, if the action of the enzyme could be effectively mimicked by small-molecule dithiols. However, the common synthetic dithiols DTT and DTE are usually not able to act as catalysts of disulfide isomerization reactions in proteins, as their low standard redox potentials (-0.33 V at pH 7^{103}) lead to preferential reduction of disulfide bridges. Furthermore, the formation of mixed disulfides of these compounds with proteins as necessary intermediates of oxidative refolding is strongly disfavored, as the system tends towards ring closure and rapid elimination of the cyclic disulfides (Scheme 2.2.3). Using DTT or DTE to buffer the redox potential in solution at values conducive to disulfide bridge formation and isomerization is quite impractical. For the oxidative refolding of reduced denatured RNase A, a well-behaved and quite stable model protein, a 1700-fold excess of oxidized over-reduced DTT was necessary.[104]

The first "small" dithiols, which were reported to promote the oxidative *in vitro* refolding of denatured proteins, were peptides derived from the sequences of naturally occurring proteins. A sequence stretch of human follicle stimulating hormone contains a CXXC motif and shows high similarity to the active center of thioredoxine. A series of hexa- to pentadecapeptides derived from this sequence were shown to catalyze the oxidative refolding of reduced denatured RNase A,[105] with the most active one, hFSH-β-(81-95), being more efficient than full-length thioredoxine.

The peptide-based approach was taken a step further by the design of cyclic hexapeptides containing the CXXC motif based on the sequence environment of

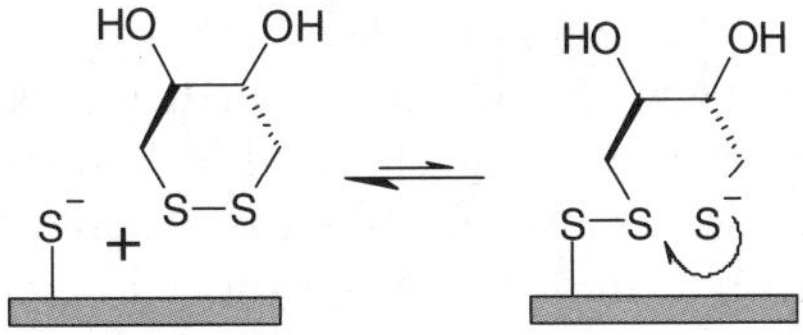

Scheme 2.2.3 Reaction of DTT with free thiols.

the cysteine pairs in glutaredoxin, thioredoxin, protein disulfide isomerase (PDI) and thioredoxin reductase.[106] Not surprisingly, two distinct thiol pK_a values could be observed for each of the four peptides. The standard redox potentials of the cysteine/cystine couple in these peptides were found to range from -0.20 V to -0.13 V. The oxidative refolding of reduced denatured RNase A in a GSH/GSSG redox buffer system proceeded with higher yield when these peptides were present in sub-stoichiometric amounts. Yield and kinetics of renaturation were redox potential-dependent, with the least reducing compound, the PDI-derived cyclic peptide, being the most effective refolding enhancer.

In an alternative approach to mimic the properties of the active center of PDIs, the synthetic disulfide ($\pm$)-*trans*-1,2-bis(2-mercaptoacetamido)cyclohexane (Vectrase-P) was developed some years ago as a reagent for *in vitro* protein refolding.[107] Its chemical structure leads to a biochemical standard redox potential $E^{0\prime}$ of 0.24 V and thiol pK_a values of 8.3 and 9.9, respectively. In this respect, Vectrase-P is much more similar to the cysteine pair in the active center of PDIs than DTT. The addition of Vectrase-P to buffers containing a GSH/GSSG redox system resulted in a significantly increased rate of renaturation of disulfide-scrambled RNase A. Since being commercially available, Vectrase-P has been successfully employed to improve the *in vitro* refolding of the thera-peutically relevant protein proinsulin.[108]

2.2.3.2 Aromatic Thiols

As mentioned above, the use of aliphatic thiol compounds like cysteine or glutathione as redox buffers limits the pH range over which *in vitro* refolding reactions may be carried out. The active species in thiol–disulfide exchange reactions is the reduced monothiol in its deprotonated form. This thiolate anion effects the nucleophilic attack at the oxidized sulfur center within a disulfide bond, at which the substitution takes place (Scheme 2.2.1). The proton acidities of aliphatic thiols, as expressed by their respective pK_a values, lie in the alkaline range; the pK_a value of the thiol proton in reduced glutathione, *e.g.*, is 8.7.[109] Consequently, at pH values below 6, the concentration of free thiolate anion, and the rate of thiol–disulfide exchange, becomes negligible. In some cases, this may represent a severe limitation. The *in vitro* refolding of the cystine knot protein bone morphogenetic protein-2, *e.g.*, proceeds quite slowly at pH 8, and necessarily at low concentrations,[110] due to the limited solubility of the protein at pH values above 6. The conformational transition that accompanies the formation of spider silk protein fibers requires acidic conditions *in vivo*[111] and *in vitro*.[112] These two examples of highly application-relevant proteins demon-strate that reagents enabling thiol-disulfide exchange reactions at moderately acidic pH values would be potentially very valuable tools for *in vitro* refolding.

Due to the electron-withdrawing effect of the more electronegative sp_2-hybridized carbons in unsaturated or aromatic systems, thiol groups attached to such an aromatic carbon should be more stable in their deprotonated form, hence their pK_a values should be lowered. This effect might be additionally enhanced by attaching electron-withdrawing substituents *ortho* or *para* to mercapto groups bound to aromatic systems. Therefore, aromatic thiol compounds should be able to sustain higher free thiolate concentrations at lower pH values than aliphatic thiols, which might make them potentially useful reagents for the catalysis of thiol–disulfide exchange reactions under these conditions.

This idea was first explored in a study where the time constants for the reduction of 2-pyridinyl dithioethanol and of insulin, respectively, by a series of mercaptobenzene derivatives were measured and compared with the rates obtained in presence of glutathione.[113] A systematic correlation between the reaction rates for the thiol–disulfide exchange reactions and the thiol pK_a values was observed. This encouraging observation led the authors of this study to continually expand the list of mercaptobenzene derivatives over the last five years, and to test their potential as enhancers of *in vitro* refolding. In all cases, the renaturation of disulfide-scrambled RNase A served as the model system.[114–117] The activity of a PDI-based refolding system for RNase A was also enhanced by the presence of *ortho*-substituted mercaptobenzene derivatives.[118]

The heteroaromatic thiol compounds 2- and 4-mercaptopyridine were first used in protein biochemistry in their oxidized disulfide forms for titrating the active cysteine of papain.[119] Recently, the potential of some heteroaromatic compounds as redox catalysts for the *in vitro* refolding of proteins was investigated by our group.[62] The model protein was a single-chain antibody fragment, which was refolded *in vitro* from its mixed disulfide with glutathione. Although the tested compounds increased the refolding yields, an acceleration of the refolding kinetics was not observed. These observations were tentatively ascribed to the low nucleophilicity of heteroaromatic thiols due to electron withdrawing effects of the nitrogen heteroatoms within the aromatic ring structures. Furthermore, tautomeric structures exist that are considerably more stable than the thiol/thiolate structures of these compounds, which drastically reduces the amount of thiolate that is available for the catalysis of renaturation by thiol–disulfide exchange.

2.2.3.3 Matrix-assisted Oxidative Refolding

Matrix-assisted *in vitro* refolding of proteins has been successfully used for the renaturation of a number of disulfide-containing proteins. As early as 1962, the reversible reduction and on-column air oxidation of immobilized RNase and trypsin was demonstrated in one of Anfinsen's ground-breaking contributions to the field of protein folding.[120] Early systematic studies on the influence of different redox systems on oxidative refolding were carried out with agarose-immobilized trypsin and trypsinogen.[39] These works demonstrated that, under the right conditions, denatured proteins bound to a chromatography matrix generally retain their ability to form the correct pattern of disulfide bonds and to fold into the native state. Matrix-assisted refolding of proteins that were reversibly bound to an ion-exchange resin and could therefore be recovered after renaturation was first demonstrated more than two decades ago.[121] In another early example, denaturant removal by high-pressure hydrophobic interaction chromatography was explored as a means of protein refolding.[122] Denatured α-glucosidase that had been reversibly bound to an ion-exchange matrix was successfully refolded on the column and could be recovered in soluble, active form in high yield.[123] However, the model systems for protein refolding used in these early studies did not require the formation of disulfide bonds. Recent advances enabling the oxidative matrix-assisted refolding of disulfide bridge-containing proteins include the development of a system for the continuous renaturation of the model protein α-lactalbumin,[124] and the controlled oxidative refolding of BSA in high concentrations[125] on an ion-exchange column. In the latter case, stringent control of the redox conditions was emphasized.

In all examples mentioned so far, a chromatography matrix was used to bind the denatured protein during the renaturation process in order to suppress the aggregation of unfolded and intermediate protein species, while disulfide-bond formation was effected by soluble redox systems. In the context of this chapter on strategies for the oxidative *in vitro* refolding of disulfide bridge-containing proteins, it is of interest to review those matrix-assisted methods where the matrix plays an active role in the redox chemistry of the renaturation reaction.

In a first step towards creating a redox-active chromatographic matrix for *in vitro* refolding, PDI was immobilized on agarose. The feasibility of the oxidative refolding in the presence of a glutathione redox buffer of RNase A and of lysozyme on this material was demonstrated more than a decade ago.[126] The renaturation of lysozyme could be performed with reasonable yields of more than 60% at high concentrations of up to $1.6\,mg\,ml^{-1}$ protein. Matrix-immobilized PDI-like proteins, in particular the bacterial disulfide isomerases DsbA and DsbC, have also been combined with stepwise dialysis systems for oxidative *in vitro* refolding.[127] A considerable advantage of this approach is the ease with which the folding helper proteins can be separated from the refolded product and recycled.

A sophisticated system for oxidative refolding chromatography, consisting of the mini-chaperone GroEL, peptidyl prolyl isomerase, and DsbA co-immobilized on agarose beads was used for the successful oxidative refolding of the two highly challenging target proteins scorpion toxin Cn5[128] and human CD1d-glycolipid tetramer.[129] The production of the active transmembrane complex of CD1d and β-microglobulin from inclusion body material in the latter example represents a major achievement. Still, given the cost and effort that has to be invested in the production and purification of the chaperones and PDI-like proteins for the matrix, oxidative refolding chromatography using immobilized folding helper enzymes will probably stay reserved for the production of disulfide-bridged proteins with high value for research that cannot be obtained by simpler means.

In an effort to develop simpler matrix-assisted systems for the oxidative refolding of proteins, thiol-modified polymer microspheres were introduced some years ago.[130] In this work, DTT was tethered to the surface of poly-(styrene/glycidyl methacrylate) microspheres. The resulting polymer latex suspension was shown to promote the renaturation of disulfide-scrambled RNase A, when applied in batch mode. Microspheres with an intermediate surface concentration of thiol groups of approx $0.2\,nm^{-2}$ were found to be the most effective catalysts of disulfide-bond isomerization. The same system was later used for the oxidative *in vitro* refolding of reduced denatured RNase A.[131] In this case, however, microspheres with a lower surface concentration of thiol residues were the more effective refolding enhancers. The report on this study does not mention the source for the oxidizing power for the renaturation of reduced RNase, so it can only be assumed that a significant portion of the immobilized DTT was present on the microsphere surface in its oxidized, di-sulfide form under the aerobic conditions of the experiment. In another approach to develop redox-active materials for the matrix-assisted oxidative refolding of proteins, the trithiol tris(2-mercaptoacetamidoethyl)amine was coupled to bromine-activated Tentagel™ resin *via* one of its free thiol groups.[132] The resulting Tentagel™-immobilized dithiol exhibited an apparent redox potential of -0.21 V, close to the value for PDI, and readily reacted in disulfide exchange reactions with, *e.g.*, DTNB. However, this material was a relatively poor catalyst for the renaturation of disulfide-scrambled RNase A, possibly due to sterical hindrance of the reaction of the unfolded protein within the matrix. When functionalized poly(styrene/glycidyl methacrylate) microspheres were used as support for immobilizing the trithiol instead of the Tentagel™ resin, the resulting material was found to be a relatively effective refolding enhancer. The yields of renaturation of disulfide-scrambled RNase were increased by approx. 50% compared to an experiment in which the protein was refolded in the presence of DTT-modified microspheres.

2.2.3.4 Other Oxidizing Compounds

The chemistry of the chalcogens sulfur and selenium is relatively similar, as expected from their positions in the periodic system of elements. Both elements have been recruited by living systems and are incorporated into proteins.[133–134,150] Selenocysteine is a rare, naturally occurring amino acid that is essential to a number of intracellular redox processes in nearly all kingdoms of life. It was recognized early that selenium is required for certain enzymatic activities,[135] and selenocysteine was discovered in proteins more than three decades ago.[136] Only recently, the potential of selenium compounds for the oxidative *in vitro* refolding of proteins has been explored for the first time. Selenoglutathione (γ-Glu-Sec-Gly) was shown to be an effective reagent for the renaturation of reduced denatured RNase A and BPTI,[137] in spite of its very reducing standard redox potential $E^{0\prime}$ of -0.41 V. Apparently, in these cases the free energy of protein folding was sufficiently large to overcome the unfavorable energy balance of the diselenide to disulfide exchange reaction (Scheme 2.2.4). The oxidative refolding of RNase A in the presence of oxidized selenoglutathione proceeded two times faster than when GSSG was used as oxidizing agent, which represents a significant advantage.

Reactive oxygen species are naturally occurring by-products of cellular redox processes. These species are generally quite unspecific oxidants and tend to be pernicious for living cells, when their concentration becomes too high. *Inter alia*, the formation of disulfide bridge-cross-linked protein aggregates has been

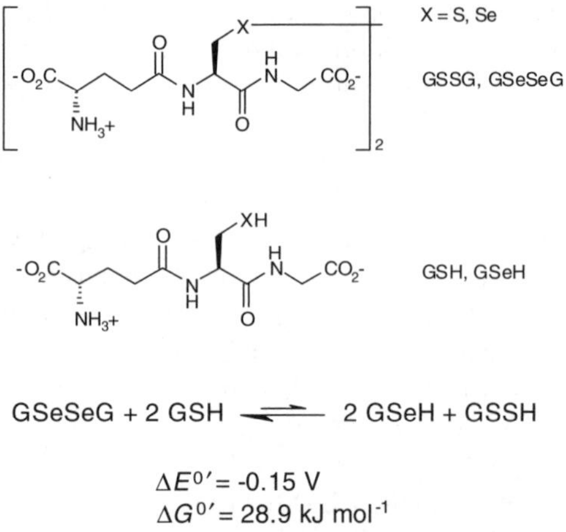

$$GSeSeG + 2\,GSH \rightleftharpoons 2\,GSeH + GSSH$$

$$\Delta E^{0\prime} = -0.15\ V$$
$$\Delta G^{0\prime} = 28.9\ kJ\ mol^{-1}$$

Scheme 2.2.4 Thiol–diselenide exchange between glutathione and selenoglutathione.

reported.[138] Cells dispose of a number of defense mechanisms against oxidative damage.[139,140] In this context, the Zn^{2+}-dependent direct oxidation of specific cysteine residues to disulfide bridges by hydrogen peroxide has been shown for the specialized *E. coli* heat shock chaperone Hsp33.[141] This shows that it is possible to use hydrogen peroxide as a selective oxidizing reagent for protein disulfide-bond formation, at least in principle, and it might be of interest to observe whether oxidative *in vitro* refolding methods will become available that harness the potential of this easily and economically available reagent in the future.

2.2.3.5 Electrochemical Oxidation

Reversible electrochemistry of glutathione and of thioredoxins on a lipid-coated gold electrode was demonstrated more than 15 years ago.[142] Direct electrochemical conversion of reduced proteins, or the electrochemical recycling of spent reduced thiols that may act as redox mediators, could potentially enable substantial improvements for oxidative *in vitro* refolding processes. However, to our knowledge no practical renaturation process making use of electrochemical oxidation has been developed so far.

Improved electrodes for the direct electrochemical oxidation of cysteine were reported recently. One of these was a fluorosurfactant-coated gold electrode;[143] while the other one consisted of a mixture of graphite with the ionic liquid octylpyridinum hexafluorophosphate.[144] The latter approach may prove interesting in the future, in light of recent progresses in the use of ionic liquids as enhancers of *in vitro* refolding[145,146] and as stabilizing solvents for native proteins.[147–149] These developments might open avenues for viable *in vitro* refolding methods based on the electrochemical oxidation of reduced denatured proteins.

2.2.4 Chemical Modification of Cysteine Residues *in vitro*

Cysteine-specific labeling of proteins with, *e.g.*, fluorophores or spin labels is a well-established and frequently employed technique for biophysical studies. The targeted cysteines are generally present in their free thiol form and located on the surface of the studied proteins. Native chemical ligation and expressed protein ligation as established methods for protein modification and semi-synthesis also make use of a thiol/thioester coupling chemistry.[150,151] An extensive review of these methods, however, is beyond the scope of this chapter. In the following, a technologically relevant *in vitro* protein modification method that specifically targets disulfide bridges will be briefly discussed.

Modification of proteins with polyethylenglycol, or pegylation, has been shown to be beneficial for the stability, immunogenicity and pharmacokinetics of therapeutically relevant proteins,[152–154] and has become an important commercial process. Pegylation is often performed by random coupling to surface lysines. The modification of cysteine residues has emerged as a more specific alternative. For example, the F_{ab}' fragment of a humanized antibody was site-specifically pegylated at a free hinge cysteine by reaction with N-pegylated maleimide derivatives.[155] The modifications led to prolonged serum half-lives, while the antigen binding affinity of the parent IgG was retained. In a similar system, the coupling efficiency for the pegylation of F_{ab}' fragments could be increased by employing harsher reaction conditions that guaranteed the full availability of the target cysteine in its reduced form for the reaction with the maleimide group.[156] These conditions, however, led to the disruption of the interchain disulfide bond in the antibody fragment. Despite this harsh treatment, the F_{ab}' fragments used in this study retained their structural stability and their antigen binding ability. Another strategy for the site-directed pegylation of disulfide bond-containing proteins was presented recently. In this approach, native disulfide bonds in the protein are reduced and then replaced with a pegylated three-carbon dithiol bridge.[157,158] The coupling proceeds by the addition of the free thiols to a conjugated double bond system in the coupling reagent (Scheme 2.2.5). Examples for proteins modified by disulfide-bridging

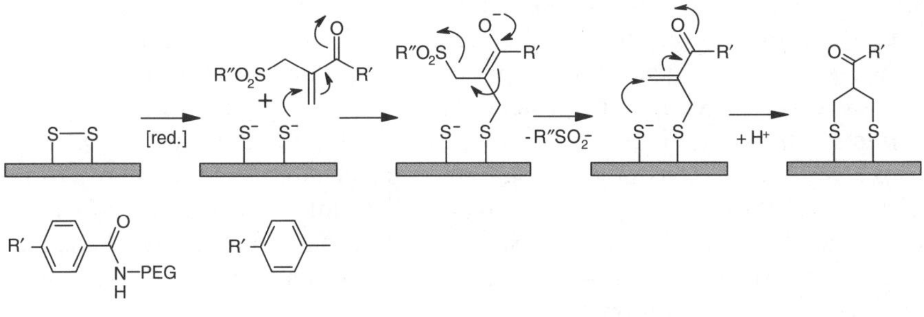

Scheme 2.2.5 Disulfide-bridging pegylation.[160]

pegylation include interferon α2 and an anti-CD4$^+$ F$_{ab}$ antibody fragment.[159] Antiviral activity and target binding, respectively, were retained after the modification in both cases. Breaking a disulfide bond naturally compromises the stability of the modified protein, and the preservation of the stability of the native state has to be kept in view during the modification process that is somewhat analogous to reductive denaturation followed by a special case of oxidative *in vitro* renaturation. Recently, an algorithm for the identification of solvent-exposed disulfide bonds that are accessible for the disulfide bridge modification, and for the prediction of the structural consequences of this modification, was presented.[160]

The developed methods for the site-specific modification of proteins at disulfide bridges are in principle not restricted to the introduction of polyethylenglycol moieties, and it will be interesting to observe which novel therapeutic molecules will be created in the future using this or a similar approach.

2.2.5 Cell-free Expression Systems

The following section will not directly deal with the *in vitro* refolding of inclusion body proteins, but with an alternative *in vitro* method for the production of active proteins. The expression of proteins in cell-free systems represents an option when the production of a given protein in its active form from microorganisms or cell culture has failed. As mentioned before, most therapeutically relevant proteins are extracellular or secreted and potentially contain specific disulfide bonds. This fact makes control of the thiol redox chemistry an important aspect of the setup of cell-free expression systems. In principle, insoluble expression in cell-free systems followed by *in vitro* refolding is conceivable. This case is not very much different from bacterial over-expression into inclusion bodies. On the other hand, the stringent control over the expression parameters possible in an *in vitro* system should allow for the direct expression of correctly folded, active protein with the desired disulfide-bridge pattern.

In the first successful examples for the production of disulfide bridge-containing proteins in cell-free expression systems, single-chain antibody F$_v$ fragments could be obtained in final concentrations in the order of 10^{-2} mg ml^{-1}.[161,162] The best results were obtained when disulfide-bond formation was promoted by the presence of bovine liver PDI and glutathione redox buffers.

Cell-free expression systems derived from *E. coli* cells contain disulfide reductase activity, which obviously presents a major problem for the production of disulfide bridge-containing proteins. However, after pre-treatment of the cell extracts with iodoacetamide, the expression of the serine protease domain of urokinase[163] and of recombinant plasminogen activator[164] in their native forms and in significant yields in the order of 10^{-1} mg ml^{-1} became possible, again only when bacterial disulfide isomerase DsbC and a glutathione redox buffer were added to the reaction mixture. This requirement for a PDI-like protein confirmed the observations made with the earlier example of single-chain antibody fragments. Similarly, the challenging cell-free expression of soluble and functional intact IgG molecules[165] in detectable concentrations was only possible in the presence of human PDI and DsbC, respectively, along with the addition of specific chaperones and a careful optimization of the redox buffer.

As mentioned above, the inactivation of free thiol groups by irreversible modification with iodoacetamide abolishes disulfide reductase activity in the cell-free extracts and allows for significantly improved yields of expression. On the downside, this treatment is rather harsh and also suppresses desirable enzymatic activities like, *e.g.*, glyceraldehyde 3-phosphate dehydrogenase, which is necessary when glucose is to be used as the energy source for the protein synthesis instead of more expensive compounds like phosphoenol pyruvate or acetyl phosphate. This problem was recently overcome by a cell-free expression system based on extracts of the extensively genetically modified *E. coli* strain KGK10.[166] This strain was derived from the KC6 cell extract source strain by deletion of the glutathione reductase-mediated disulfide reductase gene, and by modification of the thioredoxin reductase gene with a hemagglutinin affinity tag. However, removal of the thioredoxin reductase by affinity chromatography was not sufficient to abolish all disulfide reductase activity in the *E. coli* cell extracts, and a treatment with a 20-fold reduced concentration of iodoacetamide was still necessary. This relatively mild treatment left the glycolytic pathway intact and enabled the cell-free expression of the complex disulfide bridge-containing proteins urokinase and murine granulocyte macrophage-colony stimulating factor, respectively, in glucose-driven reactions.

2.2.6 Conclusions

A range of well-established methods for the oxidative *in vitro* refolding of disulfide bridge-containing proteins exists. Several recent developments hold perspectives for future improvements. In this chapter, we have focused on a review of the available chemical systems for the formation of disulfide bridges *in vitro*, which naturally only represents one aspect of the art of recombinant protein production. For any given recombinant protein that is to be obtained by oxidative *in vitro* refolding from inclusion body material, success will be dependent on the total process design that has to balance and integrate the protein engineering, expression in the microbial host organism, isolation of the inclusion body material, solubilization, renaturation, purification and for-mulation of the protein product.

References

1. K. Graumann and A. Premstaller, *Biotechnol. J.*, 2006, **1**, 164–186.
2. G. Walsh, *Nat. Biotechnol.*, 2006, **24**, 769–776.
3. R. C. Fahey, J. S. Hunt and G. C. Windham, *J. Mol. Evol.*, 1977, **10**, 155–160.
4. J. M. Thornton, *J. Mol. Biol.*, 1981, **151**, 261–287.

5. C. S. Sevier and C. A. Kaiser, *Nat. Rev. Mol. Cell. Biol.*, 2002, **3**, 836–847.
6. H. Kadokura, F. Katzen and J. Beckwith, *Annu. Rev. Biochem.*, 2003, **72**, 111–135.
7. F. Banyex and M. Mujacic, *Nat. Biotechnol.*, 2004, **22**, 1399–1408.
8. B. Fahnert, H. Lilie and P. Neubauer, *Adv. Biochem. Eng. Biotechnol.*, 2004, **89**, 93–142.
9. E. García-Fruitós, M. Martínez-Alonso, N. González-Montalbán, M. Valli, D. Mattanovich and A. Villaverde, *J. Mol. Biol.*, 2007, **374**, 195–205.
10. T. Kiefhaber, R. Rudolph, H.-H. Kohler and J. Buchner, *Bio/Technology*, 1991, **9**, 825–829.
11. T. U. Gerngross, *Nat. Biotechnol.*, 2004, **22**, 1409–1414.
12. M. Butler, *Appl. Microbiol. Biotechnol.*, 2005, **68**, 283–291.
13. V. Gomord, P. Chamberlain, R. Jefferis and L. Faye, *Trends Biotechnol.*, 2005, **23**, 559–565.
14. J. H. Choi and S. Y. Lee, *Appl. Microbiol. Biotechnol.*, 2004, **64**, 625–635.
15. G. Georgiou and L. Segatori, *Curr. Opin. Biotechnol.*, 2005, **16**, 538–545.
16. P. Martineau, J.-M. Betton and G. Winter, *J. Mol. Biol.*, 1998, **280**, 117–127.
17. H. Hellebust, M. Murby, L. Abrahmsén, M. Uhlén and S.-O. Enfors, *Bio/Technology*, 1989, **7**, 165–168.
18. E. De Bernardez Clark, E. Schwarz and R. Rudolph, *Methods Enzymol.*, 1999, **309**, 217–236.
19. A. P. Middelberg, *Trends Biotechnol.*, 2002, **20**, 437–443.
20. L. F. Vallejo and U. Rinas, *Microb. Cell Fact.*, 2004, **3**, 11.
21. C. Lange and R. Rudolph, in *Protein Folding Handbook Pt. II*, J. Buchner and T. Kiefhaber, eds., Wiley-VCH, Weinheim, 2005, p. 1215–1250.
22. E. De Bernardez Clark, D. Hevehan, S. Szela and J. Maachupalli-Reddy, *Biotechnol. Prog.*, 1998, **14**, 47–54.
23. J. Buchner, I. Pastan and U. Brinkmann, *Anal. Biochem.*, 1992, **205**, 263–270.
24. B. Fischer, B. Perry, I. Sumner and P. Goodenough, *Protein Eng.*, 1992, **5**, 593–596.
25. R. Cecil and J. R. McPhee, *Adv. Protein Chem.*, 1959, **14**, 255–389.
26. A. J. Cooper, *Annu. Rev. Biochem.*, 1983, **52**, 187–222.
27. C. Jacob, G. I. Giles, N. M. Giles and H. Sies, *Angew. Chem., Int. Ed.*, 2003, **42**, 4742–4758.
28. T. E. Creighton, *J. Mol. Biol.*, 1977, **113**, 275–293.
29. T. E. Creighton, *J. Mol. Biol.*, 1977, **113**, 329–341.
30. J. S. Weissman and P. S. Kim, *Proc. Natl. Acad. Sci. USA*, 1992, **89**, 9900–9904.
31. Y. H. Zhang, X. Yan, C. S. Meyer, M. I. Schimerlik and M. L. Deinzer, *Biochemistry*, 2002, **41**, 15495–15504.
32. N. M. Jarrett, L. Djahadi-Ohaniance, R. C. Willson, H. Tachibana and M. E. Goldberg, *Protein Sci.*, 2002, **11**, 2584–2595.
33. F. H. White, *J. Biol. Chem.*, 1961, **236**, 1353–1360.
34. C. B. Anfinsen and E. Haber, *J. Biol. Chem.*, 1961, **236**, 1361–1363.
35. C. B. Anfinsen, E. Haber, M. Sela and F. H. White, *Proc. Natl. Acad. Sci. USA*, 1961, **47**, 1309–1314.

36. A. K. Ahmed, S. W. Schaffer and D. B. Wetlaufer, *J. Biol. Chem.*, 1975, **250**, 8477–8482.
37. S. A. Charman, K. L. Mason and W. N. Charman, *Pharm. Res.*, 1993, **10**, 954–962.
38. H. G. Menzella, H. C. Gramajo and E. A. Ceccarelli, *Prot. Expr. Purif.*, 2002, **25**, 248–255.
39. N. K. Sinha and A. Light, *J. Biol. Chem.*, 1975, **250**, 8624–8629.
40. V. P. Saxena and D. B. Wetlaufer, *Biochemistry*, 1970, **9**, 5015–5023.
41. J.-Y. Chang, *Biochem. J.*, 1994, **300**, 643–650.
42. E. Welker, W. J. Wedemeyer, M. Narayan and H. A. Scheraga, *Biochemistry*, 2001, **40**, 9059–9064.
43. L. F. Vallejo and U. Rinas, *Biotechnol. Bioeng.*, 2004, **85**, 601–609.
44. F. Pecorari, A. C. Tissot and A. Plückthun, *J. Mol. Biol.*, 1999, **285**, 1831–1834.
45. J. Buchner and R. Rudolph, *Bio/Technology*, 1991, **9**, 157–162.
46. J. Honda, H. Andou, T. Mannen and S. Sugimoto, *J. Biosci. Bioeng.*, 2000, **89**, 582–589.
47. K. Tsumoto, K. Shinoki, H. Kondo, M. Uchikawa, T. Juji and I. Kumagai, *J. Immunol. Methods*, 1998, **219**, 119–129.
48. M. Umetsu, K. Tsumoto, M. Hara, K. Ashish, S. Goda, T. Adschiri and I. Kumagai, *J. Biol. Chem.*, 2003, **278**, 8979–8987.
49. J. S. Weissmann and P. S. Kim, *Cell*, 1992, **71**, 841–851.
50. Y. Hidaka, M. Ohno, B. Hemmasi, O. Hill, W.-G. Forssmann and Y. Shimonishi, *Biochemistry*, 1998, **37**, 8498–8507.
51. A. Rattenholl, M. Ruoppolo, A. Flagiello, M. Monti, F. Vinci, G. Marino, H. Lilie, E. Schwarz and R. Rudolph, *J. Mol. Biol.*, 2001, **305**, 523–533.
52. A. Hauburger, M. Kliemannel, P. Madsen, R. Rudolph and E. Schwarz, *FEBS Lett.*, 2007, **581**, 4159–4164.
53. S. Hober, G. Forsberg, G. Palm, M. Hartmanis and B. Nilsson, *Biochemistry*, 1992, **31**, 1749–1756.
54. K. R. Hejnaes, S. Bayne, L. Nørskov, H. H. Sørensen, J. Thomsen, L. Schäffer, A. Wollmer and L. Skriver, *Protein Eng.*, 1992, **5**, 797–806.
55. S. Hober, A. Hansson, M. Uhlén and B. Nilsson, *Biochemistry*, 1994, **33**, 6758–6761.
56. E. Haber and C. B. Anfinsen, *J. Biol. Chem.*, 1962, **237**, 1839–1844.
57. L. Kanarek, R. A. Bradshaw and R. L. Hill, *J. Biol. Chem.*, 1965, **240**, 2755–2757.
58. M. Ruoppolo and R. B. Freedman, *Biochemistry*, 1995, **34**, 9380–9388.
59. T. W. Odorzynski and A. Light, *J. Biol. Chem.*, 1979, **254**, 4291–4295.
60. A. Light, C. T. Duda, T. W. Odorzynski and W. G. Moore, *J. Cell. Biochem.*, 1986, **31**, 19–26.
61. C. T. Duda and A. Light, *J. Biol. Chem.*, 1982, **257**, 9866–9871.
62. G. Patil, R. Rudolph and C. Lange, *J. Biotechnol.*, 2008, **134**, 218–221.
63. J. M. Swan, *Nature*, 1957, **180**, 643–645.
64. J. L. Bailey and R. D. Cole, *J. Biol. Chem.*, 1959, **234**, 1733–1739.
65. W. W.-C. Chan, *Biochemistry*, 1968, **7**, 4247–4254.

66. P. G. Katsoyannis and A. Tometsko, *Proc. Natl. Acad. Sci. USA*, 1966, **55**, 1554–1561.

67. D. J. Cowley and R. B. Mackin, *FEBS Lett.*, 1997, **402**, 124–130.

68. R. V. Tikhonov, S. E. Pechenov, I. A. Belacheu, S. A. Yakimov, V. E. Klyushnichenko, E. F. Boldireva, V. G. Korobko, H. Tunes, J. E. Thiemann, L. Vilela and A. N. Wulfson, *Prot. Expr. Purif.*, 2001, **21**, 176–182.

69. R. V. Tikhonov, S. E. Pechenov, I. A. Belacheu, S. A. Yakimov, V. E. Klyushnichenko, H. Tunes, J. E. Thiemann, L. Vilela and A. N. Wulfson, *Prot. Expr. Purif.*, 2002, **26**, 187–193.

70. H. Ishikawa, J. Kawaguchi, Y. Yao, H. Tamaoki, T. Ono, F. Fukui and H. Yoshikawa, *J. Biosci. Bioeng.*, 1999, **87**, 296–301.

71. J. Hoppe, H. A. Weich and W. Eichner, *Biochemistry*, 1989, **28**, 2956–2960.

72. J. Hoppe, H. A. Weich, W. Eichner and D. Tatje, *Eur. J. Biochem.*, 1990, **187**, 207–214.

73. E. E. DiBella, M. C. Maurer and H. A. Scheraga, *J. Biol. Chem.*, 1995, **270**, 163–169.

74. R. F. Goldberger, C. J. Epstein and C. B. Anfinsen, *J. Biol. Chem.*, 1963, **238**, 628–635.

75. M.D. Schwaller and H. F. Gilbert, in *Encyclopedia of Life Sciences*, John Wiley & Sons, Ltd., 2001, doi: 10.1038/npg.els.0003021.

76. F. Hatahet and L. W. Ruddock, *FEBS J.*, 2007, **274**, 5223–5234.

77. N. J. Darby and T. E. Creighton, *Biochemistry*, 1995, **34**, 16770–16780.

78. K. W. Walker, M. M. Lyles and H. F. Gilbert, *Biochemistry*, 1996, **35**, 1972–1980.

79. T. Chivers, M. C. Laboissière and R. T. Raines, *EMBO J.*, 1996, **15**, 2659–2667.

80. M. Wunderlich, A. Otto, R. Seckler and R. Glockshuber, *Biochemistry*, 1993, **32**, 12251–12267.

81. H. I. Alanen, K. E. H. Salo, A. Pirneskoski and L. W. Ruddock, *Antiox. Redox Signal.*, 2006, **8**, 283–291.

82. H. Lilie, S. McLaughlin, R. Freedman and J. Buchner, *J. Biol. Chem.*, 1994, **269**, 14290–14296.

83. J. Yang, R. Raju, R. Sharma and J. Xiang, *Hum. Antibodies Hybridomas*, 1995, **6**, 129–136.

84. K. Maskos, M. Huber-Wunderlich and R. Glockshuber, *J. Mol. Biol.*, 2003, **325**, 495–513.

85. O. Buczek, B. R. Green and G. Bulaj, *Peptide Sci.*, 2006, **88**, 8–19.

86. M. Mayer, U. KIes, R. Kammermeier and J. Buchner, *J. Biol. Chem.*, 2000, **275**, 29421–29425.

87. M. Mayer, S. Frey, P. Koivunen, J. Myllyharju and J. Buchner, *J. Mol. Biol.*, 2004, **341**, 1077–1084.

88. H. Okudo, H. Kato, Y. Arayaki and R. Urade, *J Biochem. (Tokyo)*, 2005, **138**, 773–780.

89. T. P. Primm, K. W. Walker and H. F. Gilbert, *J. Biol. Chem.*, 1996, **271**, 33664–33669.

90. J. Winter, P. Klappa, R. B: Freedman, H. Lilie and R. Rudolph, *J. Biol. Chem.*, 2002, **277**, 310–317.
91. M. A. Watt, R. Y. Lo and A. Mellors, *Cell Stress Chaperones*, 1997, **2**, 180–190.
92. A. R. Frand and C. A. Kaiser, *Mol. Cell*, 1998, **1**, 161–170.
93. M. G. Pollard, K. J. Travers and J. S. Weissman, *Mol. Cell*, 1998, **1**, 171–182.
94. C. S. Sevier, J. W. Cuozzo, A. Vala, F. Åslund and C. A. Kaiser, *Nat. Cell Biol.*, 2001, **3**, 874–882.
95. T. Kobayashi, S. Kishigami, M. Sone, H. Inokuchi, T. Mogi and K. Ito, *Proc. Natl. Acad. Sci. USA*, 1997, **94**, 11857–11862.
96. M. Bader, W. Muse, D. P. Ballou, C. Gassner and J. C. Bardwell, *Cell*, 1999, **98**, 217–227.
97. J. D. Hayes and L. I. McLellan, *Free Radic. Res.*, 1999, **31**, 273–300.
98. G. Bánhegyi, M. Csala, A. Szarka, M. Varsányi, A. Benedetti and J. Mandl, *Biofactors*, 2003, **17**, 37–46.
99. V. G. Janolino and H. E. Swaisgood, *J. Biol. Chem.*, 1975, **250**, 2532–2538.
100. V. G. Janolino and H. E. Swaisgood, *Arch. Biochem. Biophys.*, 1987, **258**, 265–271.
101. K. L. Hoober, B. Joneja, H. B. White and C. Thorpe, *J. Biol. Chem.*, 1996, **271**, 30510–30516.
102. K. L. Hoober, S. L. Sheasley, H. F. Gilbert and C. Thorpe, *J. Biol. Chem.*, 1999, **274**, 22147–22150.
103. W. W. Cleland, *Biochemistry*, 1964, **3**, 480–482.
104. D. M. Rothwarf and H. A. Scheraga, *Biochemistry*, 1993, **32**, 2880–2889.
105. P. Grasso, T. A. Santa-Coloma, J. J. Boniface and L. E. Reichert, *Mol. Cell. Endocrinol.*, 1991, **78**, 163–170.
106. C. Cabrele, S. Fiori, S. Pegoraro and L. Moroder, *Chem. Biol.*, 2002, **9**, 731–740.
107. K. J. Woycechowsky, K. D. Wittrup and R. T. Raines, *Chem. Biol.*, 1999, **6**, 871–879.
108. J. Winter, H. Lilie and R. Rudolph, *FEMS Microbiol. Lett.*, 2002, **213**, 225–230.
109. J. Houk, R. Singh and G. M. Whitesides, *Methods Enzymol.*, 1987, **143**, 129–140.
110. F. Hillger, G. Herr, R. Rudolph and E. Schwarz, *J. Biol. Chem.*, 2005, **280**, 14974–14980.
111. C. Dicko, F. Vollrath and J. M. Kenney, *Biomacromolecules*, 2004, **5**, 704–710.
112. C. Dicko, J. M. Kenney, D. Knight and F. Vollrath, *Biochemistry*, 2004, **43**, 14080–14087.
113. T. V. DeCollo and W. J. Lees, *J. Org. Chem.*, 2001, **66**, 4244–4249.
114. J. D. Gough, R. H. Williams, A. E. Donofrio and W. J. Lees, *J. Am. Chem. Soc.*, 2002, **124**, 3885–3892.
115. J. D. Gough, J. M. Gargano, A. E. Donofrio and W. J. Lees, *Biochemistry*, 2003, **42**, 11787–11797.

116. J. D. Gough and W. J. Lees, *J. Biotechnol.*, 2005, **115**, 279–290.
117. J. D. Gough, E. J. Barrett, Y. Silva and W. J. Lees, *J. Biotechnol.*, 2006, **125**, 39–47.
118. J. D. Gough and W. J. Lees, *Bioorg. Med. Chem. Lett.*, 2005, **15**, 777–781.
119. K. Brocklehurst and G. Little, *Biochem. J.*, 1973, **133**, 67–80.
120. C. J. Epstein and C. B. Anfinsen, *J. Biol. Chem.*, 1962, **237**, 2175–2179.
121. T. E. Creighton, in *UCLA Symposia on Molecular and Cellular Biology -* New series Vol. 39, D. L. Oxender, ed., Academic Press, New York, 1985, p. 249–258.
122. X. Geng and X. Chang, *J. Chromatogr. A*, 1992, **599**, 185–194.
123. G. Stempfer, B. Höll-Neugebauer and R. Rudolph, *Nat. Biotechnol.*, 1996, **14**, 329–334.
124. C. Machold, R. Schlegl, R. W. Buchinger and A. Jungbauer, *J. Chromatogr. A*, 2005, **1080**, 29–42.
125. M. Langenhof, S. S. Leong, L. K. Pattenden and A. P. Middelberg, *J. Chromatogr. A*, 2005, **1096**, 195–201.
126. N. A. Morjana and H. F. Gilbert, *Protein Expr. Purif.*, 1994, **5**, 144–148.
127. K. Tsumoto, M. Umetsu, H. Yamada, T. Ito, S. Misawa and I. Kumagai, *Protein Eng.*, 2003, **16**, 535–541.
128. M. M. Altamirano, C. García, L. D. Possani and A. R. Fersht, *Nat. Biotechnol.*, 1999, **17**, 187–191.
129. A. Karadimitris, S. Gadola, M. Altamirano, D. Brown, A. Woolfson, P. Klenerman, J.-L. Chen, Y. Koezuka, I. A. G. Roberts, D. A. Price, G. Dusheiko, C. Milstein, A. R. Fersht, L. Luzzato and V. Cerundolo, *Proc. Natl. Acad. Sci. USA*, 2001, **98**, 3294–3298.
130. H. Shimizu, K. Fujimoto and H. Kawaguchi, *Colloids Surf., A*, 1999, **153**, 421–427.
131. H. Shimizu, K. Fujimoto and H. Kawaguchi, *Biotechnol. Prog.*, 2000, **16**, 248–253.
132. K. J. Woycechowsky, B. A. Hook and R. T. Raines, *Biotechnol. Prog.*, 2003, **19**, 1307–1314.
133. M. Birringer, S. Pilawa and M. Flohé, *Nat. Prod. Rep.*, 2002, **19**, 693–718.
134. L. A. Wessjohan, A. Schneider, M. Abbas and W. Brandt, *Biol. Chem.*, 2007, **388**, 997–1006.
135. J. Pinsent, *Biochem. J.*, 1954, **57**, 10–16.
136. L. Flohé, W. A. Gunzler and H. H. Schock, *FEBS Lett.*, 1973, **32**, 132–134.
137. J. Beld, K. J. Woycechowsky and D. Hilvert, *Biochemistry*, 2007, **46**, 5382–5390.
138. A. T. McDuffee, G. Senisterra, S. Huntley, J. R. Lepock, K. R. Sekhar, M. J. Meredith, M. J. Borrelli, J. D. Morrow and M. L. Freeman, *J. Cell Physiol.*, 1997, **171**, 143–151.
139. D. J. Jamieson, *Yeast*, 1998, **14**, 1511–1527.
140. N. Fedoroff, *Ann. Bot.*, 2006, **98**, 289–300.
141. U. Jakob, W. Muse, M. Eser and J. C. Bardwell, *Cell*, 1999, **96**, 341–352.
142. Z. Salamon, F. K. Gleason and G. Tollin, *Arch. Biochem. Biophys.*, 1992, **299**, 193–198.

143. Z. Chen, H. Zheng, C. Lu and Y. Zu, *Langmuir*, 2007, **23**, 10816–10822.
144. N. Maleki, A. Safavi, F. Sedaghati and F. Tajabadi, *Anal. Biochem.*, 2007, **369**, 149–153.
145. C. A. Summers and R. A. Flowers, *Protein Sci.*, 2000, **9**, 2001–2008.
146. C. Lange, G. Patil and R. Rudolph, *Protein Sci.*, 2005, **14**, 2693–2701.
147. K. Fujita, D. R. MacFarlane and M. Forsyth, *Chem. Commun.*, 2005, 4804–4806.
148. N. Byrne, L.-M. Wang, J.-P. Belieres and C. A. Angell, *Chem. Commun.*, 2007, 2714–2716.
149. D. Constantinescu, H. Weingärtner and C. Hermann, *Angew. Chem.*, 2007, **119**, 9044–9046.
150. D. Macmillan, *Angew. Chem., Int. Ed. Engl.*, 2006, **45**, 7668–7672.
151. Z. Machova and A. Beck-Sickinger, *Methods Mol. Biol.*, 2005, **298**, 105–130.
152. C. Monfardini and F. M. Veronese, *Bioconjugate Chem.*, 1998, **9**, 418–450.
153. J. M. Harris and R. B. Chess, *Nat. Rev. Drug Discov.*, 2003, **2**, 214–221.
154. R. Haag and F. Kratz, *Angew. Chem. Int. Ed. Engl.*, 2006, **45**, 1198–1215.
155. A. P. Chapman, P. Antoniw, M. Spitali, S. West, S. Stephens and D. J. King, *Nat. Biotechnol.*, 1999, **17**, 780–783.
156. D. P. Humphreys, S. P. Heywood, A. Henry, L. Ait-Lhadj, P. Antoniw, R. Palframan, K. J. Greenslade, B. Carrington, D. G. Reeks, Leigh Bowering, S. West and H. A. Brand, *Protein Eng. Des. Sel.*, 2007, **20**, 227–234.
157. S. Brocchini, S. Balan, A. Godwin, J.-W. Choi, M. Zloh and S. Shaunak, *Nat. Protoc.*, 2006, **1**, 2241–2252.
158. S. Brocchini, A. Goldwin, S. Balan, J.-W. Choi, M. Zloh and S. Shaunak, *Adv. Drug Deliv. Rev.*, 2008, **60**, 3–12.
159. S. Shaunak, A. Godwin, J.-W. Choi, S. Balan, E. Pedone, D. Vijayarangam, S. Heidelberger, I. Teo, M. Zloh and S. Brocchini, *Nat. Chem. Biol.*, 2006, **2**, 312–313.
160. M. Zloh, S. Shaunak, S. Balan and S. Brocchini, *Nat. Protoc.*, 2007, **2**, 1070–1083.
161. L. A. Ryabova, D. Desplancq, A. S. Spirin and A. Plückthun, *Nat. Biotechnol.*, 1997, **15**, 79–84.
162. H. Merk, W. Stiege, K. Tsumoto, I. Kumagai and V. A. Erdmann, *J. Biochem. (Tokyo)*, 1999, **125**, 328–333.
163. D. M. Kim and J. R. Swartz, *Biotechnol. Bioeng.*, 2004, **85**, 122–129.
164. G. Yin and J. R. Swartz, *Biotechnol. Bioeng.*, 2004, **86**, 188–195.
165. S. Frey, M. Haslbeck, O. Hainzl and J. Buchner, *Biol. Chem.*, 2008, **389**, 37–45.
166. K. G. Knapp, A. R. Goerke and J. R. Swartz, *Biotechnol. Bioeng.*, 2007, **97**, 901–908.

CHAPTER 3

Redox Potentials of Cysteine Residues in Peptides and Proteins: Methods for their Determination

DALLAS L. RABENSTEIN

Department of Chemistry, University of California, Riverside, California 92521, USA

3.1 Introduction

Formation, reduction and isomerization of disulfide bonds takes place in biology in such diverse processes as the oxidative folding of proteins,[1-4] enzyme activation[5] and regulation of gene activity.[6,7] In the oxidative folding of proteins, disulfide bonds are formed by multistep, reversible thiol–disulfide exchange reactions, in which disulfide bonds are transferred from a disulfide donor to the protein.

3.2 Formation of Disulfide Bonds by Thiol-disulfide Exchange

The overall reaction for transfer of an *inter*molecular disulfide bond, *e.g.* the disulfide bond of oxidized glutathione (GSSG), to a protein by thiol–disulfide

RSC Biomolecular Sciences
Oxidative Folding of Peptides and Proteins
Edited by Johannes Buchner and Luis Moroder
© Royal Society of Chemistry 2009
Published by the Royal Society of Chemistry, www.rsc.org

exchange:

$$P_{SH}^{SH} + GSSG \leftrightarrow P_S^S + 2GSH \tag{3.1}$$

takes place in two steps:

$$P_{SH}^{SH} + GSSG \leftrightarrow [P_{SH}^{SSG} + P_{SSG}^{SH}] + GSH \tag{3.2}$$

$$[P_{SH}^{SSG} + P_{SSG}^{SH}] \leftrightarrow P_S^S + GSH \tag{3.3}$$

P_{SH}^{SH} and P_S^S represent a peptide or protein with two cysteines in the reduced (dithiol) and oxidized (disulfide) forms. In the first step, a thiol group of P_{SH}^{SH} reacts with the disulfide bond of GSSG to form one of the two possible mixed disulfide intermediates, which undergo an *intra*molecular thiol–disulfide exchange reaction to form the peptide or protein disulfide bond in the second step.

The overall reaction for transfer of an intramolecular disulfide bond, *e.g.* from one protein to another, also takes place in two steps *via* mixed disulfide intermediates:

$$P_{SH}^{SH} + R_S^S \leftrightarrow [P_{SH}^{SSRSH} + P_{SSRSH}^{SH}] \leftrightarrow P_S^S + R_{SH}^{SH} \tag{3.4}$$

Thiol–disulfide exchange reactions are nucleophilic displacement reactions that result in oxidation of the thiol groups of one thiol/disulfide pair and reduction of the disulfide group of another. Each thiol/disulfide pair is formally a redox couple that can be represented as:

$$P_{SH}^{SH} \leftrightarrow P_S^S + 2H^+ + 2e^- \tag{3.5}$$

for a peptide or protein that forms an intramolecular disulfide bond. The standard electrode, or redox, potential for the redox couple, $E_{peptide}^o$ or $E_{protein}^o$, is a measure of the tendency of the thiol and disulfide groups to react as electron donors and acceptors. The more positive the redox potential, the greater the tendency of the disulfide bond to be an electron acceptor (to transfer a disulfide bond) in a thiol–disulfide exchange reaction; the more negative the redox potential, the greater the tendency of the thiol group to be an electron donor (to reduce a disulfide bond) in a thiol-disulfide exchange reaction. The reduction potential of a thiol-disulfide redox couple in solution, E, and thus its tendency to be an electron donor or electron acceptor, is concentration dependent according to the Nernst equation:

$$E = E_{peptide/protein}^o - \frac{RT}{nF} \ln \left(\frac{[P_{SH}^{SH}]}{[P_S^S][H^+]^2} \right) \tag{3.6}$$

where R is the gas constant, T the temperature, n the number of electrons transferred and F the Faraday constant. Redox potentials for peptides and proteins are measured in buffer solutions, generally at neutral pH, and are expressed as conditional (apparent) redox potentials, $E^{o'}_{\text{peptide}}$ or $E^{o'}_{\text{protein}}$, which incorporate [H$^+$]. The Nernst equation, expressed in terms of $E^{o'}_{\text{peptide}}$ or $E^{o'}_{\text{protein}}$, is:

$$E = E^{o'}_{\text{peptide/protein}} - \frac{RT}{nF} \ln \left(\frac{[\text{P}^{\text{SH}}_{\text{SH}}]}{[\text{P}^{\text{S}}_{\text{S}}]} \right) \tag{3.7}$$

The thiolate anion is the reactive species in thiol-disulfide exchange reactions. Mechanistically, thiol–disulfide exchange takes place by an S_N2 displacement reaction in which a thiolate anion approaches the disulfide bond along its S-S axis, forming a new bond with one sulfur atom and displacing the other.[8–11] The reaction goes through a transition state in which the thiolate negative charge is partially centered on each of the three sulfur atoms. The rate of the reaction depends on the Brønsted basicity of the incoming thiolate, the central sulfur in the transition state and the leaving thiolate.[8–11]

Redox potentials for peptide and protein cysteines are determined indirectly from equilibrium constants for thiol–disulfide exchange reactions with reference thiol/disulfide redox couples of known redox potential, including the GSH/GSSG redox couple (Equation (3.1)) and proteins that transfer an intramolecular disulfide bond (Equation (3.4)).

3.3 Redox Potentials of Mixed Disulfide Bonds

Mixed disulfides are formed as intermediates in the transfer of disulfide bonds to peptides and proteins (Equations (3.2) and (3.4)).[12–16] Mixed disulfides can also be formed by peptides and proteins that contain a single cysteine or more than one cysteine but not located in sufficiently close proximity to form an intramolecular disulfide bond.[17] The equilibrium constant for formation of a mixed disulfide with GSH is defined as:

$$K_{\text{mix}} = \frac{[\text{PSSG}][\text{GSH}]}{[\text{PSH}][\text{GSSG}]} \tag{3.8}$$

where [GSH] and [PSH] represent the total concentrations of GSH and PSH. The conditional redox potential for the mixed disulfide can be obtained from K_{mix} and $E^{o'}$ of the reference redox couple, in this case $E^{o'}_{\text{GSH}}$, with the equation:

$$E^{o'}_{\text{PSSG}} = E^{o'}_{\text{GSH}} - \frac{RT}{nF} \ln(K_{\text{mix}}) \tag{3.9}$$

K_{mix} and $E^{o'}_{\text{PSSG}}$ for several amino acids, peptides and proteins are listed in Table 3.1. The $E^{o'}$ values in Table 3.1, and in the following tables, are relative to $E^{o'}_{\text{GSH}}$. A range of values has been reported for $E^{o'}_{\text{GSH}}$ (Section 3.7). To facilitate comparison with majority of the literature, a value of $E^{o'}_{\text{GSH}} = -0.240\,\text{V}$ was used for calculating the apparent redox potentials.[18] Several of

Table 3.1 Equilibrium constants and redox potentials for mixed disulfides with glutathione.[a]

Amino acid, peptide or protein	K_{mix}	$E^{o'}_{PSSG}$, V	Reference
Cysteine	1.11	−0.241	11
Homocysteine	2.46	−0.252	11
Penicillamine	2.95	−0.254	11
Arginine Vasopressin[b]	4.17	−0.258	13
Oxytocin[b]	3.85	−0.257	13
Somatostatin (Cys[3])[c]	2.86	−0.253	14
Somatostatin (Cys[14])[c]	2.04	−0.249	14
Thioredoxin (Cys[32])	1.95[d]	−0.240[d]	19
DsbA (Cys[30])	6.1×10^{-3}	−0.175	16,20
DsbC (Cys[98])	4.1×10^{-3}	−0.171	20

[a] $E^{o'}_{PSSG}$ values calculated using $E^{o'}_{GSH} = -0.240$ V
[b] K_{mix} is the sum of K_{mix} for formation of the mixed disulfides with Cys[1] and Cys[6]
[c] K_{mix} for the mixed disulfide with the indicated cysteine residue
[d] K_{mix} for the mixed disulfide with β-mercaptoethanol (RSH). $E^{o'}_{PSSG}$ calculated using $E^{o'}_{mercaptoethanol} = -0.231$ V.

the values reported in Tables 3.1–3.4 were recalculated using reported values for K_{mix} or K_{ox} and $E^{o'}_{GSH} = -0.240$ V. Reference redox couples are discussed in Sections 3.6 and 3.7.

3.4 Redox Potentials of Intramolecular Disulfide Bonds

Equilibrium constants for the formation of intramolecular peptide and protein disulfide bonds by the thiol–disulfide exchange reactions in Equations (3.1) and (3.4) are defined by Equations (3.10) and (3.11):

$$K_{ox} = \frac{[P_S^S][GSH]^2}{[P_{SH}^{SH}][GSSG]} \tag{3.10}$$

$$K_{ox} = \frac{[P_S^S][R_{SH}^{SH}]}{[P_{SH}^{SH}][R_S^S]} \tag{3.11}$$

$E^{o'}_{peptide}$ and $E^{o'}_{protein}$ can be calculated with the Nernst equation:

$$E^{o'}_{peptide/protein} = E^{o'}_{reference\ thiol} - \frac{RT}{nF}\ln(K_{ox}) \tag{3.12}$$

Representative values of K_{ox} and $E^{o'}$ are listed in Tables 3.2 and 3.3 for selected peptides and proteins, respectively.

Octapeptides 1, 3, 5 and 7 in Table 3.2 are active-site containing segments of the thiol/disulfide oxidoreductases thioredoxin (Trx), glutaredoxin (Grx),

Table 3.2 Equilibrium constants and redox potentials for intramolecular disulfide bonds in peptides.^{a,b}

Peptide		K_{ox}, M	$E^{o\prime}_{peptide(SS)}$, V	Reference
1	Ac-WCGPCKHI-NH$_2$^{c}	0.016	−0.190	12
2	c[WCGPCK-]	4.2×10^{-3d}	−0.152	21
3	Ac-GCPYCVRA-NH$_2$	0.123	−0.213	12
4	c[GCPYCV-]	9.4×10^{-3}	−0.179	21
5	Ac-ACATCDGF-NH$_2$	0.142	−0.215	12
6	c[ACATCD-]	5.2×10^{-2}	−0.201	21
7	Ac-WCGHCKAL-NH$_2$	0.079	−0.205	12
8	c[WCGHCK-]	8.0×10^{-4d}	−0.130	21
9	Arginine-Vasopressin (Cys1-Cys6)	6.9×10^{-2}	−0.206	13
10	Oxytocin (Cys1-Cys6)	2.8×10^{-2}	−0.194	13
11	Somatostatin (Cys3-Cys14)	4.0×10^{-2}	−0.199	14

^{a}K_{ox} for oxidation of dithiol form of the peptide by GSSG.
^{b}$E^{o\prime}_{peptide(SS)}$ calculated using $E^{o\prime}_{GSH} = -0.240$ V.
^{c}Met^{35} of thioredoxin was replaced by His.
^{d}Relative to the cysteine/cystine redox couple, $E^{o\prime}_{cysteine/cystine} = -0.223$ V, adapted from the value of $E^{o\prime}_{cysteine/cystine}$ in reference 11 using $E^{o\prime}_{GSH} = -0.240$ V.

thioredoxin reductase (Trr) and protein disulfide isomerase (PDI).[12] Hexapeptides 2, 4, 6 and 8 contain the same Cys-Xaa-Yaa-Cys sequences, respectively, but are conformationally constrained by backbone cyclization, which causes the disulfide bonds to become significantly more oxidizing.[21] The cyclic peptides are effective adjuvants in the refolding of reduced RNase A, with the yield and rate increasing with the oxidizing power of the hexapeptide.[21] Peptides 9–11 are peptide hormones that contain 20- (peptides 9 and 10) and 38-membered (peptide 11) disulfide-bridged rings.[13,14]

The proteins listed in Table 3.3 are all thiol-disulfide oxidoreductases (proteins 1, 14–16, 21, 22), or variants thereof (proteins 2–13, 17–20), that are involved in the formation, reduction or isomerization of disulfide bonds during the oxidative folding of proteins.[1–4] As discussed below, the data reported for the variants of DsbA and Trx illustrate how $E^{o\prime}_{protein}$ can be tuned over a wide range by substitution of the Xaa-Yaa dipeptide between the two active-site cysteine residues. The engineered DsbA and Trx variants are useful as reference redox couples for determination of $E^{o\prime}_{protein}$ for other proteins by direct protein-protein equilibration.

3.5 Measurement of Equilibrium Constants for Thiol-disulfide Exchange

Redox potentials of cysteine residues in peptides and proteins are determined by reaction of the peptide or protein with a reference thiol/disulfide redox couple. The equilibrium constant for the thiol–disulfide exchange reaction is obtained from the equilibrium concentrations of the thiol and disulfide species, and the redox potential is then calculated using Equation (3.9) or (3.12).

Table 3.3 Equilibrium constants and redox potentials for intramolecular disulfide bonds in proteins.[a,b]

Protein		K_{ox}, M	$E^{o\prime}_{protein(ss)}$, V	Reference
	Wild Type DsbA			
1	(Cys30-Pro-His-Cys33)	8.1×10^{-5}	-0.120	16
		1.2×10^{-4}	-0.122	24
		1.31×10^{-4}	-0.125	22
2	DsbA(Gly-His)[c]	7.3×10^{-4}	-0.147	24
3	DsbA(Ala-Thr)[c]	1.4×10^{-3}	-0.156	24
4	DsbA(Pro-Tyr)[c]	1.8×10^{-3}	-0.159	24
5	DsbA(Gly-Pro)[c]	1.4×10^{-1}	-0.215	24
6	DsbA(Pro-Gly)[c]	5.9×10^{-4}	-0.145	24
7	DsbA(Ser-Val)[c]	9.0×10^{-4}	-0.150	27
8	DsbA(Ser-Phe)[c]	1.2×10^{-3}	-0.154	27
9	DsbA(Pro-Leu)[c]	1.6×10^{-3}	-0.158	27
10	DsbA(Leu-Thr)[c]	3.9×10^{-3}	-0.169	27
11	DsbA(Thr-Arg)[c]	6.8×10^{-3}	-0.176	27
12	DsbA(Pro-Pro)[c]	2.0×10^{-1}	-0.190	27
13	DsbA(F26L)	4.3×10^{-5}	-0.111	22
14	DsbB			
	(Cys41-Val-Leu-Cys44)	1.7×10^{-6}	-0.069	23
	(Cys104-Cys130)		-0.186	23
15	DsbC			
	(Cys98-Gly-Tyr-Cys101)	1.95×10^{-4}	-0.130	28
16	Wild Type Trx			
	(Cys32-Gly-Pro-Cys35)	11.3, 10	-0.271, -0.270	40,41
17	Trx(Ala-Thr)[c]	0.243	-0.222	40
18	Trx(Gly-His)[c]	0.228	-0.221	40
19	Trx(Pro-His)[c]	0.061	-0.204	40
20	Trx(Pro-Tyr)[c]	0.0273	-0.194	40
21	Grx1			
	(Cys-Pro-Tyr-Cys)		-0.233	15
22	Grx3			
	(Cys-Pro-Tyr-Cys)		-0.198	15

[a] K_{ox} for reaction of the dithiol form of the protein with GSSG.
[b] Relative to $E^{o\prime}_{GSH} = -0.240$ V.
[c] The indicated dipeptide sequence was substituted for the dipeptide between the two active-site cysteines of wild-type DsbA or Trx.

Experimental variables include the reference redox couple and its concentration, pH, temperature, equilibration time, the method used to quench the reaction and the analytical method used to determine the concentrations of P^{SH}_{SH}, P^S_S and the oxidized and reduced forms of the reference redox couple. Reference redox couples are discussed in Sections 3.6 and 3.7.

Equilibrium constants for thiol-disulfide exchange reactions are generally measured in pH 7–8 phosphate buffer at temperatures in the range 20 °C–40 °C. The fraction of thiol in the reactive thiolate form depends on pH, and thus K_{mix} and K_{ox} will depend on pH and the pK_as of the two thiols if their pK_as are different.[8,11,29] The pH and temperature should be the same as those at which $E^{o\prime}$ of the reference redox couple was measured.

To exclude air oxidation, solutions should be deoxygenated, *e.g.* by bubbling with oxygen-scrubbed argon or nitrogen, the reaction should be run under an oxygen-free atmosphere and EDTA can be added to sequester metals that catalyze oxidation of thiols by dissolved oxygen. The length of time required for the system to reach equilibrium will depend on pH, temperature, the pK_as of the thiols and the concentrations and reactivities of the thiols and disulfides. The reaction mixture should be analyzed at several time intervals and equilibrium should be approached from both directions to ensure that a true equilibrium has been reached.[30]

Analytical methods used to determine the concentrations of thiols and disulfides in exchange reaction mixtures include HPLC, NMR, fluorescence spectroscopy and radioactive labeling. HPLC with UV detection is the most widely used method; the components of the reaction mixture are separated on the HPLC column and the concentration of each component is determined from its peak area.[30] Because some thiol–disulfide exchange reactions are relatively fast, *e.g.* intramolecular thiol–disulfide exchange reactions of protein mixed disulfide intermediates (Equation (3.3)), reaction mixtures must be quenched to ensure that equilibrium does not shift during the separation.[15,30] Reaction mixtures can be quenched by alkylation of thiol groups with iodoacetamide or iodoacetate.[30] To avoid a shift of the equilibrium during the alkylation reaction, all peptide/protein and reference thiol groups should react with the alkylation reagent at similar rates. Reaction mixtures can also be quenched by rapidly lowering the pH to ~2. If an acid quench is used, a low pH mobile phase must also be used for the HPLC separation to prevent further reaction when the components are separated and no longer at equilibrium.[30] Typically water/acetonitrile mobile phases containing 0.1% trifluoroacetic acid are used, in either isocratic or gradient mode. With UV detection, the detector response must be calibrated for each component of the exchange reaction, generally by using pure samples of the peptide or protein and the reference in their thiol and disulfide forms. Detector response for mixed disulfides can be determined by isolating the mixed disulfides, raising the pH to allow them to convert to disulfide, re-chromatographing the solution and then comparing the mixed disulfide and disulfide peak areas.[14]

NMR has been used in several studies to measure thiol-disulfide exchange equilibrium constants.[11,31] An advantage of NMR is that the concentration of each species can be determined directly *in situ*, provided that resolved resonances can be observed for all reactants and products. However, the requirement of resolved resonances has limited the measurement of thiol-disulfide equilibrium constants by NMR to small molecules, including amino acids and small peptides.[11,31]

K_{ox} for proteins can be determined by fluorescence spectroscopy if the fluorescent properties of the dithiol and disulfide forms of the protein or the reference protein are different.[22–24,29,33–35] Disulfide bonds are effective quenchers of tryptophan fluorescence,[16,22,32–34] and reduction of the Cys-Xaa-Yaa-Cys disulfide bonds in Trx, DsbA, DsbB and TlpA causes a strong increase in their tryptophan fluorescence.[23,29,33,34] A value of 1.31×10^{-4} M was determined for

K_{ox} of DsbA with GSH/GSSG redox buffer at pH 7.0 and 25 °C by exploiting the 3.2-fold increase in tryptophan fluorescence upon reduction of its Cys30-Cys33 disulfide bond, from which a value of –0.122 V was calculated for $E_{DsbA}^{o'}$ (Table 3.3).[22,29] A value of 1.70×10^{-6} M was determined for K_{ox} for formation of the Cys41-Cys44 disulfide bond of DsbB by reaction with GSH/GSSG redox buffer by using the 1.5-fold increase in fluorescence upon reduction of the disulfide bond.[23] This corresponds to a redox potential of –0.069 V, the most strongly oxidizing redox potential that has been reported for a protein disulfide bond.[23] The redox potential of the Cys104-Cys130 disulfide bond of DsbB, which does not affect the tryptophan fluorescence of DsbB, was determined to be −0.186 V by direct protein-protein equilibration with a fluorescent reference protein.[23]

Redox potentials of variants of green fluorescent protein (GFP) engineered to contain two redox-active cysteine residues have been determined by fluorescence spectroscopy.[35,36] The chromophore of GFP is a cyclic tripeptide; the two cysteine residues are introduced at positions on adjacent β-strands so that, in the disulfide form, fluorescence is decreased by distortion of the immediate environment of the chromophore.[35] $E^{o'}$ for the S147C/Q204C variant of GFP was determined to be –0.288 V from $K_{ox} = 0.070$ for reaction with reduced DTT/oxidized DTT redox buffers.[36] $E^{o'}$ for the N149C/S202C variant of the yellow fluorescent (YFP) variant of GFP was determined to be −0.261 V from $K_{ox} = 5$ M for reaction with GSH/GSSG redox buffers.[35] The redox-active variants were engineered for use as indicators of redox potential in living cells; the redox potential of the cytoplasm of *E. coli* was determined with the redox-active variant of YFP and the redox potential of the matrix space of HeLa cell mitochondria was measured with the redox-active variant of GFP.[35,36]

K_{ox} can be determined by radioactivity measurements by quenching equilibrium mixtures with [^{14}C]- and [^{3}H]-labeled alkylation reagents. The redox potential of the two Cys-Gly-His-Cys active-site disulfides of protein disulfide isomerase (PDI) was determined by quenching the reaction of PDI with GSH/GSSG redox buffer and the reaction of PDI with *E. coli* (P34H)-Trx with [^{14}C]- and [^{3}H]iodoacetic acid.[37] The alkylated thiols were separated by gel filtration, and their concentrations determined by radioactivity measurements. A K_{ox} of 3.1×10^{-3} M and $E_{PDI}^{o'}$ of –0.175 V were obtained with GSH/GSSG redox buffer, and a K_{ox} of 0.032 and $E_{PDI}^{o'}$ of –0.190 V were obtained with the (P34H)-Trx reference redox couple. The difference of 0.015 V was attributed to different redox scales for the GSH/GSSG and (P34H)-Trx reference redox couples. Values ranging from −0.110 V to −0.190 V have been reported for $E_{PDI}^{o'}$.[25–26,35–39]

3.6 Reference Redox Couples

Redox potentials reported for thiol–disulfide redox couples in peptides and proteins cover a wide range, from the strongly oxidizing potential of −0.069 V for the Cys41-Cys44 disulfide of DsbB[23] to the strongly reducing potential of −0.270 V for the Cys32-Cys35 thiols of Trx[40,41] for the family of thiol-disulfide

Table 3.4 Redox potentials for reference redox couples.

Thiol	pH or pD	T, °C	$K_{ox}{}^a$	$E^{o\prime}$, V	Reference
Cysteine (CSH)	7.0	25	0.265	−0.223	11
β-Mercaptoethanol	7.0	25	0.486	−0.231	11
Dithiothreitol (DTT)	7.0	25	210	−0.323	31
Lipoic acid	7.0	25		−0.288	31
Glutathione (GSH)	7.0	25		−0.244	46
Glutathione (GSH)	7.4	40		−0.240	18

^{a}For oxidation by GSSG.

oxidoreductases that catalyze formation, reduction and isomerization of di-sulfide bonds in proteins to the even more strongly reducing potentials of structural disulfide bonds in proteins, *e.g.* −0.335 V and −0.448 V for Cys30-Cys51 ($K_{ox} = 1.6 \times 10^3$ M)[42] and Cys5-Cys55 ($K_{ox} = 1.1 \times 10^7$ M)[42,43] of bovine trypsin inhibitor. To cover this wide range, reference redox couples that have a wide potential range have been used, including GSH, β-mercaptoethanol, cysteine, dithiothreitol (DTT), lipoic acid (Table 3.4) and protein redox couples, *e.g.* DsbA and Trx (Table 3.3). The first three transfer an inter-molecular disulfide bond; DTT, lipoic acid and the proteins transfer an intra-molecular disulfide bond. A disadvantage of reference redox couples that transfer an *inter*molecular disulfide is the tendency for accumulation of mixed disulfides at equilibrium; there is less tendency for accumulation of mixed disulfides with DTT, lipoic acid and protein redox buffers because of the intramolecular nature of the second step of the overall thiol–disulfide exchange reaction (Equation (3.3)).[15,30,40]

The GSH/GSSG redox couple is the most widely used reference redox couple for the measurement of redox potentials of cysteines in peptides and proteins (Section 3.7).[24,30,40] The DTT and lipoic acid redox systems can be used to determine $E^{o\prime}$ values at the more negative (strongly reducing) end of the redox potential scale, *i.e.* for peptides and proteins that form the more stable disulfide bonds.[30] When small molecule thiol/disulfide systems are used as the reference redox couple, the concentrations of both the reduced and oxidized forms (*e.g.* GSH and GSSG) are generally present in large excess to buffer the redox potential of the solution. The potential of the redox buffer, which depends on the ratio of the reduced and oxidized forms, should be adjusted so that the ratio of P_{SH}^{SH} and P_S^S is near one for the most precise determination of K_{ox}.[15] When a protein is used as the reference redox couple, the reference and test protein concentrations are generally similar and the reduced or oxidized form of the reference protein is reacted with the oxidized or reduced form of the test protein, respectively, *i.e.* the solution is not redox buffered by the reference redox couple.[15] For the most precise determination of K_{ox}, $E^{o\prime}$ of the reference protein should be close to $E^{o\prime}$ of the test protein.[23,37]

DsbA and Trx have been used as reference redox couples for measurement of more oxidizing and more reducing protein redox potentials, respec-tively.[15,23,27,37,44] In addition to their tendency not to accumulate mixed disulfides, DsbA and Trx offer other advantages as reference redox couples. $E^{o\prime}$ of DsbA and Trx can be tuned to be close to the redox potential of the test protein by varying the dipeptide between the two cysteine residues[22,24,40,45] and K_{ox} can be determined *in situ* by exploiting the 3.2-fold and 3.5-fold increases in tryptophan fluorescence upon reduction of their Cys30-Cys33 and Cys32-Cys35 disulfide bonds, respectively.[23,40] The redox potential of DsbA can be tuned to be less oxidizing, while that of Trx can be tuned to be less reducing, by replacing the dipeptides between their Cys30-Cys33 and Cys32-Cys35 catalytic residues, respectively, with other dipeptides.[15,23,24,40,45] K_{ox} and $E^{o\prime}$ for wild-type DsbA and eleven variants of DsbA and for wild-type Trx and four variants of Trx are listed in Table 3.3.[24,40,45] The dipeptides between Cys30 and Cys33 in the first four variants of DsbA (proteins 2–5) are the dipeptides between the two cysteine residues at the catalytic centers of PDI, Trr, Grx and Trx, while the dipeptides in the four variants of Trx (proteins 17–20) are the dipeptides between the two cysteine residues at the catalytic centers of Trr, PDI, DsbA and Grx, respectively. K_{ox} for reaction of DsbA and the DsbA variants in Table 3.3 in which the Xaa-Yaa dipeptide between the Cys30-Cys33 catalytic residues is replaced with another dipeptide varies from 1.2×10^{-4} M for wild-type DsbA to 2.0×10^{-1} M for DsbA(Pro-Pro) ($E^{o\prime}$ values from -0.120 V to -0.190 V), while that for Trx and its variants varies from 11.3 to 0.0273 ($E^{o\prime}$ values from -0.270 V to -0.194 V). It is also interesting to note that, opposite to the changes in $E^{o\prime}$ that result from substitution of the Xaa-Yaa dipeptide, the F26L variant of DsbA is more oxidizing than wild-type DsbA.[22]

3.7 The GSH/GSSG Reference Redox Couple

Advantages of the GSH/GSSG redox couple are that both GSH and GSSG are readily available, both are relatively soluble and thus both can be present in excess to buffer the redox potential of the solution. A disadvantage is the greater tendency for accumulation of mixed disulfides, as compared to reference redox buffers or proteins that transfer an intramolecular disulfide bond.[15] Air oxidation and contamination of GSH with GSSG limits the maximum $[GSH]^2/[GSSG]$ ratio (Equation (3.10)) that can be attained, and thus the range of redox potentials that can be measured with the GSH/GSSG redox buffer.[29] GSH is typically contaminated with 0.3–1.0% GSSG; the GSSG content can be determined directly by ^{1}H NMR.[46]

Values ranging from -0.16 V to -0.28 V have been reported for $E^{o\prime}_{GSH}$.[8,15,18,46–48] The most frequently used value is -0.240 V, which was determined at pH 7.4 and 40 °C.[18] The reported values for $E^{o\prime}_{GSH}$ were all determined indirectly from equilibrium constants for the GSH/GSSG redox system equilibrated with other redox couples, including the NADH/NAD$^+$ and NADPH/NADP$^+$ redox couples:

$$GSSG + NADPH + H^+ \leftrightarrow 2GSH + NADP^+ \tag{3.13}$$

The wide range of values reported for $E_{\mathrm{GSH}}^{o\prime}$ is due, in part at least, to limitations of the experimental procedures used to determine the conditional equilibrium constants from which $E_{\mathrm{GSH}}^{o\prime}$ was calculated; in most cases, the concentration of only one or two of the species involved in the equilibrium reaction was determined by direct measurement, and the concentrations of the other species were calculated by difference. The one exception is the determination of $E_{\mathrm{GSH}}^{o\prime}$ by ^{1}H NMR; the concentration of each species at equilibrium was determined by direct measurement, from which a value of 139 ± 21 M was obtained for the conditional equilibrium constant, defined by Equation (3.14), at pH 7.07 and 25 °C:[46]

$$K_{\mathrm{eq}}^{\mathrm{c}} = \frac{[\mathrm{GSH}]^2 [\mathrm{NADP}^+]}{[\mathrm{GSSG}][\mathrm{NADPH}]} \tag{3.14}$$

Using this value for $K_{\mathrm{eq}}^{\mathrm{c}}$, a value of -0.315 ± 0.002 V for $E_{\mathrm{NADP+/NADPH}}^{o\prime}$ at pH 7.0 and 30 °C,[49] and correcting for the temperature and pH differences, a value of -0.244 ± 0.002 V is obtained for $E_{\mathrm{GSH}}^{o\prime}$ at pH 7.00 and 25 °C. Given the small difference between this value and the widely used value of -0.240 V, it is recommended that the value of -0.240 V continue to be used for $E_{\mathrm{GSH}}^{o\prime}$. A value of $E_{\mathrm{GSH}}^{o\prime} = -0.240$ V has also been confirmed independently through multiple pairwise linkages of protein-protein redox equilibria to $E_{\mathrm{NADP+/NADPH}}^{o\prime} = -0.315$ V.[15]

It is also recommended that researchers report both the experimentally measured quantities K_{mix} or K_{ox} as well as the derived values for $E_{\mathrm{peptide}}^{o\prime}$ or $E_{\mathrm{protein}}^{o\prime}$. The relative stabilities of disulfide bonds and their redox potentials can be compared directly with K_{mix} and K_{ox}, independent of the redox scale of the reference redox buffer; the larger the value of K_{mix} or K_{ox}, the greater the relative stability of the disulfide bond formed by the thiol–disulfide exchange reaction. Each factor of 10 increase in K_{ox} at 25 °C corresponds to a 0.0295 V more negative $E^{o\prime}$ for the disulfide bond formed by the thiol-disulfide exchange reaction.

3.8 Determination of Redox Potentials with GSH/GSSG1 Redox Buffers: an Example

The biologically active form of the tetradecapeptide hormone somatostatin contains a Cys3-Cys14 disulfide bond. The thiol/disulfide equilibria involved in the reaction of somatostatin with the GSH/GSSG redox system are shown in Figure 3.1. The somatostatin disulfide bond is formed in two steps *via* the mixed disulfide intermediates $P_{\mathrm{SH}}^{\mathrm{SSG}}$ and $P_{\mathrm{SSG}}^{\mathrm{SH}}$, which can also interconvert by intramolecular thiol/disulfide exchange reactions.[14] In addition to the reactions in Figure 3.1, the mixed disulfides can react with another molecule of GSSG to form the double mixed disulfide $P_{\mathrm{SSG}}^{\mathrm{SSG}}$. The equilibrium constants defined by Equations (3.15) to (3.20) were determined at 25 °C in pH 7.0 phosphate buffer by HPLC analysis of equilibrium mixtures.[14]

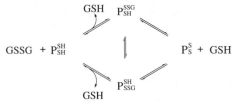

Figure 3.1 Thiol–disulfide exchange reactions for formation of the Cys3-Cys14 disulfide bond of somatostatin by reaction with GSSG.[14] In addition to the reactions shown, the two mixed disulfides can also react with another molecule of GSSG to form the double mixed disulfide P_{SSG}^{SSG}.

$$K_{mix}^3 = \frac{[P_{SH}^{SSG}][GSH]}{[P_{SH}^{SH}][GSSG]} \tag{3.15}$$

$$K_{mix}^{14} = \frac{[P_{SSG}^{SH}][GSH]}{[P_{SH}^{SH}][GSSG]} \tag{3.16}$$

$$K_{intra}^3 = \frac{[P_S^S][GSH]}{[P_{SH}^{SSG}]} \tag{3.17}$$

$$K_{intra}^{14} = \frac{[P_S^S][GSH]}{[P_{SSG}^{SH}]} \tag{3.18}$$

$$K_{intra} = \frac{[P_{SH}^{SSG}]}{[P_{SSG}^{SH}]} \tag{3.19}$$

$$K_{ox} = \frac{[P_S^S][GSH]^2}{[P_{SH}^{SH}][GSSG]} \tag{3.20}$$

The mixed disulfides P_{SH}^{SSG} and P_{SSG}^{SH} were isolated by HPLC and the location of the disulfide bond in each was established by ^{1}H NMR, making possible determination of K_{mix} for each of the mixed disulfides by HPLC. Equilibrium mixtures were acid quenched, and to establish that a true equilibrium was reached, equilibrium was approached from both directions. The GSH reagent was found to be 99.7% GSH and 0.3% GSSG by ^{1}H NMR. The values obtained for the equilibrium constants are $K_{mix}^3 = 2.86$, $K_{mix}^{14} = 2.04$, $K_{intra}^3 = 1.41 \times 10^{-2}$ M, $K_{intra}^{14} = 2.00 \times 10^{-2}$ M, $K_{intra} = 1.41$ and $K_{ox} = 4.05 \times 10^{-2}$ M. Using this value for K_{ox} and $E_{GSH}^{o'} = -0.240$ V, a value of -0.199 V is calculated for $E_{somatostain}^{o'}$.

3.9 Determination of Redox Potentials by the Direct Protein–Protein Equilibration Method: an Example

E. coli Grx1 and Grx3 have identical active-site sequences (Cys-Pro-Tyr-Cys), 33% amino acid sequence identity and highly conserved secondary structure elements and overall fold.[50] However, they have significantly different activities as reductants of ribonucleotide reductase.[28,51,52] To determine if their different activities as reductants are due to differences in their redox properties, the redox potentials of Grx1 and Grx3 were measured by the direct protein–protein equilibration method.[15]

$E^{o\prime}_{Grx1}$ and $E^{o\prime}_{Grx3}$ were determined from equilibrium constants for reaction of Grx1 and Grx3 with the P34H mutant of Trx; the P34H mutant was used because it has a redox potential closer to those of Grx1 and Grx3 than does Trx.[15,38] The redox reactions were run by adding one protein in the reduced form to the other in the oxidized form in degassed and N_2-purged solutions of 100 mM pH 7.0 phosphate buffer and 1 mM EDTA. The initial concentration of both the test protein and the reference protein was 50 μM. To ensure that a true equilibrium was reached, irrespective of the redox state of the initial mixture, reactions were run with both the reduced/oxidized and oxidized/reduced combinations of the test protein/reference protein pairs. Equilibrium mixtures were acid quenched, and the oxidized and reduced forms of each protein were separated and quantitated by HPLC. Redox potentials of $E^{o\prime}_{Grx1} = -0.233$ V and $E^{o\prime}_{Grx3} = -0.198$ V were obtained from the respective equilibrium constants using the Nernst equation and a reference value of -0.235 V for $E^{o\prime}_{Trx(P34H)}$.[38] The reference value for $E^{o\prime}_{Trx(P34H)}$ was determined from the equilibrium constant for the Trr-catalyzed reaction of Trx(P34H) with NADPH; a value of -0.315 V was used for $E^{o\prime}_{NADPH}$.[49] The difference of 35 mV between $E^{o\prime}_{Grx1}$ and $E^{o\prime}_{Grx3}$ was confirmed independently by measuring the equilibrium constant for the direct reaction of Grx1 with Grx3.[15]

The above values for $E^{o\prime}_{Grx1}$ and $E^{o\prime}_{Grx3}$ were each confirmed by using the thermodynamic linkage relationship between the stability of the disulfide bond and the stability of the protein to denaturation:

$$
\begin{array}{ccc}
N^{SH}_{SH} & \xrightleftharpoons{\ K^{N}_{ox}\ } & N^{S}_{S} \\[1em]
\Big\updownarrow K^{red}_{u} & & \Big\updownarrow K^{ox}_{u} \\[1em]
U^{SH}_{SH} & \xrightleftharpoons{\ K^{U}_{ox}\ } & U^{S}_{S}
\end{array}
\tag{3.21}
$$

where N and U represent the folded and unfolded states and K^{red}_{u} and K^{ox}_{u} are the equilibrium constants for unfolding reduced and oxidized protein, respectively.[15] $E^{o\prime}_{Grx1}$ and $E^{o\prime}_{Grx3}$ for unfolded Grx1 and Grx3 were determined by measuring equilibrium constants for the reaction of Grx1 and Grx3 with

GSH/GSSG redox buffer in pH 7.0 buffered 6 M guanidine hydrochloride (GdnHCl) solution. Using $E^{o\prime}_{GSH} = -0.240$ V, $E^{o\prime}_{Grx1}$ and $E^{o\prime}_{Grx3}$ for the unfolded proteins were found to be nearly identical, -0.217 V and -0.218 V, respectively, which are similar to redox potentials for the denatured state of DsbA, DsbC and for the individual Trx-like domains of PDI.[15,16,20,45] This suggests a restricted range of redox potentials for the Cys-Xaa-Yaa-Cys catalytic domain of thiol/disulfide oxidoreductases in the unfolded state, as has also been found for model peptides of the active-site regions of thiol/disulfide oxidoreductases (peptides 1, 3, 5 and 7 in Table 3.2.).[12] The Gibbs free energy changes for unfolding the oxidized and reduced forms of Grx1 and Grx3 were determined by denaturation experiments using GdnHCl and monitored by CD spectroscopy.

In a similar study of Trx, using the GSH/GSSG redox buffer, the linkage relationship between disulfide-bond formation and protein stability was confirmed.[41] The ratio of K_{ox} for the folded and unfolded protein gives, by the linkage relationship, the difference in free energy for unfolding with and without the disulfide bond. The method can be used to evaluate the stabilizing or destabilizing effect of a natural or genetically engineered disulfide bond and to evaluate the effect of amino acid substitutions on disulfide-bond formation in the native and unfolded state of a protein.[41] In the case of Trx, the disulfide bond was found to stabilize the folded structure by 3.5 ± 0.3 kcal mol^{-1}.

References

1. C. S. Sevier and C. A. Kaiser, *Nat. Rev. Mol. Cell Biol.*, 2002, **3**, 836–847.
2. H. Nakamoto and J. C. A. Bardwell, *Biochim. Biophys. Acta*, 2004, **1694**, 111–119.
3. B. P. Tu and J. S. Weissman, *J. Cell Biol.*, 2004, **164**, 341–346.
4. J. T. Tan and J. C. A. Bardwell, *ChemBioChem.*, 2004, **5**, 1479–1487.
5. M. Hirasawa, E. Ruelland, I. Schepens, E. Issakidis-Bourguet, M. Miginiac-Maslow and D. B. Knaff, *Biochemistry*, 2000, **39**, 3344–3350.
6. M. Zheng, F. Åslund and G. Storz, *Science*, 1998, **279**, 1718–1722.
7. U. Jakob, W. Muse, M. Eser and J. C. A. Bardwell, *Cell*, 1999, **96**, 341–352.
8. R. P. Szajewski and G. M. Whitesides, *J. Am. Chem. Soc.*, 1980, **102**, 2011–2026.
9. D. Hupe and D. Wu, *J. Org. Chem.*, 1980, **45**, 3100–3103.
10. J. M. Wilson, R. J. Bayer and D. Hupe, *J. Am. Chem. Soc.*, 1977, **99**, 7922–7926.
11. D. A. Keire, E. Strauss, W. Guo, B. Noszál and D. L. Rabenstein, *J. Org. Chem.*, 1992, **57**, 123–127.
12. F. Siedler, S. Rudolph-Böhner, M. Doi, H.-J. Musiol and L. Moroder, *Biochemistry*, 1993, **32**, 7488–7495.
13. D. L. Rabenstein and P. L. Yeo, *J. Org. Chem.*, 1994, **59**, 4223–4229.
14. D. L. Rabenstein and K. H. Weaver, *J. Org. Chem.*, 1996, **61**, 7391–7397.

15. F. Åslund, K. D. Berndt and A. Holmgren, *J. Biol. Chem.*, 1997, **272**, 30780–30786.
16. A. Zapun, J. C. A. Bardwell and T. E. Creighton, *Biochemistry*, 1993, **32**, 5083–5092.
17. D. W. Walters and H. F. Gilbert, *J. Biol. Chem.*, 1986, **33**, 15372–15377.
18. J. Rost and S. Rapoport, *Nature*, 1964, **201**, 185.
19. R. Wynn, M. J. Cocco and F. M. Richards, *Biochemistry*, 1995, **34**, 11807–11813.
20. A. Zapun, D. Missiakas, S. Raina and T. E. Creighton, *Biochemistry*, 1995, **34**, 5075–5089.
21. C. Cabrele, S. Fiori, S. Pegoraro and L. Moroder, *Chem. Biol.*, 2002, **9**, 731–740.
22. J. Hennecke, A. Sillen, M. Huber-Wunderlich, Y. Engelborghs and R. Glockshuber, *Biochemistry*, 1997, **36**, 6391–6400.
23. U. Grauschopf, A. Fritz and R. Glockshuber, *EMBO J.*, 2003, **22**, 3503–3513.
24. M. Huber-Wunderlich and R. Glockshuber, *Folding Des.*, 1998, **3**, 161–171.
25. N. J. Darby and T. E. Creighton, *Biochemistry*, 1995, **34**, 16770–16780.
26. M. M. Lyles and H. F. Gilbert, *Biochemistry*, 1991, **30**, 613–619.
27. N. Reckenfelderbäumer and R. L. Krauth-Siegel, *J. Biol. Chem.*, 2002, **277**, 17548–17555.
28. A. Jordan, E. Pontis, M. Atta, M. Krook, I. Gibert, J. Barbe and P. Reichard, *Proc. Natl. Acad. Sci. USA*, 1994, **91**, 12892–12896.
29. M. Wunderlich and R. Glockshuber, *Protein Sci.*, 1993, **2**, 717–726.
30. H. F. Gilbert, *Methods Enzymol.*, 1995, **251**, 8–28.
31. W. J. Lees and G. M. Whitesides, *J. Org. Chem.*, 1993, **58**, 642–647.
32. R. W. Cowgill, *Biochim. Biophys. Acta*, 1967, **140**, 37–44.
33. A. Holmgren, *J. Biol. Chem.*, 1972, **247**, 1992–1998.
34. H. Loferer, M. Wunderlich, H. Hennecke and R. Glockshuber, *J. Biol. Chem.*, 1995, **270**, 26178–26183.
35. H. Østergaard, A. Henriksen, F. G. Hansen and J. R. Winther, *EMBO J.*, 2001, **20**, 5853–5862.
36. G. T. Hanson, R. Aggeler, D. Oglesbee, M. Cannon, R. A. Capaldi, R. Y. Tsien and S. J. Remington, *J. Biol. Chem.*, 2004, **279**, 13044–13053.
37. J. Lundstöm and A. Holmgren, *Biochemistry*, 1993, **32**, 6649–6655.
38. G. Krause, J. Lundstöm, J. L. Barea, C. P. de la Cuesta and A. Holmgren, *J. Biol. Chem.*, 1991, **266**, 9494–9500.
39. H. C. Hawkins, M. de Nardi and R. B. Freedman, *Biochem. J.*, 1991, **275**, 341–348.
40. E. Mössner, M. Huber-Wunderlich and R. Glockshuber, *Protein Sci.*, 1998, **7**, 1233–1244.
41. T.-Y. Lin and P. S. Kim, *Biochemistry*, 1989, **28**, 5282–5287.
42. H. F. Gilbert, in *Mechanisms of Protein Folding*, R. H. Pain, ed., Oxford University Press, Oxford, 1994, pp. 104–135.

43. T. E. Creighton, in *Functions of Glutathione: Biochemical, Physiological, Toxicological and Chemical Aspects*, A. Larsson, S. Orrenius, A. Holmgren and B. Mannervik, eds., Raven Press, New York, 1983, p. 205–213.
44. H. Schmidt and R. L. Krauth-Siegel, *J. Biol. Chem.*, 2003, **278**, 46329–46336.
45. U. Grauschopf, J. R. Winther, P. Korber, T. Zander, P. Dallinger and J. C. A. Bardwell, *Cell*, 1995, **83**, 947–955.
46. K. K. Millis, K. H. Weaver and D. L. Rabenstein, *J. Org. Chem.*, 1993, **58**, 4144–4146.
47. L. W. Mapson and F. A. Isherwood, *Biochem. J.*, 1963, **86**, 173–191.
48. R. L. Veech, L. V. Eggleston and H. A. Krebs, *Biochem. J.*, 1969, **115**, 609–619.
49. W. M. Clark, *Oxidation Potentials of Organic Systems*, Williams and Wilkins, Baltimore, 1960.
50. F. Åslund, K. Nordstrand, K. D. Berndt, M. Nikkola, T. Bergman, H. Ponstingl, H. Jörnvall, G. Otting and A. Holmgren, *J. Biol. Chem.*, 1996, **271**, 6736–6745.
51. F. Åslund, B. Ehn, A. Miranda-Vizuete, C. Pueyo and A. Holmgren, *Proc. Natl. Acad. Sci. USA*, 1994, **91**, 9813–9817.
52. A. Porat, C. H. Lillig, C. Johansson, A. P. Fernandes, L. Nilsson, A. Holmgren and J. Beckwith, *Biochemistry*, 2007, **46**, 3366–3377.

CHAPTER 4

Engineered Disulfide Bonds for Protein Design

LUIS MORODER,[a] HANS-JÜRGEN MUSIOL[a] AND CHRISTIAN RENNER[b]

[a] Max Planck Institute of Biochemistry, Am Klopferspitz 18, D-82152 Martinsried, Germany; [b] Deutsche Forschungsgemeinschaft, Kennedyallee 40, 53175 Bonn/Bad-Godesberg, Germany

4.1 Introduction

Disulfide bonds are thought to stabilize proteins primarily by reducing the conformational entropy of the unfolded state.[1–4] When treating the unfolded state of proteins as a random coil, polymer theory would predict that disulfide bonds increase the free energy of the unfolded state by decreasing its conformational entropy to an extent that, in crude approximation, is proportional to the logarithm of the number of residues between the disulfide-bridged cysteines.[5,6] If this entropic effect is exploited for stabilization of proteins, the specific mutations required to introduce the novel disulfide bonds have to be designed with great care on the basis of known spatial relationships in order to prevent both perturbations in the folded state and rather drastic decreases in folding efficiency. Indeed, the energetic contribution of additional disulfides is not only entropic in nature, but also includes enthalpic and native-state effects.[7,8] The mixed success of attempts to increase protein stability by introducing novel disulfides illustrates the engineering challenge. T4 lysozyme, which lacks disulfides in its native form, is one of the proteins most extensively studied in this respect. Some engineered

RSC Biomolecular Sciences
Oxidative Folding of Peptides and Proteins
Edited by Johannes Buchner and Luis Moroder
© Royal Society of Chemistry 2009
Published by the Royal Society of Chemistry, www.rsc.org

disulfides were found to enhance its stability by increasing the melting temperature (T_m), while others destabilized the protein.[2,9] For industrially important proteases of broad specificity such as subtilisin[10-12] and the thermolysin-like proteases[13] most attempts to enhance stability by introducing additional disulfide bonds have failed. These disappointing results were mainly attributed to side effects resulting from the individual Xaa/Cys mutations and/or to the strain introduced by suboptimal geometries of the disulfide bridges. However, extreme stabilization of a thermolysin-like protease has also been reported by placing the additional disulfide bond in an area of the molecule that is involved in partial unfolding processes and thus determines its thermal stability.[14]

Although most disulfide bonds assist the folding of proteins and stabilize their three-dimensional structures,[15] there are also functional disulfides which can confer redox activity, as in the case of the thiol-protein oxidoreductases,[16] or serve as allosteric switches, which control the function of proteins by mediating conformational changes upon reduction or oxidation.[17,18] Correspondingly, new disulfide bonds may be engineered into proteins not only to enhance stability but also to introduce new functional properties as exemplarily shown in the case of a green fluorescent protein (GFP) variant with a disulfide group installed in close proximity to the chromophore. The disulfide acts as a redox switch and thus allows monitoring of redox potentials *via* the significant fluorescence quenching induced by the disulfide moiety in the dithiol in the oxidized compared to the reduced protein form.[19]

There are two basic strategies for protein design: i) the development of new macromolecules by expanding or improving an existing function or generating new biological properties in a natural protein scaffold, and ii) the *de novo* construction of isolated secondary structural modules and assembling them into compact structures with defined folds and, at times, functions. The first strategy often relies on disulfide-stabilized proteins and cystine-rich peptides as robust scaffolds amenable to exchanges of larger, more or less-structured native sequence segments with others borrowed from biomolecules possessing particular properties or with artificially designed peptide chains that encode the desired structural and functional properties.[20-28] Progress on the design and production of functional miniature proteins by the use of disulfide-stabilized scaffolds is exhaustively reviewed in Chapter 7.

The second approach, *de novo* protein design, critically tests our fundamental understanding of how proteins fold and function.[29-32] Inspection of three-dimensional structures of proteins suggests that complex tertiary folds can be decomposed, in principle, into a limited number of secondary structure elements such as turns, strands and helices, which assemble into compact folds *via* loosely structured loops. The stability of the specific fold is determined by hydrophobic and electrostatic interactions between residues within the modules and, in the case of cystine-containing polypeptide chains, by the disulfide bonds, too. By exploiting optimal residue patterning and stereochemical restraints, secondary structure modules can be generated, which may preferentially self-assemble into controlled tertiary folds or be induced to do so by tethering individual modules with more or less flexible linkers or with disulfide bonds.

An additional promising use of engineered disulfide bonds can be envisaged in the induction and stabilization of quaternary structures as a higher level organization of protein systems. This has been exemplarily documented with the ring-shaped hexameric Hcp1 protein of *P. aeruginosa*.[33] By selective mutation of a glycine at the bottom and of an arginine on top of the ring structures a supramolecular self-assembly of the hexameric protein into extended nanotubes stabilized by inter-ring disulfides was achieved.[34]

The purpose of the present chapter is to review progress achieved in recent years using engineered disulfide bonds to stabilize secondary structure elements that can serve as programmed tectons[35] for self-assembly into higher ordered structures.

4.2   Helices

Systematic studies of isolated α-helices have resulted in the formulation of a set of *de novo* design rules based on the use of i) residues with high helix-forming propensities, ii) capping groups to prevent unfavorable interactions with the helix dipole and iii) stabilizing hydrogen bonding, ionic and hydrophobic interactions between residues separated by a turn of the helix.[36–38]

4.2.1   Disulfide-stabilized Helices

Lactam bridges or olefinic cross-links between the i and $i+4$ positions of an α-helix have been exploited to "lock in" the desired helical conformation.[36–39] In contrast, disulfide-crossbridging of two cysteine residues in the $i/i+4$ positions is not compatible with an α-helix structure, although a combination of D- and L-half-cystines should allow such conformational constraint.[40] This has been experimentally confirmed with two short helical segments of the antimicrobial thionin from *Pyrularia pubera*, which both contain a pair of Cys residues in $i/i+4$ positions. As for all thionins with the typical "Γ" structural motif,[41] these two antiparallel helices that form the long arm are cross-linked by two disulfide bonds. Interchain cross-linking of the two helical thionin fragments with the two disulfide bonds by regioselective synthetic procedures induces stabilization of the two helices, but replacement of one L-Cys with a D-Cys residue and subsequent intramolecular disulfide-bond formation also stabilizes the helical conformation of the two excised fragments to a significant extent.[40] In a similar manner, α-helix stabilization has been achieved by crossbridging D- and L-2-amino-6-mercaptohexanoic acid in the i and $i+7$ positions, respectively, by a disulfide bond.[42] This type of disulfide bond was found to lock in two turns of the helix and to induce propagation of the helical conformation to neighboring residues. Cysteine residues spaced $i/i+3$ apart have been disulfide bonded to obtain cyclic peptides, which fold into a 3_{10} helix.[43] Such Cys-Xaa-Yaa-Cys motifs, located at the N-terminus of α-helices, form the redox active sites of bacterial and eukaryotic disulfide-forming enzymes, which all show the characteristic thioredoxin fold.[44] The tripeptide

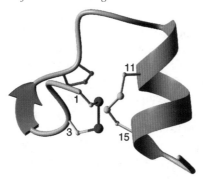

Figure 4.1   The $i/i+4$ disulfide-bridged α-helix of Sec[1,3]-apamin.[45]

portion Xaa-Yaa-Cys forms a 3_{10}-helical loop at the N-terminal end of the α-helix with minimal structural variation between the oxidized and reduced state and with the N-terminal Cys residue exposed at the surface of the molecule.[44]

Air oxidation of the Sec[1,3], Cys[11,15] analog of apamin afforded a Sec[1,3]-diselenide/Cys[11,15]-disulfide framework.[45] NMR conformational analysis of this apamin analog clearly revealed the presence of the characteristic C-terminal apamin α-helix, although this structure was partially distorted and compressed as a result of the geometrically unfavorable disulfide constraint between the i and $i+4$ half-cystine residues (Figure 4.1). Based on these findings, an isomer with Cys I-II/III-IV connectivities was proposed as a productive intermediate in the folding mechanism of apamin, as its formation should be highly favored by the proximity rule and the distorted α-helix should promote the disulfide exchange reactions that establish the energetically more favored I-III/II-IV disulfide framework.[46] The latter disulfide connectivities allow apamin to assume the classical cystine-stabilized α-helix motif with Cys-11 and Cys-15 in the i and $i+4$ positions of an α-helical turn, disulfide-paired with the N-terminal Cys-1 and Cys-3 residues, respectively; the two resulting disulfides form the hydrophobic core of a globular fold (see Section 6.3.2.3).[47] The strong α-helix-inducing and -stabilizing effects of the apamin disulfide framework have been exploited for the design of chimeric α-helical peptides in which the helical portion of apamin was replaced by a variant of the S-peptide of RNase S that contains two Cys residues positioned at i and $i+4$ positions within its α-helical stretch.[48] This construct exploits the intrinsic ability of the S-peptide to assume an α-helical conformation[49,50] to generate an artificial cystine-stabilized α-helix motif.

4.2.2   Helical Bundles

α-Helices can self-associate to form helical hairpins, coiled-coils or helical bundles and such associations are governed by burial of hydrophobic surfaces and formation of ion pairs, hydrogen bonds and/or aromatic interactions between the monomer units. From extensive studies of numerous *de novo* designed α-helical and coiled-coil polypeptides the crucial role of specific interactions between the seven-residue geometric (heptad) repeats, which constitute the individual helices of the bundles, has been recognized.[51–56] This information has been used to construct analogs of natural or fully artificial parallel and anti-parallel helical bundles, some of which were structurally characterized by X-ray crystallography and NMR conformational analysis.[36] In order to eliminate the monomer-oligomer equilibrium and the effect of peptide concentration on the stability of the associates, as well as to dictate the reciprocal helix orientations and to establish heterodimers, individual helices have been connected by more or less flexible linkers as reported *e.g.* in references 38,52,57–60. The effect of linker length on the compactness of helix-loop-helix motifs suggested a $(Gly)_7$ sequence as optimal for fold stabilization.[61] An alternative approach utilizes disulfide bridges to tether peptide modules for induction and stabilization of local folding[62–64] or for further stabilization of helix-loop-helix motifs.[65] For the *de novo* design of an up-and-down four-helix bundle, which is the most common α-helical tertiary structure,[66] cysteine residues were incorporated into helices I and IV. The covalent link was expected to stabilize the folded helix bundle and at the same time provide a simple probe for the folded state of this artificial designed protein, called Felix.[67] Indeed, after expression and purification of the 79 residue long polypeptide chain, oxidation produced a monomer which possessed α-helical structure as confirmed by spectroscopic measurements. However, the protein was not amenable to detailed structural analysis.

From the observation that most naturally occurring disulfide bonds in proteins are buried,[4] it has been proposed that introduction of these cross-links into the most closely packed portion of a protein core should provide the greatest stability.[15] However, such stabilizing effects can be offset by disruption of preexisting interactions; moreover, maintaining the strict stereochemical requirements of disulfide bonds in such packed regions of folded structures is extremely difficult.[9] Therefore, a systematic study was performed of the effect of a disulfide bond on the stability of a two-stranded parallel α-helical coiled-coil. A single disulfide was introduced into each of the five *a* and five *d* positions of a pentaheptad polypeptide chain (Figure 4.2).[68]

A disulfide bond at the N-terminal *a*2 or C-terminal *d*33 position of the coiled-coil was found to significantly increase structural stability.[64,68] This fully agrees with the observation that the fraying ends of coiled-coils are less stable and more flexible than internal regions.[69] Disulfides at the terminal *d*5 and *d*33 positions induced similar stabilities, but disulfides could also be placed in the internal *d*12, *d*19 and *d*26 positions of the coiled-coil without disruption of the structure, while in the internal *a*9, *a*16, *a*23 and *a*30 positions significant destabilization was observed. This systematic study clearly demonstrated that disulfide bonds can be installed even in the hydrophobic interface between two α-helices to enhance structural stability considerably without changing the overall folded structure, provided that appropriate positions are selected for the Xaa → Cys replacements.

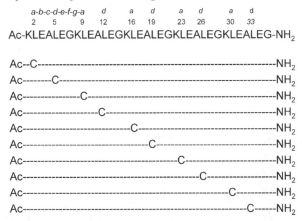

Figure 4.2 Scanning of *a* and *d* positions of an α-helical coiled-coil peptide consisting of five heptads to identify optimal positions for disulfide bonds.[68]

The mitochondrial protein p8[MTCP1] consists of three α-helices associated with an unusual cystine motif.[70] It exhibits a classical antiparallel α-helical hairpin motif stabilized by two disulfide bridges without any distortion of either the helices or the disulfide geometries. However, the steric constraints imposed by the disulfide impose an angle of about 5° between the helix axes instead of the typical value of 20° observed in a helical supercoil. The third helix is oriented roughly parallel to the hairpin plane forming an angle of about 60° with the main axis of this motif. In the antiparallel helix-loop-helix portion the Cys residues are located at positions *a*, *d* and *a'*, *d'* of helix I and helix II, respectively. Regular spacing of the cysteine residues according to the sequence pattern -Cys-(Xaa)$_9$-Cys-(Xaa)$_m$-Cys-(Xaa)$_9$-Cys- leads to a ladder-type cystine framework upon oxidative folding with the interstrand cross-links located at the *a-d'* and *a'-d* positions (see Figure 4.3A). The disulfide bonds appear to be in a favorable geometry for structure stabilization, since they are buried in the hydrophobic interface of the antiparallel helix-loop-helix motif. Oxidative folding of a reduced synthetic replicate of the excised helical hairpin produced the correct disulfide pairing in high yield, and the helical structure of this protein fragment as determined by NMR spectroscopic analysis was found to be identical to that of the α-hairpin in the intact protein.[71] The neurotoxin B-IV from the marine worm *Cerebratulus lacteus*, which also contains an α-hairpin with the correct helical geometry, possesses a different regular Cys spacing, -Cys-(Xaa)$_7$-Cys-(Xaa)$_m$-Cys-(Xaa)$_7$-Cys-, which generates two disulfide bonds, cross-linking the *d* and *d'* positions of the heptad. As a result, the disulfide bonds are located on the same face of the hairpin and exposed to the solvent.[72]

As an alternative to direct disulfide bridging of helices, linkers of the type Cys-Gly-Gly at the N- or Gly-Gly-Cys at the C-termini confer complete flexibility to the peptide chains, allowing them to adopt their most stable conformation (Figure 4.3B). Both parallel and antiparallel helical orientations can be induced

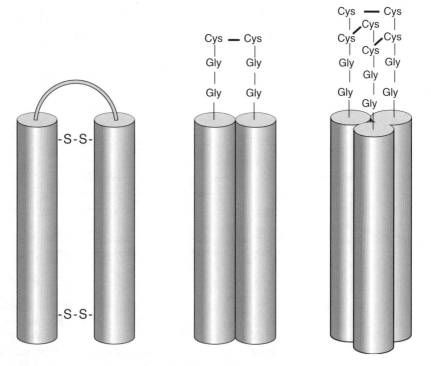

Figure 4.3   Disulfide-bridged helix-loop-helix motifs (A); homodimeric (B) and homotrimeric (C) disulfide-bridged α-helical coiled-coils. Bold lines between Cys residues denote disulfide bonds.

and stabilized by this strategy.[62,63,73,74] This principle has been extended to the design of homotrimeric coiled-coils by N-terminal extension of the polypeptide chain with Cys-Cys-Gly-Gly motifs, which upon oxidation form three disulfide bonds as shown in Figure 4.3C, whose connectivities have not been determined.[75] This type of interchain cystine knot is reminiscent of homotrimeric triple-helical collagen peptides containing collagen type III cystine knots (see Section 6.5.2.1). The disulfide-crossbridged homotrimeric coiled-coil showed an extraordinary thermodynamic stability with a $T_m > 90\,°C$. Introduction of the flexible disulfide linker at both the N- and C-terminus of an α-helical peptide was exploited recently for the design of a dimeric parallel coiled-coil that could be switched into a monomeric intramolecular antiparallel coiled-coil through oxidation of the terminal cysteine residues.[76]

4.3   β-Turns

β-Turns are the smallest element of secondary structure and their role in folding and stability of peptides and proteins has been extensively investigated.[77–79] The idealized β-turn consists of four consecutive residues with a strictly defined stereochemistry, which are joined by a hydrogen bond between the carbonyl of the first turn residue (*i*) and the amide of the last residue $(i+3)$.[80] A classical approach for stabilizing such chain reversals involves incorporation of disulfide bridges at optimal positions. Systematic studies on small disulfide-bridged loops revealed a significantly enhanced tendency to form the desired macrocyclic structure for Cys-$(Xaa)_m$-Cys peptides when *m* is an even not an odd number,[81] but also a marked effect of the β-turn propensities of the intervening $(Xaa)_m$ peptide sequences.[82–84] The alternating odd-even pattern also appears in statistics of disulfide loops in naturally occurring proteins, where cysteine pairs with 2 or 4 intervening residues have a greater probability of forming loops than sequences with $m = 1$, 3 or 5. This pattern disappears when $m > 5$ due to competition between configurational entropy, loop strain and loop size.

4.4 β-Sheets

α-Helices and β-sheets are equally abundant in proteins. However, the principles behind β-sheet formation were much less understood until more recently because of the difficulty in developing simple model systems for quantifying the different factors contributing to β-sheet stability. The basic unit of a β-sheet is a β-strand with the peptide backbone almost fully extended. β-Strands are aligned adjacent to each other such that hydrogen bonds can form between the carbonyls of one peptide chain and the amide groups of the other. The peptide strands can be aligned in parallel or antiparallel orientation, with each arrangement characterized by a distinct hydrogen bonding pattern. The nucleation event in the assembly of an antiparallel β-sheet is the formation of a β-hairpin defined by a chain reversal, *i.e.* the β-turn, flanked by two antiparallel strands.[85]

4.4.1 β-Hairpins

The most common β-hairpins found in the protein database have been classified according to their hydrogen bonding and chain reversal patterns.[38,86] From recent studies of isolated β-hairpins drawn largely from native proteins, but also of structures obtained by *ab initio* design, considerable insight has been obtained into the factors that contribute to the conformational energetics of these systems. Cross-strand interactions are key to the stability of β-hairpins and these include aromatic-aromatic,[87] charge-charge electrostatic[88] and hydrophobic interactions.[89,90] The importance of the β-turn in hairpin stability has been extensively reviewed.[36,38] The observation that the nucleating β-turns in protein β-hairpins are predominantly type I' or type II' β-turns has fostered intensive research in the design of such structures.[91] From these studies, the D-Pro-Gly loop was shown to strongly promote formation of small β-sheets in aqueous solution.[91–93] In general, type I' and II' β-turns are less stable than type I and II β-turns for segments containing L-residues.[78] Analogous to D-Pro-Gly, Asn-Gly can also adopt the optimal geometry that is needed to nucleate hairpins. This segment is the most likely of all proteinogenic sequences to be involved in a type I′ turn, but its ability to stabilize β-hairpin structures is less than that of D-Pro-Gly.[89,94]

In addition to using turn sequences with a high tendency to induce and stabilize β-hairpin folds, disulfide bonds have been exploited to determine the tendency of different sequences to adopt this specific supersecondary structure. In the search for reference compounds that can be used to quantitate double-stranded β-sheet conformations, *i.e.* the β-hairpin fold, of flexible peptides in aqueous solution, a 12-residue peptide containing a D-Pro-Gly loop was backbone-cyclized *via* an additional D-Pro-Gly segment or disulfide-bridged by cysteines placed at the N- and C-termini.[95] NMR conformational analysis indicated that the two cyclization modes promote almost identical structures except for the disulfide- or backbone-cyclized portion of the molecules. Both strategies can stabilize the antiparallel β-sheet sufficiently to provide a "pure" folded state in aqueous solution. This observation fully agrees with the finding that side-chain linkage involving cysteine residues and backbone cyclization are among the best strategies for promoting and stabilizing double-stranded anti-parallel β-sheets in natural bioactive peptides, particularly antimicrobial peptides (see also Section 6.3.).[96] In such peptides, the number of disulfide bonds can vary from one to four and the interstrand cross-links are crucial for maintaining the bioactive β-hairpin fold (Figure 4.4). In backbone-cyclized

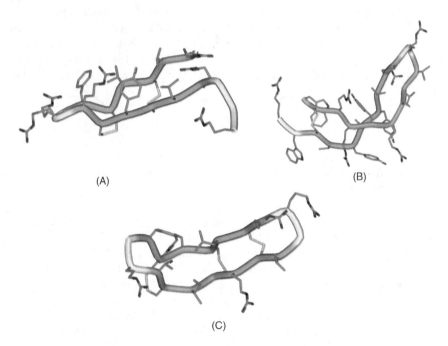

(A)

(B)

(C)

Figure 4.4 Examples of antimicrobial peptides with interstrand disulfide-bonded β-hairpin structures: (A) arenicin (2jsh)[97] (B) protegrin-1 (1pg1)[98] and (C) RTD-1 (1hvz).[99]

defensins, *i.e.* the θ-defensins, the hairpin structure is retained even upon reduction of the disulfide bonds.

By analyzing β-sheets from a set of 928 non-redundant protein structures,[100] the mean C^β–C^β distances between hydrogen-bonded and non-hydrogen-bonded pairs of residues in adjacent strands were found to be 4.82 ± 0.58 and 5.37 ± 0.56 Å, respectively, while the average C^β–C^β distance in disulfide-bonded Cys residues is 3.84 Å.[101] Therefore, the C^β atoms of opposing residues on antiparallel strands are normally too far apart for disulfide formation. Nonetheless, disulfide bonds are found between cysteine residues in the non-hydrogen-bonded register of β-sheets.[91,102] Generally the disulfides pack tightly against hydrophobic side-chains of residues located two positions before one or both Cys residue(s). Extrapolating from this observation, the model peptide Ac-CTXEGNKLTC-NH$_2$ was cross-linked with a disulfide, and the effect of residue X and different loop sequences on the stability of the β-hairpin was analyzed by monitoring thiol-disulfide exchange reactions with GSH/GSSG.[101] Tryptophan was found to be the most stabilizing X residue. Of the loop sequences analyzed, Gly-Asn (type II′ turn) was less stabilizing than Asn-Gly (type I′ turn), while D-Pro-containing turns (type II′) enhance hairpin stability considerably. Although the hairpin stem of this model peptide was very small, combining disulfide bonding of the two strands with the cross-strand tertiary contacts yielded a small, well-structured β-hairpin, clearly indicating that turns and strand-strand interactions contribute in an independent and additive manner to the stability of β-sheet structures.

Disulfides have also been used to tether two β-strands in the absence of connecting loops and thereby induce formation of two-stranded β-sheets.[103,104] As the cystine disulfide bond does not have the ideal geometry to hold two adjacent strands in a β-sheet conformation,[66] a comparative study was performed with a 9-residue peptide modeled after the β-sheet segment of the Met J repressor.[105] A cysteine or an L-2-amino-6-mercaptohexanoic acid (Amh) residue was incorporated into the central position of the nonapeptide, the latter to avoid any structural strain. In the absence of disulfide cross-links both peptides were unstructured, whereas dimerization induced β-sheet formation. Moreover, the strain-free Amh-based dimer showed a significantly enhanced β-sheet population compared to the cystine-containing peptide.

In order to gain better insight into the position-dependent nature of the enthalpic and entropic contributions of a disulfide bond to β-hairpin stability, a systematic study was performed with a 15-residue peptide in which different pairs of non-hydrogen-bonded and hydrogen-bonded residues were replaced with cysteines (see Figure 4.5).[106] Peptide **1** served as the starting point for these experiments. It had been previously shown to adopt a monomeric 3:5 β-hairpin with a type I + G1 β-bulge turn in aqueous solution.[107] This type of turn, also called an α-turn,[108] encompasses three residues in place of the two corner residues of a β-turn. The hydrogen bond is retained but is now between residues i and $i + 4$; the additional residue (in the $i + 3$ position) is often glycine, due to its ability to sterically access the required conformational space, although small polar residues such as Asp and Asn are also sometimes found.

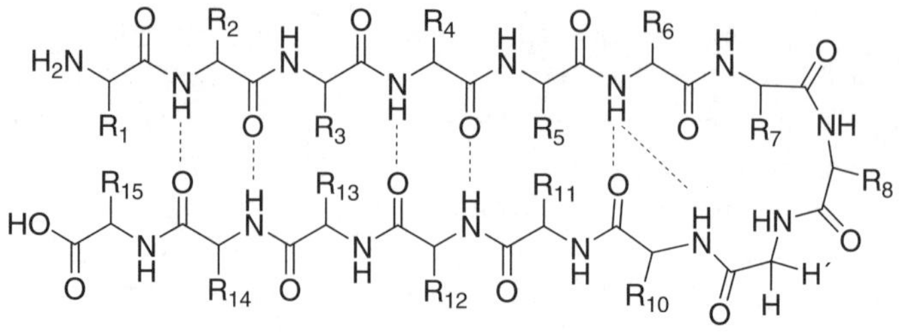

	R$_1$	R$_2$	R$_3$	R$_4$	R$_5$	R$_6$	R$_7$	R$_8$	R$_9$	R$_{10}$	R$_{11}$	R$_{12}$	R$_{13}$	R$_{14}$	R$_{15}$
1	Ser	Glu	Ser	Tyr	Ile	Asn	Ser	Asp	Gly	Thr	Trp	Thr	Val	Thr	Glu
2	Cys	---	---	---	---	---	---	---	---	---	---	---	---	---	Cys
3	---	---	Cys	---	---	---	---	---	---	---	---	---	Cys	---	Val
4	---	---	Ile	---	Cys	---	---	---	---	---	Cys	---	Trp	---	Val
5	---	Cys	---	---	---	---	---	---	---	---	---	---	---	Cys	---

Figure 4.5　In reference peptide **1** different pairs of residues were replaced by Cys residues in non-hydrogen-bonded positions and at different distances from the R$_6$-R$_{10}$ ($i/i+4$) turn region. The sequences of peptides **2**, **3** and **4** were designed in a way that pairs of cross-strand facing residues were the same. In peptide **5** the Cys residues were placed at hydrogen-bonded sites. The observation that reduced peptide **5** exhibits a higher population of β-hairpin than the parent peptide **1** suggests unfavorable cross-strand side-chain/side-chain interactions between Glu-2 and Thr-14 and that a pair of facing cysteine residues at hydrogen-bonded positions is compatible with β-hairpin formation, particularly if favorable packing with the side-chains of surrounding residues can occur.[106]

NMR conformational analysis confirmed β-hairpin structures for peptides **1**–**5**. The β-hairpin populations generally differed for the reduced and oxidized states, except in the case of peptide **5**, where the difference in both states is very small. As expected, disulfide bonds were found to stabilize the β-hairpin folds mainly by decreasing the loss of entropy upon folding, but only when the Cys residues were placed at non-hydrogen-bonded sites. When the disulfide was located at hydrogen-bonded sites, the entropic contribution was offset by the strain generated by the suboptimal geometry of the disulfide bond for cross-linking the antiparallel strands. Moreover, experiments at different temperatures clearly revealed that larger ring sizes better stabilized the β-hairpin, leading to a rank order of β-hairpin populations of **2** > **3** > **4**. On average, the conformational stability of the β-hairpin is increased by about 1.0–1.2 kcal mol^{-1} when the disulfide bond was inserted at a non-hydrogen-bonded site. Moreover, regardless of its location, the stability provided by the interstrand cross-link depends upon which pair of facing residues is replaced by the half-cystines as well as on the packing of the disulfide moiety with other side-chains in the β-sheet face. Apart from entropic considerations, the lower β-hairpin stability of peptide **4** relative to peptide **3** might also be explained by the packing of the Ile-3/Trp-13/disulfide, which is less favorable than the disulfide/Trp-11/Ile-5 packing.

4.4.2 Multi-stranded β-Sheets

In principle, a multi-stranded β-sheet in which D-Pro-Xaa sequences are used as the corner residues of the reversed turns should automatically fold into a compact structure stabilized by cooperative hydrogen bonding. Indeed four- to eight-stranded systems with various degrees of structural stability have been reported.[109–111] Disulfide cross-links were exploited in these model systems to further enhance conformational stability. Incremental cooperative effects on stability were also observed upon addition of an extra β-strand to both two- and three-stranded β-sheets,[109] although the effects were quantitatively small. Association of hydrophobic side-chains rather than creation of a strong hydrogen-bonding network was found to drive formation of these simple multi-stranded β-sheet structures.

A variety of larger β-sandwich proteins have also been successfully designed *de novo*, including the β-sandwich protein betadoublet. The latter was constructed entirely from proteinogenic amino acids and thus employed Asp and Gly for the $i+1$ and $i+2$ positions of type I' β-turns.[112] The strands were optimized to self-associate into a four-stranded β-sheet with high propensity for dimerization; a disulfide bond connecting the sheets was added to mimic the disulfide in the sandwich domains of immunoglobulins.[113] Oxidative folding by air produced the inter-β-sheet disulfide bonded sandwich structure and its thermal denaturation clearly revealed a cooperative unfolding transition. However, the free energy of unfolding of betadoublet was only 2.5 kcal mol^{-1} at 293 K, which is substantially lower than the values observed for natural proteins (5–15 kcal mol^{-1}). The properties of such *ab initio* designed proteins clearly underscore the difficulty in generating tertiary structures with the unique packing characteristics of native proteins.

4.5 Conclusions

In this brief review, recent progress using disulfide bonds to stabilize α-helical, β-turn and β-hairpin modules and their assembly into α-helical bundles and β-sheet structures has been described. The results of these studies are directly relevant to ongoing efforts to design novel functional proteins as well as to rational approaches for stabilization of biotechnologically important proteins with engineered disulfides. Although synthetic approaches to protein design have advanced only moderately in recent years because of difficulties in controlling all the complex interactions that cooperatively determine the structures and function of natural globular proteins, the steady accumulation and comparative analysis of new tertiary structures and advances in computational and experimental (*i.e.* directed evolution) approaches can be expected to provide the detailed understanding that will be necessary for real molecular engineering in the future. Presently, as reviewed in Chapter 7, more success has been achieved in the engineering of functional proteins by relying on robust natural cystine scaffolds.

Acknowledgements

The authors are grateful to Professor Donald Hilvert (ETH Zürich) for his helpful comments during the preparation of this manuscript.

References

1. S. F. Betz, *Protein Sci.*, 1993, **2**, 1551–1558.
2. M. Matsumura, G. Signor and W. Matthews, *Nature*, 1989, **342**, 291–293.
3. L. J. Perry and R. Wetzel, *Science*, 1984, **226**, 555–557.
4. J. M. Thornton, *J. Mol. Biol.*, 1981, **151**, 261–287.
5. P. J. Flory, *J. Am. Chem. Soc.*, 1956, **78**, 5222–5234.
6. C. N. Pace, G. R. Grimsley, J. A. Thomson and B. J. Barnett, *J. Biol. Chem.*, 1988, **263**, 11820–11825.
7. T. Zhang, E. Bertelsen and T. Alber, *Nat. Struct. Biol.*, 1994, **1**, 434–438.
8. A. J. Doig and D. H. Williams, *J. Mol. Biol.*, 1991, **217**, 389–398.
9. M. Matsumura and B. W. Matthews, *Methods Enzymol.*, 1991, **202**, 336–356.
10. J. A. Wells and D. B. Powers, *J. Biol. Chem.*, 1986, **261**, 6564–6570.
11. M. W. Pantoliano, R. C. Ladner, P. N. Bryan, M. L. Rollence, J. F. Wood, G. L. Gilliland, D. B. Stewart and T. L. Poulos, *Prot. Eng.*, 1987, **1**, 229–229.
12. C. Mitchinson and J. A. Wells, *Biochemistry*, 1989, **28**, 4807–4815.
13. B. van den Burg, B. W. Dijkstra, B. van der Vinne, B. K. Stulp, V. G. H. Eijsink and G. Venema, *Prot. Eng.*, 1993, **6**, 521–527.
14. J. Mansfeld, G. Vriend, B. W. Dijkstra, O. R. Veltman, B. van den Burg, G. Venema, R. Ulbrich-Hofmann and V. G. H. Eijsink, *J. Biol. Chem.*, 1997, **272**, 11152–11156.
15. T. E. Creighton, *Bioessays*, 1988, **8**, 57–63.
16. H. Nakamura, *Antioxid. Redox Signal.*, 2005, **7**, 823–828.
17. B. Schmidt, L. Ho and P. J. Hogg, *Biochemistry*, 2006, **45**, 7429–7433.
18. P. J. Hogg, *Trends Biochem. Sci.*, 2003, **28**, 210–214.
19. H. Østergaard, A. Henriksen, F. G. Hansen and J. R. Winther, *EMBO J.*, 2001, **20**, 5853–5862.
20. S. Krause, H. U. Schmoldt, A. Wentzel, M. Ballmaier, K. Friedrich and H. Kolmar, *FEBS J.*, 2007, **274**, 86–95.
21. D. J. Craik, M. Čemažar and N. L. Daly, *Curr. Opin. Drug Discovery Dev.*, 2006, **9**, 251–260.
22. R. J. Clark, N. L. Daly and D. J. Craik, *Biochem. J.*, 2006, **394**, 85–93.
23. C. J. Weston, C. H. Cureton, M. J. Calvert, O. S. Smart and R. K. Allemann, *ChemBioChem*, 2004, **5**, 1075–1080.
24. L. Martin, P. Barthe, O. Combes, C. Roumestand and C. Vita, *Tetrahedron*, 2000, **56**, 9451–9460.
25. B. Imperiali and J. J. Ottesen, *Biopolymers (Peptide Sci.)*, 1998, **47**, 23–29.

26. B. C. Cunningham and J. A. Wells, *Curr. Opin. Struct. Biol.*, 1997, **7**, 457–462.
27. P. A. Nygren and M. Uhlen, *Curr. Opin. Struct. Biol.*, 1997, **7**, 463–469.
28. C. Vita, C. Roumestand, F. Toma and A. Menez, *Proc. Natl. Acad. Sci. USA*, 1995, **92**, 6404–6408.
29. L. Baltzer, H. Nilsson and J. Nilsson, *Chem. Rev.*, 2001, **101**, 3153–3163.
30. D. A. Moffet and M. H. Hecht, *Chem. Rev.*, 2001, **101**, 3191–3203.
31. L. Baltzer and J. Nilsson, *Curr. Opin. Biotechnol.*, 2001, **12**, 355–360.
32. R. B. Hill, D. P. Raleigh, A. Lombardi and W. F. DeGrado, *Acc. Chem. Res.*, 2000, **33**, 745–754.
33. J. D. Mougous, M. E. Cuff, S. Raunser, A. Shen, M. Zhou, C. A. Gifford, A. L. Goodman, G. Joachimiak, C. L. Ordoñez, S. Lory, T. Walz, A. Joachimiak and J. J. Mekalanos, *Science*, 2006, **312**, 1526–1530.
34. E. R. Ballister, A. H. Lai, R. N. Zuckermann, Y. Cheng and J. D. Mougous, *Proc. Natl. Acad. Sci. USA*, 2008, **105**, 3733–3738.
35. E. H. C. Bromley, K. Channon, E. Moutevelis and D. N. Woolfson, *ACS Chem. Biol.*, 2008, **3**, 38–50.
36. J. Venkatraman, S. C. Shankaramma and P. Balaram, *Chem. Rev.*, 2001, **101**, 3131–3152.
37. W. F. DeGrado, C. M. Summa, V. Pavone, F. Nastri and A. Lombardi, *Annu. Rev. Biochem.*, 1999, **68**, 779–819.
38. L. Serrano, *Adv. Protein Chem.*, 2000, **53**, 49–85.
39. M. J. I. Andrews and A. B. Tabor, *Tetrahedron*, 1999, **55**, 11711–11743.
40. M. Vila-Perello, S. Tognon, A. Sanchez-Vallet, F. Garcia-Olmedo, A. Molina and D. Andreu, *J. Med. Chem.*, 2006, **49**, 448–451.
41. F. Garcia-Olmedo, A. Molina, J. M. Alamillo and P. Rodriguez-Palenzuela, *Biopolymers*, 1998, **47**, 479–491.
42. D. Y. Jackson, D. S. King, J. Chmielewski, S. Singh and P. G. Schultz, *J. Am. Chem. Soc.*, 1991, **113**, 9391–9392.
43. A. Ravi, B. V. V. Prasad and P. Balaram, *J. Am. Chem. Soc.*, 1983, **105**, 105–109.
44. J. L. Martin, *Structure*, 1995, **3**, 245–250.
45. S. Fiori, S. Pegoraro, S. Rudolph-Böhner, J. Cramer and L. Moroder, *Biopolymers*, 2000, **53**, 550–564.
46. S. Pegoraro, S. Fiori, J. Cramer, S. Rudolph-Böhner and L. Moroder, *Protein Sci.*, 1999, **8**, 1605–1613.
47. J. H. B. Pease and D. E. Wemmer, *Biochemistry*, 1988, **27**, 8491–8498.
48. J. H. B. Pease, R. W. Storrs and D. E. Wemmer, *Proc. Natl. Acad. Sci. USA*, 1990, **87**, 5643–5647.
49. P. S. Kim, A. Bierzynski and R. L. Baldwin, *J. Mol. Biol.*, 1982, **162**, 187–199.
50. R. Rocchi, G. Borin, E. F. Marchiori, L. Moroder, E. Peggion, E. Scoffone, V. Crescenzi and F. Quadrifoglio, *Biochemistry*, 1972, **11**, 50–57.
51. S. F. Betz, J. W. Bryson and W. F. DeGrado, *Curr. Opin. Struct. Biol.*, 1995, **5**, 457–463.

52. Y. Fezoui, D. L. Weaver and J. J. Osterhout, *Protein Sci.*, 1995, **4**, 286–295.

53. G. T. Dolphin, L. Brive, G. Johansson and L. Baltzer, *J. Am. Chem. Soc.*, 1996, **118**, 11297–11298.

54. N. E. Zhou, C. M. Kay and R. S. Hodges, *J. Mol. Biol.*, 1994, **237**, 500–512.

55. N. E. Zhou, C. M. Kay and R. S. Hodges, *Biochemistry*, 1992, **31**, 5739–5746.

56. F. H. C. Crick, *Acta Crystallogr.*, 1953, **6**, 689–697.

57. U. A. Ramagopal, S. Ramakumar, D. Sahal and V. S. Chauhan, *Proc. Natl. Acad. Sci. USA*, 2001, **98**, 870–874.

58. A. Lombardi, C. M. Summa, S. Geremia, L. Randaccio, V. Pavone and W. F. DeGrado, *Proc. Natl. Acad. Sci. USA*, 2000, **97**, 6298–6305.

59. S. T. R. Walsh, H. Cheng, J. W. Bryson, H. Roder and W. F. DeGrado, *Proc. Natl. Acad. Sci. USA*, 1999, **96**, 5486–5491.

60. C. E. Schafmeister, L. J. Miercke and R. M. Stroud, *Science*, 1993, **262**, 734–738.

61. N. Suzuki and I. Fujii, *Tetrahedron Lett.*, 1999, **40**, 6013–6017.

62. D. N. Marti, I. Jelesarov and H. R. Bosshard, *Biochemistry*, 2000, **39**, 12804–12818.

63. K. Wagschal, B. Tripet and R. S. Hodges, *J. Mol. Biol.*, 1999, **285**, 785–803.

64. N. E. Zhou, B. Y. Zhu, C. M. Kay and R. S. Hodges, *Biopolymers*, 1992, **32**, 419–426.

65. Y. Kuroda, T. Nakai and T. Ohkubo, *J. Mol. Biol.*, 1994, **236**, 862–868.

66. J. S. Richardson, *Adv. Protein Chem.*, 1981, **34**, 167–339.

67. M. H. Hecht, J. S. Richardson, D. C. Richardson and R. C. Ogden, *Science*, 1990, **249**, 884–891.

68. N. E. Zhou, C. M. Kay and R. S. Hodges, *Biochemistry*, 1993, **32**, 3178–3187.

69. N. E. Zhou, C. M. Kay and R. S. Hodges, *J. Biol. Chem.*, 1992, **267**, 2664–2670.

70. P. Barthe, Y. S. Yang, L. Chiche, F. Hoh, M. P. Strub, L. Guignard, J. Soulier, M. H. Stern, H. van Tilbeurgh, J. M. Lhoste and C. Roumestand, *J. Mol. Biol.*, 1997, **274**, 801–815.

71. P. Barthe, S. Rochette, C. Vita and C. Roumestand, *Protein Sci.*, 2000, **9**, 942–955.

72. K. J. Barnham, T. R. Dyke, W. R. Kem and R. S. Norton, *J. Mol. Biol.*, 1997, **268**, 886–902.

73. J. J. Skalicky, B. R. Gibney, F. Rabanal, R. J. B. Urbauer, P. L. Dutton and A. J. Wand, *J. Am. Chem. Soc.*, 1999, **121**, 4941–4951.

74. P. Lavigne, L. H. Kondejewski, M. E. Houston, F. D. Sönnichsen, B. Lix, B. D. Sykes, R. S. Hodges and C. M. Kay, *J. Mol. Biol.*, 1995, **254**, 505–520.

75. E. Bianchi, M. Finotto, P. Ingallinella, R. Hrin, A. V. Carella, X. S. Hou, W. A. Schleif, M. D. Miller, R. Geleziunas and A. Pessi, *Proc. Natl. Acad. Sci. USA*, 2005, **102**, 12903–12908.
76. M. J. Pandya, E. Cerasoli, A. Joseph, R. G. Stoneman, E. Waite and D. N. Woolfson, *J. Am. Chem. Soc.*, 2004, **126**, 17016–17024.
77. H. J. Dyson and P. E. Wright, *Annu. Rev. Biophys. Biophys. Chem.*, 1991, **20**, 519–538.
78. G. D. Rose, L. M. Gierasch and J. A. Smith, *Adv. Protein Chem.*, 1985, **37**, 1–109.
79. C. Toniolo, *CRC Crit. Rev. Biochem.*, 1980, **9**, 1–44.
80. C. M. Venkatachalam, *Biopolymers*, 1968, **6**, 1425–1436.
81. R. M. Zhang and G. H. Snyder, *J. Biol. Chem.*, 1989, **264**, 18472–18479.
82. P. J. Milburn, Y. C. Meinwald, S. Takahashi, T. Ooi and H. A. Scheraga, *Int. J. Pept. Protein Res.*, 1988, **31**, 311–321.
83. F. Siedler, S. Rudolph-Böhner, M. Doi, H. -J. Musiol and L. Moroder, *Biochemistry*, 1993, **32**, 7488–7495.
84. C. M. Falcomer, Y. C. Meinwald, I. Choudary, S. Talluri, P. J. Milburn, J. Clardy and H. A. Scheraga, *J. Am. Chem. Soc.*, 1992, **114**, 4036–4042.
85. A. V. Finkelstein, *Prot. Struct. Funct. Genet.*, 1991, **9**, 23–27.
86. B. L. Sibanda and J. M. Thornton, *Nature*, 1985, **316**, 170–174.
87. C. D. Tatko and M. L. Waters, *J. Am. Chem. Soc.*, 2002, **124**, 9372–9373.
88. S. J. Russell and A. G. Cochran, *J. Am. Chem. Soc.*, 2000, **122**, 12600–12601.
89. F. Blanco, M. Ramirez-Alvarado and L. Serrano, *Curr. Opin. Struct. Biol.*, 1998, **8**, 107–111.
90. G. Colombo, G. M. S. De Mori and D. Roccatano, *Protein Sci.*, 2003, **12**, 538–550.
91. K. Gunasekaran, C. Ramakrishnan and P. Balaram, *Protein Eng.*, 1997, **10**, 1131–1141.
92. J. F. Espinosa and S. H. Gellman, *Angew. Chem., Int. Ed.*, 2000, **39**, 2330–2333.
93. S. H. Gellman, *Curr. Opin. Chem. Biol.*, 1998, **2**, 717–725.
94. E. G. Hutchinson and J. M. Thornton, *Protein Sci.*, 1994, **3**, 2207–2216.
95. F. A. Syud, J. F. Espinosa and S. H. Gellman, *J. Am. Chem. Soc.*, 1999, **121**, 11577–11578.
96. P. Bulet, R. Stöcklin and L. Menin, *Immunol. Rev.*, 2004, **198**, 169–184.
97. J. Andrä, I. Jakovkin, J. Grötzinger, O. Hecht, A. D. Krasnosdembskaya, T. Goldmann, T. Gutsmann and M. Leippe, *Biochem. J.*, 2008, **410**, 113–122.
98. R. L. Fahrner, T. Dieckmann, S. S. L. Harwig, R. I. Lehrer, D. Eisenberg and J. Feigon, *Chem. Biol.*, 1996, **3**, 543–550.
99. M. Trabi, H. J. Schirra and D. J. Craik, *Biochemistry*, 2001, **40**, 4211–4221.
100. U. Hobohm and C. Sander, *Protein Sci.*, 1994, **3**, 522–524.

101. A. G. Cochran, R. T. Tong, M. A. Starovasnik, E. J. Park, R. S. McDowell, J. E. Theaker and N. J. Skelton, *J. Am. Chem. Soc.*, 2001, **123**, 625–632.
102. E. G. Hutchinson, R. B. Sessions, J. M. Thornton and D. N. Woolfson, *Protein Sci.*, 1998, **7**, 2287–2300.
103. H. Balaram, K. Uma and P. Balaram, *Int. J. Pept. Protein Res.*, 1990, **35**, 495–500.
104. R. Kishore, A. Kumar and P. Balaram, *J. Am. Chem. Soc.*, 1985, **107**, 8019–8023.
105. A. M. Aberle, H. K. Reddy, N. V. Heeb and K. P. Nambiar, *Biochem. Biophys. Res. Commun.*, 1994, **200**, 102–107.
106. C. M. Santiveri, E. Leon, M. Rico and M. A. Jimenez, *Chem. Eur. J.*, 2008, **14**, 488–499.
107. C. M. Santiveri, D. Pantoja-Uceda, M. Rico and M. A. Jimenez, *Biopolymers*, 2005, **79**, 150–162.
108. V. Pavone, G. Gaeta, A. Lombardi, F. Nastri, O. Maglio, C. Isernia and M. Saviano, *Biopolymers*, 1996, **38**, 705–721.
109. F. A. Syud, H. E. Stanger, H. S. Mortell, J. F. Espinosa, J. D. Fisk, C. G. Fry and S. H. Gellman, *J. Mol. Biol.*, 2003, **326**, 553–568.
110. N. Carulla, C. Woodward and G. Barany, *Protein Sci.*, 2002, **11**, 1539–1551.
111. J. Venkatraman, G. A. N. Gowda and P. Balaram, *J. Am. Chem. Soc.*, 2002, **124**, 4987–4994.
112. T. P. Quinn, N. B. Tweedy, R. W. Williams, J. S. Richardson and D. C. Richardson, *Proc. Natl. Acad. Sci. USA*, 1994, **91**, 8747–8751.
113. R. J. Poljak, L. M. Amzel, B. L. Chen, R. P. Phizackerley and F. Saul, *Proc. Natl. Acad. Sci. USA*, 1974, **71**, 3440–3444.

CHAPTER 5

Selenocysteine as a Probe of Oxidative Protein Folding

JORIS BELD, KENNETH J. WOYCECHOWSKY AND
DONALD HILVERT

Laboratory of Organic Chemistry, ETH Zürich, 8093 Zürich, Switzerland

5.1 Introduction

Selenium was discovered in 1818 by the Swedish chemist Berzelius as a poisonous contaminant in locally mined sulfur.[1] He isolated the element from the ore and named it selenium after the Greek moon goddess Selene, in analogy to tellurium which was previously named for Tellus, the earth.[2] As one of the chalcogens, selenium shares many properties with oxygen, sulfur and tellurium (Table 5.1). Although selenium was regarded as a purely toxic substance for more than a century, this view changed radically some 50 years ago when it was found to be an essential micronutrient for animals, albeit one with a very narrow beneficial dosage range. Recent studies have even reported that some selenium compounds may help to prevent cancer, and nowadays dietary supplements containing selenium can be purchased in many pharmacies.[3]

In biological systems, selenium is incorporated into proteins non-specifically as selenomethionine[4] and specifically as selenocysteine (abbreviated as Sec or U).[5] Because selenium and sulfur are similar in size (Table 5.1), replacement of sulfur-containing amino acids with their selenium counterparts generally causes little or no structural perturbation in proteins. However, such substitutions can lead to significant differences in protein function. Many selenoproteins are enzymes that feature an active-site selenocysteine in order to harness the special reactive

RSC Biomolecular Sciences
Oxidative Folding of Peptides and Proteins
Edited by Johannes Buchner and Luis Moroder
© Royal Society of Chemistry 2009
Published by the Royal Society of Chemistry, www.rsc.org

Table 5.1 Characteristics of the chalcogens and their amino acids.

	Oxygen	*Sulfur*	*Selenium*	*Tellurium*
Electronegativity	3.44	2.58	2.55	2.1
Covalent radius (Å)	0.73	1.02	1.17	1.35
van der Waals radius (Å)	1.52	1.80	1.90	2.06
Bond Length, C-X (Å)	1.43	1.82	1.95–1.99	2.4
Amino acid	Serine	Cysteine	Selenocysteine	Tellurocysteine
pK_a	13	8.3	5.2	n.d.
Reduction potential (mV)	–	-238[10]	-388[11]	n.d.

Table 5.2 Overview of representative selenoproteins and their activities.[12,13]

Selenoprotein	*Catalyzed reaction*	*Role of selenocysteine*	*Effect of Sec→Cys mutation on activity*
Glutathione peroxidases[14]	Reduction of hydroperoxides	Redox	1000-fold decrease[15]
Iodothyronine deiodonase D1[16]	Modification of T3 and T4 hormones	Selenenyliodide intermediate	10-fold decrease[17]
Thioredoxin reductases[18]	Reduction of Trx	Redox	10–100-fold decrease[19,20]
Selenophosphate synthase 2[21]	Conversion of selenite to selenophosphate	Unknown	>100-fold decrease[22]
Methionine-R-sulfoxide reductase B[23]	Reduction of methione sulfoxide to methionine	Redox	>100-fold decrease[24]
Formate dehydrogenase[25]	Conversion of formate to CO_2	Metal ligand	>100-fold decrease[26]
Glycine reductase[27]	Conversion of glycine to acetyl phosphate	Nucleophile	Unknown

properties of selenols. Glutathione peroxidase, glycine reductase, iodothyronine deiodonase and thioredoxin reductase are exemplary selenoenzymes that promote an interesting range of reactions (Table 5.2). In each case, selenocysteine serves as an essential active-site residue, and even conservative replacement with cysteine diminishes activity by 10 to 1000-fold.

The reactivity differences between selenocysteine and cysteine are manifold. For example, selenols are softer S_N2-type nucleophiles than thiols because the selenium atom is more polarizable than sulfur. Moreover, the selenol of selenocysteine is more acidic (p$K_a \approx 5.2$) than the thiol of cysteine (p$K_a \approx 8.3$).[6] Thus, in contrast to cysteine, selenocysteine is ionized at physiological pH, further enhancing its reactivity. The differences in pK_a between selenocysteine and cysteine and in the electronic structure of selenium and sulfur also give rise to differences in metal coordination properties,[7] although these are not well understood.

Selenium and sulfur do share similar redox states. For instance, selenocysteine is often found in the active sites of thiol-dependent peroxidases, where the selenium cycles between the selenol, selenosulfide and seleninic acid states during catalysis.[8,9] In principle, selenocysteines can also oxidize to form diselenide bonds. While disulfide bonds between cysteine residues are found in many proteins where they confer extra stability or redox functions, diselenide cross-links seem to occur infrequently in biology. The prospect of diselenides in proteins is especially interesting because these bonds are both more thermodynamically stable and more kinetically labile than disulfides.

The first identification of a diselenide bond in a natural protein was made only recently.[28] This protein, selL, is present in aquatic organisms and possesses a thioredoxin fold with a UXXU tetrapeptide in place of the canonical CXXC motif. While its function is unknown, the similarity of selL to thioredoxin suggests a role in redox metabolism. The intramolecular diselenide of selL is extremely stable, as it could not be reduced by the strong thiol reductant DTT ($E^{o\prime} = -327 \, mV$)[i]. Because selL is localized to the cytosol (which has a redox potential of *ca.* -230 to $-270 \, mV$),[31] the mechanism for its reduction remains unclear.

The diselenide bond of selL may not be unique; the 366 amino acid long glycosylated protein selP also likely possesses diselenide bonds.[32] SelP contains 17 cysteines and at least 10 selenocysteines, depending on the organism. In human blood, up to 60% of the total selenium content can be accounted for by selP. Consequently, selP has been hypothesized to act as a storage protein involved in the removal of toxic selenium compounds from the body. SelP possesses a variety of other interesting functions, including glutathione peroxidase activity, heparin binding and heavy metal ion complexation.

Aside from its role(s) in living cells, selenium has also become a valuable spectroscopic and mechanistic probe in protein chemistry. Anomalous scattering by selenium is exploited extensively in protein X-ray crystallography to solve the phasing problem without the need for heavy metals.[33] This method has achieved widespread popularity because facile biosynthetic methods exist for globally replacing methionines and/or cysteines in a protein with selenomethionine and selenocysteine, respectively.[34] Harnessing distinct selenium isotopes has similarly benefited a variety of other applications. For example, the positron emitter [73]Se is used for non-invasive PET-studies,[35] whereas the gamma emitter ^{75}Se is valuable as a residue-specific radiolabel.[36] The [77]Se isotope (8% natural abundance) has a nuclear spin of 1/2 and is ideal for NMR spectroscopy.[37] The high sensitivity of this nucleus to its surroundings makes ^{77}Se-NMR an invaluable tool for systematically probing structure–function relationships in large and small molecules alike.

Selenium has also proved to be useful for labeling and affinity purification of proteins.[38] Proteins bearing the tetrapeptide GCUG, the so-called Sel tag, at their C-termini can be produced recombinantly and detected with selenium-specific reagents (see Section 8.2.5). Another application of selenocysteine is found in the production of artificial catalysts with novel hydrolytic and redox activities, which can be produced by introducing selenium into an appropriately configured active site.[39–43]

---

[i] The given $E^{o\prime}$ value is calculated from the published K_{eq} value[29] using $E^{o\prime} = -256 \, mV$ for the reference GSSG.[30]

In the context of this monograph, the deliberate replacement of one or more cysteines in disulfide-containing proteins with selenocysteine is particularly germane. Selenium is a sensitive probe of redox behavior and folding pathway.[44–48] After briefly reviewing general strategies for preparing selenocysteine-containing peptides and proteins, we discuss how selenium can provide insight into the process of oxidative protein folding.

5.2 Incorporation of Selenocysteine into Proteins

Selenocysteine can be introduced into proteins in a variety of ways. Depending on the specific molecule and intended application, either molecular biological or chemical strategies can be adopted. As outlined below, each approach has distinctive advantages and disadvantages.

5.2.1 Codon Suppression

Although selenocysteine can provide proteins with many special and useful properties, the evolutionary origins of selenoproteins are poorly understood and are only now being unraveled with the help of comparative genomics.[49–51] The mechanism by which selenium is incorporated into natural selenoproteins was elucidated in pioneering work by Böck and co-workers.[52,53] They discovered that selenium is inserted cotranslationally into proteins as the amino acid selenocysteine, rather than *via* posttranslational modification. The amber stop codon UGA, which normally signals truncation of the message, is used to genetically encode this amino acid. For this reason, selenocysteine is frequently referred to as the 21st proteinogenic amino acid.

In bacteria, reassignment of the UGA stop to selenocysteine is achieved by an overcoding mechanism: the UGA is read as an amino acid when it is immediately followed by a specific mRNA stem loop structure, called a selenocysteine insertion sequence (SECIS) element (Figure 5.1).[54] Four additional gene products – selA, selB, selC and selD – are needed to implement this expansion of the standard genetic code. Selenocysteine is synthesized directly from serine loaded onto a suppressor tRNA (selC), whose anticodon is complementary to the UGA codon.[55] This reaction is catalyzed by the enzyme selenocysteine synthase (selA) and exploits selenophosphate,[56] generated by selenophosphate synthase (selD),[57] as a nucleophile. A special GTP-dependent elongation factor (selB) is also required for proper decoding and delivers the selenocysteinyl-tRNA to the active site of the ribosome after binding to the SECIS element.[58] In the absence of either an appropriate SECIS element or the selB · selenocysteinyl-tRNA complex, protein synthesis is prematurely truncated when the ribosome reaches the UGA codon.

Arnér and others have shown that it is possible to exploit the selenocysteine incorporation machinery in *E. coli* for the heterologous overproduction of several natural selenoproteins,[59] including formate dehydrogenase,[60,61] thioredoxin reductase R,[53,59] glutathione peroxidase[62] and methionine sulfoxide reductase B.[24,63] In addition, it has been possible to produce some artificial

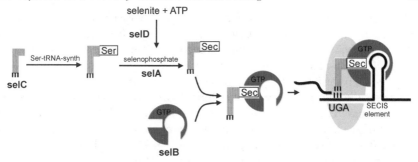

Figure 5.1 Pathway of selenium incorporation in bacteria. First, the tRNA encoded by the selC gene is loaded with the amino acid serine. The selenocysteine synthase selA converts the loaded serine to selenocysteine using selenophosphate generated by the selenophosphate synthase selD. The charged tRNA must form a complex with the elongation factor selB before entering the ribosome. Upon recognition of the UGA codon and the SECIS element of the mRNA, the polypeptide chain is elongated by transfer of the selenocysteine from the tRNA.

selenoproteins,[64,65] particularly if the selenocysteine is located near the C-terminus, as in the case of proteins bearing the Sel tag.[38] Since the SECIS element immediately follows the UGA codon in bacteria, the production of artificial selenoproteins often requires careful design of this sequence to minimize undesired coding changes while preventing truncation.

In eukaryotes, selenocysteine incorporation into proteins is less well understood. As in bacteria, serine is loaded onto the suppressor tRNA by seryl-tRNA synthetase.[66] Phosphoseryl-tRNA kinase phosphorylates the alcohol side-chain of the amino acid, and the resulting phosphate monoester is replaced by selenophosphate to give, after hydrolysis of the phosphate group, selenocysteinyl-tRNA.[3] However, in contrast to prokaryotes, the SECIS element needed to decode the UGA codon is located in the 3′-untranslated region of the mRNA, sometimes kilobases distant from the site of suppression. The decoding process is complicated and involves at least five different proteins (SBP2, EFSec, L30, Secp43 and SLA). Simplified, SECIS binding protein 2 (SBP2) binds to the SECIS element, which in turn is recognized by a complex between the selenocysteinyl-tRNA and the elongation factor Efsec.[12,67] Another stem-loop structure in the mRNA, called the selenocysteine redefinition element (SRE), was recently found to modulate selenocysteine insertion.[68] The SRE is located in the coding region, directly downstream of the UGA codon, much like the bacterial SECIS element.

5.2.2 Codon Reassignment

As an alternative to stop codon suppression, the similarity of cysteine and selenocysteine can be successfully exploited to produce some artificial selenoproteins in microorganisms that are auxotrophic for cysteine. In this approach, an *E. coli* strain that cannot produce its own cysteine is starved for cysteine in a medium supplemented with high concentrations of selenocysteine.[69] The endogenous cysteinyl-tRNA synthetase recognizes selenocysteine and loads it onto tRNA^{Cys} in place of cysteine, leading to global insertion of selenocysteine into proteins in response to the cysteine codon. As proof-of-principle, the two cysteine residues in the redox protein thioredoxin were successfully replaced by selenocysteine. The chief advantage of this method is that it enables incorporation of selenocysteine at multiple sites throughout a protein. Nevertheless, the extent of substitution is typically only 75–80%, and removal of contaminating variants containing mixtures of cysteines and selenocysteines is generally difficult. Moreover, site-selective replacement of a single cysteine in a protein with multiple cysteines is not possible by this approach. Using a similar strategy, an auxotrophic *E. coli* strain has become a standard tool for the global replacement of methionine by selenomethione, in order to produce heavy atom derivatives that can be used to solve the phasing problem in protein X-ray crystallography.[34]

5.2.3 Post-translational Modification

If a protein contains a uniquely reactive residue, chemical modification can be an effective method for introducing selenium into proteins. Experiments on the serine protease subtilisin illustrate this approach. The catalytic serine residue (Ser221) was converted post-translationally to selenocysteine, taking advantage of the specific reaction between Ser221 and phenylmethane sulfonyl fluoride (Figure 5.2). Treatment of the resulting adduct with hydrogen selenide leads to selenosubstilisin in good yields. The artificial selenoenzyme hydrolyzes activated esters, but not amides. Moreover, the acyl–enzyme intermediate that is formed shows high selectivity for aminolysis over hydrolysis (a 14 000-fold increase over the wild type, and 20-fold over thiolsubtilisin, the corresponding cysteine-containing enzyme).[70] Even more striking is the observation that incorporation of selenocysteine confers a completely new activity on the active site.[39] Selenosubtilisin efficiently catalyzes the reduction of hydroperoxides by thiols, in analogy with the natural selenoenzyme glutathione peroxidase. A detailed picture of the catalytic cycle for the novel peroxidase activity has

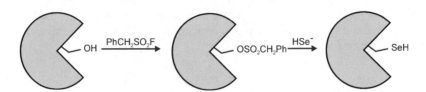

Figure 5.2   Selenium incorporation by chemical modification of the active site serine in subtilisin. The highly reactive hydroxyl group of the enzyme is first activated by phenylmethanesulfonyl fluoride and then displaced by hydrogen selenolate.

been developed based on extensive characterization of selenosubtilisin by X-ray crystallography,[71] NMR spectroscopy,[72] kinetic analysis[73] and mutagenesis.[74] A similar strategy has been exploited to create selenium-containing derivatives of trypsin[42] and an antibody.[41] While powerful, this approach is generally limited to systems in which the residue at the desired site of modification is unusually reactive.

5.2.4 Peptide Synthesis

Chemical synthesis is perhaps the most general approach to artificial seleno-peptides and selenoproteins. Both Boc/Bzl and Fmoc/tBu chemistries can be employed (see Sections 8.3 and 8.4).[75]

Selenocysteine derivatives, suitably protected for standard solid phase peptide synthesis have been prepared by several routes.[25] For example, the hydroxyl group of serine can be activated by tosylation or halogenation,[76] followed by nucleophilic displacement by M_2Se_2 or tetraethylammonium tetra-selenotungstate.[76,77] Nucleophilic attack of Li_2Se_2 on a serine-derived lactone[78] or indium iodide-catalyzed attack of hydrogen diselenide on N-Boc-aziridine[79] are additional possibilities. Recently, cysteine has also been transformed into selenocysteine, although in low yield.[80] For more detailed information on synthetic strategies, the reader is directed to Sections 8.3 and 8.4.

In analogy to native chemical ligation,[81] selenocysteine-mediated ligations provide access to longer selenoproteins.[82–84] C-Terminal peptide thioesters react efficiently with peptide fragments containing an N-terminal selenocysteine to afford a selenoester intermediate that subsequently rearranges to give an amide bond. The utility of this procedure has been demonstrated by the synthesis of selenium-containing derivatives of bovine pancreatic trypsin inhibitor (BPTI),[84] a C-terminal fragment of ribonucleotide reductase,[82] and sele-nocysteine-containing analogs of glutaredoxins 1[43] and 3,[85] among other peptides and proteins (for more details see Section 8.5.1). An extension of this methodology, in which selenocysteine is replaced with homoselenocysteine and the selenol is chemoselectively methylated after ligation, provides access to peptides containing unique selenomethionine residues.[86]

Even larger proteins can be produced by expressed protein ligation. In this method, a protein thioester, produced recombinantly, is coupled with the synthetic selenocysteine-containing fragment.[87] Selenocysteine-containing variants of RNase A[83] and the copper-binding protein azurin[7] have been generated in this way. Recently, the natural selenoprotein thioredoxin reduc-tase, containing a modified active-site sequence, was produced by expressed protein ligation of a 487 amino acid recombinant fragment with the synthetic tripeptide CUG.[88] The semisynthetic enzyme had similar activity to the wild type. The combination of solid-phase peptide synthesis,[89] (selenocysteine-mediated) native chemical ligation[81,84] and expressed protein ligation[87] can afford a wide range of interesting selenocysteine-containing proteins.

5.3 Oxidative Protein Folding

In biology, many secreted proteins contain disulfide bonds. The covalent cross-linking of two cysteine residues is important for the structure, stability and function of these molecules. Often, disulfides influence both the kinetics and the thermodynamics of protein folding. Once formed, native disulfide bonds usually fix the protein in the properly folded conformation. However, these bonds can also be transient when directly involved in protein function, serving as (allosteric) activity switches[90] or providing redox capabilities.[91]

The process by which a fully reduced cysteine-containing protein attains its oxidized native state is called oxidative protein folding. Thiols are oxidized to disulfide bonds by a well-studied two electron transfer mechanism.[91] *In vitro*, disulfide bonds can form spontaneously, using molecular oxygen as the electron acceptor. However, this process usually requires a slightly alkaline pH and the presence of an intermediary (such as a transition metal ion) to overcome the kinetically slow direct reaction of O_2 with protein thiolates.[92] Nevertheless, isomerization of incorrectly formed disulfide cross-links is often the rate-determining step in the overall folding process.[93] *In vivo*, oxidoreductase enzymes facilitate oxidative protein folding by catalyzing disulfide-bond formation as well as thiol-disulfide exchange reactions.[94,95]

5.3.1 Selenium as a Folding Probe

Protein folding pathways can be defined by describing populations of intermediates.[96] The study of oxidative protein folding is facilitated by the possibility of isolating intermediates that differ in their covalent structures. The characterization of covalently trapped intermediates can illuminate general protein folding mechanisms by providing information that is not experimentally accessible for proteins whose folding involves purely non-covalent interactions.[97] Pioneering studies by Creighton,[98] Anfinsen and Scheraga,[99] Weissman and Kim,[100] and others have illuminated the pathways by which several proteins proceed from their reduced, unfolded states to their native structures. Oxidative protein folding mechanisms lie along a spectrum. At one extreme, folding occurs *via* a limited number of distinct (mainly) native-like species. At the other extreme, the intermediates are highly heterogeneous mixtures containing native and non-native disulfide bonds.

Replacement of one or more cysteines with selenocysteine in disulfide-containing proteins can provide insight into folding mechanisms.[44-48] Diselenides and selenosulfides are substantially more stable and more rapidly formed than disulfides, which can alter the partitioning of folding intermediates. While diselenides are apparently rare in nature, peptides and proteins containing diselenides have been produced in the laboratory.

Numerous structural studies on selenocysteine-containing proteins have established that selenium is an essentially isomorphic replacement for sulfur.[101] For instance, experiments with endothelin-1, a 21 amino acid long peptide with potent vasoconstrictor activity, have shown that diselenides can replace both structural disulfides forming native-like cross-links over non-native selenosulfides.[42] Replacement of a native cysteine pair with selenocysteines yielded a peptide that adopts a three-dimensional structure that is, by NMR and CD spectroscopy, indistinguishable from that of the wild-type hormone and exhibits identical activity in biochemical assays.[42]

Building on this result, peptides and proteins have been deliberately stabilized by targeted incorporation of diselenides in place of disulfides. In one striking example, pairwise substitution of the four cysteines in the 12 amino acid long peptide α-conotoxin ImI substantially increased its stability with respect to reduction and scrambling (see Section 8.2.7).[102] As in the case of endothelin, the incorporation of two selenocysteines in place of individual native cysteine pairs did not affect the yield of properly folded product. Further, a variant containing four selenocysteines attained a native-like arrangement of diselenide bonds between residues 2–8 and 3–12.[102] Apparently, the lower free energy of the native state favors the observed arrangement of intramolecular bridges. However, it is worth noting that the α-conotoxin ImI variant with two diselenide bonds displays significant structural deviations from the wild type (Figure 5.3, bottom right), but still retains full biological activity.

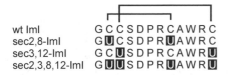

wt ImI	G C C S D P R C A W R C
sec2,8-ImI	G U C S D P R U A W R C
sec3,12-ImI	G C U S D P R C A W R U
sec2,3,8,12-ImI	G U U S D P R U A W R U

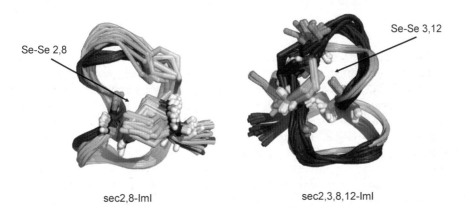

Figure 5.3 Four synthesized variants of α-conotoxin ImI containing various disulfide and diselenide bonds. Top: primary structures of the variants showing the native disulfide connections. Bottom: NMR structures of variants containing two (left) or four (right) selenocysteines (orange), each overlaid with the wild type structures (blue).

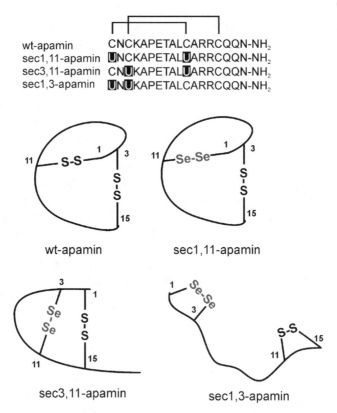

Figure 5.4   Four synthesized variants of apamin containing various disulfide and diselenide bonds. Top: primary structures of the variants showing the native disulfide connections. Bottom: the cross-linking patterns observed upon air oxidation of each variant.

The effect of diselenide bonds on oxidative protein folding has been systematically examined with apamin, an 18 amino acid peptide containing two disulfide bonds.[46] Cysteine residues in this peptide were replaced pairwise by selenocysteines (Figure 5.4). For all variants, diselenide bonds formed spontaneously upon air oxidation regardless of whether such a cross-link was native or non-native. In this system, the modest stability of the native fold is apparently insufficient to overcome the higher stability of (non-native) diselenide bonds relative to (native) selenosulfide bonds. Thus, these apamin analogs represent thermodynamically stable models of kinetically unstable intermediates in the folding pathway of the wild-type peptide.[45] Targeted incorporation of diselenides into disulfide-containing proteins in this way thus provides an attractive tool to study oxidative protein folding mechanisms.

Disulfide-bond stability is linked to conformational stability. In proteins, structural disulfides can be exceptionally stable, having reduction potentials

Table 5.3 Conformational stabilities of disulfide-containing proteins.

Protein	Disulfide bonds	ΔG $(kcal\,mol^{-1})$
Apamin	2	-4.5^a
Tendamistat	2	-9.0^b
RNase T1	2	-9.0^b
BPTI	3	-10.6^b
Papain	3	-22.4^b
Hirudin	3	-4.8^c
hEGF	3	-16.0^d
RNase A	4	-8.7^e
Lysozyme	4	-13.8^b

[a]Free energy of folding at pH 7.0 and 20 °C[103]
[b]Free energy of folding at pH 7.0 and 25 °C[104]
[c]Free energy of folding at pH 7.0 and 25 °C[105]
[d]Free energy of folding at pH 7.0 and 25 °C[106]
[e]Free energy of folding at pH 8.0 and 20 °C[107]

that range from -350 to $-470\,mV$.[91] The difference in energy relative to disulfides in the corresponding unfolded proteins or simple model disulfides is derived from the free energy of folding, which typically lies between -5 and $-20\,kcal\,mol^{-1}$ (Table 5.3). When selenocysteine is incorporated into proteins and larger peptides, the preferential stabilization of native cross-links should often be sufficient to offset the intrinsically higher stability of a non-native diselenide bond. Therefore, the strategy used to trap apamin folding intermediates may only be applicable to modestly stable proteins. Dropping the pH of the folding reaction could maximize the chances that non-native diselenides can trap folding intermediates, as the stability difference between diselenides and disulfides increases under acidic conditions.

In line with the energetic considerations discussed above, selenocysteine incorporation does not disrupt the proper folding of BPTI. This 58 amino acid long protein contains six cysteines, which form three disulfide bonds. Its folding mechanism (Figure 5.5) has been extensively investigated.[100,108] A distinctive feature of the BPTI folding pathway is the accumulation of a non-productive intermediate, N*, which contains a native disulfide bond between Cys5 and Cys55 that has to be broken to form the productive N' intermediate that leads to the correctly oxidized protein.[100] The diselenide-containing Cys5Sec-Cys55Sec-BPTI variant was prepared by selenocysteine-mediated chemical ligation to examine whether the N* intermediate could be enriched.[109] Folding of the reduced protein at neutral pH in air overnight gave a molecule that had the same HPLC retention time and circular dichroism spectrum as native BPTI. Moreover, like the native protein, it stoichiometrically inhibits trypsin. Although detailed kinetic studies must still be performed, it is clear that the enhanced stability of the 5–55 diselenide bond does not present an insurmountable barrier to reaching the native state.

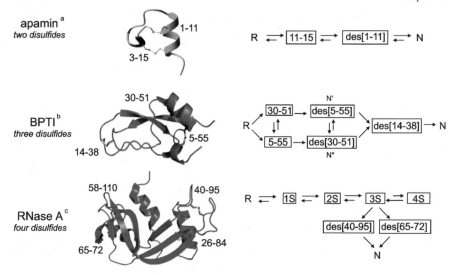

Figure 5.5 Different folding pathways of small disulfide containing proteins. a) apamin, structure adapted from reference 110; b) bovine pancreatic trypsin
inhibitor (BPT1, pdb: 1pit); c) ribonuclease A (RNase A, pdb: 7rsa).

5.3.2 Selenium as a Folding Catalyst

Early *in vitro* studies of oxidative protein folding were typically performed in
air,[111–114] although copper ions[115] and microsomal enzyme preparations[116]
were sometimes used to catalyze the renaturation process. While a variety of
heavy metals,[117,118] DMSO[119] and several enzymes[120–122] have been subsequently shown to catalyze the oxidation of thiols, the natural tripeptide glutathione (γ-Glu-Cys-Gly), in its reduced and oxidized forms (GSH and GSSG,
respectively) has found particularly widespread practical application as a redox
reagent for *in vitro* protein folding. Saxena and Wetlaufer introduced this now
common thiol/disulfide redox buffer in 1970, providing improvements in both
rate and yield compared to simple air oxidation.[123]

Oxidative protein folding is a complex process, and *in vitro* folding conditions can vary widely between different proteins. In addition to the choice of
redox buffer, some of the factors that influence the success of a folding
experiment include the overall folding kinetics of the target protein, temperature, pH, ionic strength, denaturants, various additives (cosolvents, chaotropic
agents, detergents and osmolytes) and the presence of catalysts.[124,125] Despite
these complicating factors, optimal redox buffers typically contain a mixture of
GSH and GSSG, at millimolar concentrations similar to those observed *in
vivo*.[92,126] During protein folding, GSSG directly oxidizes the protein and GSH
enhances disulfide-bond isomerization. The initial oxidation of a reduced
protein is a relatively fast process; the isomerization of partially oxidized species to the native fold is generally rate determining. Immobilized folding catalysts (thiols,[127] disulfides[128] and enzymes[129]) and oxidative refolding

chromatography[130] are two recent innovations with potentially practical bio-technological applications.

Given their similarity to thiols, selenols (and the corresponding diselenides) have attracted interest as alternative catalysts for thiol–disulfide exchange reactions during oxidative protein folding. As early as 1965, selenocystine was shown to catalyze the O_2-oxidation of thiol groups,[131] and combinations of cysteine, cystine and catalytic amounts of selenocystine were found to activate thiol-enzymes like papain and glyceraldehyde-3-phosphate dehydrogenase.[132] Later, Singh and Whitesides studied selenol-catalyzed thiol-disulfide inter-change in considerable detail.[133] They showed that selenocystamine promotes the reaction of dithiols, like DTT and dihydroasparagusic acid, with oxidized β-mercaptoethanol. Selenocystamine also catalyzes the reduction of disulfides in immunoglobins and α-chymotrypsinogen by DTT.[134] Although sub-stoichiometric amounts of selenol afford only modest rate enhancements, excess selenol can accelerate the interchange reaction up to 100-fold.

Inspired by the use of glutathione as the redox reagent of choice for *in vitro* protein folding, selenoglutathione, an analog of glutathione that contains a diselenide bond in place of the natural disulfide, was synthesized and investi-gated as a protein folding catalyst (Figure 5.6).[30] Surprisingly, the diselenide promotes oxidative protein folding with high efficiency, even though its redox potential is 150 mV lower than that of the disulfide bond in glutathione. For example, selenoglutathione efficiently oxidizes BPTI, giving >90% yields of properly folded product under anaerobic conditions. Evidently, the stabiliza-tion of the disulfide bonds in native BPTI is sufficient to drive the reduction of the diselenide.

The folding of BPTI by selenoglutathione is also faster compared to folding with natural glutathione. This enhanced rate is somewhat surprising since the most stable disulfide that has been used for oxidative protein folding, oxidiza

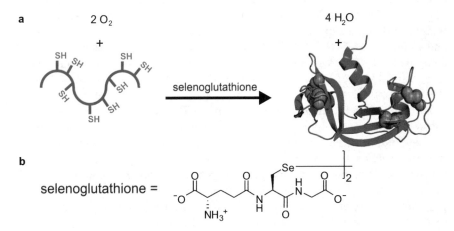

Figure 5.6 The folding catalyst selenoglutathione. a) Schematic of RNase A folding by molecular oxygen, catalyzed by selenoglutathione; b) chemical struc-ture of selenoglutathione.

DTT, exhibits a long lag phase and low rate.[135] Diselenides are presumably superior to disulfides as non-specific oxidants for protein thiols because they are more rapidly reduced by thiols. Further, the selenols generated during the reaction are better catalysts of disulfide isomerization.[30]

The oxidative folding of BPTI involves several one- and two-disulfide-bonded intermediates (Figure 5.5). In the presence of glutathione, one fast and one slow route to the native state are observed, with half of the reduced protein quickly reaching the native state, and the other half becoming trapped as the stable, native-like intermediate (N*) which lacks the native 30–51 disulfide.[85] Conversion of N* to the native protein requires disulfide-bond isomerization. In the presence of selenoglutathione, a similar partitioning between the two folding pathways is observed, but conversion of N* to the native state is significantly accelerated compared to glutathione, reflecting improved catalysis of isomerization by the selenol.[30] The ability of the selenium-based redox buffer to rescue non-productive folding intermediates is notable and potentially of considerable practical utility.

The folding proficiency of selenoglutathione extends to other proteins as well. For example, the folding of ribonuclease A (RNase A) by the diselenide occurs with a two-fold higher rate and a shorter lag phase compared to sulfur glutathione.[30] During RNase A folding, a scrambled ensemble of one-, two-, three- and four-disulfide-bonded intermediates is rapidly formed (Figure 5.5).[93] The rate-limiting step of folding involves disulfide bond isomerization within the 3S ensemble to give either des[40-95] or des[65-72], and these native-like intermediates are rapidly oxidized to the native state. Although RNase A folding with selenoglutathione was carried out aerobically,[30] the observed rate enhancement suggests that the selenols generated *in situ* may promote disulfide isomerization. However, selenols rapidly reoxidize in air, which could also push protein oxidation forward for less stable proteins.

The efficient regeneration of selenoglutathione in air enables true catalysis of oxidative protein folding. *In vitro* this process is usually performed in redox buffers containing superstoichiometric concentrations of both oxidant (GSSG) and reductant (GSH).[92,126] We have developed a novel redox buffer that is composed of oxidized selenoglutathione and reduced sulfur glutathione. While using ten-fold lower concentrations of oxidant and reductant compared to the standard optimized glutathione redox buffer, our redox buffer affords the same rate and yield for the aerobic folding of RNase A.[136] Even lower – and substoichiometric – concentrations of selenoglutathione show substantial protein folding activity in air. In contrast to the direct stoichiometric oxidation of proteins by disulfide reagents, selenoglutathione promotes electron transfer between molecular oxygen and protein thiols giving water and disulfides as products while not being consumed itself. Furthermore, the lower pK_a values of selenols significantly broaden the pH range for carrying out oxidative protein folding experiments to acidic regimes where glutathione is inactive.

Diselenide reagents, such as selenoglutathione, have high potential utility for oxidative protein folding applications. More detailed investigations into the mechanism of selenoglutathione-catalyzed folding should aid the design of new and useful diselenides, and might productively focus on protein disulfide-bond isomerization, which is often the rate-limiting step in oxidative folding. For example, engineering of the electrostatic environment around the diselenide bond may allow tuning of reactivity and the creation of novel redox reagents with tailored properties. The greater flexibility in the choice of solution conditions could be useful for *in vitro* folding of traditionally troublesome proteins, such as antibodies and growth factors.

5.4 Perspectives

By expanding the genetic code to include selenocysteine, nature has successfully capitalized on the special properties of selenium. The selenol group extends the functional capabilities of proteins in a variety of interesting ways, as evidenced by the sophisticated catalysts and other selenoproteins that have been studied to date. Based on the number of putative selenoproteins that have been identified from genomic analyses that remain to be isolated and characterized, much new chemistry can be expected in the future.

Technical advances in chemistry and molecular biology have made it possible to produce natural selenoproteins for detailed study. These methods have also provided the means to create artificial selenopeptides and selenoproteins with properties not (yet) found in nature. As we have seen, selenium has been successfully incorporated into a variety of polypeptide scaffolds, where it can serve as a structural element, a mechanistic probe or a catalytic prosthetic group. In these systems, the peptide/protein environment modulates the intrinsic reactivity of the selenium atom, opening up interesting possibilities for practical applications.

In particular, the ability of selenopeptides and selenoproteins to cycle between their selenol and diselenide states, mediated by facile reaction of the former with molecular oxygen (and other oxidants) and the latter with thiols, makes them excellent antioxidants and ideal candidates for the development of novel redox buffers. Because selenols promote disulfide shuffling, it may be possible to accelerate protein folding by targeted insertion of selenocysteines into proteins containing multiple disulfides. Such modifications can be expected to provide control over folding pathways, favoring specific reaction channels over others. The greater robustness of the final folded product, arising from the enhanced stability of selenosulfides and diselenides compared to disulfides, promises to be an added biotechnological benefit.

The properties of simple compounds like selenoglutathione exemplify the utility of selenium derivatives as true folding catalysts. While such first generation molecules are unlikely to be optimal for all practical applications, their small size and modular structure will facilitate efforts to engineer their properties in a systematic fashion. Given the apparent benefits of a diselenide over a disulfide, it is interesting that selenocysteine-containing "foldases" have never been found in nature. By replacing the catalytic CXXC motif in enzymes like protein disulfide isomerase (PDI) and DsbA, which catalyze oxidative protein folding *in vivo*, with redox active selenosulfides and diselenides, it may be possible to create a range of novel protein-based catalysts for diverse applications in living organisms.

References

1. J. J. Berzelius, *Afh. Fys. Kemi och Mineralog.*, 1818, **6**, 42.
2. J. Wisniak, *Chem. Educ.*, 2000, **5**, 343–350.
3. L. V. Papp, J. Lu, A. Holmgren and K. K. Khanna, *Antioxid. Redox Signal.*, 2007, **9**, 775–806.
4. G. N. Schrauzer, *J. Nutr.*, 2000, **130**, 1653–1656.
5. T. C. Stadtman, *Annu. Rev. Biochem.*, 1996, **65**, 83–100.
6. R. E. Huber and R. S. Criddle, *Arch. Biochem. Biophys.*, 1967, **122**, 164–173.
7. S. M. Berry, M. D. Gieselman, M. J. Nilges, W. A. van der Donk and Y. Lu, *J. Am. Chem. Soc.*, 2002, **124**, 2084–2085.
8. G. Casi and D. Hilvert, *J. Biol. Chem.*, 2007, **282**, 30518–30522.
9. H. E. Ganther and R. J. Kraus, *Methods Enzymol.*, 1984, **107**, 593–602.
10. P. C. Jocelyn, *Eur. J. Biochem.*, 1967, **2**, 327–331.
11. T. Nauser, S. Dockheer, R. Kissner and W. H. Koppenol, *Biochemistry*, 2006, **45**, 6038–6043.
12. S. Gromer, J. K. Eubel, B. L. Lee and J. Jacob, *Cell. Mol. Life Sci.*, 2005, **62**, 2414–2437.
13. L. Johansson, G. Gafvelin and E. S. J. Arnér, *Biochim. Biophys. Acta*, 2005, **1726**, 1–13.
14. F. Ursini, M. Maiorino, R. Brigelius-Flohé, K. D. Aumann, A. Roveri, D. Schomburg and L. Flohé, *Methods Enzymol.*, 1995, **252**, 38–53.
15. C. Rocher, J. L. Lalanne and J. Chaudière, *Eur. J. Biochem.*, 1992, **205**, 955–960.
16. J. Köhrle, *Cell. Mol. Life Sci.*, 2000, **57**, 1853–1863.
17. M. J. Berry, L. Banu and P. R. Larsen, *Nature*, 1991, **349**, 438–440.
18. L. Zhong, E. S. J. Arnér and A. Holmgren, *Proc. Natl. Acad. Sci. USA*, 2000, **97**, 5854–5859.
19. L. Zhong and A. Holmgren, *J. Biol. Chem.*, 2000, **275**, 18121–18128.
20. S.-R. Lee, S. Bar-Noy, J. Kwon, R. L. Levine, T. C. Stadtman and S. G. Rhee, *Proc. Natl. Acad. Sci. USA*, 2000, **97**, 2521–2526.
21. X.-M. Xu, B. A. Carlson, R. Irons, H. Mix, N. Zhong, V. N. Gladyshev and D. L. Hatfield, *Biochem. J.*, 2007, **404**, 115–120.
22. I. Y. Kim, M. J. Guimarães, A. Zlotnik, J. F. Bazan and T. C. Stadtman, *Proc. Natl. Acad. Sci. USA*, 1997, **94**, 418–421.
23. H.-Y. Kim and V. N. Gladyshev, *Mol. Biol. Cell*, 2004, **15**, 1055–1064.
24. S. Bar-Noy and J. Moskovitz, *Biochem. Biophys. Res. Commun.*, 2002, **297**, 956–961.

25. L. A. Wessjohann, A. Schneider, M. Abbas and W. Brandt, *Biol. Chem.*, 2007, **388**, 997–1006.
26. M. J. Axley, A. Böck and T. C. Stadtman, *Proc. Natl. Acad. Sci. USA*, 1991, **88**, 8450–8454.
27. M. Wagner, D. Sonntag, R. Grimm, A. Pich, C. Eckerskorn, B. Söhling and J. R. Andreesen, *Eur. J. Biochem.*, 1999, **260**, 38–49.
28. V. A. Shchedrina, S. V. Novoselov, M. Y. Malinouski and V. N. Gladyshev, *Proc. Natl. Acad. Sci. USA*, 2007, **104**, 13919–13924.
29. W. J. Lees and G. M. Whitesides, *J. Org. Chem.*, 1993, **58**, 642–647.
30. J. Beld, K. J. Woycechowsky and D. Hilvert, *Biochemistry*, 2007, **46**, 5382–5390.
31. C. Hwang, A. J. Sinskey and H. F. Lodish, *Science*, 1992, **257**, 1496–1502.
32. R. F. Burk and K. E. Hill, *Ann. Rev. Nutr.*, 2005, **25**, 215–235.
33. W. A. Hendrickson, *Science*, 1991, **254**, 51–58.
34. M.-P. Strub, F. Hoh, J.-F. Sanchez, J. M. Strub, A. Böck, A. Aumelas and C. Dumas, *Structure*, 2003, **11**, 1359–1367.
35. M. Faßbender, D. de Villiers, M. Nortier and N. van der Walt, *Appl. Radiat. Isot.*, 2001, **54**, 905–913.
36. S. Müller, J. Heider and A. Böck, *Arch. Microbiol.*, 1997, **168**, 421–427.
37. H. Duddeck, *Prog. Nucl. Magn. Reson. Spectrosc.*, 1995, **27**, 1–323.
38. L. Johansson, C. Chen, J.-O. Thorell, A. Fredriksson, S. Stone-Elander, G. Gafvelin and E. S. J. Arnér, *Nat. Methods*, 2004, **1**, 61–66.
39. Z.-P. Wu and D. Hilvert, *J. Am. Chem. Soc.*, 1990, **112**, 5647–5648.
40. H.-J. Yu, J.-Q. Liu, A. Böck, J. Li, G.-M. Luo and J.-C. Shen, *J. Biol. Chem.*, 2005, **280**, 11930–11935.
41. G.-M. Luo, Z.-Q. Zhu, L. Ding, G. Gao, Q.-A. Sun, Z. Liu, T.-S. Yang and J.-C. Shen, *Biochem. Biophys. Res. Commun.*, 1994, **198**, 1240–1247.
42. J.-Q. Liu, M.-S. Jiang, G.-M. Luo, G.-L. Yan and J.-C. Shen, *Biotechnol. Lett.*, 1998, **20**, 693–696.
43. G. Casi, G. Roelfes and D. Hilvert, *ChemBioChem*, 2008, **9**, 1623–1631.
44. D. Besse, F. Siedler, T. Diercks, H. Kessler and L. Moroder, *Angew. Chem. Int. Ed. Engl.*, 1997, **36**, 883–885.
45. S. Fiori, S. Pegoraro, S. Rudolph-Böhner, J. Cramer and L. Moroder, *Biopolymers*, 2000, **53**, 550–564.
46. S. Pegoraro, S. Fiori, J. Cramer, S. Rudolph-Böhner and L. Moroder, *Protein Sci.*, 1999, **8**, 1605–1613.
47. S. Pegoraro, S. Fiori, S. Rudolph-Böhner, T. X. Watanabe and L. Moroder, *J. Mol. Biol.*, 1998, **284**, 779–792.
48. D. Besse and L. Moroder, *J. Pept. Sci.*, 1997, **3**, 442–453.
49. P. R. Copeland, *Genome Biol.*, 2005, **6**, 221.
50. Y. Zhang, H. Romero, G. Salinas and V. N. Gladyshev, *Genome Biol.*, 2006, **7**, R94.

51. D. E. Fomenko, W. Xing, B. M. Adair, D. J. Thomas and V. N. Gladyshev, *Science*, 2007, **315**, 387–389.
52. A. Böck, K. Forchhammer, J. Heider and C. Baron, *Trends Biochem. Sci.*, 1991, **16**, 463–467.
53. E. S. J. Arnér, H. Sarioglu, F. Lottspeich, A. Holmgren and A. Böck, *J. Mol. Biol.*, 1999, **292**, 1003–1016.
54. M. Thanbichler and A. Böck, *EMBO J.*, 2002, **21**, 6925–6934.
55. S. Commans and A. Böck, *FEMS Microbiol. Rev.*, 1999, **23**, 335–351.
56. K. Forchhammer, W. Leinfelder, K. Boesmiller, B. Veprek and A. Böck, *J. Biol. Chem.*, 1991, **266**, 6318–6323.
57. W. Leinfelder, K. Forchhammer, B. Veprek, E. Zehelein and A. Böck, *Proc. Natl. Acad. Sci. USA*, 1990, **87**, 543–547.
58. K. Forchhammer, K. P. Rücknagel and A. Böck, *J. Biol. Chem.*, 1990, **265**, 9346–9350.
59. E. S. J. Arnér, *Methods Enzymol.*, 2002, **347**, 226–235.
60. G. T. Chen, M. J. Axley, J. Hacia and M. Inouye, *Mol. Microbiol.*, 1992, **6**, 781–785.
61. J. Heider and A. Böck, *J. Bacteriol.*, 1992, **174**, 659–663.
62. S. Hazebrouck, L. Camoin, Z. Faltin, A. D. Strosberg and Y. Eshdat, *J. Biol. Chem.*, 2000, **275**, 28715–28721.
63. G. V. Kryukov, R. A. Kumar, A. Koc, Z. Sun and V. N. Gladyshev, *Proc. Natl. Acad. Sci. USA*, 2002, **99**, 4245–4250.
64. H.-Y. Kim, D. E. Fomenko, Y.-E. Yoon and V. N. Gladyshev, *Biochemistry*, 2006, **45**, 13697–13704.
65. H.-Y. Kim and V. N. Gladyshev, *PLoS Biology*, 2005, **3**, 2080–2089.
66. X.-M. Xu, B. A. Carlson, H. Mix, Y. Zhang, K. Saira, R. S. Glass, M. J. Berry, V. N. Gladyshev and D. L. Hatfield, *PLoS Biology*, 2006, **5**, 0096–0105.
67. D. L. Hatfield, B. A. Carlson, X.-M. Xu, H. Mix and V. N. Gladyshev, *Prog. Nucleic Acid Res.*, 2006, **81**, 97–142.
68. M. T. Howard, M. W. Moyle, G. Aggarwal, B. A. Carlson and C. B. Anderson, *RNA*, 2007, **13**, 912–920.
69. S. Müller, H. Senn, B. Gsell, W. Vetter, C. Baron and A. Böck, *Biochemistry*, 1994, **33**, 3404–3412.
70. Z.-P. Wu and D. Hilvert, *J. Am. Chem. Soc.*, 1989, **111**, 4513–4514.
71. R. Syed, Z.-P. Wu, J. M. Hogle and D. Hilvert, *Biochemistry*, 1993, **32**, 6157–6164.
72. K. L. House, R. B. Dunlap, J. D. Odom, Z.-P. Wu and D. Hilvert, *J. Am. Chem. Soc.*, 1992, **114**, 8573–8579.
73. I. M. Bell, M. L. Fisher, Z.-P. Wu and D. Hilvert, *Biochemistry*, 1993, **32**, 3754–3762.
74. D. Hilvert, *Biochemistry*, 1995, **34**, 6616–6620.
75. L. Moroder, *J. Pept. Sci.*, 2005, **11**, 187–214.
76. E. M. Stocking, J. N. Schwarz, H. Senn, M. Salzmann and L. A. Silks, *J. Chem. Soc., Perkin Trans. 1*, 1997, 2443–2447.

77. R. G. Bhat, E. Porhiel, V. Saravanan and S. Chandrasekaran, *Tetrahedron Lett.*, 2003, **44**, 5251–5253.
78. A. Schneider, O. E. D. Rodrigues, M. W. Paixão, H. R. Appelt, A. L. Braga and L. A. Wessjohann, *Tetrahedron Lett.*, 2006, **47**, 1019–1021.
79. A. L. Braga, P. H. Schneider, M. W. Paixão, A. M. Deobald, C. Peppe and D. P. Bottega, *J. Org. Chem.*, 2006, **71**, 4305–4307.
80. M. Iwaoka, C. Haraki, R. Ooka, M. Miyamoto, A. Sugiyama, Y. Kohara and N. Isozumi, *Tetrahedron Lett.*, 2006, **47**, 3861–3863.
81. P. E. Dawson, T. W. Muir, I. Clark-Lewis and S. B. H. Kent, *Science*, 1994, **266**, 776–779.
82. M. D. Gieselman, L. Xie and W. A. van der Donk, *Org. Lett.*, 2001, **3**, 1331–1334.
83. R. J. Hondal, B. L. Nilsson and R. T. Raines, *J. Am. Chem. Soc.*, 2001, **123**, 5140–5141.
84. R. Quaderer, A. Sewing and D. Hilvert, *Helv. Chim. Acta*, 2001, **84**, 1197–1206.
85. N. Metanis, E. Keinan and P. E. Dawson, *J. Am. Chem. Soc.*, 2006, **128**, 16684–16691.
86. G. Roelfes and D. Hilvert, *Angew. Chem. Int. Ed. Engl.*, 2003, **42**, 2275–2277.
87. T. W. Muir, D. Sondhi and P. A. Cole, *Proc. Natl. Acad. Sci. USA*, 1998, **95**, 6705–6710.
88. B. Eckenroth, K. M. Harris, A. A. Turanov, V. N. Gladyshev, R. T. Raines and R. J. Hondal, *Biochemistry*, 2006, **45**, 5158–5170.
89. B. Merrifield, *Science*, 1986, **232**, 341–347.
90. B. Schmidt, L. Ho and P. J. Hogg, *Biochemistry*, 2006, **45**, 7429–7433.
91. H. F. Gilbert, *Adv. Enzymol. Relat. Areas Mol. Biol.*, 1990, **63**, 69–172.
92. D. B. Wetlaufer, P. A. Branca and G.-X. Chen, *Protein Eng.*, 1987, **1**, 141–146.
93. M. Narayan, E. Welker, W. J. Wedemeyer and H. A. Scheraga, *Acc. Chem. Res.*, 2000, **33**, 805–812.
94. L. Ellgaard, *Biochem. Soc. Trans.*, 2004, **32**, 663–667.
95. C. S. Sevier and C. A. Kaiser, *Nat. Rev. Mol. Cell Biology*, 2002, **3**, 836–847.
96. J. L. Arolas, F. X. Aviles, J.-Y. Chang and S. Ventura, *Trends Biochem. Sci.*, 2006, **31**, 292–300.
97. C. C.-J. Lin and J.-Y. Chang, *Biochemistry*, 2007, **46**, 3925–3932.
98. T. E. Creighton, *J. Mol. Biol.*, 1974, **87**, 563–577.
99. C. B. Anfinsen and H. A. Scheraga, *Adv. Protein Chem.*, 1975, **29**, 205–300.
100. J. S. Weissman and P. S. Kim, *Science*, 1991, **253**, 1386–1393.
101. D. Besse, N. Budisa, W. Karnbrock, C. Minks, H.-J. Musiol, S. Pegoraro, F. Siedler, E. Weyher and L. Moroder, *Biol. Chem.*, 1997, **378**, 211–218.

102. C. J. Armishaw, N. L. Daly, S. T. Nevin, D. J. Adams, D. J. Craik and P. F. Alewood, *J. Biol. Chem.*, 2006, **281**, 14136–14143.
103. C. E. Dempsey, *Biochemistry*, 1986, **25**, 3904–3911.
104. G. I. Makhatadze and P. L. Privalov, *Adv. Protein Chem.*, 1995, **47**, 307–425.
105. A. Otto and R. Seckler, *Eur. J. Biochem.*, 1991, **202**, 67–73.
106. L. A. Holladay, C. R. Savage Jr, S. Cohen and D. Puett, *Biochemistry*, 1976, **15**, 2624–2633.
107. C. N. Pace, D. V. Laurents and J. A. Thomson, *Biochemistry*, 1990, **29**, 2564–2572.
108. T. E. Creighton and D. P. Goldenberg, *J. Mol. Biol.*, 1984, **179**, 497–526.
109. R. Quaderer, Ph.D. thesis, ETH Zürich, 2001.
110. Y. Kobayashi, H. Takashima, H. Tamaoki, Y. Kyogoku, P. Lambert, H. Kuroda, N. Chino, T. X. Watanabe, T. Kimura, S. Sakakibara and L. Moroder, *Biopolymers*, 1991, **31**, 1213–1220.
111. C. B. Anfinsen, E. Haber, M. Sela and F. H. White Jr., *Proc. Natl. Acad. Sci. USA*, 1961, **47**, 1309–1314.
112. T. Isemura, T. Takagi, Y. Maeda and K. Imai, *Biochem. Biophys. Res. Commun.*, 1961, **5**, 373–377.
113. F. H. White Jr, *J. Biol. Chem.*, 1960, **235**, 383–389.
114. F. H. White Jr, *J. Biol. Chem.*, 1961, **236**, 1353–1360.
115. J.-P. Perraudin, T. E. Torchia and D. B. Wetlaufer, *J. Biol. Chem.*, 1983, **258**, 11834–11839.
116. R. F. Goldberger, C. J. Epstein and C. B. Anfinsen, *J. Biol. Chem.*, 1963, **238**, 628–635.
117. M. Arisawa, C. Sugata and M. Yamaguchi, *Tetrahedron Lett.*, 2005, **46**, 6097–6099.
118. M. Kirihara, K. Okubo, T. Uchiyama, Y. Kato, Y. Ochiai, S. Matsushita, A. Hatano and K. Kanamori, *Chem. Pharm. Bull.*, 2004, **52**, 625–627.
119. J. P. Tam, C.-R. Wu, W. Liu and J.-W. Zhang, *J. Am. Chem. Soc.*, 1991, **113**, 6657–6662.
120. D. A. Hillson, N. Lambert and R. B. Freedman, *Methods Enzymol.*, 1984, **107**, 281–294.
121. R. Noiva, *Protein Express. Purif.*, 1994, **5**, 1–13.
122. Y. Akiyama, S. Kamitani, N. Kusukawa and K. Ito, *J. Biol. Chem.*, 1992, **267**, 22440–22445.
123. V. P. Saxena and D. B. Wetlaufer, *Biochemistry*, 1970, **9**, 5015–5023.
124. R. Rudolph and H. Lilie, *FASEB J.*, 1996, **10**, 49–56.
125. G. Bulaj, *Biotechnol. Adv.*, 2005, **23**, 87–92.
126. M. M. Lyles and H. F. Gilbert, *Biochemistry*, 1991, **30**, 613–619.
127. K. J. Woycechowsky, B. A. Hook and R. T. Raines, *Biotechnol. Prog.*, 2003, **19**, 1307–1314.
128. I. Annis, L. Chen and G. Barany, *J. Am. Chem. Soc.*, 1998, **120**, 7226–7238.

129. M. M. Altamirano, C. García, L. D. Possani and A. R. Fersht, *Nat. Biotechnol.*, 1999, **17**, 187–191.
130. X. Geng and C. Wang, *J. Chromatogr. B*, 2007, **849**, 69–80.
131. K. A. Caldwell and A. L. Tappel, *Arch. Biochem. Biophys.*, 1965, **112**, 196–200.
132. R. C. Dickson and A. L. Tappel, *Arch. Biochem. Biophys.*, 1969, **131**, 100–110.
133. R. Singh and G. M. Whitesides, *J. Org. Chem.*, 1991, **56**, 6931–6933.
134. R. Singh and L. Kats, *Anal. Biochem.*, 1995, **232**, 86–91.
135. D. M. Rothwarf and H. A. Scheraga, *J. Am. Chem. Soc.*, 1991, **113**, 6293–6294.
136. J. Beld, K. J. Woycechowsky and D. Hilvert, *Biochemistry*, 2008, **47**, 6985–6987.

CHAPTER 6

Oxidative Folding of Peptides in vitro

CHAPTER 6.1

Oxidative Folding of Single-stranded Disulfide-rich Peptides

GRZEGORZ BULAJ[a] AND ALEKSANDRA WALEWSKA[b]

[a] Department of Medicinal Chemistry, University of Utah, Salt Lake City, Utah 84108, USA; [b] Faculty of Chemistry, University of Gdansk, 80-952 Gdansk, Poland

6.1.1 Introduction

6.1.1.1 Molecular Diversity of Disulfide-rich Peptides

Molecular diversity of disulfide-rich peptides is reflected in hundreds of thousands of unique sequences found primarily in plant and animal kingdoms. Cysteine-rich peptides were employed early in evolution; bacteria produce two-disulfide-containing toxins, such as heat-stabilized enterotoxin B or ST1b. In contrast, fungi produce antimicrobial peptides, such as highly knotted sillucin. In the course of evolution, an explosion in disulfide-rich peptide diversity occurred when spiders, scorpions and cone snails evolved numerous families of

RSC Biomolecular Sciences
Oxidative Folding of Peptides and Proteins
Edited by Johannes Buchner and Luis Moroder
© Royal Society of Chemistry 2009
Published by the Royal Society of Chemistry, www.rsc.org

short, 10–50 residue long peptides. The members of these families are characterized by a hypervariability of primary amino acid sequences in between otherwise highly conserved cysteines. Arguably, the largest diversity of amino acid sequences among short disulfide-rich peptides may exist in venoms of spiders; current estimates are that 38 000 known spider species may have evolved from 1.5 to even 19 million different toxins. Recently reported toxins from a venomous turrid snail *Lophiotoma olangoensis* (representing one out of approximately 10 000 species) may signal a new source of unprecedented molecular diversity, by far exceeding that observed in *Conus* snails.[1] Furthermore, plants from the *Rubiaceae* and *Violaceae* families evolved a vast repertoire of Cys-rich peptides with a cyclic backbone termed cyclotides. Thus, structurally and functionally diverse disulfide-rich peptides are produced from bacteria to humans.

As illustrated in Table 6.1.1 disulfide-rich peptides vary in size and content of Cys residues. They can be as short as 13 amino acid residues containing four Cys (25% Cys content) such as in α-conotoxins, 12 residues of tx3a peptide with six Cys (50% Cys content) or as long as 41 residues in σ-GVIIIA conotoxin with ten Cys residues (24% Cys content). Hepcidin, a 25-membered peptide, contains eight Cys yielding a high 33% Cys content. Equally impressive to the molecular diversity found in Cys-rich peptides is their functional diversity. The most well-known examples are toxins, venom components of spiders, scorpions and cone snails primarily used for prey capture. Other functions of disulfide-rich peptides include host-defense of plants from insects, such as cyclotides, antimicrobial defense of animals from bacteria and regulation by endogenous regulatory peptides, such as hepcidin, guanilin or endothelin. The structural and functional diversity of Cys-rich peptides is described further in Chapter 6.3.

6.1.1.2 Oxidative Folding Problem

One of the most fundamental questions in biology is the protein folding problem: how does the primary amino acid sequence determine the three-dimensional structure of a protein? Tremendous progress has been made in understanding how polypeptides acquire their native conformations.[2,3] Protein engineering, stopped-flow spectroscopy, NMR and computational methods all help to define roles of individual amino acid residues in the formation of folding intermediates and transitional states in several small proteins. Key findings are that the denatured/unfolded state is characterized by native-like intramolecular interactions, and that only a limited number of conformations are searched by a polypeptide chain. Several folding mechanisms have been proposed for various proteins, ranging from "framework", "hydrophobic collapse", to "nucleation-condensation" as possible means to effectively fold. Interestingly, although proteins necessarily evolved with the ability to efficiently fold into the bioactive conformation, there are many examples when refolded in either recombinant systems or in a test tube they failed to acquire functional activity.

Table 6.1.1   Structural diversity of disulfide-rich peptides.

Two-disulfide bridge peptides

Apamin
(venom bee neurotoxin)

CNCKAPETALCARRCQQH#

Guanylin
(intestinal, human hormone)

PGTCEICAYAACTGC

Endothelin-1
(mammalian peptide)

CSCSSLMDKECVYFCHLDIIW

α-GI
(conotoxin, *Conus geographus*)

ECCNPACGRHYSC#

Three-disulfide bridge peptides

ω-MVIIA
(conotoxin, *Conus magus*)

CKGKGAKCSRLMYDCCTGSCRSGKC#

μ-GIIIA
(conotoxin, *Conus geographus*)

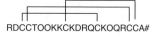
RDCCTOOKKCKDRQCKOQRCCA#

tx3a
(conotoxin, *Conus textile*)

CCSWDVCDHPSCTCCG

CMTI-I
(trypsin inhibitor, pumpkin,
  Curcurbita maxima)

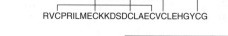
RVCPRILMECKKDSDCLAECVCLEHGYCG

Insect defensin
(fruit fly, *Drosophila*)

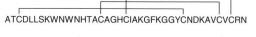
ATCDLLSKWNWNHTACAGHCIAKGFKGGYCNDKAVCVCRN

α-Defensin-1
(human neutrophils)

ACYCRIPACIAGERRYGTCIYQGRLWAFCC

β-Defensin-3
(human psoriatic scales)

GIINTLQKYYCRVRGGRCAVLSCLPKEEQIGKCSTRGRKCCRRKK

Kalata B1
(cyclotide, Möbius subfamily,
  Oldenlandia affinis)

CGETCVGGTCNTPGCTCSWPVCTRNGLPV

Four-disulfide bridge peptides

Maurotoxin
(scorpion toxin, *Scorpio maurus
  palmatus*)

VSCTGSKDCYAPCRKQTGCPNAKCINKSCKCYGC#

Hepcidin 25
(human urine and plasma)

DTHFPICIFCCGCCHRSKCGMCCKT

C-terminus amidated.

The protein folding problem has a natural spin-off: the oxidative folding problem.[4,5] Since disulfide bridges stabilize most secreted proteins, their folding is coupled to the formation of the native disulfide bonds (see Chapter 2.1.1.). Thus, oxidative folding is a complex interplay between non-covalent interactions and formation of the covalent links that restrict the search of conformational space during folding. Formation of the disulfide bridges is a useful probe for conformational changes that lead to the formation of the native state. Oxidative folding intermediates can be chemically trapped (by either protonation or alkylation), allowing a more detailed structural characterization of folding intermediates, thereby helping to define folding pathways. This approach defined roles of many individual amino acid residues in the folding mechanism of bovine pancreatic trypsin inhibitor (BPTI), including their role in stabilizing folding intermediates and in determining the transition state for forming the native conformation.[6,7] Results from extensive oxidative folding studies on BPTI, lactalbumin, lysozyme, RNase A, hirudine and other polypeptides indicated a diversity of folding mechanisms.[8] At least three folding mechanisms could be distinguished: (1) a specific/selective/ordered and hierarchic formation of the correctly folded species *via* a limited number of native-like intermediates, the "framework model", (2) non-specific and heterogeneous formation of multiple folding species (scrambled) – equivalent to the "collapse model" and (3) a mixed model in which the formation of the native conformations proceeds through a combination of the framework and scrambled mechanisms.

The oxidative folding problem also applies to short, disulfide-rich peptides. A key question is how the formation of the native disulfide bridges is determined given the hypervariability of the primary amino acid sequences between otherwise highly conserved Cys residues. How much do non-cysteine residues determine the formation of native disulfide bridges? The hypervariability of primary amino acid sequences in Cys-rich peptides seen in the same gene families (like conotoxins, cyclotides and scorpion/spider toxins) particularly poses a challenge in defining what factors govern the formation of the native disulfide bridges. As illustrated in Figure 6.1.1A, peptides significantly varying in *primary amino acid sequences* may still fold into the same disulfide framework. How much do cysteine patterns determine formation of appropriate disulfide scaffolds? As described later, *identical cysteine patterns* can produce *distinct connectivities* of the native disulfide bridges. A particularly notable example of the oxidative folding problem for small peptides is what was previously defined as the "conotoxin folding puzzle".[9]

The conotoxin folding puzzle emerged as a result of discovering hundreds of sequences of conotoxins that belong to the same gene family.[12] Conotoxins belonging to a single family always share identical cysteine patterns (and, it was presumed, disulfide connectivity), but can differ dramatically in their primary amino acid sequences. Thus, if no apparent consensus of primary amino acid sequences exists, how much sequence information is utilized during the formation of the native disulfide bridges? In addition to many reports on efficient folding of conotoxins, there are several reports showing very low folding yields for some of these peptides.[12–16] The same folding problem applies

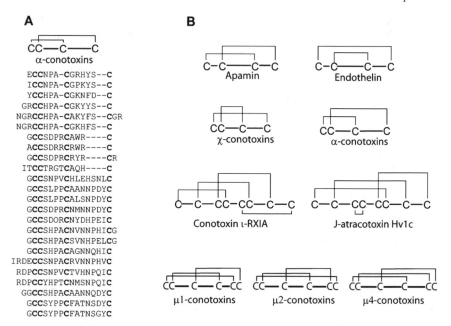

A α-conotoxins

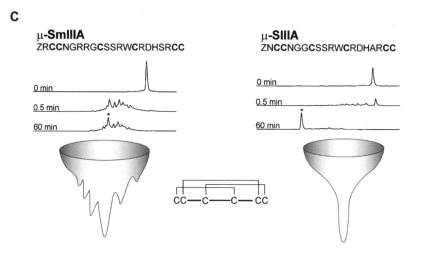

Figure 6.1.1 The oxidative folding problem for short cysteine-rich peptides. A) Despite hypervariability of the primary amino acid sequences, all α-conotoxins fold into the same disulfide scaffold (all but two α-conotoxin sequences are amidated at the C-terminus). B) Despite identical cysteine patterns, these peptides fold into distinct disulfide scaffolds. C) Despite identical cysteine patterns and similar primary amino acid sequences, these two peptides exhibit two distinct folding mechanisms. SmIIIA uses the "collapse" model whereas SIIIA folds according to the "framework" model. Folding funnels further emphasize the distinct folding mechanism, despite identical disulfide scaffold. HPLC traces reprinted with permission from references 10 and 11.

to various scorpion and spider toxins. Some scorpion toxins, such as four-disulfide-bridged toxin Cn5, are known to fold with extremely low folding yields.[17]

Another aspect of the conotoxin folding puzzle is illustrated in Figure 6.1.1C which summarizes the oxidative folding of two closely related conotoxins in the same family. These two peptides exhibit dramatically different folding mechanisms despite sharing identical topology of cysteine residues, identical connectivity of disulfide bridges and almost identical primary amino acid sequence information. Striking differences in conformational and energetic changes during folding, as expressed by the folding funnels, further emphasize that neither sequence nor cysteine patterns alone determine the mechanism of oxidative folding for these small peptides. Thus, further investigation is needed to assess what structural features govern the folding mechanism, and whether these two peptides fold differently in the cell, as they do in a test tube. Several other peptides have been shown to fold by either the framework mechanism (conotoxins GIIIA, RIIIK, GI, trypsin inhibitors EETI-II, MCoTI-II, maurotoxin or apamin) or the collapse mechanism (conotoxin PIIIA, MrVIB, inhibitors PCI and AAI). It remains unclear how much cysteine patterns and non-covalent interactions determine mechanisms by which these peptides acquire the native conformation.

6.1.1.3 Scope of the Chapter

The main objective of this chapter is to summarize studies on the mechanisms of the oxidative folding of disulfide-rich peptides, with an emphasis on small (< 50 amino acid residues) peptides; an excellent review on the mechanism of the oxidative folding of polypeptides larger than 50 amino acid residues has been recently published.[8] As outlined in section 6.1.1.1, these small peptides comprise spider, scorpion and cone snail toxins, plant-derived cyclotides or proteinase inhibitors and endogenous regulatory peptides. Many research groups contributed to defining factors that affect the formation of the native disulfide bridges in these peptides; we attempt to compile and review the major findings in the next two sections. In Section 6.1.2 we discuss how the reactivity of the cysteine thiols, cysteine patterns and primary amino acid residues affect the formation of the disulfide bridges, whereas in Section 6.1.3 we review studies relevant to *in vivo* folding. Despite well-defined rules for forming disulfide bridges in model peptides, how these rules apply to oxidative folding *in vivo* is still poorly understood.

6.1.2 Mechanisms of *in vitro* Oxidative Folding

In this section, we review mechanisms of *in vitro* oxidative folding of the disulfide-rich peptides. First, we discuss a role of cysteine thiol reactivity, cysteine patterns, primary amino acid sequences and post-translational modifications in the formation of the native disulfide bridges.

6.1.2.1   Thiol/Disulfide Exchange Reactions in Peptides

Formation of the native conformation of a peptide during oxidative folding is driven by a combination of the conformational folding and thiol/disulfide exchange reactions. The formation of the covalent disulfide bond depends on multiple factors, including the reactivity of the cysteine thiols. Thus, the kinetics and thermodynamics of the oxidative folding depends on both conformational and chemical components.

A key player in the thiol–disulfide exchange reaction is the ionized, thiolate form of the thiol group.[18] Thiol–disulfide exchange is a reversible, two-step reaction. In the first step, a nucleophilic thiolate reacts with a disulfide of an oxidizing reagent (such as DTT, GSSG, cystine) to form a mixed disulfide. This intermolecular part of the reaction depends on the reactivity of the cysteine thiolate. In the second, intramolecular step, the second peptide thiolate attacks the mixed disulfide bond, yielding the peptide disulfide bond and freeing the thiol of the oxidizing reagent. In this step, proximity of the second peptide thiolate (determined by peptide conformation and the sequence position of the Cys residue relative to the mixed disulfide) dictates the overall rate for forming the intramolecular disulfide bridge.

Thiol/disulfide exchange reactions of the cysteine thiols in peptides are influenced by neighboring amino acid residues if they carry an electric charge.[19] The presence of positively charged Lys/Arg residues increases the reactivity of adjacent thiols due to decrease of their pK_a (thus favoring a thiolate form at a neutral pH), whereas negatively charged Asp/Glu have the opposite effect. The thiol–disulfide exchange rates between the peptide thiolates and the negatively charged glutathione are significantly affected if the charged residues are neighboring Cys residues. Consequently, electrostatic interactions were shown to play a role in the early folding of BPTI.[20] The electrostatic effects on the thiol–disulfide exchange rates can be further modulated by a presence of organic cosolvents; coulombic interactions are favored in 50% methanol and at low ionic strength.[21] Although engineering the reactivity of the cysteine thiols *via* electrostatic interactions was validated,[22] there have been no studies reported on engineering oxidative folding of small peptides using this strategy.

6.1.2.2   Cysteine Patterns and Loop Sizes

A striking feature of Cys-rich peptides is a high conservation of cysteine patterns (cysteine pattern is defined here as a distribution of cysteine residues within a primary amino acid sequence). Since the conserved Cys patterns are in stark contrast to the hypervariability of primary amino acid sequences in the inter-cysteine loops, the appropriate positioning of cysteine residues might be considered a key factor determining the connectivities of the disulfide bridges. Indeed, dozens or even hundreds of primary amino acid sequences sharing identical Cys patterns fall into the same disulfide scaffold (disulfide scaffold is defined here as a three-dimensional topology of a peptide cross-linked by disulfide bridges). Furthermore, some scaffolds, like the inhibitory cystine knot

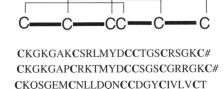

ω-MVIIA	CKGKGAKCSRLMYDCCTGSCRSGKC#
ω-MVIIC	CKGKGAPCRKTMYDCCSGSCGRRGKC#
δ-TxVIA	CKQSGEMCNLLDQNCCDGYCIVLVCT
δ-SVIE	EACSSGGTFCGIHPGLCCSEFCFLWCITFID

Figure 6.1.2   Distinct folding properties of conotoxins sharing identical cysteine patterns. All ω- and δ-conotoxins share an identical cysteine pattern that forms inhibitory cystine knot motif (ICK). ω-Conotoxin MVIIA folds very efficiently (yield 50%), as compared to ω-conotoxin MVIIC (yield 16%).[13,14] δ-Conotoxin SVIE folds with extremely low folding yield (less than 1%), whereas δ-TxVIA can form the native disulfide bridges with yield approximately 7%.[15,16]

(ICK) motif are highly conserved across plant and animal kingdoms.[23] Despite this conservation, folding mechanisms of the peptides sharing the ICK motif can significantly differ: for example comparing folding of ω-conotoxins MVIIA to MVIIC,[14] or folding of δ-conotoxins TxVIA to SVIE[15] (Figure 6.1.2). Although Cys patterns do play an important role in defining the native topology of the disulfide bridges, folding studies suggest the Cys patterns alone are insufficient to determine native disulfide bridges.

As shown in Figure 6.1.1B, identical Cys patterns can produce distinct connectivities of the native disulfide bridges. In one extreme case, the same Cys pattern found in the M-superfamily of conotoxins yielded three distinct disulfide scaffolds for μ1-, μ2- and μ4-conotoxins.[24,25] It has been speculated that differences in the inter-cysteine loop sizes could in part determine the connectivity of the native disulfide bridges. In the case of scorpion toxins belonging to the α-KTx6 family, mauritoxin has a distinct connectivity of the native disulfide bridges, compared to other members of the family (HsTx1 or Pi1 toxins); this difference could be accounted for by a presence of two proline residues in maurotoxin.[26] A combination of Cys patterns and some non-cysteine residues are dominant factors in forming the native disulfide bridges.

Identical Cys patterns yielding different native bridges were found in mammalian endothelin-1 and bee apamin. Interestingly, these two peptides not only have the same pattern of four cysteine residues, but also share an identical number of amino acid residues in the inter-cysteine loops. In endothelin-1, disulfide bonds are formed between Cys^I-Cys^{IV} and Cys^{II}-Cys^{III},[27] while in apamin, between Cys^I-Cys^{III} and Cys^{II}-Cys^{IV}.[28] The cysteine connectivities in apamin significantly contribute to the stability of the native peptide; the high stability of apamin is not affected by extreme pH, temperature or the addition of denaturant. Furthermore, synthetic apamin folds to almost 100% of correctly formed peptide,[29] whereas folding of synthetic endothelin-1 gives 75% of the native form.[27] The differences in the primary amino acid sequences can account for distinct connectivities of the native disulfide bridges. As illustrated in

Figure 6.1.1B, two families of cone snail toxins: α-conotoxin and χ-conotoxin have identical Cys patterns, but their native conformations are stabilized by a distinct framework of the disulfide bridges. Similarly, conotoxin ι-RXIA, which belongs to the I_1-superfamily, has an ICK motif with an additional disulfide bridge between Cys^V-Cys^{VIII},[30] whereas the same cysteine pattern, C-C-CC-CC-C-C, found in the spider toxin J-atracotoxin Hv1c contains different disulfide connectivities in which Cys^{III} and Cys^{IV} form a vicinal disulfide bond[31] (Figure 6.1.1B). In addition, α- and β-defensins from human and animals have six cysteine residues in a following pattern: C-C-C-C-CC, but the disulfide bridges in α-defensins are between cysteine Cys^I-Cys^{VI}, Cys^{II}-Cys^{IV} and Cys^{III}-Cys^V, while the pairing in β-defensins is Cys^I-Cys^V, Cys^{II}-Cys^{IV}, Cys^{III}-Cys^{VI}.[32,33] Thus, predicting disulfide bridging based on Cys patterns may leave a high level of uncertainty, in particular where there are differences in the size of inter-cysteine loops.

Formation of a disulfide bridge greatly depends on the number of the intervening amino acid residues. Zhang and Snyder defined rules governing the thermodynamics and kinetics of forming the disulfide bridges based on positions of Cys residues.[34,35] For model peptides containing two cysteines and a different number of intervening residues, the oxidation rates varied over a 50-fold range and were faster for shorter loops. On the other hand, equilibria for forming the disulfide bridges were dependent on whether there was an odd or even number of intervening amino acid residues. The same authors showed that the equilibria for forming the native disulfide bridges in two-disulfide bridged α-conotoxin were primarily determined by the position of the Cys residues, rather than by primary amino acid sequences of the non-cysteine residues. These rules applied to formation of the first disulfide bridge during early steps of folding conotoxin ω-MVIIA, where no preference for forming the native disulfide bridges was observed.[36] Many Cys patterns contain two vicinal cysteine residues forming a diad "-Cys-Cys-", but the vicinal disulfide bridge is rarely present in polypeptides.[31,37]

While the formation of the first disulfide bridge is largely governed by positions of Cys residues and loop sizes, the formation of the second bridge is significantly affected by the topology of the pre-existing disulfide bridge within the one-disulfide folding intermediate.[35,36] During formation of two-disulfide folding intermediates in MVIIA that exhibited so-called overlapping or enclosed loops, the pre-existing loop generally increased effective concentrations for forming the second disulfide bridge.[36] During folding of α-conotoxin GI, formation of the first disulfide bridge increases an effective concentration for forming the second bridge.[38] All of these studies strongly suggest that positions of cysteine residues and loop sizes play a major role in forming the disulfide bridges.

6.1.2.3 Amino Acid Sequences and Non-covalent Interactions

Anfinsen proposed that the native conformation of a polypeptide is encoded in its primary amino acid sequence. How much sequence information is utilized during the oxidative folding of small peptides, given the extreme hypervariability of their primary amino acid sequences? This structure–folding relationship paradox becomes apparent when analyzing alignments of sequences of conotoxins; an example is shown in Figure 6.1.1A. Thus, the key question is what role non-cysteine residues play in the formation of the native disulfide bridges.

Importance of non-cysteine residues in oxidative folding is evident from many studies. Human β-defensins, hBD1, hBD2 and hBD3, share the same spacing between cysteine residues, but only hBD3 could not correctly form the folded isomer under various oxidative conditions.[39] These differences could be accounted for by the positively charged residues in hBD3 (hBD1 net charge $+5$, hBD2 $+7$, hBD3 $+11$). Similarly, four conotoxins that share an identical disulfide scaffold, but significantly differed in the primary amino acid sequences exhibited distinct folding properties.[10] On the other hand, folding studies of various α-conotoxins (Figure 6.1.1A) indicated the overall folding yields in the range from 40–70%. As described in Section 6.1.2.2, apamin and endothelin are an excellent example where primary amino acid sequence, rather than Cys pattern is important for forming native disulfide bridges.

The role of individual non-cysteine residues in oxidative folding was established in many peptides. In minicollagen-1, N-terminal and C-terminal domains have identical cysteine patterns and identical loop sizes, but their disulfide frameworks are different.[40] Oxidative folding of the N-terminal domain Mcoll1-[1-33]-NH$_2$ leads to an accumulation of fully oxidized intermediates, in which the cysteine paring is Cys^I-Cys^{II}, Cys^{III}-Cys^{IV} and Cys^V-Cys^{VI}, followed by a rearrangement to the native form containing Cys^I-Cys^{IV}, Cys^{II}-Cys^{VI} and Cys^{III}-Cys^V connectivities. During this rearrangement, Pro24 plays an important role with a slow conversion from *trans* to *cis* isomer. On the other hand, the C-terminal domain Mcoll-[107-130] folds efficiently to the major isomer ($>95\%$) without forming any transient intermediates. Furthermore, the disulfide bond framework of the C-terminal domain (Cys^I-Cys^V, Cys^{II}-Cys^{IV}, Cys^{III}-Cys^{VI}) is different from that of the N-terminal domain, and is probably constrained by the *trans* conformation of the Val117-Pro118 bond while in the N-terminal domain the related Ala23-Pro24 bond is in the *cis* conformation. Structure-folding relationship studies on potato carboxypeptidase inhibitor (PCI) identified individual amino acid residues at the N- and C-termini that significantly affected folding yield.[41] In the trypsin inhibitor CMTI-I, the conservative replacement of Met8 by Leu dramatically changed the folding mechanism and yields.[42] In α-conotoxin ImI, replacement of the conserved Pro6 by Lys resulted in favoring an alternative disulfide scaffold.[43] The same authors showed the reverse effect of replacing Lys in MrIA by Pro: this substitution significantly favored α-conotoxin scaffold.[43] Replacement of two Pro residues in maurotoxin resulted in a change of the disulfide scaffold to one characteristic of other scorpion toxins.[26]

In late stage folding, non-covalent interactions most likely contribute to the formation of the native disulfide bridges. Folding of several peptides appears to be sensitive to the presence of denaturant. For example, the folding yield of ω-MVIIA in the presence of 8 M urea was reduced, likely by disrupting hydrophobic effects.[14] Similarly, later stages of folding of the inhibitor PCI were sensitive to denaturants, such as urea and guanidinium chloride.[44] The denaturing solvent influenced the accumulation of the correctly folded isomer of α-conotoxin GI (folding yield decreased from 71% to 47%).[35] Furthermore, the addition of 4 M urea to the folding mixture of μ-conotoxin GIIIA shows a negative effect on the accumulation of the native form of this peptide.[10] However, the oxidative folding of other peptides, such as SmIIIA or apamin, is insensitive to the presence of denaturing conditions.[10] Similarly, in folding of the short two-disulfide bridged peptides ImI or MrIA denaturants did not affect the formation of the native disulfide bridges, suggesting no major role of non-covalent interactions in the forming of the native disulfide bridges.[43,45]

6.1.2.4   A Case Study – Folding of ω-Conotoxin MVIIA

Studies on the oxidative folding of ω-conotoxin MVIIA have provided insight into how folding is governed by primary amino acid sequence, non-covalent interactions and pre-formed disulfide bridges in the folding inter-mediates.[14,36,46,47] The early folding events were determined by the Cys patterns, but the later folding stages were determined by non-covalent interactions. MVIIA was shown to fold with relatively high yields in the presence of oxidized and reduced glutathione.[14] ω-Conotoxin MVIID, which had identical loop sizes to MVIIA, folded with lower yield, whereas other ω-conotoxins with different loop sizes folded with higher yields. These data suggest that the Cys pattern alone does not determine the folding yield.

In the early stages of folding, there is no preference in forming any of the native disulfide bonds from the unfolded conformation.[47] Equilibrium constants for forming individual disulfide bridges were measured and those confirmed no preference in forming native *vs.* non-native disulfide bridges.[36] The effective concentrations of the Cys pairs forming the disulfide bonds were largely dependent on the spacing between the cysteine residues, and were independent of denaturing conditions. Two-disulfide-bridged folding intermediates were found to be very unstable.[47] Once the two native disulfide bonds were formed, however, there was a high degree of cooperativity in forming the third native disulfide bond: the effective concentration for forming any of the third native disulfide bonds was denaturant-sensitive and increased several orders of magnitude. The presence of the C-terminal glycine also significantly improved the folding yield by providing stabilizing interactions within the native conformation of the peptide.[46,48] The folding yield for the MVIIA was sensitive to urea, further confirming the role of non-covalent interactions in the formation of native disulfide bridges.[14]

6.1.2.5   Role of Post-translational Modifications

Many Cys-rich peptides contain various post-translational modifications; the best known examples are the conotoxins and the cyclotides. The most abundant posttranslational modification in conotoxins is C-terminal amidation. The role of this modification in oxidative folding was studied in conotoxin ω-MVIIA (C-terminus amidated) and its precursor form, containing an additional C-terminal glycine residue with a free carboxyl group.[46] The C-terminal glycine improved folding yield from 50% (in the C-terminal amidated ω-MVIIA) to 85%. In contrast, conotoxin α-ImI with C-terminal glycine or the free carboxylic acid led to lower folding yields than the C-terminally amidated conotoxin α-ImI. Nevertheless, there were no significant structural differences between these α-ImI analogs, as judged from NMR spectra.[49] The role of C-terminal amidation in folding was also evaluated for μ-conotoxin SmIIIA, but the results indicated that C-terminal amidation did not influence folding of SmIIIA.[10] Another post-translational modification, γ-carboxylation of glutamate (Gla) in the conotoxin tx9a, accelerated formation of disulfide bonds in the presence of Ca^{2+}.[50] The authors suggested that both Ca^{2+} and γ-carboxyglutamate residues may stabilize the correctly folded conotoxin, and/or hinder the formation of non-native disulfide bridges.

In cyclotides, the cyclic cystine knot (CCK) results from a head-to-tail cyclic backbone (N-to-C cyclization).[51] In this motif, the disulfide bonds between Cys^I-Cys^{IV} and Cys^{II}-Cys^V form an embedded ring, which is threaded by a third Cys^{III}-Cys^{IV} disulfide bridge. *In vitro* oxidative folding of cyclotides can be carried out using two different strategies: (1) oxidation prior to cyclization and (2) cyclization prior to oxidation.[52] The second strategy appeared significantly better, providing high yields of the correctly folded cyclotide kalata B1. For the first strategy, the presence of organic cosolvents improved the folding yields. The mechanism of the oxidative folding of cyclotides has been extensively studied, and an excellent review on this topic was recently published.[53]

6.1.2.6 Oxidative Folding Conditions – Practical Considerations

Formation of the native disulfide bridges in peptides is often sensitive to oxidative folding conditions.[54] Changing the cysteine oxidation methods (air oxidation, redox buffers, DMSO) may significantly improve final folding yields. Factors such as temperature, pH and buffer composition were also shown to affect the formation of the native disulfide bridges.[55] Table 6.1.2 provides several examples of oxidative folding conditions. The optimization of the oxidative folding is an empirical process. For that reason, applying various folding additives or immobilized systems to change the ionic strength or temperature remains a good approach to improve folding yields.

Organic cosolvents appeared very effective in improving the oxidative folding of small peptides. As described earlier, the oxidative folding of cyclotide kalata B1 carried out in the presence of acetonitrile, methanol, ethanol or isopropanol improved final folding yields.[52] Similarly, the presence of organic cosolvents increased the folding yield of conotoxins ω-TxVII, μO-MrVIB or α-ImI.[45,58,59] Adding non-ionic detergents, such as Tween, to the folding mixture was very

Table 6.1.2   Examples of oxidative conditions used for folding of disulfide-rich peptides.

Peptide	Folding conditions	Yield
Apamin	Peptide was dissolved in 20 mM Tris-HCl, pH 8 and air-oxidized, at RT, 72 hrs.[56]	~100%
Guanylin	Reduced peptide was dissolved in 100 mM ammonium bicarbonate, 2 mM EDTA, pH 8.3 and air-oxidized, at 20 °C.[57]	5%
Endothelin-1	Peptide was resuspended in 0.1% aqueous ammonia containing 8 M urea and subjected to air oxidation, at RT, 3 hrs.[27]	75%
α-GI	Reduced peptide was oxidized with solvents containing 20 mM Tris-HCl, 2 mM EDTA at pH 8.5, air oxidation, 2 hrs.[35]	71%
ω-MVIIA	Peptide was dissolved in 10 mM HCl (final concentration 40 μM), and mixed with an equal volume of 0.2 M MOPS, pH 7.3, 0.4 M KCl, 2 mM EDTA, 1 mM GSSG, 2 mM GSH, at RT, 90 min.[14]	50%
μ-GIIIA	Reduced peptide was resuspended in 0.01% TFA (v/v) and added to the folding mixture containing 0.1 M Tris-HCl, buffer pH 7.5, 0.1 mM EDTA, and mixture of 1 mM GSH and 1 mM GSSG, at RT, 1 hr.[10]	30%
CMTI-I	Peptide was dissolved in 0.1 M Tris-HCl, pH 8.7, 0.2 M KCl, 1 Mm EDTA and air-oxidized, at RT, overnight.[42]	~100%
Kalata B1	Linear peptide was dissolved in 0.1 M ammonium bicarbonate, (pH 7.8, 8.5, 9) and mixed with 1 mM GSH, at RT, 1–5 days.	2%
	Above conditions and in the presence of 50% 2-propanol.[52]	28%

efficient in improving the yield of very hydrophobic peptides such as δ-cono-toxins SVIE or PVIA.[15,16] Increasing ionic strength enhanced the folding yields for μ-SmIIIA as well as ω-MVIIC.[10,13] The oxidative folding of the latter peptide was even better in the presence of high salt concentration and low temperature.[13] Oxidative folding in immobilized systems is a relatively new concept.[17] When scorpion toxin Cn5 was oxidized in the presence of immobilized mini-chaperone (a short fragment of GroEL chaperone) the folding yields increased dramatically.[17] Studies by Barany and Darlak[60,61] suggested that oxidative folding in the presence of immobilized Ellman's reagent (commercial name Clear-Ox™) provides a beneficial pseudo-dilution effect. This promotes the formation of intramolecular, rather than intermolecular, disulfide bridges; the latter led to aggregation and precipitation. Indeed, when Clear-Ox™ was applied to the oxidative folding of two- and three-disulfide bridged conotoxins (α-GI and μ-PIIIA) the folding yields were significantly higher at increased peptide concentrations.[62] This and other folding strategies are important factors for manufacturing disulfide-rich peptides on a commercial scale.

6.1.3 Biosynthetic Aspects of the Oxidative Folding

The relatively detailed understanding of the mechanism of *in vitro* oxidative folding is in stark contrast to our poor knowledge of how this process may occur during biosynthesis in a cell. Several important factors that affect the formation of the native disulfide bridges include: (1) the presence of precursor sequences, (2) folding catalysts and molecular chaperones, (3) a dynamic chemical environment and (4) macromolecular crowding. Most of the experiments relevant to biosynthetic oxidative folding focused on elucidating the role of propeptide sequences and protein disulfide isomerase in the formation of the disulfide bonds.

6.1.3.1 Precursor Sequences

As molecules targeted for secretion, disulfide-rich peptides are biosynthesized as larger precursors containing the mature peptide sequence and, at the minimum, an N-terminal signal sequence. Additionally, a propeptide region may be present that is N- or C-terminally located, relative to the mature sequence. Thus, *in vivo* oxidative folding occurs as the precursor is being post-translationally processed. While propeptides play a key role in the folding of larger proteins, it is still unknown whether any precursor sequences are important for oxidative folding of smaller, Cys-rich peptides. The relevant studies on evaluating the role of propeptides are summarized in Table 6.1.3. These studies showed that propeptides either directly facilitated or were important for the formation of the native disulfide framework. Based on a chaperone role of prosequences in folding of larger proteins, Buczek *et al.* proposed a similar classification for small, disulfide-rich peptides.[63] Thus, prosequences that acted as intramolecular chaperones belong to class I propeptides, whereas those that did not directly affect oxidative folding were called class II propeptides.

Class I precursor sequences appear critical for forming native disulfide bridges in several peptides, including proguanylin. For this two-disulfide-containing peptide, the folding yields increased from 7% to 95% in the presence of the N-terminal propeptide.[57] Similar studies on the 24-residue-long guanylyl cyclase-activating peptide, GCAP-II, showed that two amino acids in the N-terminal fragment of the propeptide were directly involved in a chaperone-like function during folding.[64,65] In endothelin-1, the N-terminal addition of two positively charged residues that are present in the precursor sequence dramatically improved folding yield; this effect was accounted for by electrostatic interactions between the propeptide fragment and the mature peptide sequence, as examined by NMR and molecular dynamics simulations.[66] The role of precursor sequences in oxidative folding was also studied in the case of macrophage inhibitory cytokine-1, MIC-1 and nerve growth factor, hNGF.[67,68] For oxidative folding of BPTI, the propeptide substantially increased folding yields and improved folding kinetics through an additional N-terminal cysteine residue naturally present in the propeptide fragment.[69] There have been no studies reported that evaluate how class I propeptides may

Table 6.1.3 Role of propeptide in oxidative folding of disulfide-rich peptides. Modified from reference 63.

Name	Propeptide	Mature peptide
Class I: Propeptides as intramolecular chaperones		
Proguanylin	VTVQDGNFSFSLESVK-KLKDLQEPQEPRVGKLR-NFAPIPGEPVVPILCSNPN-FPEELKPLCKEPNAQE-ILQRLEEIIAED	PGTCEICAYAACTGC
Pro-GCAP-II	VYIQYQGFRVQLESMK-KLSDLEAQWAPSPRL-QAQSLLPAVCHH-PALPQDLQPVCAS-QEASSI	FKTLRTIANDDCELCVN-VACTGCL
Pro-defensin	EPLQARADEVAAAPEQIA-ADIPEVVVSLAWDESL-APKHPGSRKNM	ACYCRIPACIAGERRYG-TCIYQGRLWWAFCC
Pro-BPTI	TPGCDTSNQAKAQ	RPDFCLEPPYTGPCK-ARIIRYFYNAKAGLCQTF-VYGGCRAKRNNFKSAE-DCMRTCGGA
Class II: Propeptides without apparent chaperone activity		
Pro-MVIIA	DDSRGTQKHRALRST-TKLSTSTR	CKGKGAKCSRLMYDC-CTGSCRSGKC
Pro-GI	FPSERASDGRDDTAKD-EGSDMEKLVEKK	ECCNPACGRHYSC
Pro-PVIA	DDSKNGLENHFWKARDE-MKNREASKLDKK	EACYAOGTFCGI-KOGLCCSEFCLP-GVCFG
Pro-PCI	AHDNSFYSTKIHVMAQD-VVLPTVTKLF	EQHADPICNKPCKTH-DDCSGAWFCQACWNSA-RTCGPYVC

improve the oxidative folding of those Cys-rich peptides that naturally contain class II propeptides (such as conotoxins).

For disulfide-rich peptides containing class II propeptides, neither folding kinetics nor thermodynamics was significantly affected by the precursor sequences, as judged by studies on three conotoxins MVIIA, GI, PVIA and the inhibitor PCI.[9,46,63,70] Folding yields of conotoxin precursor pro-GI and the mature toxin GI were comparable under a variety of folding conditions.[9] A potential chaperone role of a conotoxin propeptide was rigorously tested using the highly hydrophobic δ-conotoxin PVIA.[63] This peptide is prone to aggregation and was previously shown to fold with extremely low yields. The 58-membered precursor of PVIA, pro-PVIA, was synthesized using native chemical ligation. However, the folding yield for the pro-PVIA was also very low and comparable for the precursor and the mature toxin δ-PVIA. Despite a lack of intramolecular chaperone activity of the class II propeptides, these sequences may play an important role in the oxidative folding catalyzed by protein disulfide isomerase, and perhaps other folding catalysts and molecular chaperones.

6.1.3.2 Protein Disulfide Isomerase

PDI is an ER-resident oxidoreductase that catalyzes *in vivo* oxidative folding of polypeptides (see Chapter 1.5). This enzyme was shown to be sufficient for promoting oxidative folding, even in the absence of glutathione. PDI plays three important functions in folding: (1) it promotes disulfide-bond formation, (2) it catalyzes the rearrangement of disulfide bridges and (3) it provides a chaperone function and minimizes aggregation. The multimodality in function of PDI was studied in detail in the oxidative folding of proinsulin and cyclotides.[71,72] PDI's role in the formation of the native disulfide bridges in small peptides was examined in greater detail in the oxidative folding of cyclotides and conotoxins.

PDI is a key enzyme that catalyzes the oxidative folding of Cys-rich peptides. Interestingly, PDIs from organisms that specialize in producing disulfide-rich peptides have recently been discovered; *Conus* PDI and OaPDI were cloned and characterized from cone snails and the *Rubiacea* plant family that produces cyclotides.[50,72,73] Strikingly, no reports on PDIs from spiders or scorpions, organisms that both evolved millions of Cys-rich peptides, exist. In plants that express cyclotides, OaPDI appeared to exhibit similar isomerases with better chaperone activity, when compared to human PDI. Furthermore, OaPDI significantly improved the oxidative folding of kalata B1 by increasing the folding yield by at least one order of magnitude (as mentioned earlier, this peptide does not fold with high yields and requires organic co-solvents to increase folding yields). The first functional characterization of *Conus* PDI suggested that this enzyme has similar properties to human PDI with respect to oxidation, reduction and isomerization activities.

The mechanism of the PDI-catalyzed folding of small Cys-rich peptides was investigated for several peptides including proinsulin, maurotoxin, conotoxins and cyclotides. PDI increased folding rates (conotoxins GIIIA or tx3a, maurotoxin),[10,73,74] and in some cases also folding yields (cyclotides, proinsulin).[71,72] Interestingly, PDI was more efficient in catalyzing oxidative folding of conotoxin GI when covalently linked to the N-terminal propeptide fragment.[9] As illustrated in Figure 6.1.3, the appearance of correctly folded GI was approximately three times faster for the precursor compared to the mature toxin. Based on this work, it was suggested that class II propeptides modulate substrate properties of the Cys-rich peptides for PDI-assisted folding. The role of propeptides in the modulation of substrate specificity and kinetics for PDI-catalyzed folding requires further detailed studies.

6.1.3.3 Macromolecular Crowding

Since the cellular milieu is very crowded, oxidative folding must occur in the context of: (1) multiple *intermolecular* interactions with ER-resident

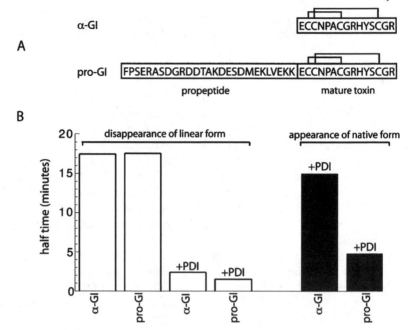

Figure 6.1.3 Effects of propeptide and PDI on the oxidative folding of α-conotoxin GI. A) Structure and disulfide scaffold of the α-GI propeptide and mature toxin. The mature α-GI is amidated at the C-terminus. B) Folding of α-GI and pro-GI PDI-catalyzed and uncatalyzed. Bar graphs reprinted with permission from reference 9.

macromolecules, and (2) the excluded volume effect caused by an extremely high concentration of macromolecules that typically exceed $100\,mg\,mL^{-1}$. Oxidative folding in the presence of macromolecular crowding agents was previously examined in great detail for the lysozyme. These studies suggest that crowding promotes aggregation and decreases folding yields; as well as that PDI acts as a chaperone that prevents crowding-induced aggregation.[75,76] Theoretical calculations predicted that the excluded volume should affect the unfolded conformation of Cys-rich peptides as short as 25 residues.[77] Interestingly, when these predictions were tested in a folding experiment, no apparent excluded volume effects were observed during the oxidative folding of three conotoxins, ranging in size from 13 to 46 amino acid residues.[78] This suggested that peptides of that size are not prone to aggregation. In the same study, the authors noted that the commonly used macromolecular crowding agent, bovine serum albumin, promoted the oxidative folding of the peptides *via* a thiol–disulfide exchange mechanism. Based on these observations, macromolecular crowding does not affect PDI-assisted folding of small Cys-rich peptides. More folding and structural studies are needed to evaluate how crowding may affect early events in forming disulfide bridges in these peptides.

6.1.3.4 Oxidative Folding in the Endoplasmic Reticulum

Formation of the disulfide bonds during oxidative folding in the ER has been well characterized (see Chapters 1.5 and 1.6).[79,80] However, little is known about how small, disulfide-rich peptides form the native disulfide bridges in the ER. Understudied aspects of the *in vivo* folding are how molecular chaperones and folding catalysts affect the formation of the native disulfide bridges, and how much energy is utilized during the oxidative folding of thermodynamically less stable peptides (such as δ-conotoxins). Since ATP was demonstrated to be required for the oxidative folding of larger proteins,[81] it is likely that the formation of native disulfide bridges in smaller peptides is energy-dependent. Furthermore, ATP-dependent oxidative folding catalyzed by BiP and PDI was shown for antibodies.[82]

The role of molecular chaperones and folding catalysts in promoting the oxidative folding of short peptides is not well understood. It is unknown whether additional chaperones and folding catalysts, excluding PDI, are involved in the formation of native disulfide bridges in peptides. For example, the oxidative folding of maurotoxin was most efficient in the presence of combined PDI and PPI.[74] Another unknown aspect is how energy landscapes for the *in vitro* and *in vivo* oxidative folding of small, disulfide-rich peptides differ from each other. This issue is exemplified in Figure 6.1.4, where pro-peptide, folding enzymes and chaperones modulate the reactivity of cysteine thiols and non-covalent interactions of a nascent polypeptide. Differences between *in vitro* and *in vivo* folding were discussed for larger proteins,[83] but never directly addressed for small Cys-rich peptides.

With regard to the oxidative folding of short peptides, it is unclear how the ER quality control system can discriminate between the native and misfolded conformations. What are key structural features of the misfolded conforma-

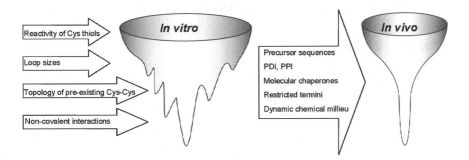

Figure 6.1.4 Differences between the *in vitro* and *in vivo* oxidative folding of small peptides. Key factors affecting the *in vitro* folding are shown. How cellular factors may affect the *in vivo* folding of these peptides remains largely unexplored. A provocative question exemplified by the folding funnels is whether peptides that fold *in vitro* using the "collapse" mechanism can undergo the oxidative folding in the cell by a framework mechanism.

tions that are recognized by the ERAD proteins? Studies from the Wittrup group suggest that the thermodynamic stability of the Cys-rich polypeptides (BPTI) correlated with the secretion efficiency and was coupled to the ER quality control.[84,85] It appears the ER-based quality control must determine the fate of a folded species, directing it to either secretion or degradation.

6.1.4   Conclusions and Outlook

Numerous studies defined basic principles for forming the native disulfide bridges in small, Cys-rich peptides. At the early stages of folding, the key factors are the reactivity of the cysteine thiols and the cysteine patterns, whereas, at later stages, non-covalent interactions may play an increasingly important role. Precursor sequences can either act directly as intramolecular chaperones, or facilitate PDI-assisted folding. A dual role for PDI, both as a folding catalyst and as a molecular chaperone has been established, but how this enzyme affects the oxidative folding of small peptides in the ER milieu is not fully understood. Thus, despite identifying key intra- and intermolecular factors that direct the oxidative folding, how this process occurs in the cell will remain a subject for many future studies.

Cysteine-rich peptides continue to be an attractive source of current and future therapeutics;[86] oxidative folding will remain an important aspect of their discovery and subsequent structure-activity-relationship studies. There is a continuing need to improve the folding strategies that can be applied both for research material as well as on a commercial manufacturing scale. Novel immobilized folding strategies may be developed based on the inspiration from studying *in vivo* folding. For example, a combination of artificial chaperones and folding catalyst mimetics acting in the iterative annealing manner may one day revolutionize the synthesis of cysteine-rich peptides. Until then, all incremental improvements of the oxidative folding methods introduced by chemists and biochemists are important for studying disulfide-rich peptides.

Acknowledgements

We would like to thank Professor Baldomero M. Olivera for his continuous support of the conotoxin folding research, including NIH Program Project Grant 48677. GB acknowledges support from the University of Utah Startup Funds and R21 grant NS055845. We thank Ashley Chadwick for her invaluable help with the manuscript. We would like to apologize to many investigators whose studies on the oxidative folding of peptides have not been acknowledged or referenced due to space constraints of this monograph.

References

1. M. Watkins, D. R. Hillyard and B. M. Olivera, *J. Mol. Evol.*, 2006, **62**, 247–256.

2. K. A. Dill, S. B. Ozkan, T. R. Weikl, J. D. Chodera and V. A. Voelz, *Curr. Opin. Struct. Biol.*, 2007, **17**, 342–346.
3. V. Daggett and A. Fersht, *Nat. Rev. Mol. Cell Biol.*, 2003, **4**, 497–502.
4. W. J. Wedemeyer, E. Welker, M. Narayan and H. A. Scheraga, *Biochemistry*, 2000, **39**, 4207–4216.
5. E. Welker, W. J. Wedemeyer, M. Narayan and H. A. Scheraga, *Biochemistry*, 2001, **40**, 9059–9064.
6. D. P. Goldenberg, *Trends Biochem. Sci.*, 1992, **17**, 257–261.
7. G. Bulaj and D. P. Goldenberg, *Nat. Struct. Biol.*, 2001, **8**, 326–330.
8. J. L. Arolas, F. X. Aviles, J. Y. Chang and S. Ventura, *Trends Biochem. Sci.*, 2006, **31**, 292–301.
9. O. Buczek, B. M. Olivera and G. Bulaj, *Biochemistry*, 2004, **43**, 1093–1101.
10. E. Fuller, B. R. Green, P. Catlin, O. Buczek, J. S. Nielsen, B. M. Olivera and G. Bulaj, *FEBS J.*, 2005, **272**, 1727–1738.
11. B. R. Green, P. Catlin, M. M. Zhang, B. Fiedler, W. Bayudan, A. Morrison, R. S. Norton, B. J. Smith, D. Yoshikami, B. M. Olivera and G. Bulaj, *Chem. Biol.*, 2007, **14**, 399–407.
12. G. Bulaj and B. M. Olivera, *Antioxid. Redox Signal.*, 2008, **10**, 141–156.
13. S. Kubo, N. Chino, T. Kimura and S. Sakakibara, *Biopolymers*, 1996, **38**, 733–744.
14. M. Price-Carter, W. R. Gray and D. P. Goldenberg, *Biochemistry*, 1996, **35**, 15537–15546.
15. G. Bulaj, R. DeLaCruz, A. Azimi-Zonooz, P. West, M. Watkins, D. Yoshikami and B. M. Olivera, *Biochemistry*, 2001, **40**, 13201–13208.
16. R. DeLa Cruz, F. G. Whitby, O. Buczek and G. Bulaj, *J. Pept. Res.*, 2003, **61**, 202–212.
17. M. M. Altamirano, C. Garcia, L. D. Possani and A. R. Fersht, *Nat. Biotechnol.*, 1999, **17**, 187–191.
18. H. F. Gilbert, *Adv. Enzymol. Relat. Areas Mol. Biol.*, 1990, **63**, 69–172.
19. G. Bulaj, T. Kortemme and D. P. Goldenberg, *Biochemistry*, 1998, **37**, 8965–8972.
20. G. Bulaj and D. P. Goldenberg, *Protein Sci.*, 1999, **8**, 1825–1842.
21. G. H. Snyder, *J. Biol. Chem.*, 1984, **259**, 7468–7472.
22. R. E. Hansen, H. Ostergaard and J. R. Winther, *Biochemistry*, 2005, **44**, 5899–5906.
23. P. K. Pallaghy, K. J. Nielsen, D. J. Craik and R. S. Norton, *Protein Sci.*, 1994, **3**, 1833–1839.
24. G. P. Corpuz, R. B. Jacobsen, E. C. Jimenez, M. Watkins, C. Walker, C. Colledge, J. E. Garrett, O. McDougal, W. Li, W. R. Gray, D. R. Hillyard, J. Rivier, J. M. McIntosh, L. J. Cruz and B. M. Olivera, *Biochemistry*, 2005, **44**, 8176–8186.
25. Y. H. Han, Q. Wang, H. Jiang, L. Liu, C. Xiao, D. D. Yuan, X. X. Shao, Q. Y. Dai, J. S. Cheng and C. W. Chi, *FEBS J.*, 2006, **273**, 4972–4982.
26. E. Carlier, Z. Fajloun, P. Mansuelle, M. Fathallah, A. Mosbah, R. Oughideni, G. Sandoz, E. Di Luccio, S. Geib, I. Regaya, J. Brocard, H.

Rochat, H. Darbon, C. Devaux, J. M. Sabatier and M. de Waard, *FEBS Lett.*, 2001, **489**, 202–207.

27. S. Kumagaye, H. Kuroda, K. Nakajima, T. X. Watanabe, T. Kimura, T. Masaki and S. Sakakibara, *Int. J. Pept. Protein Res.*, 1988, **32**, 519–526.
28. J. H. Pease and D. E. Wemmer, *Biochemistry*, 1988, **27**, 8491–8498.
29. M. H. Chau and J. W. Nelson, *Biochemistry*, 1992, **31**, 4445–4450.
30. O. Buczek, D. Wei, J. J. Babon, X. Yang, B. Fiedler, P. Chen, D. Yoshikami, B. M. Olivera, G. Bulaj and R. S. Norton, *Biochemistry*, 2007, **46**, 9929–9940.
31. X. Wang, M. Connor, R. Smith, M. W. Maciejewski, M. E. Howden, G. M. Nicholson, M. J. Christie and G. F. King, *Nat. Struct. Biol.*, 2000, **7**, 505–513.
32. Y. Q. Tang and M. E. Selsted, *J. Biol. Chem.*, 1993, **268**, 6649–6653.
33. R. I. Lehrer, A. K. Lichtenstein and T. Ganz, *Annu. Rev. Immunol.*, 1993, **11**, 105–128.
34. R. M. Zhang and G. H. Snyder, *J. Biol. Chem.*, 1989, **264**, 18472–18479.
35. R. M. Zhang and G. H. Snyder, *Biochemistry*, 1991, **30**, 11343–11348.
36. M. Price-Carter, G. Bulaj and D. P. Goldenberg, *Biochemistry*, 2002, **41**, 3507–3519.
37. O. Carugo, M. Čemažar, S. Zahariev, I. Hudáky, Z. Gáspári, A. Perczel and S. Pongor, *Protein Eng.*, 2003, **16**, 637–639.
38. A. Kaerner and D. L. Rabenstein, *Biochemistry*, 1999, **38**, 5459–5470.
39. Z. Wu, D. M. Hoover, D. Yang, C. Boulègue, F. Santamaria, J. J. Oppenheim, J. Lubkowski and W. Lu, *Proc. Natl. Acad. Sci. USA*, 2003, **100**, 8880–8885.
40. C. Boulègue, A. G. Milbradt, C. Renner and L. Moroder, *J. Mol. Biol.*, 2006, **358**, 846–856.
41. G. Venhudova, F. Canals, E. Querol and F. X. Aviles, *J. Biol. Chem.*, 2001, **276**, 11683–11690.
42. I. Zhukov, L. Jaroszewski and A. Bierzynski, *Protein Sci.*, 2000, **9**, 273–279.
43. T. S. Kang, Z. Radic, T. T. Talley, S. D. Jois, P. Taylor and R. M. Kini, *Biochemistry*, 2007, **46**, 3338–3355.
44. J. Y. Chang, F. Canals, P. Schindler, E. Querol and F. X. Aviles, *J. Biol. Chem.*, 1994, **269**, 22087–22094.
45. J. S. Nielsen, P. Buczek and G. Bulaj, *J. Pept. Sci.*, 2004, **10**, 249–256.
46. M. Price-Carter, W. R. Gray and D. P. Goldenberg, *Biochemistry*, 1996, **35**, 15547–15557.
47. M. Price-Carter, M. S. Hull and D. P. Goldenberg, *Biochemistry*, 1998, **37**, 9851–9861.
48. D. P. Goldenberg, R. E. Koehn, D. E. Gilbert and G. Wagner, *Protein Sci.*, 2001, **10**, 538–550.
49. T. S. Kang, S. Vivekanandan, S. D. Jois and R. M. Kini, *Angew. Chem. Int. Ed. Engl.*, 2005, **44**, 6333–6337.
50. G. Bulaj, O. Buczek, I. Goodsell, E. C. Jimenez, J. Kranski, J. S. Nielsen, J. E. Garrett and B. M. Olivera, *Proc. Natl. Acad. Sci. USA*, 2003, **100, S2**, 14562–14568.

51. D. J. Craik, N. L. Daly, J. Mulvenna, M. R. Plan and M. Trabi, *Curr. Protein Pept. Sci.*, 2004, **5**, 297–315.
52. N. L. Daly, S. Love, P. F. Alewood and D. J. Craik, *Biochemistry*, 1999, **38**, 10606–10614.
53. M. Čemažar, C. W. Gruber and D. J. Craik, *Antioxid. Redox Signal.*, 2008, **10**, 103–112.
54. G. Bulaj, *Biotechnol. Adv.*, 2005, **23**, 87–92.
55. T. Kimura, in *Houhen Weyl, Methods of Organic Chemistry, Synthesis of Peptides and Peptidomimetics*, M. Goodman, A. Felix, L. Moroder and C. Toniolo, eds., Georg Thieme Verlag, Stuttgart, 2002, Vol. E22b, pp. 142–161.
56. B. F. Volkman and D. E. Wemmer, *Biopolymers*, 1997, **41**, 451–460.
57. A. Schulz, U. C. Marx, Y. Hidaka, Y. Shimonishi, P. Rosch, W. G. Forssmann and K. Adermann, *Protein Sci.*, 1999, **8**, 1850–1859.
58. T. Sasaki, Z. P. Feng, R. Scott, N. Grigoriev, N. I. Syed, M. Fainzilber and K. Sato, *Biochemistry*, 1999, **38**, 12876–12884.
59. G. Bulaj, M. M. Zhang, B. R. Green, B. Fiedler, R. T. Layer, S. Wei, J. S. Nielsen, S. J. Low, B. D. Klein, J. D. Wagstaff, L. Chicoine, T. P. Harty, H. Terlau, D. Yoshikami and B. M. Olivera, *Biochemistry*, 2006, **45**, 7404–7414.
60. I. Annis, L. Chen and G. Barany, *J. Am. Chem. Soc.*, 1998, **120**, 7226–7238.
61. K. Darlak, D. Wiegandt Long, A. Czerwinski, M. Darlak, F. Valenzuela, A. F. Spatola and G. Barany, *J. Pept. Res.*, 2004, **63**, 303–312.
62. B. R. Green and G. Bulaj, *Protein Pept. Lett.*, 2006, **13**, 67–70.
63. P. Buczek, O. Buczek and G. Bulaj, *Biopolymers*, 2005, **80**, 50–57.
64. Y. Hidaka, M. Ohno, B. Hemmasi, O. Hill, W. G. Forssmann and Y. Shimonishi, *Biochemistry*, 1998, **37**, 8498–8507.
65. Y. Hidaka, C. Shimono, M. Ohno, N. Okumura, K. Adermann, W. G. Forssmann and Y. Shimonishi, *J. Biol. Chem.*, 2000, **275**, 25155–25162.
66. A. Aumelas, L. Chiche, S. Kubo, N. Chino, H. Tamaoki and Y. Kobayashi, *Biochemistry*, 1995, **34**, 4546–4561.
67. A. R. Bauskin, H. P. Zhang, W. D. Fairlie, X. Y. He, P. K. Russell, A. G. Moore, D. A. Brown, K. K. Stanley and S. N. Breit, *EMBO J.*, 2000, **19**, 2212–2220.
68. A. Rattenholl, M. Ruoppolo, A. Flagiello, M. Monti, F. Vinci, G. Marino, H. Lilie, E. Schwarz and R. Rudolph, *J. Mol. Biol.*, 2001, **305**, 523–533.
69. J. S. Weissman and P. S. Kim, *Cell*, 1992, **71**, 841–851.
70. S. Bronsoms, J. Villanueva, F. Canals, E. Querol and F. X. Aviles, *Eur. J. Biochem.*, 2003, **270**, 3641–3650.
71. J. Winter, P. Klappa, R. B. Freedman, H. Lilie and R. Rudolph, *J. Biol. Chem.*, 2002, **277**, 310–317.
72. C. W. Gruber, M. Čemažar, R. J. Clark, T. Horibe, R. F. Renda, M. A. Anderson and D. J. Craik, *J. Biol. Chem.*, 2007, **282**, 20435–20446.
73. Z. Q. Wang, Y. H. Han, X. X. Shao, C. W. Chi and Z. Y. Guo, *FEBS J.*, 2007, **274**, 4778–4787.

74. E. di Luccio, D. O. Azulay, I. Regaya, Z. Fajloun, G. Sandoz, P. Mansuelle, R. Kharrat, M. Fathallah, L. Carrega, E. Esteve, H. Rochat, M. De Waard and J. M. Sabatier, *Biochem. J.*, 2001, **358**, 681–692.
75. B. van den Berg, R. J. Ellis and C. M. Dobson, *EMBO J.*, 1999, **18**, 6927–6933.
76. B. van den Berg, R. Wain, C. M. Dobson and R. J. Ellis, *EMBO J.*, 2000, **19**, 3870–3875.
77. D. P. Goldenberg, *J. Mol. Biol.*, 2003, **326**, 1615–1633.
78. O. Buczek, B. R. Green and G. Bulaj, *Biopolymers*, 2007, **88**, 8–19.
79. B. P. Tu, S. C. Ho-Schleyer, K. J. Travers and J. S. Weissman, *Science*, 2000, **290**, 1571–1574.
80. B. P. Tu and J. S. Weissman, *J. Cell. Biol.*, 2004, **164**, 341–346.
81. A. Mirazimi and L. Svensson, *J. Virol.*, 2000, **74**, 8048–8052.
82. M. Mayer, U. Kies, R. Kammermeier and J. Buchner, *J. Biol. Chem.*, 2000, **275**, 29421–29425.
83. P. L. Clark, *Trends Biochem. Sci.*, 2004, **29**, 527–534.
84. J. M. Kowalski, R. N. Parekh, J. Mao and K. D. Wittrup, *J. Biol. Chem.*, 1998, **273**, 19453–19458.
85. J. M. Kowalski, R. N. Parekh and K. D. Wittrup, *Biochemistry*, 1998, **37**, 1264–1273.
86. C. E. Beeton, G. A. Gutman and K. G. Chandy, in *Handbook of Biologically Active Peptides*, A. J. Kastin, ed., Academic Press, 2005, pp. 403–414.

CHAPTER 6.2

Regioselective Disulfide Formation

KNUT ADERMANN[a] AND KLEOMENIS BARLOS[b]

[a] VIRO Pharmaceuticals GmbH & Co. KG, Feodor-Lynen-Strasse 31,
D-30625 Hannover, Germany; [b] Department of Chemistry, University
of Patras, 26010 Patras, Greece

6.2.1 Introduction

Multiple disulfide bonds between cysteine residues form disulfide frameworks, which represent an important structural feature in many peptides and proteins secreted from cells. Among these, numerous peptide hormones, cytokines, protease inhibitors and toxins are found. Starting with chemically synesized precursors containing the Cys residues in their thiol form, some of the multiple disulfide-bonded polypeptides are produced in satisfying-to-good yields in oxidative folding processes under experimental conditions that partly mimic the environment of intracellular compartments as described in the preceding Chapter 6.1. However, in other cases where insufficient structural information is encoded in the amino acid sequence to generate the correct disulfide framework, application of oxidative folding proved to be inexpedient or even impossible. In addition, for many research purposes such as structure-function studies, modifications of the natural amino acid sequence are required, and often, in contrast to the parent molecules, the analogs cannot be correctly oxidized. Indeed, more or less complex mixtures of disulfide isomers are formed, which are difficult to resolve by chromatographic techniques. The complexity of disulfide isomer mixtures depends on the number of disulfide bonds: from a peptide

RSC Biomolecular Sciences
Oxidative Folding of Peptides and Proteins
Edited by Johannes Buchner and Luis Moroder
© Royal Society of Chemistry 2009
Published by the Royal Society of Chemistry, www.rsc.org

containing four reduced Cys residues, three intrachain disulfide isomers are theoretically formed with the Cys^I-Cys^{II}/Cys^{III}-Cys^{IV}, Cys^I-Cys^{III}/Cys^{II}-Cys^{IV} or Cys^I-Cys^{IV}/Cys^{II}-Cys^{III} disulfide connectivities, while from a peptide with six Cys residues the maximum number of disulfide isomers theoretically rises to 15. In addition, oligomers are formed at variable extents. Since the early days of peptide chemistry these facts have been recognized and protection strategies for regioselective disulfide formation have been one of the most challenging tasks. Great advances have been achieved in the field as is well assessed by the numerous highly efficient syntheses of multiple-disulfide peptides reported so far. These progresses have been extensively reviewed in recent years.[1–4]

The scope of the present review is to elaborate on the available sets of orthogonal thiol-protecting groups (Section 6.2.2) the main strategies evolved for the synthesis of Cys-rich peptides: i) stepwise regioselective Cys pairings (Sections 6.2.3 and 6.2.4) and ii) a combination of oxidative folding with regioselective disulfide formation (Section 6.2.5). Selected elegant synthetic examples reported during recent years are discussed to illustrate the state of the art in the field.

6.2.2   Thiol-protecting Groups

Among the natural amino acids, suitable protection of the nucleophilic thiol group of cysteine represents a particular challenge, and only a restricted set of available protecting groups fulfills the requirements of orthogonality in terms of the overall protection strategy in the polypeptide chain assembly, *i.e.* the 9-fluorenylmethoxycarbonyl/*tert*-butyl (Fmoc/tBu) or *tert*-butoxycarbonyl/ benzyl (Boc/Bzl) chemistry, and of reciprocal selectivity for subsequent regio- selective disulfide formation. In Table 6.2.1 the most important thiol protecting groups are reported, which applied for pairs of Cys residues largely satisfy the criteria of orthogonality. These protecting groups can be classified into those which are removed by both acidic and oxidative procedures, and those which are cleaved with other reagents, *e.g.* the *tert*-butylthio (StBu) group requiring reduction with thiols or phosphines for cleavage. All the groups are compatible with the Fmoc/tBu chemistry except the Xpys group (substituted 2-thiopyridyl). The latter activated disulfide-type protecting groups tend to disproportionate during base treatment, but upon post-synthetic introduction can be useful in regioselective disulfide-bond formation.

Because of the acid-sensitivity of most thiol-protecting groups (Table 6.2.1) and thus incompatibility with the Boc/Bzl strategy, the Fmoc/tBu chemistry allows for a larger combinatorial diversity as required for multiple-disulfide peptides. The major difficulty arising from the Fmoc/tBu chemistry are the repetitive base treatments in the chain elongation steps, which are known to provoke side reactions such as Cys racemization or β-elimination. Racemiza- tion resulting from repeated base treatments, *i.e.* piperidine in Fmoc/tBu-based procedures and triethylamine in Boc/Bzl syntheses, is enhanced when a cysteine is directly esterified to the support as C-terminal residue. Since not only the

Table 6.2.1 Structure and conditions for cleavage of common thiol-protecting groups.

Protecting group	Structure	Deprotection	Oxidative deprotection	Ref.
StBu	H$_3$C–C(CH$_3$)$_2$–S–	thiols, phosphines	–	5–7
Xan	xanthyl structure	1% TFA	I$_2$, Tl^{3+}	8
Mmt	4-methoxytrityl structure (—OCH$_3$)	1% TFA	I$_2$, Tl^{3+}	9
Trt	trityl structure	>25% TFA	I$_2$, Tl^{3+}	10,11
tBu	H$_3$C–C(CH$_3$)$_3$	TFMSA, Hg^{2+}	I$_2$, Tl^{3+}, Ph$_2$SO/ MeSiCl$_3$/TFA	12,13
Mob	H$_3$CO–C$_6$H$_4$–CH$_2$–	HF, TFMSA/ TFA/cresol	Tl^{3+}, Ph$_2$SO/ MeSiCl$_3$	14
MeBzl	H$_3$C–C$_6$H$_4$–CH$_2$–	HF, TFMSA, TFA/Me$_3$SiBr/ thioanisole	Tl^{3+}, Ph$_2$SO/ MeSiCl$_3$	15
Acm	H$_3$C–C(O)–NH–CH$_2$–	TFA/Npys-Cl	I$_2$, Tl^{3+}, Ph$_2$SO/ MeSiCl$_3$	16
Xpys	X-substituted 2-pyridinesulfenyl structure	thiols	–	17–19
Thz	H$_2$C=< thiazolidine	methoxyamine x HCl	–	20

StBu: *tert*-butylthio; Xan: xanthyl; Mmt: 4-methoxytrityl; Trt: trityl; tBu: *tert*-butyl; Mob: 4-methoxybenzyl; MeBzl: 4-methylbenzyl; Acm: acetamidomethyl; Xpys: substituted 2-pyridinesulfenyl. TFA: trifluoroacetic acid; TFMSA: trifluoromethane sulfonic acid; HF: hydrogen fluoride; Me$_3$SiBr: trimethylsilyl bromide; Npys: 3-nitro-2-pyridinesulfenyl.

electron-donating character of the protecting group but also its bulkiness and the steric hindrance of the resin play a crucial role in this side reaction, the 2-chlorotrityl resin represents the solid support of choice in such cases.[21] In addition, this resin effectively suppresses piperidinylalanine formation when Cys(Trt) is esterified onto the resin. This side reaction, which is caused by β-elimination followed by Michael addition of piperidine, is often observed with Cys(Acm) or Cys(StBu) directly linked to the resin. Conversely, with Cys(StBu) esterified onto the Wang resin, the Cys-peptides could not be obtained at all.[2] Again, electron-donating and bulky protecting groups such as Trt and Xan effectively suppress the piperidinylalanine formation.[22]

An additional difficulty encountered in the synthesis of Cys-peptides derives from the electrophilic nature of intermediate species formed during the acidic resin cleavage/deprotection step, which are likely to alkylate the free thiol groups acting as effective internal scavengers for these reactive species. When such intermediates are generated in the linker moiety on the resin, the Cys-peptide is irreversibly grafted to the support. Many sophisticated scavenger mixtures were proposed to bypass this problem: a particularly effective combination is reagent K, a mixture of TFA/H_2O/phenol/thioanisole/ethanedithiol (82.5:5:5:5:2.5). Very efficient in these terms are trialkylsilanes, which form silylthioethers with the free thiol functions[9] and prevent irreversible alkylation. The silyl groups are readily removed by hydrolysis or alcoholysis during the work-up procedure.[9,23] Silanes are even better than thiols as scavengers as reduction of previously formed disulfide bonds occurs to lesser extents under the strongly acidic conditions, and the addition of H_2O to such cleavage mixtures is not necessary. For instance, good deprotection/cleavage results are obtained with the mixture TFA/CH_2Cl_2/triethylsilane (70:25:5).[24]

A special protection of Cys residues is its conversion to thiazolidines (Thz) by reaction with formaldehyde and ketones.[25–27] This cysteine precursor has found recent application in convergent chemical ligation reactions when an N-terminal Cys residue is required. Examples for such an application are the synthesis of crambin[28] and lysozyme.[29] Intermediate protection of a Cys residue as Thz allowed for regeneration of the required N-terminal cysteine with methoxyamine hydrochloride.

6.2.3   Regioselective Disulfide Formation in Solution

Regioselective disulfide formation is usually carried out in solution following solid-phase assembly of the peptide chain on suitable resins. Some work, however, has been carried out investigating disulfide formation on the solid support (see Section 6.2.4). As mentioned above, in many cases the thiol-protecting groups and synthetic strategy applied, *i.e.* Fmoc/tBu or Boc/Bzl chemistry, affect the degree of homogeneity of the crude synthetic peptide, and in this way the chromatographic separation of intermediates, *e.g.* peptides with one or two pairs of protected Cys residues and with one or two disulfide bonds. Therefore, the potential combinations of pairwise Cys protections strongly depend both on the chemistry applied in the chain assembly and in the regioselective disulfide formation. Moreover, because of their hydrophobic character, the thiol-protecting groups markedly affect the solubility of Cys-containing peptides in solvents required for the regioselective Cys pairings.

6.2.3.1 Peptides with Two Disulfides

For the synthesis of peptides with two disulfide bonds by Fmoc/tBu chemistry, the most widely used combination of thiol-protecting groups is Trt/Acm, which can be considered as the standard strategy. While the Trt groups are cleaved under acidic conditions such as TFA, the Acm groups remain stable. Correspondingly, the precursor peptides with two free thiols are generated by TFA-mediated resin cleavage/deprotection, and the first disulfide is formed under mild oxidative conditions such as by exposure to air oxygen, DMSO or $K_3[Fe(CN)_6]$. Under these conditions, the Cys(Acm) derivatives are stable. For the Boc/Bzl chemistry, a pairwise combination of Cys(MeBzl) and Cys(Acm) derivatives has found large application. After HF-mediated cleavage/deprotection, peptides are generated that contain a pair of thiol groups and of Acm-protected Cys residues, as these Cys derivatives are largely stable to liquid HF. After oxidative formation of the first disulfide, the two Acm groups are generally cleaved under oxidative conditions with slight excesses of I_2 and concomitant formation of the second disulfide bond.

An alternative orthogonal pair of thiol protections are the Cys(Acm)/Cys(StBu) derivatives. After acidic resin cleavage/deprotection, the peptide is generated with the Cys residues fully protected and thus allowing its purification prior to the disulfide formation. After reductive cleavage of the two StBu groups, regioselective disulfide formation follows the route described above.

6.2.3.1.1 Syntheses with the Trt/Acm Protection Scheme

Among the large number of two-disulfide peptides, also guanylin and uroguanylin, gastrointestinal peptide hormones binding to the membrane receptor guanylate cyclase type C and involved in electrolyte transport,[30] have been synthesized by this strategy.[31] Indeed, oxidative folding was unsuccessful as almost none of the correct disulfide isomer with the required Cys^I-Cys^{III}/Cys^{II}-Cys^{IV} connectivities was formed (Figure 6.2.1).[32] Consisting of 15 and 16 amino acid residues, respectively, these peptides are very cysteine-rich. For guanylin in particular, it was found that combinations other than Trt/Acm for Cys protection led to unsatisfactory low overall yields. In this case the positioning of the Acm derivatives relative to the Trt groups did not significantly affect the purity and yield of the final product.[31,33] However, careful positioning of the Acm and Trt groups should be considered since an efficient chain assembly may depend upon their location. Similarly important is the choice of reaction conditions for oxidation of the Cys(Acm) residues such as solvent and pH.[16,34] Moreover, by titration with I_2 instead of using excesses of reagent side reactions can be minimized.

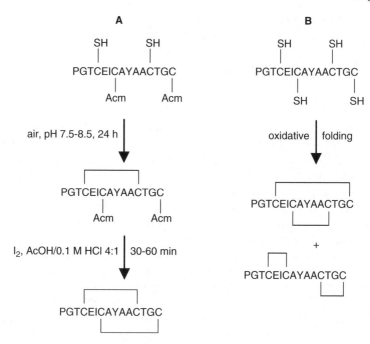

Figure 6.2.1   (A) Synthetic route for regioselective disulfide formation of guanylin; (B) oxidative folding of guanylin which results predominantly in non-native isomers.

Additional examples of successful regioselective disulfide formation are endothelin, a mammalian vasoconstrictor,[35] with the orthogonal Trt/Acm[36] or the MeBzl/Acm[37] protection. The versatility of the Trt/Acm approach was confirmed by an Fmoc-based convergent solid-phase synthesis of endothelin, *i.e.* by assembling the peptide chain in solution using fully protected fragments. Immer *et al.* prepared the endothelin fragment 1-12 with a pair of Cys(Trt) residues and the third Cys protected as an Acm derivative.[38] Based on earlier systematic work on the solvent-dependent iodolysis of the Acm group,[34] the disulfide bond was formed between Cys-3 and Cys-11, while the Acm-protected Cys-1 remained intact during treatment with I_2 in CH_2Cl_2/trifluoroethanol (9:1). The monocyclic fragment was then coupled to the protected endothelin fragment 13-21 containing the second Cys(Acm) residue. Formation of the second disulfide bond was achieved by I_2 in methanol. Under these mild conditions, almost no by-products, especially those concerning His and Trp side-chain modification, were observed. A similar solvent-dependency of the Trt and Acm cleavage rates was exploited for the synthesis of orexin A,[39] a 33-mer neuropeptide involved in food intake.[40] A precursor peptide with free thiols at Cys-7 and Cys-14, and with Acm-protected Cys residues in positions 6 and 12, was disulfide-cross-linked by a 10-fold excess of I_2 in AcOH within 60 min at room temperature. Upon a 4:1 dilution of the reaction mixture with water,

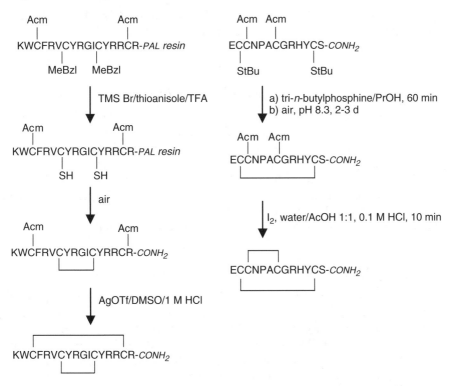

Figure 6.2.2 Regioselective disulfide introduction in tachyplesin I (left) and α-conotoxin GI (right). Both tachyplesin I and α-conotoxin GI exhibit the identical I-IV/II-III disulfide connectivity but, for tachyplesin I, the inner disulfide was formed first, while for α-conotoxin it was the outer disulfide.

cleavage/oxidation of the two Cys(Acm) derivatives was achieved within an additional 90 min. This one-pot procedure carries the advantage that it can be performed within a short time without chromatographic work-up of the monocyclic intermediate, at least in the case of orexin A. The reported yield (30–35%) can be considered excellent for such types of peptides.

In some cases, over-oxidation or incomplete cleavage of Cys(Acm) was observed. An advantage of Acm is that it can be removed not only oxidatively by I_2, but also by heavy metal ions in the presence of oxidizing agents yielding the corresponding disulfide-peptide. This procedure was applied for endothelin-1 and tachyplesin I, a small antibacterial peptide from horseshoe crabs,[41] using silver trifluoromethanesulfonate (AgOTf) (Figure 6.2.2). The solvent system consisting of DMSO in 1 M HCl caused the immediate formation of the corresponding disulfide.[42] This method apparently produced fewer side products compared to I_2-mediated disulfide formation. Hg(II) salts have also been applied, but it has been observed by mass spectrometry that the metal ion is difficult to remove.

Among the many other Cys-rich peptides synthesized by this protection scheme the α-conotoxin GI[43] was also obtained in good yields.[44] An alternative protection scheme for this synthesis is discussed in the following section.

6.2.3.1.2   Syntheses with the StBu/Acm Protection Scheme

The synthetic route of α-conotoxin GI by the StBu/Acm combination is shown in Fig 6.2.2.[45] After TFA-mediated resin cleavage/deprotection the peptide containing a pair of Cys(StBu) and Cys(Acm) derivatives was isolated chromatographically as protection of the Cys residues allows for easier purification of the intermediate prior to disulfide formation steps. Subsequent cleavage of the StBu groups was achieved by tributylphosphine, thereby generating the dithiol-peptide for the first disulfide formation by exposure to air oxygen, followed by I_2 oxidation to generate the second disulfide bond.

6.2.3.1.3   Syntheses with Alternative Protection Schemes

In a one-pot synthesis of α-conotoxin SI the temperature-dependent different lability of the tBu and MeBzl thiol-protecting groups was exploited.[46] While the first disulfide bond was formed with TFA/DMSO at room temperature, a condition under which MeBzl is stable, the second disulfide was generated at 70 °C. However, it is clear that such harsh reaction conditions favor side reactions, *e.g.* oxidation of sensitive amino acids (Trp, Met), cleavage of acid-sensitive peptide bonds (Asp-Pro)[47] and aspartimide formation.[48] Although this procedure reveals limitations, it has also been applied in combination with the Trt group for a successful synthesis of the heat-stable enterotoxin ST,[49] a very Cys-rich peptide containing three disulfide bonds in the 14-residue sequence.[50] A combination of Cys(tBu) and Cys(Acm) derivatives was applied in the synthesis of α-conotoxin MI by Boc/tBu chemistry.[51] After introduction of the first disulfide by I_2 oxidation, oxidative cleavage of the tBu groups with $MeSiCl_3/Ph_2SO$[52] was applied for generation of the second disulfide bond. For the synthesis of α-conotoxin GI, oxidative cleavage of the tBu groups with this reagent led to significant scrambling of the preformed disulfide, thus strongly questioning the general efficiency of this procedure.[53]

6.2.3.1.4   Topological Isomers

In the case of the guanylin-type peptides (see Section 6.2.3.1.1), the rare phenomenon of topological isomerism has been observed. Both peptides, when containing the correct disulfide-bond pattern (Cys^I-Cys^{III}/Cys^{II}-Cys^{IV}), occur as two topological isomers (Figure 6.2.3).[31,54] The other two disulfide isomers with the Cys^I-Cys^{IV}/Cys^{II}-Cys^{III} and Cys^I-Cys^{II}/Cys^{III}-Cys^{IV} connectivities do not show this isomerism. Detailed investigations have demonstrated that the isoforms interconvert into each other without disulfide rearrangement following a sterically controlled process.[33] The isoforms differ in their HPLC

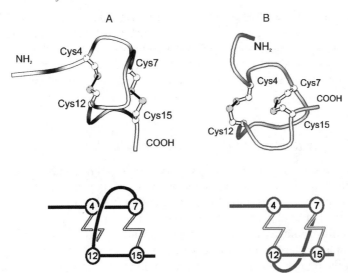

Figure 6.2.3 3D structure of the topological isomers A and B of human uroguanylin. Top: lowest energy solution structures. Bottom: schematic view of the backbone fold. Only isomer A is biologically active.

retention time and 3D structures as assessed by NMR conformational analysis. Such isomerism has also been observed for peptides of the conotoxin[55] and contryphan[56] families. It is highly probable that this isomerism may emerge in the synthesis of other Cys-rich peptides,[57] in particular of those containing two overlapping disulfides. It is, however, difficult to distinguish such topological isomers from positional disulfide isomers. Interestingly, only one of the two isomers is biologically active.

6.2.3.2 Peptides with Three Disulfides

Many naturally occurring peptides with three intrachain disulfide bonds can be oxidatively folded into the correct disulfide framework (see Chapter 6.1), *e.g.* charybdotoxin and many functionally related channel-blocking toxins,[58–61] or members of the widespread defensin family of antimicrobial peptides.[62,63] However, among defensin peptides, in particular the mammalian β-defensins, some peptides do not fold to the native disulfide pattern as the β-defensins hBD-3 and hBD-28,[63–65] and thus regioselective procedures are essential.

6.2.3.2.1 *Syntheses with the Trt/Acm/Mob Protection Scheme*

By combining the Cys(Trt)/Cys(Acm) derivatives with the third Fmoc-compatible orthogonal Cys(Mob) protection, the regioselective formation of three disulfide bonds of α-conotoxin MVIIA was successful.[66] Following the standard air and I_2 oxidation steps described in Section 6.2.3.1.1 for transformation of the Cys(Trt) and Cys(Acm) pairs, a TFMSA/TFA mixture was used in the presence of *p*-cresol as scavenger to cleave the Mob groups, followed by a DMSO oxidation without affecting the two already preformed disulfide bonds. However, this Mob deprotection method failed in the synthesis of sapecin,[67] an insect defensin.[68] Instead, a mixture of TFA/MeSiCl₃/Ph₂SO/anisole was required for simultaneous cleavage and oxidation.

6.2.3.2.2   Syntheses with the Trt/Acm/tBu Protection Scheme

Less troublesome is the Cys(Trt)/Cys(Acm)/Cys(tBu) combination as well documented in the one-pot synthesis of an EGF-like domain.[69] Upon resin cleavage/deprotection the dithiol-peptide was subjected to air oxidation followed by deprotection/oxidation of the Cys(tBu) pair with a TFA/DMSO/anisole treatment for 5 min at room temperature. Under these conditions the Acm groups remained largely intact as assessed by mass spectrometry. By extending the reaction time to 90 min, the third disulfide was installed (Figure 6.2.4).

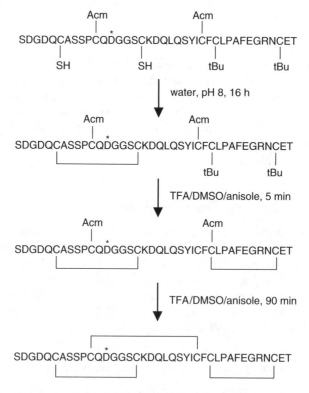

Figure 6.2.4   Synthetic scheme for an analog of the EGF-like domain of human coagulation factor VIII. For coupling the aspartic acid residue marked with an asterisk, the C-terminally protected Fmoc-Asp-OtBu with a free β-carboxy function was used, leading to the desired β-peptide bond at this position.

The lability of the Acm group at room temperature was unexpected as in earlier work elevated temperatures were required for cleavage of the Acm by this acid treatment (see 6.2.1.3).[46,50] Such quasi-orthogonal protections of two pairs of Cys residues may not be of general use, but is an attractive strategy in particular cases.

In contrast to this one-pot introduction of the last two disulfide bonds, Gali *et al.* used the combination Trt/Acm/tBu with cleavage and oxidation of the tBu groups as the final transformation step for the synthesis of variants of *E. coli* heat-stable enterotoxin N-terminally modified with DOTA, a macro-cyclic chelating group for radionuclides.[70] The first disulfide was formed within 60 min in the presence of 2,2'-dithiopyridine (DPS) instead of air oxidation. In this reaction, a free thiol reacts initially with DPS forming a mixed peptide-thiopyridyl disulfide which is then attacked by the second thiol resulting in the intrachain disulfide-bond formation. After I_2 oxidation of the two Acm groups, the tBu groups were cleaved/oxidized with TFA/Ph$_2$SO/MeSiCl$_3$/thioanisole.[51–52,71]

In other cases, this protection scheme was less satisfying. For the synthesis of the β-defensin hBD-28 with three overlapping disulfides, the TFA/DMSO/anisole mixture applied for longer reaction times in the formation of the third disulfide bond at room temperature (Figure 6.2.5) led to over-oxidation of the Cys residues.[63] More appropriate was a short treatment at 60 °C, while depro-tection by the chlorosilane procedure[51,52,71] was not advantageous. Formation of the third disulfide bond of hBD-28 from the Cys(tBu) pair was found to strongly depend on the excess of DMSO and, in particular, the peptide concentration. However, at a conveniently low concentration, the synthesis of hBD-3[63,65] can hardly be scaled up without loss in the chemical performance, product purity and yield. Some defensin peptides are readily accessible by using oxidative folding techniques, *e.g.*, the human β-defensins hBD-2 and hBD-27 with their I-V/II-IV/III-VI connectivities. However, others such as the above-mentioned hBD-3 and hBD-28 can satisfactorily be synthesized only using regioselective approaches. Thereby juxtaposition of the Cys protecting groups and thus the order of disulfide-bond formation was found to be crucial. Indeed, for hBD-28, disulfides, were formed in the order III-VI, I-V and II-IV,[63–65] while for hBD-3, the best order was II-IV, I-V and III-VI, suggesting significant conformational effects, a fact that is difficult to predict and control.

6.2.3.2.3 Syntheses with Alternative Protection Schemes

Only a few comparative synthetic studies with alternative protection schemes were reported. Kellenberger *et al.* compared the combination Trt/Acm/Mob with Mmt/Trt/Acm in the synthesis of PMP-D2, a 35-residue insect peptide with three overlapping disulfide bonds blocking calcium channels and inhi-biting serine proteases.[72] Figure 6.2.6 shows the use of the Mob protection group, where, upon standard formation of the first two disulfides, the Mob groups were cleaved with TFMSA/TFA/*p*-cresol at 0 °C within 10 min. The

hBD-3

hBD-28

LQKYYCRVRGGRCAVLSCLPKEEQIGKCSTRGRKCCRRKK   ARLKKCFNKVTGYCRKKCKVGERYEIGCLSGKLCCAN

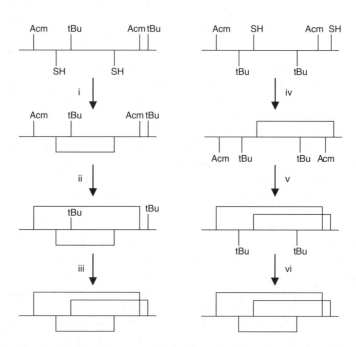

Figure 6.2.5   Synthesis of human β-defensins hBD-3 and hBD-28 with different positioning of the thiol protecting groups Trt, Acm and tBu. Reaction conditions: i) DMSO oxidation, 24 h; ii) I$_2$ in AcOH, 90 min; iii) TFA/DMSO/anisole, 90 min, room temperature; iv) air oxidation, 4 h; v) I$_2$ in AcOH, 75 min; vi) TFA/DMSO/anisole, 60 min, 60 °C.

intermediate containing two disulfide bonds and two thiol groups was isolated by HPLC and then subjected to DMSO oxidation yielding the fully disulfide-bonded product. In the alternative synthesis, where Mmt was used besides Trt and Acm for Cys protection, the resin-bound peptide was deprotected at the Cys(Mmt) residues with 2% TFA/2% triisopropylsilane in CH$_2$Cl$_2$ within 60 min. The partially deprotected peptide was then oxidized on resin in NMP/CCl$_4$/triethylamine as described elsewhere.[73] Upon resin cleavage/deprotection, two successive classical oxidation steps were performed to yield PMP-D2. Although this protection strategy was not superior to the Trt/Acm/Mob combination, it may represent an alternative for other synthetic target peptides.

6.2.3.3   Peptides with Multiple Disulfides

There are almost no reports on the synthesis of peptides with regioselective formation of more than three disulfide bonds. In the synthesis of a dimeric

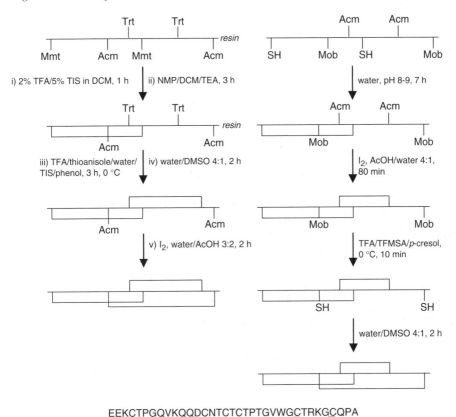

EEK**C**TPGQVKQQD**C**NT**C**T**C**TPTGVWG**C**TRKG**C**QPA

Figure 6.2.6 Regioselective disulfide formation of the insect peptide PMP-D2 by the Cys(Trt)/Cys(Acm)/Cys(Mob) (left) and Cys(Mmt)/Cys(Trt)/Cys(Acm) (right) protection schemes. TIS, tri-isopropylsilane.

α-conotoxin analog consisting of two identical conotoxin units head-to-tail linked by a Gly-Lys-Gly segment,[74] a Cys(Trt)/Cys(Acm)/Cys(tBu)/Cys(MeBzl) protection scheme was applied as shown in Figure 6.2.7. After cleavage of the crude peptide from the resin, the first disulfide bond was formed by DMSO oxidation of the two free thiols within 24 h. Dilution with acetic acid and addition of I_2 caused oxidative cleavage of the two Acm groups, forming a double disulfide-bonded intermediate with a pair of tBu- and MeBzl-protected cysteines. The third and fourth disulfide bonds were introduced by the one-pot procedure described earlier by the authors[46] (see Section 6.2.3.1.3) using 5% DMSO in TFA to form the third disulfide within 30 min at room temperature and the fourth after an additional 3 h at 45 °C. Not surprisingly, partial MeBzl cleavage at room temperature was observed, causing the formation of the wrong disulfide isomers. Although this chemistry is far from fulfilling the many requirements of regioselectivity it demonstrates that it is in principle possible to synthesize such complex targets in a regioselective manner.

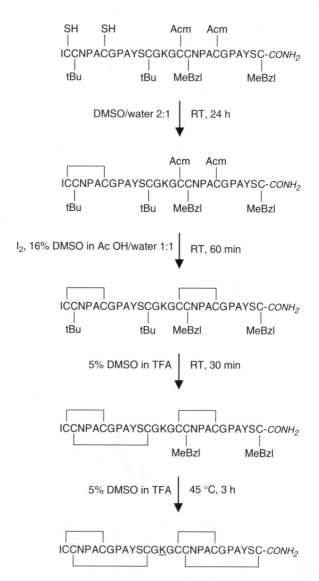

Figure 6.2.7 Regioselective synthesis of a dimeric α-conotoxin construct.

6.2.4 Disulfide Formation on the Solid Support

On-resin thiol chemistry involving Cys residues has been investigated to a minor extent, although this strategy could bear, at least in selected cases, significant advantages over solution methods. For such synthesis, additional orthogonality of the thiol protecting group relative to the resin linker is essential, since the Cys-protection group must be removed selectively without cleavage from the resin.

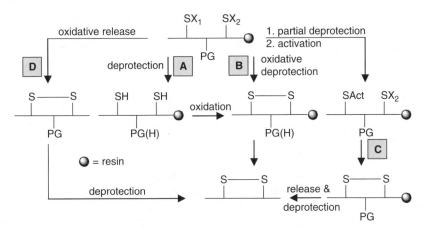

Figure 6.2.8 Different routes for on-resin synthesis of Cys-rich peptides. For explanation see text; X_1, X_2: thiol-protecting groups; Act (activating group): 3- or 5-Npys; PG: other Cys-protecting group.

From the knowledge accumulated in the synthesis of Cys-rich peptides in solution, potential synthetic routes for the on-resin disulfide formation can be elaborated. As shown in Figure 6.2.8 route A, the thiol functions or the whole peptide are deprotected and subsequently oxidized; according to route B, the Cys-protecting groups are oxidatively removed followed by deprotection and simultaneous release from the resin; route C is a non-oxidative method as one Cys residue is selectively deprotected in the presence of all other protecting groups and the resin linkage and then activated electrophilically, *e.g.* as an Npys derivative. Thereafter, the second thiol-protecting group is selectively removed and the resulting thiol reacts with the Npys derivative to produce the disulfide. By this procedure a peptide ligand for $\alpha_v\beta_3$ integrin was obtained in high yield and purity.[75] Route D represents a more recent approach for the solid-phase synthesis of Cys-containing peptides. In this variation of route B, a Cys-peptide is oxidatively deprotected and simultaneously released from the resin. The resulting protected fragment is then condensed to the final peptide. This approach is not only interesting because it allows an optimal deprotection of the peptide in solution, but is has also proved very advantageous for the synthesis of multiple intra- or intermolecular disulfide-containing peptides. This has been well demonstrated by the preparation of the A chain of insulin-like peptides by convergent synthesis.[76]

In contrast to chemistry in solution, the reactions for on-resin intra- or intermolecular oxidation of the thiol groups cannot easily be monitored, and thus the events during on-resin oxidation processes are poorly understood. But it can reasonably be assumed that a low resin loading favors intramolecular disulfide formation and that intramolecular reactions are favored in proximity of the resin-linker while distal from it, solution-type conditions may prevail and thus lower the advantages of the pseudo-dilution effect existing on resin.

6.2.5   Semi-selective Formation of Disulfide Bonds

As a hybrid approach, the semi-selective disulfide formation has been applied for the synthesis of peptides with three disulfide bonds, in some cases with considerable success and advantages over the regioselective procedures. The basic concept is to reduce the number of possible disulfide isomers generated in oxidative folding processes to three by protecting one pair of Cys residues, preferably as Acm derivatives. Correspondingly, the precursor tetra-thiol peptide with a Cys(Acm) pair is subjected to the oxidative folding reaction, which should produce the three possible isomers, possibly with the preferential formation of the correct isomer. Upon its isolation it is converted by standard methods into the native disulfide framework. Such a simple strategy can further be improved by taking into account the "proximity rule" with its two under-lying principles of folding kinetics and loop formation probability,[77] which allows a knowledge-based design of the Cys protection scheme as well illu-strated by the synthesis of EGF-like domains of larger proteins, e.g. the Cys-rich protective antigen of the malaria merozoite surface protein MSP-1[78] and the EGF-like domain of blood coagulation factor IX.[79] The latter 45-residue peptide is known to exhibit I-III/II-IV/V-VI disulfide connectivities. The syn-thetic strategy for this peptide was optimized by a prescreening of the three possible precursors with the protected Cys pair in the three different positions. Upon oxidative folding of the tetra-thiol peptides, the optimal precursor for preferential formation of the two native disulfide bonds could be identified. Only the precursor with Cys(Acm) in positions II and IV generates the I-III and V-VI disulfides as required for the production of the correct disulfide frame-work upon oxidative cleavage of the two Cys(Acm) derivatives at positions II and IV. Conversely, placing the Cys(Acm) derivatives in positions V and VI resulted in a predominantly misfolded intermediate with I-II and III-IV di-sulfide connectivities. With Cys(Acm) in positions I and II a mixture of the three possible disulfide isomers was formed. It is noteworthy that small amounts of misfolded disulfide isomers are always generated by oxidative folding. Based on these findings, the authors suggested that for such EGF-like domains, which contain two overlapping disulfide bonds (I-III and II-IV) and one isolated V-VI disulfide group, the Cys(Acm) residues should be placed at the positions of one of the overlapping disulfides. However, in the synthesis of the 52-residue α/β chimera of the EGF-like domain of neu differentiation factor α/β (NDF$_c$α/β)[80] it was reported that the precursor with the Cys(Acm) residues in positions I and III rather than in positions II and IV led to the preferential formation of the desired double-disulfide product. This fact would suggest a careful preliminary screening of the optimal protection scheme for each case.

The potential of the semi-selective approach was recently demonstrated by the synthesis of a non-sulfated hirudin variant,[24,81] a potent thrombin inhibitor containing the I-II/III-V/IV-VI disulfide pattern (Figure 6.2.9). Hirudin has been intensively investigated regarding its *in vitro* folding pathways.[82,83] A synthetic 65-residue tetra-thiol precursor with Cys(Acm) in positions III and V was subjected to oxidative folding in the presence of β-mercaptoethanol at

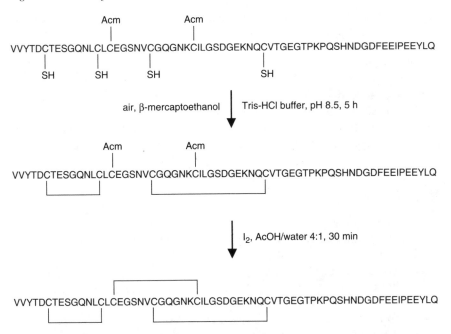

Figure 6.2.9   Synthesis of a non-sulfated hirudin variant 1 by a semi-selective procedure for disulfide-bond formation.

pH 8.5. The intermediate product with I-II and IV-VI disulfides was formed almost exclusively within 5 h. This intermediate was then oxidized by I_2 in 40% aqueous acetic acid, generating the third III-V disulfide bond and thus the desired hirudin variant with the native disulfide framework in a remarkable overall yield of 12%.[24]

These examples illustrate that the semi-selective approach facilitates the synthesis of polypeptides with three disulfide bonds. It provides an alternative pathway circumventing major problems encountered in oxidative folding (disulfide isomer mixtures difficult to separate) and three-step regioselective disulfide formation (low yields due to multiple HPLC purification steps). However, in some cases, such as for the β-defensins hBD-3 and hBD-28, the semi-selective strategy failed.[63]

6.2.6   Concluding Remarks

The quality of multiple-disulfide peptides resulting from the synthetic approaches discussed in this chapter strongly depends on the judicious choice of the resin linker, the set of thiol (and for other functional groups) protecting groups applied, the positioning of the protecting groups at the Cys residues, the coupling reagents, particularly for the Cys residues, exposure to bases, and even on the nature of the protecting groups on adjacent amino acids. In addition, the mixture of reagents used in deprotection and cleavage reactions and, of course, the chemical method for disulfide-bond formation play a crucial role. There are no general rules to support the design of such synthesis although some standard conditions such as the use of a Trt/Acm combination for double disulfide-bonded peptides are preferred. All above-mentioned experimental variables together control the synthetic efficiency in terms of homogeneity and overall yield. Correspondingly, the synthesis of each Cys-rich peptide represents an individual challenge and requires careful planning and development, particularly if syntheses have to be scaled up to gram or even kilogram amounts.

References

1. F. Albericio, I. Annis, M. Royo and G. Barany, in *Fmoc Solid Phase Peptide Synthesis*, W. C. Chan and P. White eds., Oxford University Press, Oxford, 2000, pp. 81–114.
2. L. Moroder, H.-J. Musiol, N. Schaschke, L. Chen, B. Hargittai and G. Barany, in *Houben-Weyl. Methods of Organic Chemistry. Synthesis of Peptides and Peptidomimetics*, M. Goodman, A. Felix, L. Moroder and C. Toniolo, eds., Thieme, Stuttgart, 2002, Vol. E22a, pp. 384–423.
3. K. Akaji and Y. Kiso, in *Houben-Weyl. Methods of Organic Chemistry. Synthesis of Peptides and Peptidominetics*, M. Goodman, A. Felix, L. Moroder and C. Toniolo, eds., Thieme, Stuttgart, 2002, Vol. E22b, pp. 101–141.
4. C. Boulègue, H.-J. Musiol, V. Prasad and L. Moroder, *Chem. Today*, 2006, **24**, 24–36.
5. E. Wünsch and R. Spangenberg, in *Peptides 1969*, E. Scoffone, ed., North-Holland, Amsterdam, 1971, p. 30.
6. L. Moroder, M. Gemeiner, W. Göhring, E. Jaeger, P. Thamm and E. Wünsch, *Biopolymers*, 1981, **20**, 17–37.
7. R. Eritja, J. P. Ziegler-Martin, P. A. Walker, T. D. Lee, K. Legesse, F. Albericio and B. E. Kaplan, *Tetrahedron*, 1987, **43**, 2675–2680.
8. Y. Han and G. Barany, *J. Org. Chem.*, 1997, **62**, 3841–3848.
9. K. Barlos, D. Gatos, O. Hatzi, N. Koch and S. Koutsogianni, *Int. J. Pept. Protein Res.*, 1996, **47**, 148–153.
10. L. Zervas and I. Photaki, *J. Am. Chem. Soc.*, 1962, **84**, 3887–3897.
11. L. Zervas, I. Photaki and I. Phocas, *Chem. Ber.*, 1968, **101**, 3332–3333.
12. F. M. Callahan, G. W. Anderson, R. Paul and J. E. Zimmermann, *J. Am. Chem. Soc.*, 1963, **85**, 201–207.
13. J. J. Pastuszak and A. Chimiak, *J. Org. Chem.*, 1981, **46**, 1868–1873.
14. S. Akabori, S. Sakakibara, Y. Shimonishi and Y. Nobuhara, *Bull. Chem. Soc. Jpn.*, 1964, **37**, 433–434.
15. B. W. Erickson and R. B. Merrifield, *J. Am. Chem. Soc.*, 1973, **95**, 3750–3756.
16. D. F. Veber, J. D. Milkowski, S. L. Varga, R. G. Denkewalter and R. Hirschmann, *J. Am. Chem. Soc.*, 1972, **94**, 5456–5461.

17. L. Moroder, F. Marchiori, G. Borin and E. Scoffone, *Biopolymers*, 1973, **12**, 493–505.
18. J. V. Castell and A. Tun-Kyi, *Helv. Chim. Acta*, 1979, **62**, 2507–2510.
19. R. Matsueda, T. Kimura, E. T. Kaiser and G. R. Matsueda, *Chem. Lett.*, 1981, **6**, 737–740.
20. M. Villain, J. Vizzavona and K. Rose, *Chem. Biol.*, 2001, **8**, 673–679.
21. Y. Fujiwara, K. Akaji and Y. Kiso, *Chem. Pharm. Bull. (Tokyo)*, 1994, **42**, 724–726.
22. G. Barany, Y. Han, B. Hargittai, R.-Q. Liu and J. T. Varkey, *Biopolymers*, 2003, **71**, 652–666.
23. D. A. Pearson, M. Blanchette, M. L. Baker and C. A. Guindon, *Tetrahedron Lett.*, 1989, **30**, 2739–2742.
24. S. Goulas, D. Gatos and K. Barlos, *J. Pept. Sci.*, 2006, **12**, 116–123.
25. M. D. Armstrong and V. Du Vigneaud, *J. Biol. Chem.*, 1947, **168**, 373–377.
26. F. King, J. Clark-Lewis and R. Wade, *J. Chem. Soc.*, 1957, 880–885.
27. T. Haack and M. Mutter, *Tetrahedron Lett.*, 1992, **33**, 1589–1592.
28. D. Bang and S. B. Kent, *Angew. Chem. Int. Ed. Engl.*, 2004, **43**, 2534–2538.
29. T. Durek, V. Y. Torbeev and S. B. Kent, *Proc. Natl. Acad. Sci. USA*, 2007, **104**, 4846–4851.
30. L. R. Forte Jr., *Pharmacol. Ther.*, 2004, **104**, 137–162.
31. J. Klodt, M. Kuhn, U. C. Marx, S. Martin, P. Rösch, W. G. Forssmann and K. Adermann, *J. Pept. Res.*, 1997, **50**, 222–230.
32. V. Badock, M. Raida, K. Adermann, W. G. Forssmann and M. Schrader, *Rapid Commun. Mass. Spectrom.*, 1998, **12**, 1952–1956.
33. A. Schulz, S. Escher, U. C. Marx, M. Meyer, P. Rösch, W. G. Forssmann and K. Adermann, *J. Pept. Res.*, 1998, **52**, 518–525.
34. B. Kamber, A. Hartmann, K. Eisler, B. Riniker, H. Rink, P. Sieber and W. Rittel, *Helv. Chim. Acta*, 1980, **63**, 899–915.
35. E. L. Schiffrin and R. M. Touyz, *J. Cardiovasc. Pharmacol.*, 1998, **32**(suppl. 3), S2–S13.
36. J. D. Wade, M. R. McDonald and G. W. Tregear, in *Innovation and Perspectives in Solid Phase Synthesis 1990*, R. Epton, ed., SPCC, Kingswinford, 1990, pp. 585–591.
37. S. Kumagaye, H. Kuroda, K. Nakajima, T. X. Watanabe, T. Kimura, T. Masaki and S. Sakakibara, *Int. J. Pept. Protein Res.*, 1988, **32**, 519–526.
38. H. Immer, I. Eberle, W. Fischer and E. Moser, in *Peptides 1988, Proceedings of the 20th European Peptide Symposium*, G. Jung and E. Byer, eds., de Gruyter, Berlin, New York, 1988, pp. 94–96.
39. R. Söll and A. G. Beck-Sickinger, *J. Pept. Sci.*, 2000, **6**, 387–397.
40. L. de Lecea, T. S. Kilduff, C. Peyron, X. Gao, P. E. Foye, P. E. Danielson, C. Fukuhara, E. L. Battenberg, V. T. Gautvik, F. S. Bartlett 2nd, W. N. Frankel, A. N. van den Pol, F. E. Bloom, K. M. Gautvik and J. G. Sutcliffe, *Proc. Natl. Acad. Sci. USA*, 1998, **95**, 322–327.
41. T. Nakamura, H. Furunaka, T. Miyata, F. Tokunaga, T. Muta, S. Iwanaga, M. Niwa, T. Takao and Y. Shimonishi, *J. Biol. Chem.*, 1988, **263**, 16709–16713.

42. H. Tamamura, T. Murakami, S. Horiuchi, K. Sugihara, A. Otaka, W. Takada, T. Ibuka, M. Waki, N. Yamamoto and N. Fujii, *Chem. Pharm. Bull. (Tokyo)*, 1995, **43**, 853–858.
43. C. J. Armishaw and P. F. Alewood, *Curr. Protein Pept. Sci.*, 2005, **6**, 221–240.
44. Y. Nishiuchi and S. Sakakibara, *FEBS Lett.*, 1982, **148**, 260–262.
45. E. Atherton, R. C. Sheppard and P. Ward, *J. Chem. Soc. Perkin Trans. 1*, 1985, 2065–2073.
46. A. Cuthbertson and B. Indrevoll, *Tetrahedron Lett.*, 2000, **41**, 3661–3663.
47. J. Rittenhouse and F. Marcus, *Anal. Biochem.*, 1984, **138**, 442–448.
48. J. P. Tam, M. W. Riemen and R. B. Merrifield, *Pept. Res.*, 1988, **1**, 6–18.
49. R. A. Giannella, *J. Lab. Clin. Med.*, 1995, **125**, 173–181.
50. A. Cuthbertson, E. Jamæss and S. Pullan, *Tetrahedron Lett.*, 2001, **42**, 9257–9259.
51. K. Akaji, K. Fujino, T. Tatsumi and Y. Kiso, *Tetrahedron Lett.*, 1992, **33**, 1073–1076.
52. K. Akaji, T. Tatsumi, M. Yoshisa, T. Kimura, Y. Fujiwara and Y. Kiso, *J. Am. Chem. Soc.*, 1992, **114**, 4137–4143.
53. I. Szabó, G. Schlosser, F. Hudecz and G. Mezo, *Biopolymers*, 2007, **88**, 20–28.
54. N. J. Skelton, K. C. Garcia, D. V. Goeddel, C. Quan and J. P. Burnier, *Biochemistry*, 1994, **33**, 13581–13592.
55. W. R. Gray, J. E. Rivier, R. Galyean, L. J. Cruz and B. M. Olivera, *J. Biol. Chem.*, 1983, **258**, 12247–12251.
56. R. Jacobsen, E. C. Jimenez, M. Grilley, M. Watkins, D. Hillyard, L. J. Cruz and B. M. Olivera, *J. Pept. Res.*, 1998, **51**, 173–179.
57. B. Mao, *J. Am. Chem. Soc.*, 1989, **111**, 6132–6136.
58. C. Miller, *Neuron*, 1995, **15**, 5–10.
59. T. C. Tenenholz, K. C. Klenk, D. R. Matteson, M. P. Blaustein and D. J. Weber, *Rev. Physiol. Biochem. Pharmacol.*, 2000, **140**, 135–185.
60. M. L. Garcia, Y. Gao, O. B. McManus and G. J. Kaczorowski, *Toxicon*, 2001, **39**, 739–748.
61. H. Nishio, T. Inui, Y. Nishiushi, C. L. C. De Medeiros, E. G. Rowan, A. L. Harvey, E. Katoh, T. Yamazaki, T. Kimura and S. Sakakibara, *J. Pept. Res.*, 1998, **51**, 355–364.
62. R. Garcia, A. Krause, S. Schulz, F. J. Rodriguez-Jimenez, E. Klüver, K. Adermann, U. Forssmann, A. Frimpong-Boateng, R. Bals and W. G. Forssmann, *FASEB J.*, 2001, **15**, 1819–1821.
63. A. Schulz, E. Klüver, S. Schulz-Maronde and K. Adermann, *Biopolymers*, 2005, **80**, 34–49.
64. Z. Wu, D. M. Hoover, D. Yang, C. Boulègue, F. Santamaria, J. J. Oppenheim, J. Lubkowski and W. Lu, *Proc. Natl. Acad. Sci. USA*, 2003, **100**, 8880–8885.
65. E. Klüver, S. Schulz-Maronde, S. Scheid, B. Meyer, W. G. Forssmann and K. Adermann, *Biochemistry*, 2005, **44**, 9804–9816.

66. P. Durieux and R. Nyfeler, in *Peptides. Chemistry, Structure and Biology. Proceedings of the Fourteenth American Peptide Symposium*, P. T. P. Kaumaya and R. S. Hodges, eds., Mayflower Scientific, Kingswinford, 1996, pp. 42–43.
67. M. Mergler and R. Nyfeler, in *Innovations and Perspectives in Solid Phase Synthesis & Combinatorial Libraries*, R. Epton, ed., Mayflower Scientific, Birmingham, 1996, pp. 485–486.
68. T. Kuzuhara, Y. Nakajima, K. Matsuyama and S. Natori, *J. Biochem. (Tokyo)*, 1990, **107**, 514–518.
69. M. Husbyn and A. Cuthbertson, *J. Pept. Res.*, 2002, **60**, 121–127.
70. H. Gali, G. L. Sieckmann, T. J. Hoffman, N. K. Owen, D. G. Mazuru, L. R. Forte and W. A. Volkert, *Bioconjug. Chem.*, 2002, **13**, 224–231.
71. K. Akaji, K. Fujino, T. Tatsumi and Y. Kiso, *J. Am. Chem. Soc.*, 1993, **115**, 11384–11392.
72. C. Kellenberger, H. Hietter and B. Luu, *Pept. Res.*, 1995, **8**, 321–327.
73. C. Munson and G. Barany, *J. Am. Chem. Soc.*, 1992, **115**, 10203–10210.
74. A. Cuthbertson and B. Indrevoll, *Org. Lett.*, 2003, **5**, 2955–2957.
75. A. K. Galande, R. Weissleder and C.-H. Tung, *J. Comb. Chem.*, 2005, **7**, 174–177.
76. P. Tsetseni, A. Karkantzou, S. Markos, K. Barlos, D. Gatos and K. Barlos, in *Peptides for Youth*, E. Escher, W. D. Lubell and S. Del Valle, eds., American Peptide Society, 2008, in press.
77. C. J. Camacho and D. Thirumalai, *Proc. Natl. Acad. Sci. USA*, 1995, **92**, 1277–1281.
78. J. C. Spetzler, C. Rao and J. P. Tam, *Int. J. Pept. Protein Res.*, 1994, **43**, 351–358.
79. Y. Yang, W. V. Sweeney, K. Schneider, B. T. Chait and J. P. Tam, *Prot. Sci.*, 1994, **3**, 1267–1275.
80. T. J. Zamborelli, W. S. Davidson, B. J. Harding, J. Zhang, B. D. Bennett, D. M. Lenz, Y. Young, M. Haniu, C.-F. Liu, T. Jones and M. A. Jarosinski, *J. Pept. Res.*, 2000, **55**, 359–371.
81. J. Dodt, H.-P. Müller, U. Seemüller and J.-Y. Chang, *FEBS Lett.*, 1984, **165**, 180–183.
82. J.-Y. Chang, B.-Y. Lu and P.-H. Lai, *Biochem. J.*, 2006, **394**, 249–257.
83. J. L. Arolas, F. X. Aviles, J.-Y. Chang and S. Ventura, *Trends Biochem. Sci.*, 2006, **31**, 292–301.

CHAPTER 6.3

Folding Motifs of Cystine-rich Peptides

NORELLE L. DALY AND DAVID J. CRAIK

Institute for Molecular Bioscience, University of Queensland, Brisbane 4072 QLD, Australia

6.3.1 Overview of Folding Motifs in Disulfide-rich Peptides

6.3.1.1 Sources, Activities and Structures of Disulfide-rich Peptides

As is clear from Chapter 6.1, Cys-rich peptides are produced by a large number of diverse organisms, ranging from bacteria to plants and animals. The reasons for the production of these peptides vary widely, ranging from regulatory functions to host defense or prey capture. Although Cys-rich peptides may have a highly specific function associated with the organism that produces them, there is significant scope for exploiting their activities in other applications, making them of broad general interest to chemists, biologists and medical researchers. For instance, conotoxins, a well-known class of Cys-rich peptides produced by marine cone snails for prey capture, function as ion channel blockers and based on utilizing this activity in humans, several are now being pursued as novel therapeutics for the treatment of pain.[1] In another example, Cys-rich antimicrobial peptides are being evaluated as potential anti-cancer agents.[2,3] These types of applications and others reported elsewhere in this

RSC Biomolecular Sciences
Oxidative Folding of Peptides and Proteins
Edited by Johannes Buchner and Luis Moroder
© Royal Society of Chemistry 2009
Published by the Royal Society of Chemistry, www.rsc.org

book highlight the broad interest in disulfide-rich peptides. The focus of work in our laboratory is on the structural characterization of these uniquely stable and well-defined peptides, and hence we concentrate on structural aspects in this chapter.

The three-dimensional structures of Cys-rich peptides that have been reported over the last three decades highlight the diversity of structural motifs in these peptides. Illustrative examples of such peptides with different disulfide frameworks and having different origins are shown in Figure 6.3.1. In general, the two main methods for obtaining three-dimensional structures of proteins are X-ray crystallography and NMR spectroscopy, with 86% of protein structures deposited in the Protein Data Bank determined using X-ray crystallography and 13% with NMR spectroscopy (Protein Data Bank, www.rcsb.org). The main reason for the higher proportion resolved by X-ray crystallography relates to the size limitation associated with structure determination by NMR spectroscopy, which is typically restricted to proteins of <40 kDa. However,

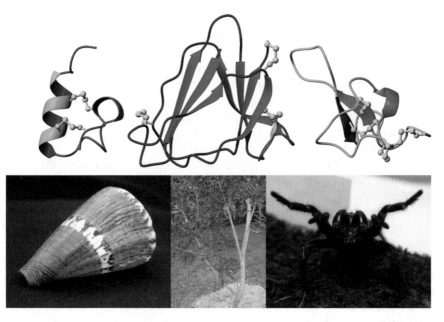

Figure 6.3.1 Cys-rich peptides. The disulfide connectivities and three-dimensional structures are shown for three cystine-rich peptides: conotoxin pl14a (PDB code 2FQC)[4]; a macadamia nut peptide, MiAMP1 (PDB code 1C01)[5]; and a spider toxin, robustoxin (PDB code 1QDP).[6] Structures were drawn using MOLMOL.[7] The disulfide bonds are shown in ball-and-stick format, the β-strands as arrows and the helical regions as thickened regions of the ribbon. Photos are shown at the bottom with the organisms that the peptides come from: (left to right) a cone snail, a macadamia nut tree and a funnel web spider. (Photographs by Dr. David Wilson, University of Queensland).

despite this limitation with respect to larger proteins NMR is highly suited to the analysis of small disulfide-bonded peptides and has the advantage of avoiding the potential difficulties associated with crystallization. The relative importance of NMR in this field is nicely exemplified by the conotoxins, for which more than 90% of the nearly 100 reported structures have been determined by NMR spectroscopy.[8]

6.3.1.2   Scope of Review

In this chapter our focus is on the structural motifs of disulfide-rich peptides. Since structural motifs in large proteins are often built on assemblies of smaller modules or domains, studies on such motifs in disulfide-rich peptides have applications to the more general field of protein structure. Disulfide-rich peptides are typically defined as proteinaceous molecules of < 100 residues, which lack an extensive hydrophobic core, and whose fold stabilization is primarily due to two or more disulfide bonds in close proximity.[9] We adopt this definition in the current review but also include peptides with single disulfide bonds as they have many structural features in common with peptides containing multiple disulfide bonds. Most of the examples we examine fall well below the 100 amino acid threshold, and indeed the majority are less than half this size.

Over recent years there has been significant effort directed towards the classification of protein structures, with the SCOP (structural classification of proteins)[10] database being one of the most comprehensive databases available in terms of classification. The most recent study that focused specifically on Cys-rich domains examined 2945 disulfide-rich protein domains, and arranged them into 41 fold groups according to structural similarity.[9] That study provided valuable information on disulfide-rich domains, highlighting structural similarities between domains that had been previously unrecognized, and is a useful precursor to this chapter. Of particular note is the fact that the analysis revealed that the cystine-knot[11] or knottin-like topology is found in nearly 40% of known Cys-rich peptide domains and is the most commonly observed structural motif. Another structural motif that is prevalent in Cys-rich peptides is the cystine-stabilized α-helix, CSH.[12] Because of this special prevalence we particularly focus on these two motifs in this chapter. We stress that our interest here is on the topologies and three-dimensional folds of the structural motifs themselves, rather than on how they fold. The oxidative folding of Cys-rich peptides is covered by Bulaj and Walewska in Chapter 6.1, which provides a complementary insight into how the structural motifs described here actually reach their final folded form.

6.3.2   Classes of Disulfide-rich Motifs

The description of any object, including a disulfide-rich motif, can in principle be made by building up successively from a description of the basic parts, quantifying the number of parts present, identifying the way they are connected, and then describing the three-dimensional assembly and function. Thus we divide this section into subsections regarding the disulfide-bond geometry, the connectivities of successively more complex disulfide frameworks and finally the functional fold classifications of these frameworks.

6.3.2.1 Geometry of the Disulfide Bond

The most basic level of description of Cys-rich peptides derives from the geometry of the component disulfide bonds, so it is useful to explicitly define the important parameters involved in disulfide-bond geometry. Classifications of geometry are based on the five χ angles that define the bond as shown in Figure 6.3.2, with the nomenclature for defining the conformation of a disulfide bond based on the signs of the χ angles. There are three basic types of disulfide bonds based on the combination of the signs of the $\chi2$, $\chi3$ and $\chi2'$ angles and these are designated spirals, hooks or staples.[13] The classification depends on the type and order of the sign; for example all positive or all negative angles are designated as spirals. Disulfide bonds are further classified as right-handed or left-handed depending on whether the sign of the $\chi3$ angle is positive or negative, respectively.[13] A shorthand nomenclature has been established to describe the type of disulfide bond; for example a right-handed spiral is designated RHSpiral. Hogg and co-workers recently included the $\chi1$ and $\chi1'$ angles to further refine the classifications of disulfide bonds.[14] This has expanded the number of types from 6 to 20 and adds a $+$, $-$, $+/-$ or $-/+$ notation to the nomenclature (*i.e.*, $+/-$ LHSpiral).[14]

The naming convention established for disulfide bonds is proving to be more useful than just a way of classifying disulfide-rich peptides. For instance, the role of a disulfide bond within a protein/peptide structure appears to be related to its configuration. Although most disulfide bonds appear to have their major role in

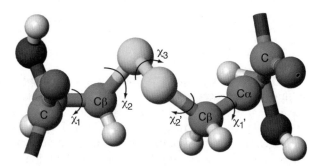

Figure 6.3.2 The geometry of a disulfide bond. The χ angles used to describe the configuration of a disulfide bond are labeled across the bond. The carbon atoms are shown in green, nitrogen in blue, oxygen in red, sulfur in yellow and hydrogens in grey.

stabilizing protein structures and are also thought to assist protein folding by decreasing the entropy of the unfolded form,[15] there are also disulfide bonds that are considered functional. The best-known role of functional disulfide bonds is as catalytic loci that mediate disulfide exchange in other proteins.[16] However, it has recently emerged that there is another type of functional disulfide bond referred to as an allosteric disulfide bond.[14,17] In this case, the disulfide bond controls the function of proteins by mediating conformational changes upon oxidation or reduction. A recent analysis of 6874 disulfide bonds from the protein data bank revealed that the hallmark of allosteric disulfide bonds is the –RHStaple.[14] In general, structural disulfide bonds are spirals and, with very few exceptions, catalytic disulfide bonds are $+/-$RHHooks.[14] Thus, analysis of disulfide-bond configuration may provide clues into the role of those bonds in a particular protein or peptide.

6.3.2.2 Disulfide-bond Frameworks

The next level of classification of Cys-rich peptides involves the number of disulfide bonds present in the peptides, *i.e.*, one-disulfide, two-disulfide, *etc*. Within these categories there are further distinctions, including the spacings between cystine residues and the connectivity of the disulfide bonds, which together define the cystine "framework". Additional features can also add complexity to Cys-rich frameworks, including post-translational modifications such as cyclization and intramolecular cleavage. The sequences of a selection of Cys-rich peptides with differing numbers of disulfide bonds, connectivities and post-translational modifications are shown in Figure 6.3.3. The topologies of these peptides are also schematically represented to highlight the diversity of structural motifs of disulfide-rich peptides. Although it is beyond the scope of a single chapter to mention all Cys-rich peptides, examples from some of the major classes are discussed below. We note that this set of examples is not exhaustive and is biased somewhat by examples with which we are more familiar as they have been studied in our laboratory, but we hope that they will provide readers with illustrative examples.

6.3.2.2.1 One-disulfide-bond Frameworks

The simplest of the Cys-rich peptides contain only one disulfide bond and generally have limited structural complexity, typically having turn or β-strand structures. One of the most significant classes of single-disulfide-bond peptides is the vasopressin-like peptide family, members of which are isolated from a range of sources, including most recently from cone snail venom,[18] and comprise only nine residues.[19] These peptides have crucial biological functions, including the regulation of water balance, the control of blood pressure, and contraction of uterine smooth muscle and mammary myoepithelium.[20] Despite their small size, these peptides have well-defined structures comprising a loop region formed by the disulfide bond between residues 1 and 6.[21–24] Oxytocin,

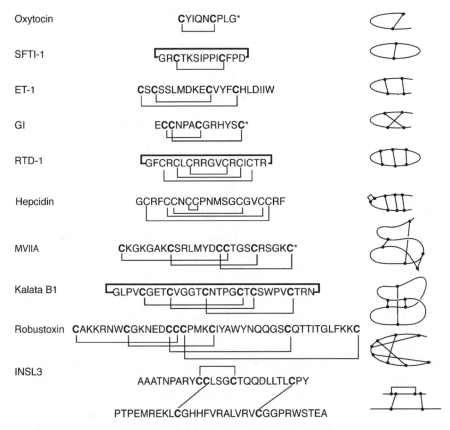

Oxytocin	CYIQNCPLG*	
SFTI-1	GRCTKSIPPICFPD	
ET-1	CSCSSLMDKECVYFCHLDIIW	
GI	ECCNPACGRHYSC*	
RTD-1	GFCRCLCRRGVCRCICTR	
Hepcidin	GCRFCCNCCPNMSGCGVCCRF	
MVIIA	CKGKGAKCSRLMYDCCTGSCRSGKC*	
Kalata B1	GLPVCGETCVGGTCNTPGCTCSWPVCTRN	
Robustoxin	CAKKRNWCGKNEDCCCPMKCIYAWYNQQGSCQTTITGLFKKC	
INSL3	AAATNPARYCCLSGCTQQDLLTLCPY	
	PTPEMREKLCGHHFVRALVRVCGGPRWSTEA	

Figure 6.3.3 Sequences and topologies of Cys-rich peptides. A selection of the sequences of representative Cys-rich peptides containing from one to four disulfide bonds is displayed.

shown in Figure 6.3.3, is a particularly well-studied member of this family and is well known for its use as an agent to induce labor and also appears to affect social behaviors such as young-parent bonding in mammals.[25–27]

A peptide that is topologically similar to the vasopressin-like peptide family, albeit with a cyclic backbone, was discovered a decade ago in sunflower seeds.[28] Sunflower trypsin inhibitor (SFTI-1) contains 14 residues and is one of the most potent trypsin inhibitors known. The three-dimensional structure comprises two antiparallel β-strands, which are braced by a disulfide bond and a cyclic backbone.[28,29] Interestingly, removal of the disulfide bond does not have a significant effect on the structure or activity. It appears the cyclic backbone and a strong network of hydrogen bonds maintain structure and activity in the absence of the disulfide bond.[30] Thus, SFTI-1 is an example where the fold is not driven by the disulfide bond, but rather stabilized by it.

A recently discovered peptide (arenicin-1) from the marine *Polychaeta* lugworm[31] is somewhat larger than the vasopressin and SFTI peptides at 21 residues but also contains a single disulfide bond. Similarly to SFTI-1 the native peptide contains two β-strands but in this case removal of the disulfide bond results in a random coil structure,[31] in contrast to the maintained structure of SFTI-1 in the absence of its disulfide bond. Thus, for arenicin-1 the disulfide bond plays a much greater role in defining the structure than that observed for SFTI-1. This is not unexpected as the head-to-tail cyclization of the backbone in SFTI-1 provides an additional degree of tethering that remains in place after removal of the disulfide bond.

6.3.2.2.2   *Two-disulfide-bond Frameworks*

Examples of well-studied two-disulfide-bond frameworks include apamin, the endothelin family (*e.g.* ET-1, Figure 6.3.3) and the α- and χ-conotoxins. Apamin, a bee venom peptide,[32] is a potassium channel blocker and was one of the first peptides of animal origin discovered to have this activity.[33] Apamin also represents one of the few cases where it has been possible to produce non-peptidic analogs of high potency[34] and it has been suggested that such compounds may have potential in the treatment of disorders involving neuronal hyperexcitability.[35] The endothelin family, and the structurally related sarafotoxin snake toxins, are potent vasoconstrictors. Their structures have been known since the early 1990s[36] following the isolation of the peptides from the culture supernatant of porcine aortic endothelial cells[37] and from the venom of the Israeli burrowing asp, *Atractaspis engaddensis*,[38,39] respectively. The α- and χ-conotoxins are isolated from the venom of *Conus* species and are potent antagonists of the nicotinic acetylcholine receptor and noradrenaline transporter, respectively.[40,41] Their primary function is as toxic venom components used by the *Conus* snails in the capture of prey but they have received significant attention from biomedical researchers as they are valuable pharmacological probes and have potential as therapeutic agents in the treatment of pain.[41] Figure 6.3.3 shows the sequence of α-conotoxin GI to illustrate the typical α-conotoxin framework.

Both the endothelin family and apamin have helical regions that are stabilized by the two-disulfide bonds. Indeed, the endothelins and apamin fit into a very common structural motif known as the cystine-stabilized α-helix[12,42] (see Section 6.3.2.3). The α-conotoxins also contain a helical region stabilized by two disulfide bonds but their connectivity is different from that of the endothelin and similar to that of apamin peptides.[43] This differing connectivity is illustrated in Figure 6.3.3, which highlights the so-called globular disulfide connectivity (Cys^{I}-Cys^{III}, Cys^{II}-Cys^{IV}) of apamin and of the prototypic α-conotoxin GI and the "ribbon" connectivity of the prototypic endothelin, ET-1 (Cys^{I}-Cys^{IV}, Cys^{II}-Cys^{III}). The third possible disulfide connectivity for a two-disulfide framework, the beads connectivity (Cys^{I}-Cys^{II}, Cys^{III}-Cys^{IV}), is not found to our knowledge in natural peptides but has been synthetically made, for example in the case of conotoxin GI analogs to explore the role of disulfide connectivity on structure.[44] The χ-conotoxins have the same "ribbon" disulfide connectivity as the endothelin peptides but their secondary structure comprises mainly a β-hairpin structure and no helical elements are present.[45,46] This comparison illustrates the importance of the sequence of Cys-rich peptides, because the same disulfide connectivity does not guarantee a similar structural motif.

In relation to the study of non-native synthetic disulfide frameworks mentioned above, we note that our focus in this chapter is on the structural aspects of disulfide motifs, rather than on their synthesis, which is covered in Chapters 6.2, 6.4 and 6.5. It is nevertheless important to note that the delineation of the structures of disulfide-rich peptides has depended in large part on the development of efficient synthetic methods to produce correctly folded peptides. The two-disulfide framework has been a particularly useful model system in this regard, because it is the simplest system containing more than one disulfide bond. The reader is referred to a number of excellent articles and reviews on the topic of the synthesis of these molecules.[47–50]

6.3.2.2.3 Three-disulfide-bond Frameworks

In a recent study[9] of disulfide-rich protein domains, involving a set of nearly 3000 peptides arranged into 41 fold groups, the average number of disulfide bonds was 3 ± 1, making the category of three-disulfide containing peptides particularly important. Conotoxins feature prominently in this category and the main structural motif found in many of these peptides is the cystine knot, which comprises two disulfide bonds (Cys^I-Cys^{IV}, Cys^{II}-Cys^V) and their connecting backbone segments forming a ring through which a third disulfide bond threads (Cys^{III}-Cys^{VI}). This motif is not limited to conotoxins and in fact is present in a very diverse range of peptides with vastly different sequences and activities.[51,52] These include, for example, a range of inhibitory molecules from plants, including α-amylase inhibitor[53] and squash trypsin inhibitors,[54] the fungal peptide AVR9,[55] spider toxins (*e.g.* ACTX-Hi:OB4219[56]) and scorpion toxins (*e.g.* maurocalcine[57]), as well as the plant cyclotides.[58] In the case of the cyclotides, the cystine-knot motif is coupled with a head-to-tail cyclized peptide backbone and the combined structural motif has been termed the cyclic cystine knot.[52] As might be expected, owing to its highly cross-braced and cyclized structure, this motif is particularly stable and is highly resistant to enzymes, chemical chaotropes and temperature extremes.[59]

Although the cystine-knot motif is very common in biologically active peptides not all three-disulfide-bond containing peptides exhibit this motif. As the name suggests, the cystine-knot motif involves a very compact knotted structure with a complex topology, but at the other end of the topological complexity scale, the cystine-ladder motif (Cys^I-Cys^{VI}, Cys^{II}-Cys^V, Cys^{III}-Cys^{IV}) can be considered as the least topologically complex of this class of peptides.[60] Examples of peptides containing the cystine-ladder motif are the mammalian θ-defensins RTD-1[61,62] and the retrocyclins.[60,63] The sequence of RTD-1 is given in Figure 6.3.3. Like the cyclotides these peptides contain a head-to-tail cyclized backbone, but there are a number of examples of laddered disulfide-rich peptides that have a conventional, acyclic peptide backbone, including the tachyplesins and the protegrins.[64,65] In the case of the protegrins, synthetic cyclized versions have been produced and shown to have advantages over their linear counterparts.[66]

Between these extremes of simple (*i.e.*, laddered) and complex (*i.e.*, knotted) disulfide connectivities a range of other topologies are seen in three-disulfide containing peptides. One peptide that has been particularly well studied is BPTI, which contains a Cys^I-Cys^{VI}, Cys^{II}-Cys^{IV} and Cys^{III}-Cys^V connectivity.[67] This molecule has been widely used as a model protein for deciphering the role of individual disulfide bonds in the folding process.[68–70] Other model peptides for studying folding processes include tick anticoagulant peptide, a factor Xa specific inhibitor that is homologous to BPTI[71] and hirudin,[72,73] a thrombin-specific inhibitor from the leech *Hirudo medicinalis*.[74]

6.3.2.2.4 Multiple-disulfide-bond Frameworks

Many peptides contain four or more disulfide bonds, with the majority being inhibitors, toxins or defense related molecules. Possibly the most disulfide-rich example is hepcidin, originally isolated from human urine and plasma ultra-filtrate and found to have antimicrobial activity,[75] but now known to be involved in iron regulation.[76] Hepcidin is only 25 residues in size, yet contains four disulfide bonds connected in a laddered arrangement, including an unusual vicinal disulfide bond (Figure 6.3.3). With a cysteine content of 8/25 residues (*i.e.*, 32% Cys), hepcidin is shaded only by the θ-defensins retrocyclin and RTD-1 in terms of cysteine content (6/18 residues = 33% Cys for these cases). The structures of hepcidins from several species have been reported and consist primarily of two β-strands[77,78] that are cross-linked by the disulfide bonds.

Significantly more complicated structural motifs are present in a range of other Cys-rich peptides, including defensins, snake and spider toxins and anti-fungal peptides.[79,80] Some of the more widely studied categories include the three-finger snake toxins with four disulfide bonds[81] and a range of plant defense proteins. An example of the latter is a series of chymotrypsin and trypsin inhibitors (C1, C2, T1, T2, T3, T4) from the stigma of the ornamental tobacco, *Nicotiana alata*. These inhibitors are involved in protecting the female reproductive tissue of the plant from predation by caterpillars.[82–86] They are ~ 50 amino acids in size and contain four disulfide bonds with a Cys^I-Cys^{VII}, Cys^{II}-Cys^V, Cys^{III}-Cys^{VI}, Cys^{IV}-Cys^{VIII} disulfide connectivity, as illustrated in Figure 6.3.4. As in a range of other inhibitory peptides, two of the disulfide bonds appear to play a role in stabilizing a somewhat flexible active site binding loop, *i.e.*, maintaining a balance between too much flexibility if it were not constrained and too little flexibility that might prevent interaction with the active site. Figure 6.3.4 shows the NMR-derived structure of the trypsin inhibitor, T1, and illustrates the bracing of the active site loop by the Cys^I-Cys^{VII}

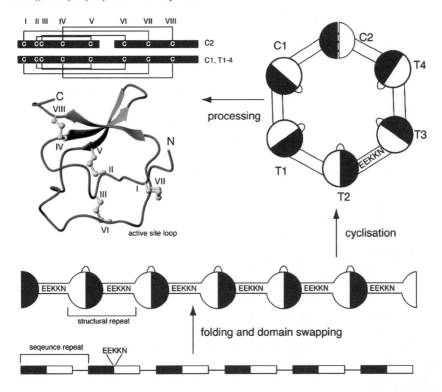

Figure 6.3.4 Structures and processing mechanism for a series of six disulfide-rich proteinase inhibitors (C1, C2, T1, T2, T3 and T4) from the ornamental tobacco *Nicotiana alata*. The gene arrangement for the *N. alata* precursor protein (NaProPI) is shown at the bottom of the diagram.[86] There are six sequence repeats separated by a linker and within each repeat is the sequence EEKKN. During folding domain swapping occurs and five structural repeats are formed whereby the EEKKN sequence now forms the linker region between the repeats. Further maturation of the precursor protein involves formation of a circle of six repeats where the chymotrypsin inhibitor C2 is formed by the joining of the half repeats present at the N- and C-termini of the precursor. The three-dimensional structure of the trypsin inhibitor domain, T1, determined using NMR spectroscopy is shown on the top left (PDB ID code 1TIH).[84] The β-strands are shown as arrows and the disulfide bonds in ball-and-stick format. The structure display was drawn with MOLMOL.[7] The disulfide connectivity of the repeats is shown above the structure with C2 comprising two chains that are disulfide linked.

and Cys[III]-Cys[VI] disulfide bonds. The structures of the other domains are similar as they differ by only a few amino acids from one another. The structures all comprise a triple-stranded β-sheet as well as the flexible binding-site loop stabilized by disulfide bonds. We return to these inhibitors in the following section because they are a case where disulfide bonds play a rather unique role not only within domains, but also in clasping domains together.

6.3.2.2.5 *Multiple-chain Frameworks*

An added complexity in disulfide-rich peptides occurs when post-translational intrachain cleavage results in multiple-chain frameworks. One of the most well known is the insulin family of peptides with three disulfide bonds, two of which are interchain and one is intrachain (see Section 6.4.2). Insulin itself has been extensively studied and was indeed one of the first proteins to be structurally characterized.[87] The relaxin family of peptides has a similar two-chain arrangement linked by disulfide bonds.[88] Our group has had a particular interest in the structure of relaxin- and insulin-like peptides over the last few years and while all maintain the basic insulin fold with significant helical structure, subtle variations in local conformations are beginning to help unravel the selectivity of individual members of the family for specific receptors.[88,89] For instance, the structure of relaxin-3 differs significantly from relaxin-2 in the C-terminal region of the B-chain and this difference provides an explanation for the differences in affinity for the relaxin receptor LGR7.[88]

Another well-characterized family of multiple-chain peptides is the 2S albumins.[90] They are 12–15 kDa in size and are generally composed of two chains with four disulfide bonds, two intrachain and two interchain.[91] 2S albumins are storage proteins that belong to the prolamin superfamily, which includes α-amylase and trypsin inhibitors.[92] The structures of several 2S albumins have been determined and they generally display four-helix bundles similar to other members of the prolamin superfamily.[93]

Finally, although not strictly an example of a multi-chain disulfide framework, the family of *N. alata* protease inhibitors mentioned in the preceding section demonstrates the variety of structural roles that disulfide bonds may play in defining and stabilizing domains within protein structures. Six of these inhibitors are all coded for by a single gene that expresses a precursor protein comprising six repeats that differ from each other by only a few amino acids. Interestingly, the folded domains in the precursor structure are offset from the sequence repeats, leaving two half domains at the N- and C-terminus of the 40 kDa precursor protein, as illustrated schematically at the bottom of Figure 6.3.4. In the final folded form of the precursor, these two half domains are bound together with three inter-domain disulfide bonds, and form an intact module (C2) similar to the contiguous domains (C1, T1-4). Thus, as illustrated in Figure 6.3.4, the precursor protein has a unique circular structure, clasped together with a tri-disulfide buckle.[85,86] The disulfide connectivity in the two-chain "buckle" domain is homologous to the connectivity in the internal domains that may be likened to beads on a necklace. Processing of the precursor protein involves proteolytic cleavage at charged linker regions between the domains, thus releasing the six small inhibitors.

6.3.2.3 Fold Classifications

Analysis of the different structural motifs and disulfide connectivities of Cys-rich peptides has led to classifications of motifs into a number of classes. As already

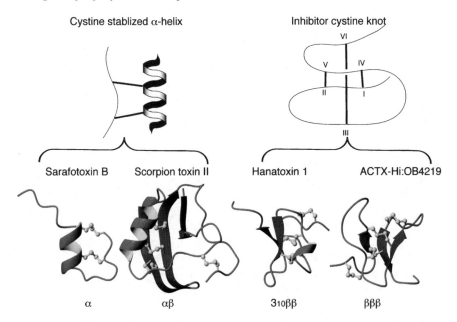

Figure 6.3.5 Two major classes of Cys-rich peptides: cystine-stabilized α-helix (CSH)[12] and the inhibitor-cystine knot (ICK).[11,52] Schematic representations of the CSH and ICK are shown at the top of the diagram with examples of structures from both classes shown at the bottom. Sarafotoxin B (PDB code 1SRB),[95] Scorpion toxin II (PDB code 1PTX),[96] Hanatoxin 1 (PDB code 1D1H),[97] ACTX-Hi:OB4219 (PDB code 1KQI).[56] Structures were drawn using MOLMOL.[7] The disulfide bonds are shown in ball-and-stick format, the β-strands as arrows and the helical regions as thickened regions of the ribbon.

noted, two of the major classes are the cystine-stabilized α-helix (CSH)[12] and the cystine-knot motif,[52] which are schematically represented in Figure 6.3.5. In addition to these types of classifications, a convenient method of describing the secondary structure has been developed whereby the occurrence of α and β structure (from N- to C-termini) is stated in the classification. For instance, the α/β scaffold[94] is quite common in a range of peptides. Examples of secondary structure types are shown in Figure 6.3.5.

The CSH motif is associated with a particular cysteine spacing in which one pair of cysteine residues is separated by three residues (CXXXC) and the other pair of cysteines is separated by one residue (CXC).[12] Two examples of CSH peptides are given in Figure 6.3.5, namely sarafotoxin B and scorpion toxin II. Despite the conserved structural motif, the overall protein folds are significantly different, with the scorpion toxin displaying the classic α/β scaffold and the sarafotoxin having a simple α fold.

Although the cysteine spacing associated with the CSH motif can lead to local structural similarity, with or without similar global folds, the presence of CXXXC and CXC segments does not guarantee the presence of a CSH motif. For example, the cystine knot can also contain the same segments of cysteine spacing but has a significantly different structure,[42] as illustrated in Figure 6.3.5. The cystine-knot motif has two topologically distinct forms termed the inhibitor cystine knot (ICK) and the growth factor cystine knot.[11,52,98] Although all cystine-knotted peptides contain the same disulfide connectivity (Cys^I-Cys^{IV}, Cys^{II}-Cys^V, Cys^{III}-Cys^{VI}), the growth factor and inhibitor cystine knots have different threading disulfide bonds and thus have been classified as two separate subfamilies. Recently it has been suggested that the cystine-knot motif originates from an ancestral motif termed the disulfide-directed β-hairpin (DDH).[99] This motif only contains two disulfide bonds and it has been suggested that the N-terminal disulfide bridge is not essential for formation of the basic ICK fold based on the structures of the cellulose binding domain of cellobiohydrolase I, which lacks this disulfide bond. Furthermore, it has been shown that the structures of the ICK containing peptides, EETI-II[100] and kalata B1,[101] are essentially unperturbed by removal of the N-terminal disulfide bond. Consequently it has been proposed that the ICK fold is a minor elaboration of the simpler ancestral DDH fold.

6.3.3 Examples and Applications of Peptide Classes with Disulfide-rich Motifs

It is clear from the discussion above that there is a wide range of disulfide-rich motifs and that even within peptides that have similar sizes and disulfide connectivities, vastly different structural motifs can arise. Conversely peptides with no sequence homology aside from cysteine residues can display very similar structures. To illustrate the range of structures and applications of disulfide-rich peptides we now elaborate on three families of Cys-rich peptides that have diverse and interesting structural motifs. As a result of their tightly folded disulfide-rich structures onto which a range of surface-exposed epitopes are displayed, they exhibit a range of biological activities that have the potential to be exploited for pharmaceutical or agricultural applications.

6.3.3.1 Cyclotides

Cyclotides are a fascinating family of macrocyclic peptides isolated from plants that have a diverse range of biological activities, including uterotonic,[102] insecticidal,[103] anti-HIV,[104] anti-fouling,[105] trypsin inhibitory[106] and cytotoxic activity.[107] On the basis of their insecticidal activity, the natural function of cyclotides in plants is considered to be as defense agents.[108] More than 100 sequences have now been reported for cyclotides and despite the diverse range of activities, all structurally characterized cyclotides display the cyclic cystine-knot motif (CCK).[108,109]

The prototypic cyclotide, kalata B1, was first reported in the early 1970s as the active ingredient in a medicinal tea used in the Congo region of Africa to facilitate childbirth.[110] In early reports the peptide was only partially characterized but was found to contain approximately 30 amino acid residues, including six cysteines.[111] The fact that it could withstand boiling to make the medicinal tea and still maintain biological (uterotonic) activity suggested that kalata B1 might have some unusual structural features, but it was some 25 years before the full sequence, the macrocyclic character and the three-dimensional structure that revealed the cystine knot were reported.[112] Several other plant-derived macrocyclic peptides of similar size were discovered in the mid 1990s in bioassay-directed screening studies,[113–115] with activities ranging from neurotensin antagonism and haemolysis to anti-HIV activity. The further discovery and structural characterization of similar macrocyclic peptides from plants over the following few years[58,116,117] led to the recognition that they all formed part of a large protein family that we named the cyclotides.[58]

Cyclotides are categorized into two main subfamilies, Möbius and bracelet, on the basis of the presence or absence of a conceptual twist in the circular backbone arising from a *cis*-geometry of the peptide bond preceding a Pro residue in loop 5 (Figure 6.3.6).[58] The subfamily names reflect the fact that a circular backbone ribbon may be regarded as a "bracelet", whereas a circular ribbon containing a 180° twist forms a topological entity known as a Möbius strip. The two macrocyclic peptides, *Momordica cochinchinesis* trypsin inhibitor I (MCoTI-I) and MCoTI-II[118] make up a third subfamily referred to as the trypsin-inhibitor cyclotides.[106,108] They are so classified on the basis of having a CCK motif,[106,119] although they are more similar in sequence to squash trypsin inhibitors of the knottin family[120,121] than they are to other cyclotides and are also referred to as cyclic knottins.[121] *Ecballium elaterium* trypsin inhibitor-II (EETI-II) is a particularly well-characterized knottin and is 44% identical to MCoTI-II. The link between the cyclic and linear cystine knot derivatives is important because much of the recent development work done on grafting foreign bioactive epitopes into cystine-knot frameworks has been done using linear derivatives (see Section 6.3.4). The structures of representative members of the three subfamilies of cyclotides are shown in Figure 6.3.6.

The *Violaceae* (violet) family is a particularly rich source of cyclotide-containing plants but they are also present in some *Rubiaceae* (coffee) family and *Cucurbitaceae* (cucurbit) plants. The sequence diversity of cyclotides resides in the loop regions between Cys residues. The six backbone loops differ in their size and sequence diversity amongst the array of known cyclotides, with some, such as loops 1 and 4, being highly conserved while others are more variable. Loops 1 and 4 form part of the embedded ring of the cystine knot and thus are presumably conserved because of their key structural role. Because they express a range of sequences on a conserved structural core, cyclotides may be regarded as a natural combinatorial template. The "diversity wheel" shown in Figure 6.3.6 highlights the variation in sequences in this combinatorial template. It is apparent that the only completely conserved residues are the six cysteine residues, emphasizing the vital role that the Cys residues and disulfide framework play in this family of molecules.

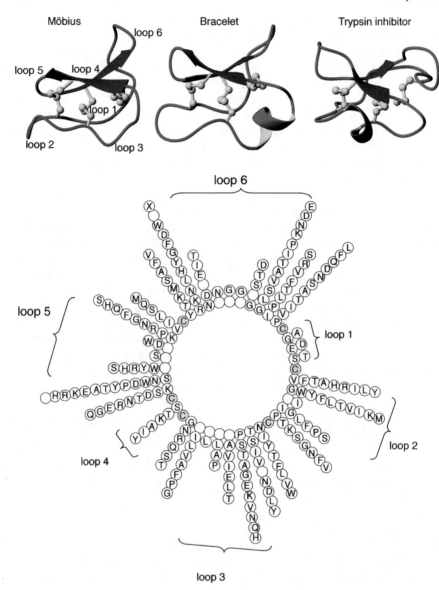

Figure 6.3.6 Subfamilies of the plant cyclotides. The three-dimensional structure of the three subfamilies (Möbius, bracelet and trypsin inhibitor) of the plant cyclotides are shown at the top of the diagram. The disulfide bonds are shown in ball-and-stick format, the β-strands as blue arrows and the helical regions as thickened regions of the ribbon. A representation of the sequence diversity of the two major classes (Möbius and bracelet) is shown at the bottom of the diagram. The Cys residues are shown in yellow and are the only absolutely conserved residues.

6.3.3.2 Conotoxin Frameworks

Conotoxins are small conformationally constrained peptides found in the venom of marine snails from the genus *Conus*.[122] They are generally Cys-rich and have exquisite potency for pharmacologically important targets, making them valuable leads in drug design applications,[123] as exemplified by a synthetic version of ω-conotoxin MVIIA that has been approved by the Food and Drug Administration for the treatment of chronic neuropathic pain.[1] Conotoxins bind their target receptors, such as ion channels and transporters, with unparalleled selectivity and specificity as a result of their high structural diversity.

Since their discovery more than two decades ago, hundreds of conotoxins have been characterized and consequently a classification scheme has been developed.[124] Conotoxins are broadly grouped into two classes, namely those with or without multiple disulfide bonds.[125] The term conotoxin is mainly used in the literature for disulfide-rich peptides from *Conus* venom in contrast to the term conopeptides for those that are not disulfide-rich, such as the conantokins.[126] Within these groupings, conotoxins are categorized in terms of pharmacological families, which have recently been grouped into superfamilies. Individual families contain peptides that have a particular disulfide framework and biological target and include the A, I, M, O, P, S and T superfamilies.[123,125,127] New sequences are regularly being reported and many have not yet been classified into superfamilies.

The sequence diversity of conotoxins is reflected in the diversity of their three-dimensional structures. Disulfide-rich conotoxins generally have well-defined structures with β-sheet, turn or α-helical elements. Figure 6.3.7 provides an indication of the range of structures seen for conotoxins, with examples of one-, two- and three-disulfide-bond containing peptides. These structures range from having a simple turn structure, to having α-helixes and β-hairpins as the major elements of secondary structure. As in the cyclotides, the disulfide-framework plays an important role as a scaffold, from which bioactive loops extend.

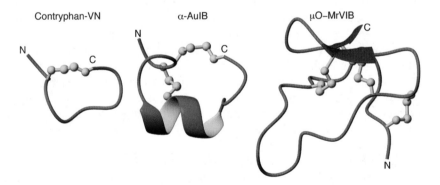

Figure 6.3.7   Conotoxin structures. The three-dimensional structures of contryphan VN (PDB code 1NXN),[128] AuIB (PDB code 1MXN),[129] and μO MrVIB (PDB code 1RMK).[130]

For example, conotoxin MrVIB contains a cystine-knot motif similar to that in conotoxin MVIIA, but in the case of MrVIB, loop 2 (the extended loop) is highly disordered, presumably as a result of flexibility in this region, in contrast to a more-defined structure in MVIIA. These topologically similar peptides target different ion channels.[131,132] MVIIA is a calcium channel blocker,[131] whereas MrVIB is a potent sodium channel blocker[132] as well as being active at molluscan calcium channels.[133] The disorder in loop 2 of MrVIB may account for this cross-channel activity.[130] Thus, although peptides can have conserved structures, there can still be significant differences in the dynamics of disulfide-rich motifs. These differences in dynamics may play critical roles in the functions of these Cys-rich peptides.

The disulfide-rich frameworks of the conotoxins accommodate a wide range of post-translational modifications. Many conotoxins have C-terminal amides, and 4-hydroxyproline and γ-carboxyglutamic acid residues are reasonably prevalent. Other examples include the presence of D-amino acids and pyro-glutamic acid.[134] Although the functions of the modified amino acids are not known in all cases, they potentially contribute to variations in activity, structure and biological stability. Again, the molecules represent a case of a defined Cys-rich core being used as a scaffold into which a diverse array of amino acids is displayed. In summary, conotoxins are a rich source of peptides with diverse structures and functions that have potential as therapeutic agents and are also useful pharmacological probes.

6.3.3.3 Defensin Frameworks

Defensins are another family of disulfide-rich peptides with a diverse range of structures. In general they are cationic antimicrobial peptides found in a wide range of species. They appear to be the only class of peptides involved in the innate immune response that is conserved between vertebrates, invertebrates and plants. Mammalian defensins are classified into three subfamilies, namely the α-, β- and θ-defensins. All contain six conserved cysteine residues but differ in their disulfide connectivities. The α-defensins, have a Cys^I-Cys^{VI}, Cys^{II}-Cys^{IV} and Cys^{III}-Cys^V, the β-defensins a Cys^I-Cys^V, Cys^{II}-Cys^{IV} and Cys^{III}-Cys^{VI}, and the θ-defensins a Cys^I-Cys^{VI}, Cys^{II}-Cys^V and Cys^{III}-$CysI^V$ connectivity.[135] The plant defensins have four disulfide bonds in a Cys^I-Cys^{VIII}, Cys^{II}-Cys^V, Cys^{III}-Cys^{VI}, Cys^{IV}-Cys^{VII} arrangement.[136] The structures of examples of each of these four classes of defensins are shown in Figure 6.3.8. Although these peptides are all antimicrobial and involved in defense, they have a diverse range of structures, ranging from the cystine-ladder arrangement of the θ-defensins to the αβ motif of the plant defensins.

The genetic origins of the disulfide-rich motifs of the defensins is quite interesting as the α-defensins are synthesized as pre-pro-peptides from single genes, but the θ-defensins are biosynthesized from two α-defensin-like pre-cursor peptides of nine amino acids that are ligated to form a head-to-tail cyclic peptide.[61,140,141] In the case of rhesus monkeys the θ-defensins are expressed in

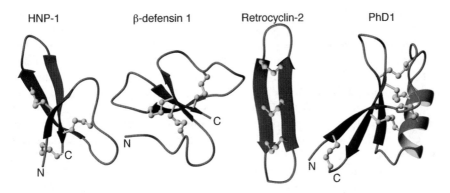

Figure 6.3.8 Three-dimensional structures of defensins from different classes. α-Defensin HNP-1 (PDB code 2PM4),[137] β-defensin 1 (PDB code 1KJ5),[138] retrocyclin-2 (PDB code 2ATG),[60] and the plant defensin PhD1 (PDB code 1N4N).[139]

leucocytes and play a role in host defence, but in the case of humans a premature stop codon before the coding sequence prevents expression of the peptides. This is an ironic twist of fate, as chemical synthesis of these retrocyclins shows them to be potent anti-HIV agents,[142] so it is intriguing that modern humans have lost the ability to synthesize these apparently useful molecules.

The defensins lend themselves to both agricultural and therapeutic applications. It has been suggested that plant defensins may be useful in generating transgenic crops with improved pathogen resistance[143] and that mammalian defensins may be useful in the prevention of infection. Indeed, the human θ-defensins, retrocyclins, have been implicated as topical agents to prevent the infection of HIV.[144–146] Furthermore, there are apparent differences in the mechanisms of defensin action and further studies aimed at understanding the structure activity relationships of defensins may identify some unifying principles that will contribute to their development as new drugs for the prevention of infection.[135]

6.3.4 Disulfide-rich Frameworks as Bioengineering Scaffolds

There is a range of potential applications of Cys-rich peptides based on their intrinsic activity, but they are also emerging as potential scaffolds in drug design. The disulfide bonds in the sequences can confer stability on the peptides and the intrinsic sequence diversity can be exploited by transferring non-native sequences into the scaffolds to confer particular biological activities. This approach has now been validated with several different frameworks, suggesting that a range of distinct activities can be accommodated and consequently a range of potential therapeutic applications can be targeted with these scaffolds. For example, the introduction of novel binding epitopes has been achieved using a range of protein based inhibitors. In one example, the α-amylase inhibitor tendamistat was expressed on the surface of phage and two loops randomized using PCR mutagenesis. Libraries were then tested for binding to a monoclonal antibody that binds endothelin. This approach was successful in generating novel binding molecules to this antibody from a scaffold that does not bind the antibody.[147] BPTI, one of the most highly structurally characterized Cys-rich peptides has also been used as a scaffold. Protease inhibitors with altered enzyme specificity towards human neutrophil elastase were generated using randomization of residues and subsequent phage display selection using the BPTI scaffold.[148,149]

As an alternative approach to using random mutagenesis to generate novel binding activities, the transfer of active site regions from larger proteins to disulfide-rich scaffolds has also proved successful. For example, a scorpion toxin scaffold was used to display the curaremimetic loop of a snake neurotoxin and the CDR2-like site of human CD4.[150] These peptides were shown to maintain the α/β structure of the parent molecule and the acetylcholine binding of the snake neurotoxin loop and the HIV-1 gp120 binding were conferred onto the scaffold.[150,151]

The concept that epitopes can be grafted onto disulfide-rich scaffolds has been further exemplified with cystine-knot peptides. For example, a recent study described the grafting of platelet aggregation inhibitory activity onto EETI-II,[152] a cystine-knot containing a peptide from the squash trypsin inhibitor family. In this case the source epitope came from disintegrins, small proteins whose bioactive RGD tripeptide sequence is directly involved in inhibition of platelet aggregation. The RGD sequence was grafted into a surface loop of EETI-II and the resulting derivative had significantly higher activity than the linear peptide.[152] More broadly, there has been significant effort placed on designing small peptidic inhibitors of platelet aggregation and Integrillin® is an example in clinical use for the treatment of patients with acute coronary syndrome.

Grafting a biologically active epitope onto a cystine-knot scaffold and enhancing activity is a significant achievement but potential drug candidates require more than just activity *in vitro* and must have appropriate pharmacokinetic properties. A recent study explored these properties for three cystine-knot miniproteins, including analogs of EETI-II.[153] The peptides permeate well through rat small intestinal mucosa relative to other peptide drugs such as insulin and bacitracin. Enzymatic digestion occurred only for a few proteases and it was suggested that this limitation may be overcome by mutating out particular cleavage sites. Overall, grafting studies have highlighted the potential of disulfide-rich scaffolds as templates in drug design and indicate that the favorable characteristics of Cys-rich peptides can be exploited in peptide engineering studies.

Although this approach of grafting bioactive sequences onto a disulfide-rich framework is built on the premise that the framework is stable, any additional improvement in stability is in principle advantageous. In this regard, the use of selenium substitution for disulfide bonds has recently been applied. This work has been built on the important synthetic methodologies developed by Moroder and colleagues.[50] In a recent example, the incorporation of selenocysteine into α-conotoxins had a significant influence in slowing the rate of disulfide scrambling and reduction compared to the native conotoxins (see Section 8.2.4).[154] Given that reduction and disulfide scrambling in biological fluids can decrease the effectiveness of a drug, these results indicate that selenoconotoxins may be useful in drug design applications.

6.3.5 Outlook

Cys-rich peptides have a diverse range of structures and NMR spectroscopy has played a vital role in determining these structures. With increasing numbers of sequences being discovered, NMR will continue to be a valuable technique in this area and with advances in technology, including higher field strengths and cryoprobes, more detailed information is likely to become available more rapidly. This structural information will be vital not only for increasing our understanding of the roles of disulfide bonds in modulating structure and activity, but will be essential for driving applied outcomes. The preliminary success stories relating to the use of Cys-rich peptides as scaffolds in drug design suggest that this area will rapidly increase in the near future. Given the structural diversity of disulfide-rich peptides, the scope for exploring these peptides as templates is enormous. The structural diversity means that a wide range of therapeutic applications can be explored by choosing scaffolds with appropriate structures for particular grafting applications.

Acknowledgements

Work in our laboratory on disulfide-rich peptides is supported by grants from the Australian Research Council (ARC) and the National Health and Medical Research Council (NHMRC). DJC is an ARC Professorial Fellow and NLD is a NHMRC Industry Fellow. We thank our colleagues Quentin Kaas, Conan Wang and Kathryn Greenwood from The University of Queensland for helpful comments on the manuscript.

References

1. G. P. Miljanich, *Curr. Med. Chem.*, 2004, **11**, 3029–3040.
2. Y. Chen, X. Xu, S. Hong, J. Chen, N. Liu, C. B. Underhill, K. Creswell and L. Zhang, *Cancer Res.*, 2001, **61**, 2434–2438.
3. N. Papo and Y. Shai, *Cell. Mol. Life Sci.*, 2005, **62**, 784–790.
4. J. S. Imperial, P. S. Bansal, P. F. Alewood, N. L. Daly, D. J. Craik, A. Sporning, H. Terlau, E. Lopez-Vera, P. K. Bandyopadhyay and B. M. Olivera, *Biochemistry*, 2006, **45**, 8331–8340.
5. A. M. McManus, K. J. Nielsen, J. P. Marcus, S. J. Harrison, J. L. Green, J. M. Manners and D. J. Craik, *J. Mol. Biol.*, 1999, **293**, 629–638.
6. P. K. Pallaghy, D. Alewood, P. F. Alewood and R. S. Norton, *FEBS Lett.*, 1997, **419**, 191–196.

7. R. Koradi, M. Billeter and K. Wüthrich, *J. Mol. Graph.*, 1996, **14**, 51–55.

8. U. C. Marx, N. L. Daly and D. J. Craik, *Magn. Reson. Chem.*, 2006, **44**, S41–50.

9. S. Cheek, S. S. Krishna and N. V. Grishin, *J. Mol. Biol.*, 2006, **359**, 215–237.

10. A. G. Murzin, S. E. Brenner, T. Hubbard and C. Chothia, *J. Mol. Biol.*, 1995, **247**, 536–540.

11. N. W. Isaacs, *Curr. Opin. Struct. Biol.*, 1995, **5**, 391–395.

12. Y. Kobayashi, H. Takashima, H. Tamaoki, Y. Kyogoku, P. Lambert, H. Kuroda, N. Chino, T. X. Watanabe, T. Kimura, S. Sakakibara and L. Moroder, *Biopolymers*, 1991, **31**, 1213–1220.

13. J. S. Richardson and D. C. Richardson, in *Prediction of Protein Structure and the Principles of Protein Conformation*, G. D. Fasman, ed., Plenum Press, New York, 1989, pp. 1–99.

14. B. Schmidt, L. Ho and P. J. Hogg, *Biochemistry*, 2006, **45**, 7429–7433.

15. J. M. Thornton, *J. Mol. Biol.*, 1981, **151**, 261–287.

16. H. Nakamura, *Antioxid. Redox Signal.*, 2005, **7**, 823–828.

17. P. J. Hogg, *Trends Biochem. Sci.*, 2003, **28**, 210–214.

18. S. Dutertre, D. Croker, N. L. Daly, A. Andersson, M. Muttenthaler, N. G. Lumsden, D. J. Craik, P. F. Alewood, G. Guillon and R. J. Lewis, *J. Biol. Chem.*, 2008, **283**, 7100–7108.

19. M. Birnbaumer, *Trends Endocrinol. Metab.*, 2000, **11**, 406–410.

20. T. A. Treschan and J. Peters, *Anesthesiology*, 2006, **105**, 599–612.

21. B. Syed Ibrahim and V. Pattabhi, *J. Mol. Biol.*, 2005, **348**, 1191–1198.

22. C. K. Wu, B. Hu, J. P. Rose, Z. J. Liu, T. L. Nguyen, C. Zheng, E. Breslow and B. C. Wang, *Protein Sci.*, 2001, **10**, 1869–1880.

23. M. Iwadate, E. Nagao, M. P. Williamson, M. Ueki and T. Asakura, *Eur. J. Biochem.*, 2000, **267**, 4504–4510.

24. E. Trzepalka, W. Kowalczyk and B. Lammek, *J. Pept. Res.*, 2004, **63**, 333–346.

25. M. Febo, M. Numan and C. F. Ferris, *J. Neurosci.*, 2005, **25**, 11637–11644.

26. C. F. Ferris, S. F. Lu, T. Messenger, C. D. Guillon, N. Heindel, M. Miller, G. Koppel, F. R. Bruns and N. G. Simon, *Pharmacol. Biochem. Behav.*, 2006, **83**, 169–174.

27. Y. Kozorovitskiy, M. Hughes, K. Lee and E. Gould, *Nat. Neurosci.*, 2006, **9**, 1094–1095.

28. S. Luckett, R. S. Garcia, J. J. Barker, A. V. Konarev, P. R. Shewry, A. R. Clarke and R. L. Brady, *J. Mol. Biol.*, 1999, **290**, 525–533.

29. M. L. Korsinczky, H. J. Schirra, K. J. Rosengren, J. West, B. A. Condie, L. Otvos, M. A. Anderson and D. J. Craik, *J. Mol. Biol.*, 2001, **311**, 579–591.

30. M. L. Korsinczky, R. J. Clark and D. J. Craik, *Biochemistry*, 2005, **44**, 1145–1153.

31. J. U. Lee, D. I. Kang, W. L. Zhu, S. Y. Shin, K. S. Hahm and Y. Kim, *Biopolymers*, 2007, **88**, 208–216.
32. E. Habermann, *Science*, 1972, **177**, 314–322.
33. B. E. Banks, C. Brown, G. M. Burgess, G. Burnstock, M. Claret, T. M. Cocks and D. H. Jenkinson, *Nature*, 1979, **282**, 415–417.
34. D. H. Jenkinson, *Br. J. Pharmacol.*, 2006, **147 S1**, S63–71.
35. J. F. Liegeois, F. Mercier, A. Graulich, F. Graulich-Lorge, J. Scuvee-Moreau and V. Seutin, *Curr. Med. Chem.*, 2003, **10**, 625–647.
36. M. Sokolovsky, *J. Neurochem.*, 1992, **59**, 809–821.
37. M. Yanagisawa, H. Kurihara, S. Kimura, Y. Tomobe, M. Kobayashi, Y. Mitsui, Y. Yazaki, K. Goto and T. Masaki, *Nature*, 1988, **332**, 411–415.
38. C. Takasaki, N. Tamiya, A. Bdolah, Z. Wollberg and E. Kochva, *Toxicon*, 1988, **26**, 543–548.
39. A. Bdolah, Z. Wollberg, G. Fleminger and E. Kochva, *FEBS Lett.*, 1989, **256**, 1–3.
40. B. G. Livett, D. W. Sandall, D. Keays, J. Down, K. R. Gayler, N. Satkunanathan and Z. Khalil, *Toxicon*, 2006, **48**, 810–829.
41. B. G. Livett, K. R. Gayler and Z. Khalil, *Curr. Med. Chem.*, 2004, **11**, 1715–1723.
42. H. Tamaoki, R. Miura, M. Kusunoki, Y. Kyogoku, Y. Kobayashi and L. Moroder, *Protein Engineer.*, 1998, **11**, 649–659.
43. E. L. Millard, N. L. Daly and D. J. Craik, *Eur. J. Biochem.*, 2004, **271**, 2320–2326.
44. J. Gehrmann, P. F. Alewood and D. J. Craik, *J. Mol. Biol.*, 1998, **278**, 401–415.
45. I. A. Sharpe, J. Gehrmann, M. L. Loughnan, L. Thomas, D. A. Adams, A. Atkins, E. Palant, D. J. Craik, D. J. Adams, P. F. Alewood and R. J. Lewis, *Nat. Neurosci.*, 2001, **4**, 902–907.
46. K. P. Nilsson, E. S. Lovelace, C. E. Caesar, N. Tynngard, P. F. Alewood, H. M. Johansson, I. A. Sharpe, R. J. Lewis, N. L. Daly and D. J. Craik, *Biopolymers (Peptide Sci.)*, 2005, **80**, 815–823.
47. S. Fiori, S. Pegoraro, S. Rudolph-Böhner, J. Cramer and L. Moroder, *Biopolymers*, 2000, **53**, 550–564.
48. S. Pegoraro, S. Fiori, J. Cramer, S. Rudolph-Böhner and L. Moroder, *Protein Sci.*, 1999, **8**, 1605–1613.
49. S. Pegoraro, S. Fiori, S. Rudolph-Böhner, T. X. Watanabe and L. Moroder, *J. Mol. Biol.*, 1998, **284**, 779–792.
50. D. Besse, N. Budisa, W. Karnbrock, C. Minks, H. J. Musiol, S. Pegoraro, F. Siedler, E. Weyher and L. Moroder, *Biol. Chem.*, 1997, **378**, 211–218.
51. R. S. Norton and P. K. Pallaghy, *Toxicon*, 1998, **36**, 573–583.
52. D. J. Craik, N. L. Daly and C. Waine, *Toxicon*, 2001, **39**, 43–60.
53. S. Lu, P. Deng, X. Liu, J. Luo, R. Han, X. Gu, S. Liang, X. Wang, F. Li, V. Lozanov, A. Patthy and S. Pongor, *J. Biol. Chem.*, 1999, **274**, 20473–20478.

54. A. Christmann, K. Walter, A. Wentzel, R. Kratzner and H. Kolmar, *Protein Engineer.*, 1999, **12**, 797–806.

55. H. W. van den Hooven, H. A. van den Burg, P. Vossen, S. Boeren, P. J. de Wit and J. Vervoort, *Biochemistry*, 2001, **40**, 3458–3466.

56. K. J. Rosengren, D. Wilson, N. L. Daly, P. F. Alewood and D. J. Craik, *Biochemistry*, 2002, **41**, 3294–3301.

57. A. Mosbah, R. Kharrat, Z. Fajloun, J. G. Renisio, E. Blanc, J. M. Sabatier, M. El Ayeb and H. Darbon, *Proteins*, 2000, **40**, 436–442.

58. D. J. Craik, N. L. Daly, T. Bond and C. Waine, *J. Mol. Biol.*, 1999, **294**, 1327–1336.

59. M. L. Colgrave and D. J. Craik, *Biochemistry*, 2004, **43**, 5965–5975.

60. N. L. Daly, Y. K. Chen, K. J. Rosengren, U. C. Marx, M. L. Phillips, A. J. Waring, W. Wang, R. I. Lehrer and D. J. Craik, *Biochemistry*, 2007, **46**, 9920–9928.

61. Y.-Q. Tang, J. Yuan, G. Ösapay, K. Ösapay, D. Tran, C. J. Miller, A. J. Ouellette and M. E. Selsted, *Science*, 1999, **286**, 498–502.

62. M. Trabi, H. J. Schirra and D. J. Craik, *Biochemistry*, 2001, **40**, 4211–4221.

63. A. M. Cole, W. Wang, A. J. Waring and R. I. Lehrer, *Curr. Protein Pept. Sci.*, 2004, **5**, 373–381.

64. T. Nakamura, H. Furunaka, T. Miyata, F. Tokunaga, T. Muta, S. Iwanaga, M. Niwa, T. Takao and Y. Shimonishi, *J. Biol. Chem.*, 1988, **263**, 16709–16713.

65. A. Aumelas, M. Mangoni, C. Roumestand, L. Chiche, E. Despaux, G. Grassy, B. Calas and A. Chavanieu, *Eur. J. Biochem.*, 1996, **237**, 575–583.

66. J. P. Tam, C. Wu and J. L. Yang, *Eur. J. Biochem.*, 2000, **267**, 3289–3300.

67. B. Kassell and M. Laskowski Sr, *Biochem. Biophys. Res. Commun.*, 1965, **20**, 463–468.

68. D. P. Goldenberg, *Trends Biochem. Sci.*, 1992, **17**, 257–261.

69. J. Y. Chang and L. Li, *Arch. Biochem. Biophys.*, 2005, **437**, 85–95.

70. J. Y. Chang, L. Li and A. Bulychev, *J. Biol. Chem.*, 2000, **275**, 8287–8289.

71. L. Waxman, D. E. Smith, K. E. Arcuri and G. P. Vlasuk, *Science*, 1990, **248**, 593–596.

72. J.-Y. Chang, *J. Biol. Chem.*, 1995, **270**, 25661–25666.

73. J.-Y. Chang, H. Grossenbacher, B. Meyhack and W. Maerki, *FEBS Lett.*, 1993, **336**, 53–56.

74. F. Markwardt and P. Walsmann, *Hoppe Seyler's Z Physiol. Chem.*, 1958, **312**, 85–98.

75. A. Krause, S. Neitz, H. J. Magert, A. Schulz, W. G. Forssmann, P. Schulz-Knappe and K. Adermann, *FEBS Lett.*, 2000, **480**, 147–150.

76. T. Ganz, *Curr. Top. Microbiol. Immunol.*, 2006, **306**, 183–198.

77. H. N. Hunter, D. B. Fulton, T. Ganz and H. J. Vogel, *J. Biol. Chem.*, 2002, **277**, 37597–37603.

78. X. Lauth, J. J. Babon, J. A. Stannard, S. Singh, V. Nizet, J. M. Carlberg, V. E. Ostland, M. W. Pennington, R. S. Norton and M. E. Westerman, *J. Biol. Chem.*, 2005, **280**, 9272–9282.

79. F. Maggio and G. F. King, *Toxicon*, 2002, **40**, 1355–1361.
80. F. Maggio and G. F. King, *J. Biol. Chem.*, 2002, **277**, 22806–22813.
81. R. M. Kini, *Clin. Exp. Pharmacol. Physiol.*, 2002, **29**, 815–822.
82. A. H. Atkinson, J. L. Lind, A. E. Clarke and M. A. Anderson, *Biochem. Soc. Symp.*, 1994, **60**, 15–26.
83. K. J. Nielsen, R. L. Heath, M. A. Anderson and D. J. Craik, *J. Mol. Biol.*, 1994, **242**, 231–243.
84. K. J. Nielsen, R. L. Heath, M. A. Anderson and D. J. Craik, *Biochemistry*, 1995, **34**, 14304–14311.
85. M. J. Scanlon, M. C. Lee, M. A. Anderson and D. J. Craik, *Structure*, 1999, **7**, 793–802.
86. M. C. Lee, M. J. Scanlon, D. J. Craik and M. A. Anderson, *Nat. Struct. Biol.*, 1999, **6**, 526–530.
87. T. L. Blundell, J. F. Cutfield, S. M. Cutfield, E. J. Dodson, G. G. Dodson, D. C. Hodgkin and D. A. Mercola, *Diabetes*, 1972, **21**, 492–505.
88. K. J. Rosengren, F. Lin, R. A. Bathgate, G. W. Tregear, N. L. Daly, J. D. Wade and D. J. Craik, *J. Biol. Chem.*, 2006, **281**, 5845–5851.
89. K. J. Rosengren, S. Zhang, F. Lin, N. L. Daly, D. J. Scott, R. A. Hughes, R. A. Bathgate, D. J. Craik and J. D. Wade, *J. Biol. Chem.*, 2006, **281**, 28287–28295.
90. R. J. Youle and A. H. Huang, *Plant Physiol.*, 1978, **61**, 13–16.
91. D. Pantoja-Uceda, M. Bruix, J. Santoro, M. Rico, R. Monsalve and M. Villalba, *Biochem. Soc. Trans.*, 2002, **30**, 919–924.
92. M. Kreis, B. G. Forde, S. Rahman, B. J. Miflin and P. R. Shewry, *J. Mol. Biol.*, 1985, **183**, 499–502.
93. D. Pantoja-Uceda, P. R. Shewry, M. Bruix, A. S. Tatham, J. Santoro and M. Rico, *Biochemistry*, 2004, **43**, 6976–6986.
94. F. Bontems, C. Roumestand, B. Gilquin, A. Menez and F. Toma, *Science*, 1991, **254**, 1521–1523.
95. A. R. Atkins, R. C. Martin and R. Smith, *Biochemistry*, 1995, **34**, 2026–2033.
96. D. Housset, C. Habersetzer-Rochat, J. P. Astier and J. C. Fontecilla-Camps, *J. Mol. Biol.*, 1994, **238**, 88–103.
97. H. Takahashi, J. I. Kim, H. J. Min, K. Sato, K. J. Swartz and I. Shimada, *J. Mol. Biol.*, 2000, **297**, 771–780.
98. P. K. Pallaghy, K. J. Nielsen, D. J. Craik and R. S. Norton, *Protein Sci.*, 1994, **3**, 1833–1839.
99. X. Wang, M. Connor, R. Smith, M. W. Maciejewski, M. E. Howden, G. M. Nicholson, M. J. Christie and G. F. King, *Nat. Struct. Biol.*, 2000, **7**, 505–513.
100. D. Le-Nguyen, A. Heitz, L. Chiche, M. El Hajji and B. Castro, *Protein Sci.*, 1993, **2**, 165–174.
101. N. L. Daly, R. J. Clark and D. J. Craik, *J. Biol. Chem.*, 2003, **278**, 6314–6322.
102. L. Gran, F. Sandberg and K. Sletten, *J. Ethnopharmacol.*, 2000, **70**, 197–203.

103. C. Jennings, J. West, C. Waine, D. Craik and M. Anderson, *Proc. Natl. Acad. Sci. USA*, 2001, **98**, 10614–10619.

104. K. R. Gustafson, T. C. McKee and H. R. Bokesch, *Curr. Protein Pept. Sci.*, 2004, **5**, 331–340.

105. U. Göransson, M. Sjogren, E. Svangard, P. Claeson and L. Bohlin, *J. Nat. Prod.*, 2004, **67**, 1287–1290.

106. M. E. Felizmenio-Quimio, N. L. Daly and D. J. Craik, *J. Biol. Chem.*, 2001, **276**, 22875–22882.

107. P. Lindholm, U. Göransson, S. Johansson, P. Claeson, J. Gulbo, R. Larsson, L. Bohlin and A. Backlund, *Mol. Cancer Ther.*, 2002, **1**, 365–369.

108. D. J. Craik, N. L. Daly, J. Mulvenna, M. R. Plan and M. Trabi, *Curr. Prot. Pept. Sci.*, 2004, **5**, 297–315.

109. D. J. Craik, S. Simonsen and N. L. Daly, *Curr. Opin. Drug Discov. Devel.*, 2002, **5**, 251–260.

110. L. Gran, *Medd. Nor. Farm. Selsk*, 1970, **12**, 173–180.

111. L. Gran, *Lloydia*, 1973, **36**, 174–178.

112. O. Saether, D. J. Craik, I. D. Campbell, K. Sletten, J. Juul and D. G. Norman, *Biochemistry*, 1995, **34**, 4147–4158.

113. T. Schöpke, M. I. Hasan Agha, R. Kraft, A. Otto and K. Hiller, *Sci. Pharm.*, 1993, **61**, 145–153.

114. K. R. Gustafson, R. C. Sowder II, L. E. Henderson, I. C. Parsons, Y. Kashman, J. H. Cardellina II, J. B. McMahon, R. W. Buckheit Jr., L. K. Pannell and M. R. Boyd, *J. Am. Chem. Soc.*, 1994, **116**, 9337–9338.

115. K. M. Witherup, M. J. Bogusky, P. S. Anderson, H. Ramjit, R. W. Ransom, T. Wood and M. Sardana, *J. Nat. Prod.*, 1994, **57**, 1619–1625.

116. U. Göransson, T. Luijendijk, S. Johansson, L. Bohlin and P. Claeson, *J. Nat. Prod.*, 1999, **62**, 283–286.

117. K. R. Gustafson, L. K. Walton, R. C. Sowder II, D. G. Johnson, L. K. Pannell, J. H. Cardellina II and M. R. Boyd, *J. Nat. Prod.*, 2000, **63**, 176–178.

118. J. F. Hernandez, J. Gagnon, L. Chiche, T. M. Nguyen, J. P. Andrieu, A. Heitz, T. T. Hong, T. T. C. Pham and D. Le Nguyen, *Biochemistry*, 2000, **39**, 5722–5730.

119. A. Heitz, J. F. Hernandez, J. Gagnon, T. T. Hong, T. T. C. Pham, T. M. Nguyen, D. Le-Nguyen and L. Chiche, *Biochemistry*, 2001, **40**, 7973–7983.

120. D. Le Nguyen, A. Heitz, L. Chiche, B. Castro, R. A. Boigegrain, A. Favel and M. A. Coletti-Previero, *Biochimie*, 1990, **72**, 431–435.

121. L. Chiche, A. Heitz, J. C. Gelly, J. Gracy, P. T. Chau, P. T. Ha, J. F. Hernandez and D. Le-Nguyen, *Curr. Protein Pept. Sci.*, 2004, **5**, 341–349.

122. B. M. Olivera, J. Rivier, J. K. Scott, D. R. Hillyard and L. J. Cruz, *J. Biol. Chem.*, 1991, **266**, 22067–22070.

123. D. J. Adams, P. F. Alewood, D. J. Craik, R. D. Drinkwater and R. J. Lewis, *Drug Dev. Res.*, 1999, **46**, 219–234.
124. B. M. Olivera, J. Rivier, C. Clark, C. A. Ramilo, G. P. Corpuz, F. C. Abogadie, E. E. Mena, S. R. Woodward, D. R. Hillyard and L. J. Cruz, *Science*, 1990, **249**, 257–263.
125. H. Terlau and B. M. Olivera, *Physiol. Rev.*, 2004, **84**, 41–68.
126. R. T. Layer, J. D. Wagstaff and H. S. White, *Curr. Med. Chem.*, 2004, **11**, 3073–3084.
127. J. M. McIntosh, S. Gardner, S. Luo, J. E. Garrett and D. Yoshikami, *Eur. J. Pharmacol.*, 2000, **393**, 205–208.
128. T. Eliseo, D. O. Cicero, C. Romeo, M. E. Schinina, G. R. Massilia, F. Polticelli, P. Ascenzi and M. Paci, *Biopolymers*, 2004, **74**, 189–198.
129. J. L. Dutton, P. S. Bansal, R. C. Hogg, D. J. Adams, P. F. Alewood and D. J. Craik, *J. Biol. Chem.*, 2002, **277**, 48849–48857.
130. N. L. Daly, J. A. Ekberg, L. Thomas, D. J. Adams, R. J. Lewis and D. J. Craik, *J. Biol. Chem.*, 2004, **279**, 25774–25782.
131. B. M. Olivera, L. J. Cruz, V. de Santos, G. W. LeCheminant, D. Griffin, R. Zeikus, J. M. McIntosh, R. Galyean, J. Varga, W. R. Gray and J. Rivier, *Biochemistry*, 1987, **26**, 2086–2090.
132. J. M. McIntosh, A. Hasson, M. E. Spira, W. R. Gray, W. Li, M. Marsh, D. R. Hillyard and B. M. Olivera, *J. Biol. Chem.*, 1995, **270**, 16796–16802.
133. M. Fainzilber, R. van der Schors, J. C. Lodder, K. W. Li, W. P. Geraerts and K. S. Kits, *Biochemistry*, 1995, **34**, 5364–5371.
134. K. J. Nielsen, M. Watson, D. J. Adams, A. K. Hammarstrom, P. W. Gage, J. M. Hill, D. J. Craik, L. Thomas, D. Adams, P. F. Alewood and R. J. Lewis, *J. Biol. Chem.*, 2002, **277**, 27247–27255.
135. M. E. Klotman and T. L. Chang, *Nat. Rev. Immunol.*, 2006, **6**, 447–456.
136. F. R. Terras, K. Eggermont, V. Kovaleva, N. V. Raikhel, R. W. Osborn, A. Kester, S. B. Rees, S. Torrekens, F. Van Leuven, J. Vanderleyden, B. P. A. Cammue and W. F. Broekaert, *Plant Cell*, 1995, **7**, 573–588.
137. G. Zou, E. de Leeuw, C. Li, M. Pazgier, C. Li, P. Zeng, W. Y. Lu, J. Lubkowski and W. Lu, *J. Biol. Chem.*, 2007, **282**, 19653–19665.
138. D. J. Schibli, H. N. Hunter, V. Aseyev, T. D. Starner, J. M. Wiencek, P. B. McCray Jr, B. F. Tack and H. J. Vogel, *J. Biol. Chem.*, 2002, **277**, 8279–8289.
139. B. J. Janssen, H. J. Schirra, F. T. Lay, M. A. Anderson and D. J. Craik, *Biochemistry*, 2003, **42**, 8214–8222.
140. L. Leonova, V. N. Kokryakov, G. Aleshina, T. Hong, T. Nguyen, C. Zhao, A. J. Waring and R. I. Lehrer, *J. Leukoc. Biol.*, 2001, **70**, 461–464.
141. D. Tran, P. A. Tran, Y. Q. Tang, J. Yuan, T. Cole and M. E. Selsted, *J. Biol. Chem.*, 2002, **277**, 3079–3084.
142. R. I. Lehrer, *Nat. Rev. Microbiol.*, 2004, **2**, 727–738.
143. B. P. Thomma, B. P. Cammue and K. Thevissen, *Planta*, 2002, **216**, 193–202.

144. C. Munk, G. Wei, O. O. Yang, A. J. Waring, W. Wang, T. Hong, R. I. Lehrer, N. R. Landau and A. M. Cole, *AIDS Res. Hum. Retroviruses*, 2003, **19**, 875–881.
145. A. Cole, W. Wang, A. J. Warring and R. I. Lehrer, *Curr. Prot. Pept. Sci.*, 2004, **5**, 373–381.
146. B. Yasin, W. Wang, M. Pang, N. Cheshenko, T. Hong, A. J. Waring, B. C. Herold, E. A. Wagar and R. I. Lehrer, *J. Virol.*, 2004, **78**, 5147–5156.
147. S. J. McConnell and R. H. Hoess, *J. Mol. Biol.*, 1995, **250**, 460–470.
148. B. L. Roberts, W. Markland, A. C. Ley, R. B. Kent, D. W. White, S. K. Guterman and R. C. Ladner, *Proc. Natl. Acad. Sci. USA*, 1992, **89**, 2429–2433.
149. B. L. Roberts, W. Markland, K. Siranosian, M. J. Saxena, S. K. Guterman and R. C. Ladner, *Gene*, 1992, **121**, 9–15.
150. C. Vita, J. Vizzavona, E. Drakopoulou, S. Zinn-Justin, B. Gilquin and A. Menez, *Biopolymers*, 1998, **47**, 93–100.
151. C. C. Huang, F. Stricher, L. Martin, J. M. Decker, S. Majeed, P. Barthe, W. A. Hendrickson, J. Robinson, C. Roumestand, J. Sodroski, R. Wyatt, G. M. Shaw, C. Vita and P. D. Kwong, *Structure*, 2005, **13**, 755–768.
152. S. Reiss, M. Sieber, V. Oberle, A. Wentzel, P. Spangenberg, R. Claus, H. Kolmar and W. Losche, *Platelets*, 2006, **17**, 153–157.
153. M. Werle, T. Schmitz, H. L. Huang, A. Wentzel, H. Kolmar and A. Bernkop-Schnurch, *J. Drug Target*, 2006, **14**, 137–146.
154. C. J. Armishaw, N. L. Daly, S. T. Nevin, D. J. Adams, D. J. Craik and P. F. Alewood, *J. Biol. Chem.*, 2006, **281**, 14136–14143.

CHAPTER 6.4

Double-stranded Cystine Peptides

JOHN D. WADE

Howard Florey Institute and School of Chemistry, University of Melbourne, Victoria 3010, Australia

6.4.1 Introduction

The elucidation of the primary structure of porcine insulin in the 1950s by Frederick Sanger and the remarkable finding that it consisted of not one but two peptide chains that were held together by two interchain and one intra-chain disulfide bonds in a distinctive arrangement remains one of the major scientific discoveries of the twentieth century.[1] Relaxin, a peptide hormone that has important actions on connective tissue during parturition and which was, like insulin, discovered in the early 1900s[2] was isolated from pregnant sows and shown by protein chemistry methods to also possess a two-chain three-disulfide bonded structure similar to insulin.[3] The two conclusions from this exciting finding were that there existed a family of such peptides which was likely to contain additional members and, secondly, such peptide architectural struc-tures, *viz.*, double-stranded cystine peptides, may in fact be more common than expected. The advent of both more sophisticated protein isolation and efficient sequencing techniques and, importantly, gene isolation and recombinant DNA sequencing has resulted in an explosive growth in the determination of the primary structures of new peptides and proteins. However the vast majority of those that contain one or more cystine cross-links are single-chain biomole-cules. The landmark sequencing of the human genome in 2003[4] has allowed an

RSC Biomolecular Sciences
Oxidative Folding of Peptides and Proteins
Edited by Johannes Buchner and Luis Moroder
© Royal Society of Chemistry 2009
Published by the Royal Society of Chemistry, www.rsc.org

unprecedented opportunity to "mine" the estimated 30 000 plus genes for additional novel double-stranded cystine peptides. Thus far, however, only one has been identified. Nevertheless, the ongoing sequencing of genomes of other mammals as well as of non-mammalian species[5,6] is likely to ultimately yield additional fascinating double-stranded cystine peptide sequences even if these should be comparatively rare. Of particular interest will be the collation of the resulting information to assist in the determination of the evolutionary pathways leading to the formation of these peptides. As by far our greatest understanding of such peptides is as a consequence of studies on insulin and its related family members, the primary emphasis of this chapter will be on such peptides.

6.4.2   Insulin and Insulin-like Peptides

6.4.2.1   Human Insulin

The primary structure of human insulin was determined soon after that of the porcine peptide and was shown also to consist of a 21-residue A-chain disulfide cross-bridged to a 30-residue B-chain.[7] The advent of recombinant DNA technology together with gene sequencing led to the rapid identification of additional human insulin-like peptides. The primary structure of human relaxin-1 was then determined by complementary DNA sequencing. The six cysteine residues that make up the three disulfide bonds were predicted to be in an identical disposition to that of insulin thus confirming the concept of the insulin superfamily.[8] Two new members of this family were then identified by conventional protein chemistry techniques but shown to be unique in that they were single chain polypeptides bearing the insulin-like disulfide cross-links. These were named insulin-like growth factors (IGFs) I and II.[9,10] Both have important mitogenic roles. A second relaxin known as relaxin-2[11] was discovered, which has subsequently been shown to be the primary mediator of birth-regulating actions in lower mammals, but to exhibit multiple physiological (pleiotropic) properties in higher mammals including humans.[12] These were followed by insulin-like peptide 3 (INSL3, also known as Leydig cell insulin-like peptide or relaxin-like factor),[13] which has recently been identified to be an important fertility regulator,[14] and insulin-like peptide 4 (placentin, early pregnancy insulin-like peptide, INSL4)[15] whose function remains unknown. More recently, the existence of two further human insulin-like peptides, INSLs 5 and 6, has been predicted from cDNA sequences.[16,17] The function of the former peptide remains uncertain, but it may be an important neurological peptide given its primary site of expression being the hypothalamic-pituitary axis of the brain.[18] In contrast, INSL6 is principally a testicular peptide with hitherto unknown actions. Finally, a novel relaxin sequence, relaxin-3, was discovered from the Celera genome database.[19] It has since been shown to be a key brain peptide with potential roles in feeding and stress regulation.[20] Further analysis of both the Celera and public

A-Chain

Relaxin-1	R	P	Y	V	A	L	F	E	K	C	C	L	I	G	C	T	K	R	S	L	A	K	Y	C		
Relaxin-2	Z	L	Y	S	A	L	A	N	K	C	C	H	V	G	C	T	K	R	S	L	A	R	F	C		
Relaxin-3	D	V	L	A	G	L	S	S	S	C	C	K	W	G	C	S	K	S	E	I	S	S	L	C		
Insulin				G	I	V	E	Q	C	C	T	S	I	C	S	L	Y	Q	L	E	N	Y	C	N		
IGF-1	~	A	P	Q	T	G	I	V	D	E	C	C	F	R	S	C	D	L	R	R	L	E	M	Y	C	A ~
IGF-2	~	R	R	S	R	G	I	V	E	E	C	C	F	R	S	C	D	L	A	L	L	E	T	L	C	A ~
INSL3	A	A	A	T	N	P	A	R	Y	C	C	L	S	G	C	T	Q	Q	D	L	L	T	L	C	P	Y
INSL4	R	S	G	R	H	R	F	D	P	F	C	C	E	V	I	C	D	D	G	T	S	V	K	L	C	
INSL5			Q	D	L	Q	T	L	C	C	T	D	G	C	S	M	T	D	L	S	A	L	C			
INSL6			G	Y	S	E	K	C	C	L	T	G	C	T	K	E	E	L	S	I	A	C				

B-chain

Relaxin-1	K	W	K	D	D	V	I	K	L	C	G	R	E	L	V	R	A	Q	I	A	I	C	G	M	S	T	W	S	
Relaxin-2	D	S	W	M	E	E	V	I	K	L	C	G	R	E	L	V	R	A	Q	I	A	I	C	G	M	S	T	W	S
Relaxin-3	R	A	A	P	Y	G	V	R	L	C	G	R	E	F	I	R	A	V	I	F	T	C	G	G	R	W			
Insulin	F	V	N	Q	H	L	C	G	S	H	L	V	E	A	L	Y	L	V	C	G	E	R	G	F	F	Y	T	P	K A
IGF-1	~	G	P	E	T	L	C	G	A	E	L	V	D	A	L	Q	F	V	C	G	D	R	G	F	Y	F	N	K	P ~
IGF-2	~	P	S	E	T	L	C	G	G	E	L	V	D	T	L	Q	F	V	C	G	D	R	G	F	Y	F	S	R	P ~
INSL3	P	T	P	E	M	R	E	K	L	C	G	H	H	F	V	R	A	L	V	R	V	C	G	G	P	R	W	S	T E A
INSL4	Z	S	L	A	A	E	L	R	G	C	G	P	R	F	G	K	H	L	L	S	Y	C	P	M	P	E	K	T	F T T T P
INSL5	S	K	E	S	V	R	L	C	G	L	E	Y	I	R	T	V	I	Y	I	C	A	S	S	R	W				
INSL6	S	D	I	S	S	A	R	K	L	C	G	R	Y	L	V	K	E	I	E	K	L	C	G	H	A	N	W	S	F R

Figure 6.4.1 Primary structure of members of the human insulin superfamily. Cysteine connectivities highlighted in shades of grey.

domain human genome databases has failed to reveal the presence of additional insulin-like peptides thus showing that membership of the human insulin superfamily is restricted to ten (Figure 6.4.1).

With the exception of IGFs I and II, each of these superfamily members are characterized by two peptide chains which are constrained by three disulfide bonds that comprise the so-called insulin structural motif. Within the A-chain, there is a single intramolecular disulfide bond between CysA6-A11 (insulin numbering). Two intermolecular disulfide bonds tether the A- and B-chains together between CysA7-B7 and CysA20-B19 (Figure 6.4.2A). It is now recognized that each peptide member of the insulin superfamily is assembled on the ribosome as a single chain pre-pro-peptide that undergoes subsequent proteolytic processing (with the exception of the IGFs) to yield the mature two-chain form.[21] Within the human insulin superfamily and with the exception of the six Cys residues and a single Gly residue within the B-chain adjacent to one of the Cys residues (B8, human insulin numbering), the primary structure homology is low. However, the tertiary structures that have thus far been determined show a very high level of conformational similarity with the A-chain containing two well-defined α-helices at its termini separated by a turn (helix-loop-helix motif). The B-chain contains a long central α-helix that ends with a β-turn at residues B20-B23 followed by an extended β-strand. The two chains are arranged in a manner such that the N-terminus of the A-chain is in close proximity to the C-terminus of the B-chain (Figure 6.4.2B). The cystines are absolutely crucial for the maintenance of this insulin-like structure and subsequent biological activity of the respective insulin-like peptides.[22]

A

A-chain

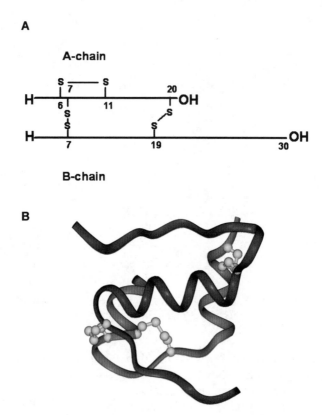

B-chain

B

Figure 6.4.2 A) General primary structure of insulin-like peptides highlighting cysteine pairings (human insulin residue numbering). B) Schematic representation of human insulin X-ray crystal structure. A-chain shown in magenta and B-chain in blue. Disulfide bonds are shown in yellow. (Figure courtesy of Dr Johan Rosengren, University of Kalmar, Sweden.)

6.4.2.2   Insulin-like Peptides from Other Species

To date, insulin, IGFs I and II have been identified in all classes of vertebrates including the dog, cat, horse and rabbit and also in amphibians, fish and birds.[23] In contrast, neither relaxin-1 nor INSL4 is present in lower mammals.[24] Relaxin-2 has been detected in all species with the exception of birds where only relaxin-3 has thus far been found. There are in fact two avian relaxin-3 genes and it has been speculated that one of these has adopted the reproductive activity of relaxin-2. Detailed phylogenetic analysis of relaxin genes shows that the ancestral relaxin is a relaxin-3-like sequence which emerged prior to the divergence of fish.[25]

In the 1980s, the brain of silkworm, *Bombyx mori*, was shown to possess an insulin-like peptide, bombyxin, that contained an A- and B-chain and the characteristic insulin disulfide network.[26] Five molecular species of bombyxins have since been identified with each differing only in primary structure. Since that time, a large number of insulin-like peptides has been reported for other invertebrates, including insects, molluscs and nematodes. These findings emphasize that insulin is an evolutionarily ancient molecule which is present in all metazoa. Remarkably, compared to vertebrate insulin genes, the invertebrate insulin-like peptides comprise large multi-gene families and are characterized by highly divergent sequences and large variations in chain length.[27] This is highlighted in *Caenorhabditis elegans*, which was shown by a combination of sequence- and structure-based algorithms to possess 37 candidate genes encoding insulin-like peptides.[28] The predicted peptide sequences have each been grouped in one of four subclasses: the native (canonical) insulin, one in which an additional intermolecular disulfide bond exists, one in which the intramolecular disulfide bond is substituted by a hydrophobic bond, and an unusual three repeats of each chain. The great majority of these invertebrate peptides are expressed in the central nervous system with each appearing to function as a mitogenic growth factor. Curiously, the newly deciphered yeast genome has failed to reveal the presence of an insulin-like peptide, which indicates that insulin-like genes have co-evolved with metazoa. Despite detailed, ongoing phylogenetic analyses of invertebrate insulin-like genes, there is yet no clear consensus regarding the origin and molecular evolution of the insulin superfamily.[23,29]

6.4.3 Other Double-stranded Cystine Peptides

6.4.3.1 From Natural Origin

Mammalian non-insulin-like double-stranded cystine proteins and peptides have been reported but these are rare. Immunoglobins (IgGs, antibodies) possess light and heavy chains that are linked by a single cystine.[30] Pairs of the light-heavy chains are folded together and restrained in a parallel alignment by both non-covalent bonds as well as by a pair of cystines within the heavy chains. Uteroglobin is a small globular protein that is formed by two identical 70-residue polypeptides linked in an antiparallel manner by two disulfide bonds.[31] In addition to α-atrial natriuretic peptide, a 28-residue peptide containing a single intramolecular disulfide bond, atrial tissue contains β-atrial natriuretic peptide (α-ANP), which is also an antiparallel disulfide dimer of α-ANP.[32] Lipophilin, isolated from tears, consists of a 69-residue A-chain linked to a 77-residue C-chain *via* three interchain disulfide bonds,[33] and human plasma apolipoprotein (apo) D exists as a disulfide-linked glycoprotein heterodimer with other lipoproteins.[34] Botrocetin, isolated from the venom of the snake *Bothrops jararaca*, is a heterodimer composed of an α-subunit (consisting of 133 amino acid residues) and a β-subunit (consisting of 125 amino acid residues) cross-linked by a single disulfide bond.[35] A novel antimicrobial peptide named halocidin was extracted from tunicate *Halocynthia aurantium*. It consists of two different peptide chains containing 18 and 15 amino acid residues, respectively, which are linked covalently by a single di-sulfide bond.[36] A somewhat similar polypeptide was extracted from the skin granular glands of the tree-frog *Phyllomedusa distincta* and called distinctin.[37] It is a 5.4 kDa heterodimeric peptide having a 22-residue A-chain and 24-residue B-chain and significant antimicrobial activity. Crustaceans have been shown to possess an androgenic gland hormone that regulates sex dif-ferentiation and which is a heterodimeric glycopeptide consisting of an A-chain of 29 amino acids bearing a single N-linked glycan and two intramolecular disulfide bonds. This chain is linked *via* two disulfides to a B-chain of 44 amino acids.[38] From the seed of a Chinese plant, a sweet-tasting protein named mabinlin II was isolated, characterized and shown to be a heterodimer con-sisting of a 33-residue A-chain and a 72-residue B-chain. The latter has two intramolecular disulfide bonds and is linked to the A-chain by two disulfide bonds.[39] It possesses extraordinary heat stability and is at least 100 times sweeter than sucrose. Three additional mabinlins have been identified with each differing from mabinlin II only in chain length or individual amino acids within the chains.[40] With further genomic analyses, additional double-stranded cystine peptides will undoubtedly be identified.

6.4.3.2 Synthetic Constructs

The literature is replete with examples of non-native homo- and heterodimeric peptides that have been assembled either chemically or by random combination of the separate chains in solution. Such peptides have a variety of uses including as immunogens, prodrugs, conformationally-constrained mimetics and recep-tor agonists or antagonists.[41–48]

6.4.4 Oxidative Folding

The study of the *in vitro* folding and oxidation of cystine-containing peptides and proteins has long been of significant interest for these provide insights into the general principles and mechanisms of protein folding. Equally important, such information allows for optimization of methods to produce adequate material for detailed biochemical analyses. With the exception of insulin, similar studies of double-stranded cystine peptides have only comparatively recently been undertaken in detail due, in part, to the complexity afforded by the two-chain nature of such peptides.

6.4.4.1 Combination of Two Chains into Double-stranded Peptides

Other than the studies on insulin, one of the earliest investigations on the oxidative folding of double-stranded cystine peptides was that of the hinge fragment of human IgG1. The two heavy chains of IgG are linked in a parallel orientation by two proximal disulfide bridges in the region of the molecule that connects the two Fab chains with the Fc segment.[49] The core of the hinge consists of a bis(cystinyl)octapeptide [H-Cys-Pro-Pro-Cys]$_2$. Oxidative folding studies in aqueous solution (pH 6.8) of an extended octapeptide monomer of the hinge peptide, H-Thr(tBu)-Cys-Pro-Pro-Cys-Pro-Ala-Pro-OH, led to the formation of an unexpected disproportionally high ratio of parallel dimer relative to the antiparallel dimer and the intramolecular disulfide monomer of 90:8:2.[50] Interestingly, similar experiments with a C-terminally extended tridecapeptide fragment of the hinge peptide yielded solely parallel-aligned dimer.[51] These important findings emphasized the likely structural role of the hinge segment in mediating correct alignment of the IgG1 chains. More recently, studies on the oxidative folding of the synthetic A- and B-chains of mabinlin II showed that the absence of a redox buffering system led to acquisition of only dimers of the individual chains themselves.[52] Addition of GSH/GSSG to the oxidation solution also failed to produce the desired products with only glutathione adducts of the chains being obtained. Successful combination of the protein was achieved in a remarkable 50% overall yield when oxidative folding was carried out at pH 8.0 and room temperature in the presence of reduced and oxidized Cys.[52] These results highlight the need for careful optimization of oxidative folding conditions on a case by case basis.

By far the greatest attention has been applied to insulin superfamily peptides, particularly insulin itself given its obvious biological and medical significance. For this reason, the following sections will deliberately focus on the oxidative folding of insulin and its homologs.

6.4.4.2 Insulin and Insulin-like Peptides

The pioneering work of Anfinsen, which showed that the disulfide bonds within native ribonuclease A could be reduced and the resulting single-chain linear protein regenerated by re-oxidation, was a milestone in both protein chemistry and structure-function relationship research.[53] It highlighted that the tertiary structure of a protein is largely dictated by its primary structure. However, it is little known that this Nobel Prize-winning study was, in fact, preceded by the successful oxidative assembly of insulin from the individual reduced A- and B-chains. In the mid 1950s and in the lead-up to the successful chemical synthesis of porcine insulin, chemists at the Shanghai Institute of Biochemistry found that insulin could be reduced with β-mercaptoethanol and then re-oxidized in high pH buffer to generate biologically active insulin.[54] However, yields were very low leading to the subsequent examination and use of the S-sulfitolysis method developed by Swan[55] to prepare S-sulfonated forms of the A- and B-chains that could be more readily handled and purified. The two chains could then be refolded and oxidized in the presence of a reducing agent (β-mercaptoethanol) in overall yield of about 10%.[56] Without being aware of this development in Shanghai, Dixon and Wardlaw published a seminal paper in 1960 that reported the reduction and reoxidation of insulin chains to generate native insulin in 1–2% yield and thus confirming the hitherto-unpublished Chinese studies.[57]

These early findings clearly showed that the primary structure of the peptide chains were the key determinants for the correct alignment and subsequent cystine formation, itself remarkable given that the number of statistically possible disulfide heterodimers of the two insulin chains is 12. There is, of course, virtually a limitless number of higher molecular weight disulfide isomers that can also be formed. These early studies also showed that the S-sulfonated chains were incapable of folding and producing disulfide-bonded insulin and that the presence of a large excess of reducing agent, typically β-mercaptoethanol, was required. It further became apparent that oxidative folding under physiological conditions (pH around 8.0, 37 °C), in fact, proceeded poorly. Exhaustive investigations by the Chinese group led to further incremental refinement of their insulin oxidative refolding method in which the A-chain was added in excess (1.2–2.0) and a lower temperature (4 °C) used for combination in high pH (9–12) buffer. These modified conditions led to a reported yield of insulin of an astonishing 50% calculated on the basis of limiting B-chain and were subsequently employed for the successful chemical synthesis of porcine insulin.[58] At about the same time, competing groups in Germany and the USA announced the successful chemical syntheses of ovine and human insulins following oxidative refolding of the synthetic A- and B-chains.[59,60]

A detailed study by Katsoyannis and Tomesko showed that significantly improved yields of insulin could be obtained if a much greater excess of A-chain was used to compensate for the resulting greater losses of this chain following S-reduction of the S-sulfonate form and the separation of the S-thiol peptide from the reducing agent (β-mercaptoethanol) by ethyl acetate extraction instead of centrifugation. A remarkable 60–80% yield of insulin was reported.[61] Despite these notable achievements, use of this approach has generally provided low and variable overall yields of insulin analogs regardless of the source of the individual chains (either chemically or recombinant DNA-derived) thus restricting progress into structure-function studies of this important hormone. Further incremental improvement in this oxidative refolding methodology was afforded by replacement of the previously used large excess of β-mercaptoethanol with a modest excess of DTT.[62] Combination of the two S-sulfonated chains is carried out in the presence of this reducing agent as its oxidized form is readily separated from the reaction products. Use of these conditions has since enabled the preparation of many insulin analogs although overall yields were generally much lower than for the native peptide chains. It is now abundantly clear that even subtle changes in the secondary structure of the chains afforded by either or both truncation or substitution of certain amino acids can significantly impact upon the efficiency of the reoxidation process.[63–65]

Curiously, the mixed S-sulfonate/thiol approach was found to work poorly for the folding and reoxidation of the A- and B-chains of relaxin.[66] The primary problem was the very poor solubility of the B-chain whether in its reduced or S-sulfonate form. This limitation was overcome by the use of only reduced peptide chains in a complex buffering system, which contained both organic solvents and denaturing agents such as guanidine hydrochloride. Upon reduction of the concentration of the latter and subsequent addition of both reduced A-chain and DTT, combination of the two chains was then able to proceed, although reaction times were variable and occasionally long.[67] Overall yields could be as high as 50% relative to the starting B-chain. Remarkably, and in contrast, no *in vitro* oxidative refolding conditions could be found under which successful combination of the two chains of human relaxin-3 could be achieved. Recourse to chemical regioselective disulfide-bond synthesis was required in order to obtain this peptide for subsequent biological study (see Section 6.4.5).[68]

Of other two-chain members of the human insulin superfamily, only INSL3 has been successfully prepared by oxidative folding and combination of the individual A- and B-chains. In contrast to relaxin, both the reduced chains of this peptide are freely soluble in the oxidation buffer and very good yields of native peptide (*ca.* 20%) were obtained using the conditions developed for relaxin oxidative refolding.[69] Again, however, the preparation of modified analogs of INSL3 is considerably more difficult and recourse to chemically directed methods is required. The oxidative refolding of both INSL5 and 6 has been examined in this author's laboratory and was found to be similar to that of relaxin in that yields were very low as a consequence of poor chain solubility.

6.4.5 Regioselective Disulfide Formation

Without question, instrumental to the acquisition and study of insulin and related peptides has been chemical peptide synthesis. Following Du Vigneaud's at-the-time benchmark assembly of the nonapeptide oxytocin in the mid 1950s by solution methods,[70] the synthesis of insulin by China's Shanghai Institute of Biochemistry, and the groups of Katsoyannis in the USA and Zahn in Germany, were landmark achievements.[58–60] However, the tedious assembly of the two chains in solution (typically requiring enormous manpower) and their subsequent random (although modestly efficient) combination were obviously severe limitations to detailed structure-function studies of this important hormone. Merrifield's development of the solid phase peptide synthesis methodology[71] is considered by many to be a turning point in the ready acquisition of peptides and, indeed, was used to successfully prepare insulin in weeks rather than many months.[72] However, as elegant as the new methodology was, the individual chains of insulin each presented significant synthetic challenges because of their inherent physicochemical properties with the result that, typically, low yields of purified peptides were obtained which, in turn, afforded very low overall yields of chain-combined product. In particular, the A-chain was recognized at a very early stage to be not only difficult to sequentially assemble due to a plethora of both cysteine and bulky, hydrophobic residues (a now-recognized "difficult" sequence), but also its very poor solubility made its purification and subsequent handling cumbersome and difficult.[61,73,74] Indeed, the A-chain was recently shown to possess the property of inducing fibrillogenesis leading to the formation of not only insoluble fibrils itself but of insulin, too.[75,76] This clearly has important implications for its clinical use in high concentration. While the solid phase peptide synthesis methodology was being further refined and optimized, proponents of the solution phase methodology continued important studies on the preparation and assay of analogs of insulin. Of particular interest, Sieber and his colleagues reported a successful and very elegant synthesis of human insulin using a fragment condensation approach in which protected peptide segments were sequentially assembled together with regioselective disulfide-bond formation.[77] Possibly the most important segment was the peptide 1-13 of the A-chain, which contained three cysteine residues, two of which are involved in the intramolecular disulfide bond. Advantage was taken of the unique chemical properties of the two thiol-protecting groups that were employed in this segment, the acid-labile trityl (Trt) and the acid-resistant acetamidomethyl (Acm), such that the pair of Trt groups on Cys6 and Cys11 were selectively removed and the intramolecular disulfide bond simultaneously formed by iodolysis in trifluoroethanol whilst the S-Acm derivative of Cys7 remained intact (for abbreviations of protecting groups see Section 6.2.2).[78] This segment was the last to be attached to the sequentially assembled insulin intermediate and the final disulfide bond, between CysA7 and B7 was formed by iodolysis of the Cys(Acm) derivatives in acetic acid. The capacity to scale up the assembly was of particular importance as was the bypassing of the random chain oxidative combination with its accompanying variable and low yields thus allowing the preparation of analogs. However, the substantial synthetic efforts required for this approach were a severe limitation.

Nevertheless, the successful use of different thiol-protecting groups for the subsequent sequential construction of disulfide bonds was an important turning point in the chemical synthesis of double-stranded cystine peptides. The strategies developed and their applications have been comprehensively reviewed elsewhere (for additional detailed information see Chapter 6.2).[79–83] A significant advance in the synthesis of insulin-like peptides was heralded by the assembly of human relaxin-2 *via* regioselective disulfide-bond formation, in which each of the three disulfide bonds was formed sequentially by stepwise removal of pairs of orthogonal S-protecting groups followed by oxidation of the resulting thiol groups.[84] In this approach, a complex eight-step strategy (Figure 6.4.3A) was followed in which the A-chain containing three different S-protecting groups was assembled by the Fmoc method. After cleavage of the

---

Figure 6.4.3 Strategies for the regioselective disulfide bond assembly of insulin-like peptides. Pairs of cysteine residues and related derivatives, which are successively disulfide cross-bridged are shown in identical colors. A) Synthesis of human relaxin-2: (I) $I_2/50\%$ acetic acid, (II) HF, (III) pH 4.5, 37 °C, 24 h, (IV) $I_2/70\%$ acetic acid, (V) OH$^-$, (VI) I$^-/90\%$ TFA. B) Synthesis of bombyxin: (I) air oxidation, pH 8.0, 4 °C, 5 days, (II) DPDS, TFMSA/TFA, (III) pH 8.0, 30 min, (IV) $I_2/95\%$ acetic acid. C) Synthesis of human insulin: (I) DPDS, 2-PrOH/acetic acid, 40 min., (II) 8 M urea, pH 8.5, 50 min, (III) $I_2/80\%$ acetic acid, (IV) $CH_3SiCl_3/PhS(O)/TFA$. Abbreviations: DPDS, 2,2′-dipyridyldisulfide; HF, hydrogen fluoride; TFA, trifluoroacetic acid; TFMSA, trifluoromethanesulfonic acid; for abbreviations of thiol-protecting groups see Chapter 6.2.

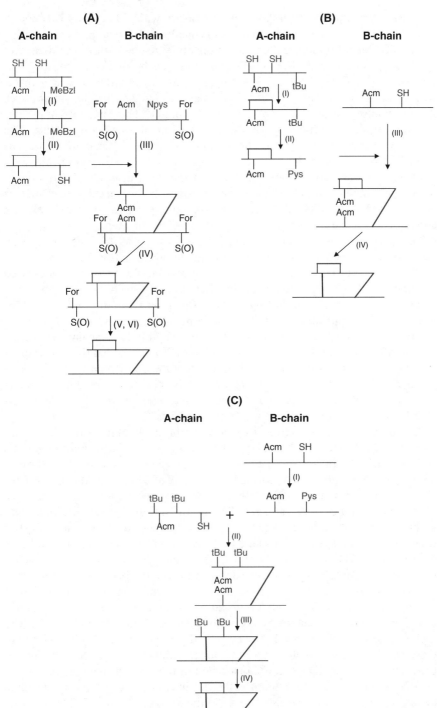

peptides from the solid support by treatment with TFA, two side-chain-deprotected cysteines were oxidized with I_2 in 50% acetic acid and the resulting intramolecular disulfide-bond-containing peptide treated with HF to deprotect the C-terminal Cys residue. The B-chain, prepared by Boc-solid phase synthesis, contained Cys(Acm) and Cys(Npys) residues, the latter of which reacts with thiols such as that present within the A-chain intermediate to produce the intermolecular disulfide bond in pH 4.5 buffer. The final intermolecular disulfide bond was formed by iodolysis of the pair of Cys(Acm) residues. The two Trp residues within the B-chain were protected at the indole group as formyl derivatives and the Met residue as sulfoxide. The formyl groups were removed without corresponding damage to the three disulfides *via* treatment with base. Finally ammonium iodide in aqueous TFA was used to back-reduce Met-sulfoxide to Met. Overall yield of native human relaxin-2 was very low as a consequence of the numerous intermediate purification steps.

The silkworm insulin-like peptide, bombyxin-IV, was the next to be successfully assembled by regioselective disulfide-bond formation between the two synthetic chains. The strategy was simpler and employed only Fmoc-based solid phase synthesis for the two chains that were selectively S-protected with a combination of Trt, Acm and *tert*-butyl (tBu) groups (Figure 6.4.3B).[85] Following A-chain intramolecular disulfide-bond formation by air-oxidation of the two free thiols, treatment of the resulting peptide intermediate with 2,2'-dipyridyldisulfide (DPDS) in trifluoromethanesulfonic acid led to displacement of the C-terminal S-tBu group and formation of the Cys(Pys) derivative, which reacted with the free thiol of Cys22 of the B-chain to form the first intermolecular disulfide. The second and final intermolecular disulfide was generated by iodolysis of the pair of Cys(Acm) residues.[86]

An important regioselective disulfide assembly of human insulin by a simpler, wholly Fmoc-based strategy was reported by Kiso and associates (Figure 6.4.3C).[87] In this synthetic route, the two chains were first linked by an intermolecular disulfide bond formed by thiolysis of B19Cys(Pys) with A20Cys. The second intermolecular disulfide bond was formed by iodolysis of the two Cys(SAcm) residues in A7 and B7. Finally, the intramolecular disulfide bond within the A-chain was generated by treatment of the corresponding Cys(tBu) residues with methyltrichlorosilane/diphenylsulfoxide. The overall yield was approximately 1% relative to the starting B-chain. This elegant approach is not suited to Trp-containing peptides such as the relaxins because this residue is destroyed by the final reaction conditions *via* chlorination of the indole ring.[88]

A detailed unpublished study in this author's laboratory on the regioselective disulfide synthesis of relaxin by the Fmoc-chemistry showed that the strategy used for the synthesis of bombyxin-IV provided the best overall yields although these remained low, largely due to the ongoing difficulty with the solubility of the B-chain intermediate. Careful placement of the thiol-protecting groups was also required in order to minimize the detrimental effect of the final iodine oxidation step used for disulfide linking of the Cys(Acm) residues. This strategy led to the first-ever acquisition of human relaxin-3, which was previously shown to be refractive to production *via* the random chain refolding approach.[68]

The quantity of peptide thus obtained enabled its tertiary structure determination by 2-D NMR spectroscopy in which it was shown to possess, not unexpectedly, the characteristic insulin-like fold.[89] A similar synthetic strategy also allowed the preparation of relaxins from other species including horse, rat, dog and mouse.[90,91] The strategy of regioselective disulfide-bond formation has also enabled a detailed structure-function relationship study to be undertaken on human INSL3 and the subsequent identification of a short span within its B-chain to be its receptor binding site.[92] The successful acquisition of human INSL5 by similar synthetic approaches has also been reported by others.[18]

As impressive as each of these achievements were, the multiple steps required for not only the separate chain syntheses but also the subsequent individual disulfide-bond formation invariably lead to overall low yields of target peptide (generally less than 10% relative to the limiting chain). This may not be critical for the preparation of analogs of insulin-like peptides for structural and functional study but presents severe limitations for the acquisition of adequate material for clinical use. Consequently, further improvements are unquestionably required in the chemical synthesis of such peptides, in particular, in still higher yields of solid phase peptide synthesis of the chains, novel S-protection that affords yet additional levels of orthogonality, improved or alternative chemical disulfide formation that avoid damage to other residues such as tryptophan and tyrosine and, finally, refined purification procedures.

6.4.6   Oxidative Folding of Single Chain Precursors

In vivo ribosomal synthesis of insulin occurs in the β cells of the pancreas as a single-chain precursor, the proinsulin, bearing a signal peptide at the N-terminus of the B-chain followed by a connecting (C-) peptide which links the C-terminus of the B-chain to the N-terminus of the A-chain (Figure 6.4.4). Following ribosomal release, the signal sequence is enzymatically cleaved in the endoplasmic reticulum and the B-C-A proinsulin sequence folds with assistance of the competent folding enzyme catalysts. The two-chain insulin is then produced following liberation of the 35-residue C-peptide (Figure 6.4.4) by prohormone convertases that cleave at the sites of pairs of basic residues, which link it at either end to the B- and A-chains and further enzymatic trimming of the terminal dibasic residues by enzymes.[93] The high efficiency of *in vitro* oxidative folding of human proinsulin (as high as 80%[94]) has prompted many to believe that C-peptide acts as a key mediator for correct native peptide production.[95,96] Remarkably, however, *in vitro* recombinant DNA expression of an "inverted" proinsulin in which the C-terminus of the A-chain is connected to the N-terminus of the B-chain by the C-peptide could be obtained in high yield (70%) when converted to the S-sulfonate form and subjected to oxidative folding at high pH in the presence of Cys·HCl.[97] This suggests that the C-peptide is acting more to hold the two chains together in near proximity for a successful folding rather than to direct the folding process itself. Consequently, there has been much investigation into the use of linear insulin analogs, either as precursors to the

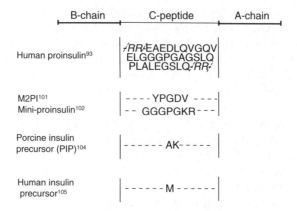

Figure 6.4.4 Proinsulin and artificial proinsulins. The proinsulin is enzymatically processed *via* cleavage at the Arg-Arg sequences (*italic*) by prohormone convertases and the terminal Arg-Arg sequences are further removed enzymatically to produce the mature double-stranded insulin.

preparation of two-chain peptides or as tools to examine the *in vitro* folding pathways.

6.4.6.1   Head-to-tail Constructs

Markussen[98] showed that insulin could be treated with trypsin to remove the C-terminal residue (Thr at position 30) of the B-chain followed by enzymatic formation of a peptide bond between B29 Lys and the A1 Gly α-amino group. It is a by-product of the enzymatic conversion of porcine to human insulin. The resulting "mini-proinsulin" surprisingly possesses native foldability but is biologically inactive despite a near-native tertiary structure, indicating that a free α-amino group at the N-terminus of the A-chain is required for activity. These findings together with the work on A- and B-chains cross-linked with bifunctional reagents and cleavable bridges such as oxaloyl-bis-Met[99,100] (Figure 6.4.4) fostered intensive research on artificial proinsulins.

6.4.6.2   Precursors with Mini-connecting Peptides

Molecular modeling studies led to the C-peptide being replaced with a short turn-forming pentapeptide (Tyr-Pro-Gly-Asp-Val), and expression of the resulting single-chain insulin analog (called M2PI) (Figure 6.4.4) by recombinant DNA methods in *E. coli* was followed by S-sulfonation and oxidative folding in the presence of β-mercaptoethanol in 50 μM glycine buffer, pH 11.5, at 4 °C.[101] The refolding yield of M2PI under these conditions was determined to be 20–40% better than that of native proinsulin indicating that the short turn-forming peptide was more effective than the much longer C-peptide. In similar studies by others,[102] the C-peptide was replaced by the turn-forming heptapeptide (Gly-Gly-Gly-Pro-Gly-Lys-Arg) (Figure 6.4.4B). The resulting mini-proinsulin analog was more potent than native proinsulin itself but less active than the 2-chain insulin. It was used in gene therapy experiments to successfully cause remission of diabetes in streptozotocin-induced diabetic rats and autoimmune diabetic rats.[102]

A porcine insulin precursor (PIP) is a single-chain analog in which B30 Ala and A1 Gly are joined by the dipeptide, Ala-Lys (Figure 6.4.4). Recombinant DNA expression of this peptide can be achieved efficiently but, as well, following S-reduction, it can undergo efficient oxidative refolding *in vitro*.[103] The resulting PIP can be converted to human insulin *via* transpeptidation *in vitro*.[104] More recently, Marglin reported preliminary studies in which the 35-residue insulin C-peptide could be substituted by a single methionine residue (Figure 6.4.4). The resulting synthetic "proinsulin" was able to fold and the Met was subsequently removed by cyanogen bromide cleavage.[105]

All these studies on short artificial linkers fully confirm that the insulin A- and B-chain contain sufficient information for the oxidative folding into the native structure and that the role of the connecting C-peptide is to bring and keep the two chains together.

Relaxin has also been obtained *via* a single-chain precursor. In contrast to proinsulin, the C-peptide of relaxin is 102 residues long. The reason for such a length is unknown although it may relate to more efficient secretory granule storage. However, it could be replaced by a short 13-residue "mini-C" sequence that was chosen principally from the comparable primary sequence in IGF-I together with appropriately selected residues at the B-C and C-A junctions. After expression of the "mini"-C prorelaxin in *E. coli* and partial purification, the reduced (not S-sulfonated) peptide could be efficiently refolded in high yield at pH 8.0 in the presence of oxidized and reduced glutathione. Removal of the "mini"-C-peptide was accomplished by sequential treatment of the refolded protein with Asp-N and Arg-C enzymes. Following further purification, the native two-chain relaxin was obtained in overall yield several fold higher than that obtained from an optimized two-chain process.[106]

6.4.7 Folding Pathways of Insulin

The past 20 years has witnessed a substantial number of studies into the mechanism of oxidative folding of not only the two insulin chains but also of single-chain proinsulins to provide insights into the principles that dictate the hierarchic processes that occur *via* a series of folded intermediates.[22,107–112] Insulin lends itself particularly well to such studies given that its small globular nature together with its readily modified disulfide bonds allow a clear determination of the sequence of events leading to association of the two insulin A- and B-chains *in vitro* with simultaneous local structure and disulfide-bond formation. By use of highly controlled conditions, the rate of oxidative folding can be sufficiently slowed so that intermediates can be trapped and subsequently analyzed to provide a descriptive pathway whereby the final folded peptide is obtained. However, such conditions are necessarily artificial and very unlikely mimic those that occur *in vivo*.

Numerous studies have been undertaken to determine the effect on insulin oxidative folding by replacement of one or more of the three disulfide bonds of insulin together with domain modification and substitution of secondary structural elements. The broad conclusions are that the intramolecular disulfide bond A6-A11 forms first during the folding process, and that this leads to local secondary structure formation which in turn directs the pairing of the two intermolecular disulfide bonds at approximately equal rates.[22,112] In the absence of the disulfide bond A6-A11, combination of the A- and B-chain occurs in unexpectedly high yield and in the correct alignment.[113] Similar results were obtained in parallel studies using single-chain insulin analogs and show that the formation of the intrachain disulfide is not an important element *per se* in the chain folding pathway. Instead, its importance lies in the acceleration of the nucleation of secondary structure elements, which does not include the N-terminal α-helix of the A-chain as its modification had little impact on folding efficiency. Of the two intermolecular disulfide bonds, replacement of A20-B19 had the greatest impact on subsequent oxidative folding indicating that it is a critical element.[22]

Similar studies on the chain combination and oxidative folding of human relaxin-2 led to somewhat similar conclusions. Unlike with insulin, however, it was postulated that the N-terminal α-helix of the A-chain is the initiation element which, upon its generation, leads to the intramolecular disulfide bond A10-A15 (relaxin numbering) forming first (like in insulin). This so-called oxidized A-chain intermediate is followed by rapid alignment and combination with the B-chain to generate native relaxin-2.[114] Such a folding pathway is able to explain the inability to produce human relaxin-3 by oxidative folding methods. Despite considerable effort, no conditions have been found under which the two chains of relaxin-3 will combine. RP-HPLC time course monitoring does not reveal the presence of an oxidized A-chain intermediate,[115] which suggested that the N-terminal A-chain α-helix was not forming as a prelude to subsequent disulfide bond linking. Analysis of the primary structure of human relaxin-3 (Figure 6.4.1) shows that the N-terminal sequence of the A-chain has a stretch of three Ser residues which is not predicted to be favorable for α-helix formation which is required to act as a folding initiation element.

Numerous studies have been undertaken to determine the *in vitro* oxidative folding pathway of single-chain insulin peptides including the IGFs.[116–118] As these were not used to subsequently produce double-stranded cystine peptides, they are beyond the scope of this review. However, it can be stated that insulin and IGF I have significantly different folding pathways that likely reflect differences in primary structures and the presence of the C-domain in the latter. However, there are also key similarities which must ultimately derive from their common evolutionary origins.

6.4.8 Concluding Remarks

The oxidative folding of double-stranded cystine peptides has been the subject of intense study for more than 50 years. It remains a source of astonishment that two separate peptide chains can generally combine *in vitro* with such high efficiency and with little evidence for formation of the numerous isomers. As the genome of various species including the human is subjected to new or further detailed analysis, it is likely that new double-stranded cystine peptides – both hetero- and homodimeric – will be discovered including additional insulin-like ligands.[119] The study of their oxidative folding will unquestionably continue unabated as an important precursor to the determination of their biological functions.

Acknowledgements

The author is grateful to Professor Geoffrey Tregear for critical appraisal of the manuscript. Reported studies undertaken in the author's laboratory were supported by NHMRC of Australia Project grants (#350245 and 350284).

References

1. A. P. Ryle, F. Sanger, L. F. Smith and K. Kitai, *Biochem. J.*, 1955, **60**, 541–556.
2. F. L. Hisaw, *Proc. Soc. Exp. Biol. Med.*, 1926, **23**, 661–663.
3. C. Schwabe and J. K. McDonald, *Science*, 1976, **197**, 914–915.
4. International Human Genome Sequencing Consortium, *Nature*, 2004, **41**, 931–945.
5. Rat Genome Sequencing Project Consortium, *Nature*, 2004, **428**, 493–521.
6. Drosophila 12 Genomes Consortium, *Nature*, 2007, **450**, 203–218.
7. D. S. Nicol and L. F. Smith, *Nature*, 1960, **187**, 483–485.
8. P. Hudson, J. Haley, M. Cronk, J. Shine and H. Niall, *Nature*, 1981, **291**, 127–131.
9. E. Rinderknecht and R. E. Humbel, *J. Biol. Chem.*, 1978, **253**, 2769–2776.
10. E. Rinderknecht and R. E. Humbel, *FEBS Lett.*, 1978, **89**, 283–286.
11. P. Hudson, M. John, R. Crawford, J. Haralambidis, D. Scanlon, J. Gorman, G. W. Tregear, J. Shine and H. Niall, *EMBO J.*, 1984, **3**, 2333–2339.
12. T. Dschietzig, C. Bartsch, G. Baumann and K. Stangl, *Pharmacol. Therapeut.*, 2006, **112**, 38–56.
13. E. Burkhardt, I. M. Adham, B. Brosig, A. Gastmann, M. G. Mattei and W. Engel, *Genomics*, 1994, **20**, 13–19.
14. K. Kawamura, J. Kumagai, S. Sudo, S.-Y. Chun, M. Pisarska, H. Morita, P. Fu, J. D. Wade, R. A. D. Bathgate and A. J. W. Hsueh, *Proc. Natl. Acad. Sci. USA*, 2004, **101**, 7323–7328.

15. D. Chassin, A. Laurent, J. L. Janneau, R. Berger and D. Bellet, _Genomics_, 1995, **29**, 465–470.
16. D. Conklin, D. E. Lofton-Day, B. A. Haldeman, A. Ching, T. E. Whitmore, S. Lok and S. Jaspers, _Genomics_, 1999, **60**, 50–56.
17. S. Lok, D. S. Johnston, D. Conklin, D. E. Lofton-Day, R. L. Adams, L. C. Jelmberg, T. E. Whitmore, S. Schrader, M. D. Griswald and S. Jaspers, _Biol. Reprod._, 2000, **62**, 1593–1599.
18. S. L. Dun, E. Brailoiu, Y. Wang, G. C. Brailoiu, L.-Y. Liu-Chen, J. Yang, J. K. Chang and N. J. Dun, _Endocrinol._, 2006, **147**, 3243–3248.
19. R. A. D. Bathgate, C. S. Samuel, T. C. D. Burazin, S. Layfield, A. A. Claasz, I. G. T. Reytomas, N. F. Dawson, C. Zhao, C. Bond, R. J. Summers, L. J. Parry, J. D. Wad and G. W. Tregear, _J. Biol. Chem._, 2002, **277**, 1148–1157.
20. B. A. McGowan, S. A. Stanley, K. L. Smith, J. S. Minnion, J. Donovana, E. L. Thompsona, M. Patterson, M. Connolly, C. R. Abbott, C. J. Small, J. V. Gardiner, M. A. Ghatei and S. R. Bloom, _Regul. Pept._, 2006, **136**, 72–77.
21. A. Zhou, G. Webb, X. Zhu and D. F. Steiner, _J. Biol. Chem._, 1999, **274**, 20745–20748.
22. Z.-Y. Guo, Z.-S. Qiao and Y.-M. Feng, _Antiox. Redox Signal._, 2008, **10**, 127–140.
23. S. J. Chan and D. F. Steiner, _Amer. Zool._, 2000, **40**, 213–222.
24. J.-I. Park, C. L. Chang and S. Y. Hsu, _Rev. Endocr. Metab. Disord._, 2005, **6**, 291–296.
25. T. C. Wilkinson, T. P. Speed, G. W. Tregear and R. A. Bathgate, _BMC Evol. Biol._, 2005, **5**, 14.
26. H. Nagasawa, H. Kataoka, A. Isogai, S. Tamura, A. Suzuki, A. Mizoguchi, Y. Fujiwara, A. Suzuki, S. Takahashi and H. Ishizaki, _Proc. Natl. Acad. Sci. USA_, 1986, **83**, 5840–5843.
27. I. Claeys, G. Simonet, J. Poels, T. Van Loy, L. Vercammen, A. De Loof and J. Van den Broeck, _Peptides_, 2002, **23**, 807–816.
28. S. B. Pierce, M. Costa, R. Wisotzkey, S. Devadhar, S. A. Homburger, A. R. Buchman, K. C. Ferguson, J. Heller, D. M. Platt, A. A. Pasquinelli, L. X. Liu, S. K. Doberstein and G. Ruvkun, _Genes Dev._, 2001, **15**, 672–686.
29. R. P. Olinsky, C. Dahlberg, M. Thorndyke and F. Hallbook, _Peptides_, 2006, **27**, 2535–2546.
30. A. Honneger, _Handb. Exp. Pharmacol._, 2008, **181**, 47–68.
31. A. B. Mukherjee, Z. J. Zhang and B. S. Chilton, _Endocr. Rev._, 2007, **28**, 707–725.
32. K. Kangawa, A. Fukuda and H. Matsuo, _Nature_, 1985, **313**, 397–400.
33. R. I. Lehrer, G. Xua, A. Abduragimov, N. N. Dinh, X.-D. Qua, D. Martin and B. J. Glasgow, _FEBS Lett._, 1998, **432**, 163–167.
34. F. Blanco-Vaca, D. P. Via, C.-Y. Yang, J. B. Massey and H. J. Pownall, _J. Lipid Res._, 1992, **33**, 1785–1790.

35. Y. Usami, Y. Fujimura, M. Suzuki, Y. Ozeki, K. Nishio, H. Fukui and K. Titani, *Proc. Natl. Acad. Sci. USA*, 1993, **90**, 928–392.
36. W. S. Jang, K. N. Kim, Y. S. Lee, M. H. Nam and I. H. Lee, *FEBS Lett.*, 2002, **521**, 81–86.
37. C. V. F. Batista, A. Scaloni, D. J. Rigden, L. R. Silva, A. Rodrigues Romero, R. Dukor, A. Sebben, F. Talamo and C. Bloch, *FEBS Lett.*, 2001, **494**, 85–89.
38. A. Okuno, Y. Hasegawa, M. Nishiyama, T. Ohira, R. Ko, M. Kurihara, S. Matsumoto and H. Nagasawa, *Peptides*, 2002, **23**, 567–572.
39. X. Liu, S. Maeda, Z. Hu, T. Aiuchi, K. Nakaya and Y. Kurihara, *Eur. J. Biochem.*, 1993, **211**, 281–287.
40. S. Nirasawa, T. Nishino, M. Katahira, S. Uesugi, Z. Hu and Y. Kurihara, *Eur. J. Biochem.*, 1994, **223**, 989–995.
41. M. Ruiz-Gayo, M. Royo, M. Fernandez, F. Alberico, E. Giralt and M. Pons, *J. Org. Chem.*, 1993, **58**, 6319–6328.
42. L. Chen, H. Bauerova, J. Slaninova and G. Barany, *Int. J. Pept. Prot. Res.*, 1996, **9**, 114–121.
43. M. C. Munson, M. Lebl, J. Slaninova and G. Barany, *J. Pept. Res.*, 2003, **6**, 155–159.
44. M. Royo, M. A. Contreras, E. Giralt, F. Alberico and M. Pons, *J. Am. Chem. Soc.*, 1998, **120**, 6639–6650.
45. D. Andreu, F. Alberico, N. A. Sole, M. C. Munson, M. Ferrer and G. Barany, *Meth. Mol. Biol.*, 1994, **35**, 91–169.
46. P. Lavigne, L. H. Kondejewski, M. E. Houston, F. D. Sönnichsen, B. Lix, B. D. Sykes, R. S. Hodges and C. M. Kay, *J. Mol. Biol.*, 1995, **254**, 505–520.
47. N. M. Jetly, K. S. Iyer, M. V. Hosur and S. D. Mahale, *J. Pept. Res.*, 2003, **62**, 269–279.
48. J. M. Fletcher and R. A. Hughes, *J. Pept. Sci.*, 2006, **12**, 515–524.
49. D. R. Burton, *Mol. Immunol.*, 1985, **22**, 161–206.
50. L. Moroder, G. Hubener, S. Göhring-Romani, W. Göhring, H.-J. Musiol and E. Wünsch, *Tetrahedron*, 1990, **46**, 3305–3314.
51. P. Terness, I. Kohl, G. Hübener, R. Battistuta, L. Moroder, M. Welschof, C. Dufter, M. Finger, C. Hain, M. Jung and G. Opelz, *J. Immunol.*, 1995, **154**, 6446–6452.
52. M. Kohmura and Y. Ariyoshi, *Biopolymers*, 1998, **46**, 215–223.
53. C. B. Anfinsen and E. Haber, *J. Biol. Chem.*, 1961, **236**, 1361–1363.
54. Y. C. Du, Y. S. Zhang, Z. X. Lu and C. L. Tsou, *Sci. Sin.*, 1961, **10**, 84–104.
55. J. M. Swan, *Nature*, 1957, **179**, 965.
56. R. Q. Jiang, Y. C. Du and C. L. Tsou, *Sci. Sin.*, 1963, **12**, 452–454.
57. G. H. Dixon and A. C. Wardlaw, *Nature*, 1960, **188**, 721–724.
58. Y. T. Kung, Y. C. Du, W. T. Huang, C. C. Chen and L. T. Ke, *Sci. Sin.*, 1965, **14**, 1710–1716.
59. J. Meienhofer, E. Schnabel, H. Bremer, O. Brinkhoff, R. Zabel, W. Sroka, H. Klostermayer, D. Brandenburg, T. Okuda and H. Zahn, *Z. Naturforschong B.*, 1963, **18**, 1120–1121.

60. P. G. Katsoyannis, K. Fukuda, A. Tomesko, K. Suzuki and M. Tilak, *J. Am. Chem. Soc.*, 1964, **86**, 930–932.

61. P. G. Katsoyannis and A. Tomesko, *Proc. Natl. Acad. Sci. USA*, 1966, **55**, 1554–1561.

62. R. E. Chance, J. A. Hoffmann, E. P. Kroeff, M. G. Johnson, E. W. Schirmer, W. W. Bromer, M. J. Ross and R. Wetzel, in *Peptides: Synthesis, Structure, Function. Proceedings of the 7th American Peptide Symposium*, D. R. Rich and E. Gross, ed., Pierce Chemical Co, 1981, pp. 721–728.

63. H. G. Gattner, G. Krail, W. Danho, R. Knorr, H. J. Wieneke, E. E. Büllesbach, B. Schartmann, D. Brandenburg and H. Zahn, *Hoppe Seyler's Z. Physiol. Chem.*, 1981, **362**, 1043–1049.

64. S. Q. Hu, G. T. Burke, G. T. Schwartz, N. Federigos, J. B. Ross and P. G. Katsoyannis, *Biochemistry*, 1993, **32**, 2631–2635.

65. M. A. Weiss, S. H. Nakagawa, W. Jia, B. Xu, Q. X. Hua, Y. C. Chu, R. Y. Wang and P. G. Katsoyannis, *Biochemistry*, 2002, **41**, 809–819.

66. P. Hudson, J. Haley, M. John, M. Cronk, R. Crawford, J. Haralambidis, G. W. Tregear, J. Shine and H. Niall, *Nature*, 1983, **301**, 628–631.

67. E. Canova-Davis, I. P. Baldonado and G. M. Teshima, *J. Chromatogr.*, 1990, **508**, 81–96.

68. R. A. Bathgate, F. Lin, N. F. Hanson, L. Otvos Jr, A. Guidolin, C. Giannakis, S. Bastiras, S. L. Layfield, T. Ferraro, S. Ma, C. Zhao, A. L. Gundlach, C. S. Samuel, G. W. Tregear and J. D. Wade, *Biochemistry*, 2006, **45**, 1043–1053.

69. P. Fu, L. Otvos, S. Layfield, T. Ferraro, H. Tomiyama, J. Hutson, G. W. Tregear, R. A. D. Bathgate and J. D. Wade, *J. Pept. Res.*, 2004, **63**, 91–98.

70. S. Gordon and V. Du Vigneaud, *Proc. Soc. Exp. Biol. Med.*, 1953, **84**, 723–725.

71. R. B. Merrifield, *J. Am. Chem. Soc.*, 1963, **85**, 2149–2154.

72. A. Marglin and R. B. Merrifield, *J. Am. Chem. Soc.*, 1966, **88**, 5051–5052.

73. S. G. Zakhariev, C. Guarnaccia, H. G. Gattner, S. Pongor and D. Brandenburg, in *Peptides 1994. Proceedings of the 23rd European Peptide Symposium*, H. L. S. Maia, ed., ESCOM, Leiden, 1995, pp. 297–298.

74. J. P. Mayer, G. S. Brooke and R. D. DiMarchi, in *Peptides: Chemistry, Structure and Biology. Proceedings of the 11th American Peptide Symposium*, J. Rivier and G. R. Marshall, eds., ESCOM, Leiden, 1990, pp. 1061–1062.

75. J. Brange, L. Andersen, E. D. Laursen, G. Meyn and E. Rasmussen, *J. Pharm. Sci.*, 1997, **86**, 517–525.

76. D.-P. Hong, A. Ahmad and A. L. Fink, *Biochemistry*, 2006, **45**, 9342–9353.

77. P. Sieber, B. Kamber, A. Hartmann, A. Johl, B. Riniker and W. Rittel, *Helv. Chim. Acta*, 1974, **57**, 2617–2621.

78. P. Sieber, B. Kamber, K. Eisler, A. Hartmann, B. Riniker and W. Rittel, *Helv. Chim. Acta*, 1976, **59**, 1489–1497.

79. L. Moroder, D. Besse, H.-J. Musiol, S. Rudolph-Böhner and F. Siedler, *Biopolymers (Pept. Sci.)*, 1996, **40**, 207–234.
80. I. Annis, B. Hargittai and G. Barany, *Meth. Enzymol.*, 1997, **289**, 198–221.
81. F. Alberico, I. Annis, M. Royo and G. Barany, in *Fmoc Solid Phase Peptide Synthesis*, W. C. Chan and P. D. White, eds., Oxford University Press, 2000, pp. 77–114.
82. C. Boulègue, H.-J. Musiol, V. Prasad and L. Moroder, *Chem. Today*, 2006, **24**, 24–36.
83. K. Akaji and Y. Kiso, in *Houben-Weyl. Methods of Organic Chemistry, Synthesis of Peptides and Peptidomimetics*, M. Goodman, A. Felix, L. Moroder and C. Toniolo, eds., Thieme, Stuttgart, 2002, Vol. E 22, pp. 101–141.
84. E. E. Büllesbach and C. Schwabe, *J. Biol. Chem.*, 1991, **266**, 10754–10761.
85. K. Maruyama, K. Nagata, M. Tanaka, H. Nagasawa, A. Isogai, H. Ishizaki and A. Suzuki, *J. Prot. Chem.*, 1992, **11**, 1–12.
86. K. Maruyama, H. Nagasawa, A. Isogai, H. Ishizaki and A. Suzuki, *J. Prot. Chem.*, 1992, **11**, 13–20.
87. K. Akaji, K. Fujino, T. Tatsumi and Y. Kiso, *J. Am. Chem. Soc.*, 1993, **115**, 11384–11392.
88. K. Akaji, T. Tatsumi, M. Yoshida, T. Kimura, Y. Fujiwara and Y. Kiso, *J. Am. Chem. Soc.*, 1992, **114**, 4137–4143.
89. K. J. Rosengren, F. Lin, R. A. Bathgate, G. W. Tregear, N. L. Daly, J. D. Wade and D. J. Craik, *J. Biol. Chem.*, 2006, **281**, 5845–5851.
90. M. A. Hossain, F. Lin, S. Zhang, T. Ferraro, R. A. D. Bathgate, G. W. Tregear and J. D. Wade, *Int. J. Pept. Res. Therap.*, 2006, **12**, 211–215.
91. C. J. Samuel, F. Lin, C. Zhao, S. Layfield, M. A. Hossain, S. Zhang, E. Giannakis, R. A. Bathgate, G. W. Tregear and J. D. Wade, *Biochemistry*, 2007, **46**, 5374–5381.
92. K. J. Rosengren, S. Zhang, F. Lin, N. L. Daly, D. J. Scott, R. A. Hughes, R. A. Bathgate, D. J. Craik and J. D. Wade, *J. Biol. Chem.*, 2006, **281**, 28287–28295.
93. D. F. Steiner, D. Cunningham, L. Spiegelman and B. Aten, *Science*, 1967, **157**, 697–700.
94. D. F. Steiner and J. L. Clark, *Proc. Natl. Acad. Sci. USA*, 1968, **60**, 622–629.
95. W. Kemmler, J. D. Peterson and D. F. Steiner, *J. Biol. Chem.*, 1971, **246**, 6786–6791.
96. J. Winter, H. Lilie and R. Rudolph, *Anal. Biochem.*, 2002, **310**, 148–155.
97. W. F. Heath, R. M. Belagaje, G. S. Brooke, R. E. Chance, J. A. Hoffmann, H. B. Long, S. G. Reams, C. Roundtree, W. N. Shaw, L. S. Slieker, K. L. Sundell and R. D. DiMarchi, *J. Biol. Chem.*, 1991, **267**, 419–425.
98. J. Markussen, *Int. J. Pept. Prot. Res.*, 1985, **25**, 431–434.
99. B. R. Srinivasa and F. H. Carpenter, *Int. J. Pept. Prot. Res.*, 1983, **22**, 214–222.
100. S. H. Nakagawa and H. S. Tager, *J. Biol. Chem.*, 1989, **264**, 272–279.

101. S.-G. Chang, D.-Y. Kim, K.-D. Choi, J.-M. Shin and H.-G. Shin, *Biochem. J.*, 1998, **329**, 631–635.

102. H. C. Lee, S.-J. Kim, K.-S. Kim, H.-G. Shin and J.-W. Yoon, *Nature*, 2000, **408**, 483–488.

103. Z.-S. Qiao, Z.-Y. Guo and Y.-M. Feng, *Biochemistry*, 2001, **40**, 2662–2668.

104. Y. Zhang, H. Hu, R. Cai, Y. Feng, S. Zhu, Q. He, Y. Tang, M. Xu, Y. Xu, X. Zhang, B. Liu and Z. Liang, *Sci. China C. Life Sci.*, 1996, **39**, 225–233.

105. A. Marglin, *Biopolymers (Pept. Sci.)*, 2008, **90**, 200–202.

106. R. J. Vandlen, J. Winslow, B. Moffat and E. Rinderknecht, in *Progress in Relaxin Research. Proceedings of the 2nd International Congress on the Hormone Relaxin*, A. H. MacLennan, G. W. Tregear and G. D. Bryant-Greenwood, eds., Global Publication Services, Singapore, 1995, pp. 59–72.

107. J.-G. Tang, C. C. Wang and C. L. Tsou, *Biochem. J.*, 1988, **255**, 451–455.

108. Y. Dai and J.-G. Tang, *Biochem. Mol. Biol. Int.*, 1994, **33**, 1049–1053.

109. S. Q. Hu, G. T. Burke and P. G. Katsoyannis, *J. Prot. Chem.*, 1993, **12**, 741–747.

110. Q. X. Hua, Y. C. Chu, W. Jia, N. F. B. Phillips, R.-Y. Wang, P. G. Katsoyannis and M. A. Weiss, *J. Biol. Chem.*, 2002, **277**, 43443–43453.

111. Q. X. Hua, S. H. Nakagawa, S.-Q. Hu, W. Jia, S. Wang and M. A. Weiss, *J. Biol. Chem.*, 2006, **281**, 24900–24909.

112. Z. S. Qiao, Z. Y. Guo and Y.-M. Feng, *Protein Pept. Lett.*, 2006, **13**, 423–429.

113. X. Y. Jia, Z. Y. Guo, Y. Wang, Y. Xu, S. S. Duan and Y.-M. Feng, *Protein Sci.*, 2003, **12**, 2412–2419.

114. J.-G. Tang, Z. H. Wang, G. W. Tregear and J. D. Wade, *Biochemistry*, 2003, **11**, 2731–2739.

115. J. D. Wade and G. W. Tregear, *Meths. Enzymol.*, 1997, **289**, 637–646.

116. Y. Yuan, Z.-H. Wang and J.-G. Tang, *Biochem. J.*, 1999, **343**, 139–144.

117. Q. X. Hua, J. P. Mayer, W. Jia, J. Zhang and M. A. Weiss, *J. Biol. Chem.*, 2006, **281**, 28131–28142.

118. Q. L. Huang, J. Zhao, Y. H. Tang, S. Q. Shao, G. J. Xu and Y.-M. Feng, *Biochemistry*, 2007, **46**, 218–224.

119. R. Manor, S. Weil, S. Oren, L. Glazer, E. D. Aflalo, T. Ventura, V. Chalifa-Caspi, M. Lapidot and A. Sagi, *Gen. Comp. Endocr.*, 2007, **150**, 326–336.

CHAPTER 6.5

Multiple-strand Cystine Peptides

MARION G. GÖTZ,[a] HANS-JÜRGEN MUSIOL[b]
AND LUIS MORODER[b]

[a] Department of Chemistry, Whitman College, 345 Boyer Ave, Walla Walla, WA 99362, USA; [b] Max Planck Institute of Biochemistry, Am Klopferspitz 18, D-82152 Martinsried, Germany

6.5.1 Introduction

The most important family of proteins that consist of multiple polypeptide chains cross-linked by disulfide knots for stabilization of the higher order structure are the collagens. These secretory proteins constitute the main components of the extracellular matrix where they provide mechanical strength and structural integrity to the connective tissues. Their stability arises from unique structural features. All collagens are formed by three polypeptide chains, which are assembled into homotrimers or heterotrimers with the latter consisting of two or three distinguishable α-chains. The common structural motif of all collagens is a rigid rod-shaped triple helix that either spans the entire molecule from the N- to the C-terminus, as in the case of fibrillar collagens, or is interspersed by non-collagenous (NC) domains resulting in increased overall structural flexibility. The (GXY) tripeptide repeats of the triple-helical segments represent a second characteristic feature. The presence of a repeating glycine residue in every third position is a steric requirement for the close packing of the three chains. The X and Y positions are most frequently occupied by proline and hydroxyproline, respectively. This imino acid-rich sequence composition favors an extended left-handed polyproline II-helix of the single α-chains, which are supercoiled in a right-handed manner around a

RSC Biomolecular Sciences
Oxidative Folding of Peptides and Proteins
Edited by Johannes Buchner and Luis Moroder
© Royal Society of Chemistry 2009
Published by the Royal Society of Chemistry, www.rsc.org

central axis into a triple helix, with a stagger of one residue between the adjacent chains.[1–3]

The biosynthesis of collagens involves a unique pathway including extensive post-translational modifications, proper chain selection and correct chain registration. The final assembly requires nucleation of the triple helix with subsequent propagation from the C- to the N-terminus, followed by additional enzymatic processing of the folded procollagens to produce the mature proteins.[4–6] The pro-α-chains contain one or more collagenous (COL) domains, which are generally flanked by two NC domains, although there are cases where one of the NC domains is missing. In most types of collagens the folded C-terminal NC domains serve as recognition modules, which first select then bind and register the three proper α-chains into homo- or heterotrimers. This process is controlled by shape complementarity, electrostatic interactions and hydrophobicity. One exception is the assembly of transmembrane collagens where the N-terminal NC domains rather than the C-terminal NC domains dictate the trimerization process and thus the folding pathway.[7,8] The homo- or heterotrimerization of the C-terminal propeptides of procollagen involves extensive intra- and interchain disulfide-bond formation catalyzed by protein disulfide isomerase.[9,10] In the case of fibrillar collagens this interchain disulfide cross-linking of the three α-chains precedes nucleation and propagation of the native triple helix in the COL domains of the molecules.[9,11–14] However, interchain disulfide-bonding, between either the C-propeptides or C-telopeptides, is not required for chain association and triple-helix formation as long as the triple-helical domains are brought together in close proximity, and are tethered at the carboxy terminus.[15–18] This is in contrast to the situation when collagens lacking C-propeptides are refolded after denaturation *in vitro*. Unless interchain disulfide bonds are present at the C-terminal end of the chains (as in the case of collagen type III),[19,20] these chains are unable to fold into the correctly aligned helices. Indeed, absence of such interchain cross-links impedes a fast and correct refolding, and results in mis-matched triple helices differing in length and stability.[21–23] As a consequence, kinetic as well as mechanistic studies of triple-helical folding have almost exclusively been performed with procollagens and related fragments, which contain cysteine residues in the N- and C-propeptides that are involved in the formation of the interchain disulfide knots, as exemplarily shown for procollagen type III in Figure 6.5.1. Type III collagen retains a cystine knot even in the mature form as a result of interchain cross-linking of two vicinal cysteine residues at the junction between the triple-helical domain and the telopeptide.[24] Therefore this collagen or fragments of its proform have served as ideal model systems for the detailed investigations of the zipper-like propagation of the triple helix as well as of the impact of the *cis*-to-*trans* isomerization of the aminoacyl-proline/hydroxyproline bonds on the folding kinetics.[19,20,25]

Collagens as well as related fragments are experimentally cumbersome, while synthetic collagenous peptides are more amenable to detailed biophysical and biological studies. Unfortunately self-association of monomeric peptides into homotrimers occurs in a concentration-dependent manner thus leading to equilibria more or less shifted to the trimeric state. Indeed the midpoint of

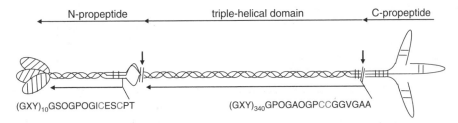

$(GXY)_{10}GSOGPOGICESCPT$ $(GXY)_{340}GPOGAOGPCCGGVGAA$

Figure 6.5.1   Schematic representation of human type III procollagen. In the trimeric C-propeptides four cysteine residues of each chain are involved in intrachain and four in interchain disulfide bonds; at the junction between the triple-helical domain and the C-telopeptide two adjacent cysteine residues form an intrachain cystine knot, which is retained in the mature type III collagen upon enzymatic processing at the two positions indicated by arrows. The N-propeptide (Col1-3) consists C-terminally of an NC domain followed by a short triple-helical segment cross-linked C-terminally by a cystine knot involving the CXYC motif and finally by a larger NC domain containing five intrachain disulfide bridges.

thermal unfolding of self-associated triple helices strongly depends on the concentration of the collagenous peptide in the entropic term $(T_m = \Delta H^0/DS^0 + R\ln(0.75c_0^2)$.[3] Moreover, hysteresis, which is more prominent for long natural collagens, is also observed for short collagen triple helices resulting in a strong time-dependency for full equilibration.[2,3] These inconvenient properties of self-associated triple helices and the early observation that collagens refold correctly when the chains are tightly packed or covalently cross-linked at the C-terminus[19,20] have fostered intensive research towards the development of artificial templates for C- and particularly N-terminal cross-linking of the three chains.[26–29] By this strategy the unfavorable entropy of self-association of the collagenous peptides into homotrimers is considerably reduced and the triple-helical fold significantly stabilized.

6.5.2   Synthesis of Disulfide Cross-linked Homotrimeric Collagenous Peptides

Unlike the synthetic scaffolds, native or *de novo* designed cystine knots have only recently found successful applications in the oxidative or regioselective assembly of trimeric collagen peptides. For recent reviews on the subject see references 29 and 30.

6.5.2.1   Oxidative Assembly of Collagenous Homotrimers with the C-Terminal Cystine Knot of Collagen Type III

A covalent homotrimeric assembly of synthetic model peptides following the design of the C-terminal cystine knot of collagen type III (see Figure 6.5.1) has been successfully pioneered by Bächinger and associates.[31,32] For this purpose collagenous peptides of higher tendency to self-associate into triple helices were C-terminally extended by the bis-cysteinyl-sequence GPCCG and then subjected to air or GSH/GSSG oxidation at lower temperature under slightly basic conditions. The disulfide-bridged homotrimers were obtained in satisfactory yields.

In the absence of specific recognition epitopes as in the case of the C-propeptides of collagens[16] a directed alignment of the chains and induced nucleation of the triple helix cannot occur. Therefore, mixtures of differently aligned triple helices, which differ in their structural stability, are expected to form at equilibrium, as shown in Figure 6.5.2. It seems that under oxidative conditions the correctly registered and thus most stable trimers are trapped by formation of the disulfide knot. This explanation is experimentally supported by the successful oxidative folding of synthetic collagenous peptides, which are capable of self-associating into triple helices, independent of a C-terminal or even N-terminal location of the bis-cysteinyl motif.[33,34] As expected, the built-in cystine knot thermodynamically induces a marked increase in the thermal stability of the triple helix, and folding/unfolding becomes a concentration-independent process. The oxidative folding, however, failed when the hydroxyproline of the native C-terminal sequence of collagen type III was replaced with proline.[35] These results support the notion that nucleation of the triple helix and thus folding of procollagens requires 4-hydroxylation of at least two C-terminal triplets,[15] and clearly suggest that prefolding of the triple helix is essential for correct oxidative formation of the cystine knot. This fact was fully confirmed by a systematic study with the synthetic collagen molecules Ac-$(POG)_n$PCCGGG-NH_2 where n was varied from 3 to 7.[36] Indeed only with $n \geq 5$ CD spectra characteristic of higher contents of triple-helical homotrimers were monitored upon equilibration at low temperature (7–8 °C) and at 1 mM concentration. Performing the oxidation experiments after pre-equilibration at a temperature below the T_m value of the self-associated homotrimers, *i.e.* at 7–8 °C, air oxidation led to product distributions consisting of the average of

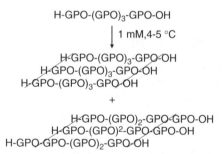

Figure 6.5.2   Self-association of collagenous peptides at low temperature and higher concentration to overcome the entropic penalty is expected to produce variously aligned homotrimers at equilibrium which differ in the number of registered triplets and thus the triple-helix stability.

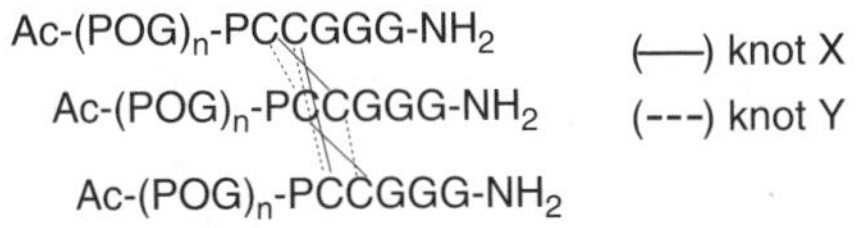

Ac-(POG)$_n$-PCCGGG-NH$_2$ (——) knot X
Ac-(POG)$_n$-PCCGGG-NH$_2$ (- - -) knot Y
Ac-(POG)$_n$-PCCGGG-NH$_2$

Figure 6.5.3 The two most plausible cystine knots derived from NMR structural analysis and MD simulations. While type Y cystine knot seems more compatible with the triple-helical conformation in terms of dihedral angles, the type X knot represents the intuitively simplest structure formed by the three disulfide bonds.

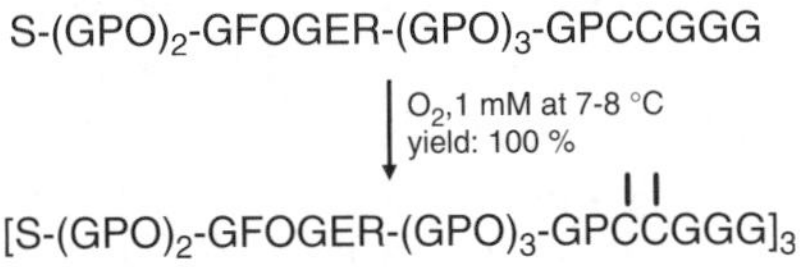

S-(GPO)$_2$-GFOGER-(GPO)$_3$-GPCCGGG

$\downarrow$ O$_2$,1 mM at 7-8 °C
yield: 100 %

| |
[S-(GPO)$_2$-GFOGER-(GPO)$_3$-GPCCGGG]$_3$

Figure 6.5.4 Oxidative folding of a synthetic peptide containing the GFOGER sequence of collagen type I as adhesion epitope and the bis-cysteinyl motif of collagen type III for formation of the C-terminal cystine knot.

approximately 70% trimer, while the oxidized monomer formed the major side product. Although chromatographic and NMR structural analysis confirmed formation of a well-defined cystine-knot isomer, the exact cysteine pairings could not be unambiguously assigned (Figure 6.5.3).[36]

The enhanced formation of a vicinal intramolecular disulfide bridge within the monomer was unexpected because disulfide bridges between adjacent cysteine residues are known to be conformationally very disfavoured.[37–40] These are rarely found in folded proteins,[41–43] although their appearance is often associated with productive intermediates that readily undergo disulfide reshuffling in order to generate the final correct cystine framework.[43–45] In the case of the collagen model peptides even operating with GSH/GSSG as redox buffer a reshuffling to the homotrimers with correspondingly higher yields of the correctly folded constructs was not observed.[36] However, oxidative folding of a collagen peptide containing the α1β1 and α2β1 integrin adhesion motif of collagen type I and the C-terminal bis-cysteinyl motif almost exclusively produced the disulfide-bridged homotrimer (Figure 6.5.4).[46] This unexpected high yield was attributed to the Arg and Glu residues that may facilitate the correct staggered alignment by electrostatic interactions, thus increasing the probability of the cysteine residues being in juxtaposition for the oxidative disulfide formation.

6.5.2.2 Oxidative Assembly of Collagenous Homotrimers with the Cystine Knot of FACIT COL1-NC1 Junctions

FACITs are collagens, which are associated with fibrils and unique for their short triple-helical domains interpersed with non-triple-helical domains.[6] All FACIT collagens are homotrimers, except for collagen type IX, which is composed of three different α-chains. Even though FACITs have significantly shorter C-terminal NC domains in comparison to fibril-forming collagens, they share a remarkable sequence homology in the first collagenous domain (COL1) and contain two strictly conserved cysteine residues separated by four residues in their COL1/NC1 junctions. These cysteine residues are responsible for an interchain disulfide knot in the completely folded trimers; however, exact disulfide connectivities have yet to be determined. *In vivo* folding has been extensively studied using different constructs of the homotrimeric collagens XII and XIV. These studies strongly support a folding mechanism that involves both the COL1 domain at the COL1/NC1 junction, which contains the first cysteine residue, and the NC1 domain containing the second conserved cysteine.[47–51] A recent work on various type IX collagen constructs showed that the α-chains can associate in the absence of COL1 and NC1 domains to form the triple helix, but that the COL1-NC1 region is important for chain specificity.[52]

Lesage *et al.*[53] have studied the intrinsic propensity of the highly conserved cystine-knot sequences of FACITs for oxidative trimerization *in vitro*. For this purpose a synthetic model peptide derived from the COL1/NC1 junction of collagen type XIV was N-terminally extended with three (GPO) repeats, the strongest triple-helix stabilizing triplets,[54–56] to replace the GSQGPAGPO sequence of type XIV human collagen. As expected, oxidation of the bis-cysteinyl peptide itself exclusively generated the intrachain-disulfide bridged monomer. Upon extension of the N-terminus with the (GPO)$_3$ sequence as well as exposing the peptide to the triple-helix favoring environment of aqueous methanol[54,57] the homotrimer was found to be the major product next to smaller amounts of the oxidized monomer and dimer (Figure 6.5.5). Increased temperatures and lower concentrations compromise the stability of the self-associated homotrimeric triple helix. As a result, the yields of homotrimer were dramatically reduced. Therefore, even with the FACIT cystine knot motif trimerization into the triple-helix register *via* the (GPO)$_3$ repeats is apparently essential for successful oxidative formation of the disulfide framework. The triple-helical structure of the homotrimer was confirmed by CD spectral analysis, however NMR structural analysis did not allow an unambiguous differentiation between the cystine knots 1 and 2 (Figure 6.5.5).[53] It can be speculated that an increase of the triple-helix stability would significantly reduce the amount of oxidized monomer as side product, thus enhancing the yields of target homotrimers.

This model study strongly supports the use of the collagen type XIV bis-cysteinyl motif for an efficient oxidative assembly of homotrimeric collagen peptides. Conversely, the CXYC motif of the N-propeptide of collagen type III (Figure 6.5.1), so far, has not been investigated for such purposes.

6.5.2.3 Assembly of Homotrimeric Collagen Peptides by Regioselective Disulfide Formation

Following essentially a reaction scheme discussed in detail in Section 6.5.3.2, Barth *et al.*[36] have compared the triple-helix stabilizing properties of the

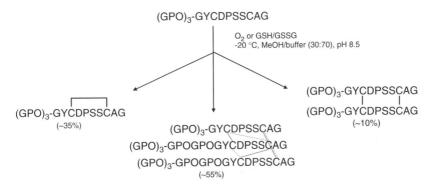

Figure 6.5.5 Oxidative folding of a synthetic collagen peptide consisting of the bis-cysteinyl motif of the COL1/NC1 junction of collagen type XIV N-terminally extended with GPO triplets. The major product was the homotrimer. The most probable disulfide connectivities are the knot 1 (—) and 2 (---) according to NMR structural analysis.

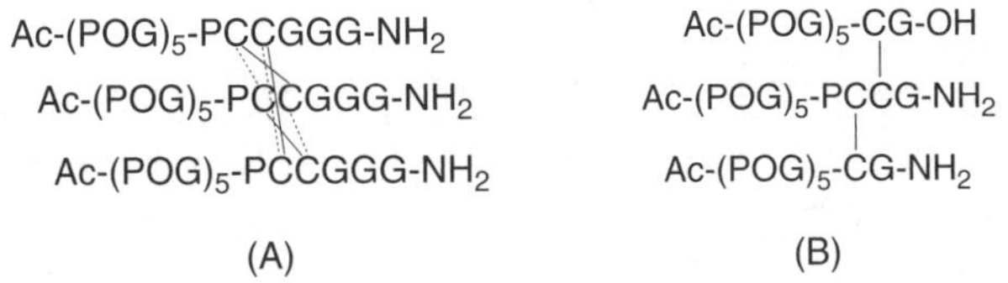

Ac-(POG)₅-PCCGGG-NH₂
Ac-(POG)₅-PCCGGG-NH₂
Ac-(POG)₅-PCCGGG-NH₂

Ac-(POG)₅-CG-OH
Ac-(POG)₅-PCCG-NH₂
Ac-(POG)₅-CG-NH₂

(A)   (B)

Figure 6.5.6 (POG)₅ trimers C-terminally cross-linked by the collagen type III C-terminal cystine knot (A) and an artificial cystine knot established by regioselective thiol chemistry.

collagen type III cystine knot (*vide supra*) with a simpler, artificially designed cystine knot shown in Figure 6.5.6. This comparative study employed the (GPO)₅ sequence as the shortest collagenous sequence known to self-associate into stable triple-helical homotrimers.[58,59]

Differently from the observations made in the regioselective disulfide assembly of trimers from collagenous peptides lacking a higher tendency to self-associate into triple helices (*vide infra*),[60–63] with an increasing propensity for triple-helix formation in aqueous solution conformation-dependent side reactions became critical in the synthesis of the heterotrimer consisting of only five (POG) triplets adjacent to the artificial cystine knot (trimer B of Figure 6.5.6).[64] Attempts to regioselectively cross-link the unprotected chains with disulfides in successive steps failed entirely. However, protecting the peptides with O-*tert*-butyl groups, prevents self-association into homotrimers, and the heterotrimer was successfully produced following the scheme outlined in Figure 6.5.7.[64] This trimer differs from the trimer A of Figure 6.5.6 only in the type of cystine knot. The simplified and thus more flexible artificial cystine framework of the trimer B leads to a thermal stability of the triple helix ($T_m = 55.7\,°C$) only slightly

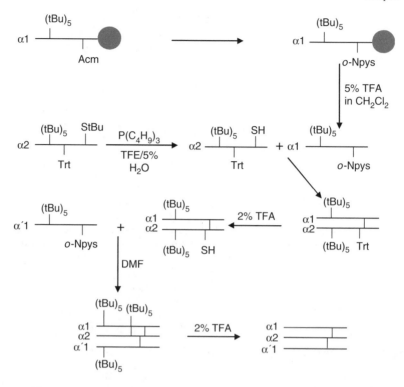

Figure 6.5.7   Regioselective disulfide-bond formation for assembly of the trimer B of Figure 6.5.6.

decreased in comparison to the more rigid native collagen type III knot of trimer A ($T_m = 66.2\,°C$).[36] NMR structural analysis of the trimer B confirmed an extension of the triple-helical hydrogen bonding network into the cross-linked cysteines. This discovery was more recently exploited in the synthesis of self-complementary collagenous peptides that are able to aggregate into collagen-like triple-helical structures of sizes mimicking those of natural collagen.[65,66]

6.5.3   Synthesis of Disulfide Cross-linked Heterotrimeric Collagenous Peptides

In contrast to the more or less efficient cysteine motifs available for the oxidative assembly of homotrimeric collagen peptides, formation of heterotrimers requires specific chain-selection and -registration by structural elements encoded mostly in the C-terminal non-collagenous domains. So far, only little attention has been paid to the identification of such recognition elements that could be exploited for heterotrimerization of *de novo* designed collagen molecules.[16]

6.5.3.1 Oxidative Assembly of Collagenous Heterotrimers with the Cystine Knot of Collagen Type IX

Results from a study analyzing the contribution of the C-terminal portion of the COL1 and the N-terminal part of the NC1 domain towards the oxidative folding of heterotrimeric constructs of FACIT collagen type IX[67] initially suggested a more or less explicit involvement of the COL1 domains in the preferred heterotrimeric assembly of this collagen. These findings, however, do not agree with the self-association and oxidation experiments performed by Bächinger and associates,[68] who employed synthetic peptides corresponding to sequences of the three α-chains at the COL1/NC1 junction and the short NC1 domains of human collagen type IX. A stoichiometric mixture of these three chains was unable to self-assemble into a heterotrimer. Therefore, the peptides were N-terminally extended with the adjacent natural (GXY)$_5$ sequences of the COL1 domains replacing all proline residues in the Y position with hydroxyprolines. Oxidative folding was found to generate a product mixture consisting of about 10% heterotrimer in addition to monomers, dimers and higher oligomers under conditions which were unfavorable for the prefolding of the N-terminal collagenous extensions into a triple helix. This observation would suggest that at least with the entire NC1 domain a prefolding does not seem to be essential for the selective heterotrimerization and cystine-knot formation. So far, this structural motif of the collagen type IX NC1 domains has not been further analyzed for its usefulness in the production of synthetic or recombinant heterotrimeric collagen models. Yields of disulfide cross-bridged triple-helical heterotrimers may possibly be enhanced by incorporation of elements that promote specific interchain interactions, such as the electrostatic interactions used recently for self-association of three different synthetic α-chains in collagenous heterotrimers.[69]

6.5.3.2 Assembly of Heterotrimeric Collagen Molecules by Regioselective Disulfide Formation

To date only trimerization procedures for the assembly of homotrimeric collagen molecules have been developed, with the exception of the COL1/NC1 junction of collagen type IX[68] discussed in the preceding Section 6.5.3.1. Therefore, for studies focusing on structural and functional properties of heterotrimeric collagens, such as collagen type I and IV, a new strategy was required that would result in regioselective cross-bridging of three peptide chains in order to establish the correct chain registration. For this purpose Ottl *et al.*[60] devised an artificial C-terminal cystine knot with minimal steric clashes within the triple helix. Since attempts to apply the information encoded in native sequences for a preferential oxidative assembly of such artificial cystine knots failed completely,[60] regioselective cysteine pairing procedures, as shown in Figure 6.5.8, were applied for cross-bridging the single chains by disulfides into defined heterotrimers. This strategy relies on regioselective

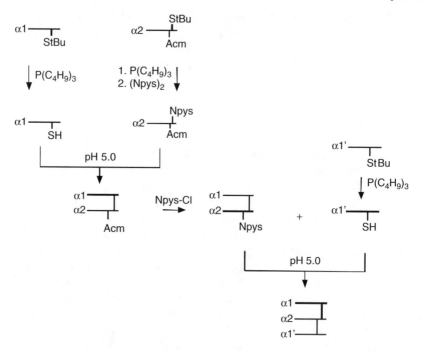

Figure 6.5.8 Reaction scheme for regioselective cross-bridging of three peptide strands into heterotrimers in aqueous solution with defined disulfide connectivities.

activation of cysteine residues as pyridyl- or nitropyridylsulfenyl derivatives followed by thiol-disulfide exchange reactions. The slightly acidic conditions prevent scrambling of already formed disulfides.[60,61]

The general scheme is outlined in Figure 6.5.8 and has been applied with some variations in the orthogonal thiol protecting groups for the successful assembly of various heterotrimers in solution[61–63,66,70,71] and on solid support.[65] Slatter et al.[72] developed a more facile synthesis using a thiol-activated species of the monocysteinyl-α1-chains in excess over the bis-cysteinyl-α2-chain, which places the incorporation of the artificial cystine knot under the control of reaction kinetics rather than regioselective disulfide chemistry. The desired heterotrimer was produced in up to 30% yield. Side products in the reaction mixture were removed using chromatographic separations in order to isolate the trimer.

The synthetic route using regioselective disulfide bridging of heterotrimers did not encounter any difficulties, as long as the peptide monomers or dimers lacked the characteristics which promote their self-assembly into triple helices. But with increasing propensity for triple-helix formation in aqueous solution conformation-dependent side reactions were encountered.[73] The percentage of side products formed reached a considerable level in the synthesis of the trimer consisting of only five (POG) triplets adjacent to the artificial cystine knot (trimer B of Figure 6.5.6).[64]

Despite the complex synthetic steps required for such regioselective strategies, important new insights into the structural properties of collagens were collected with the heterotrimeric collagen constructs. Particularly the structural plasticity of the triple helix in the regions containing recognition elements specific for interaction with proteins such as the collagenases,[74,75] integrins[76,77] or chaperons[78] was well documented.

6.5.4 Concluding Remarks

Homotrimeric collagen constructs that are cross-linked N- or more effectively C-terminally with the collagen type III cystine knot are readily accessible by synthesis and recombinant technologies, while the production of collagen heterotrimers still remains a challenging task. Possible new perspectives may be derived from the proper use of the heterotrimeric collagen type IX cystine knot, although a well-defined registration of the chains by this approach requires more detailed insights into the role of the NC1 and/or COL1 domain during chain selection. Since the registration of the α-chains has been found to significantly affect the plasticity of triple-helical functional epitopes in model heterotrimers,[75,76] a defined chain alignment can so far only be achieved with regioselective interchain disulfide formation. Heterotrimeric foldons of specific chain alignment properties could offer a valid alternative for tethering and registration of the chains, when identified in other heterotrimeric proteins. Indeed homotrimeric foldons such as the short foldon domain of bacteriophage T4 fibritin have already proven to be a highly promising alternative to the C-propeptides to bring collagen peptides in close proximity for the oxidative formation of the interchain cystine knot.[35,79]

References

1. B. Brodsky and A. V. Persikov, *Adv. Prot. Chem.*, 2005, **70**, 301–339.
2. J. Engel and H. P. Bächinger, *Top. Curr. Chem.*, 2005, **247**, 7–33.
3. H. P. Bächinger and J. Engel, in *Protein Folding Handbook*, J. Buchner and T. Kiefhaber, eds., Wiley-VCH, Weinheim, Germany, 2005, Vol. 2, pp. 1059–1110.
4. B. R. Olsen, in *Cell Biology of the Extracellular Matrix*, E. D. Hay, ed., Plenum Press, New York, 1991, pp. 177–220.
5. T. Koide and K. Nagata, *Top. Curr. Chem.*, 2005, **247**, 85–114.
6. J. Khoshnoodi, J. P. Cartailler, K. Alvares, A. Veis and B. C. Hudson, *J. Biol. Chem.*, 2006, **281**, 38117–38121.
7. S. K. Areida, D. P. Reinhardt, P. K. Müller, P. P. Fietzek, J. Köwitz, M. P. Marinkovich and H. Notbohm, *J. Biol. Chem.*, 2001, **276**, 1594–1601.
8. A. Snellman, H. M. Tu, T. Väisänen, A. P. Kvist, P. Huhtala and T. Pihlajaniemi, *EMBO J.*, 2000, **19**, 5051–5059.
9. J. Koivu and R. Myllylä, *J. Biol. Chem.*, 1987, **262**, 6159–6164.

10. J. Koivu, *FEBS Lett.*, 1987, **217**, 216–220.
11. H. P. Bächinger, L. I. Fessler, R. Timpl and J. H. Fessler, *J. Biol. Chem.*, 1981, **256**, 13193–13199.
12. P. Bruckner and E. F. Eikenberry, *Eur. J. Biochem.*, 1984, **140**, 391–395.
13. P. Bruckner, E. F. Eikenberry and D. J. Prockop, *Eur. J. Biochem.*, 1981, **118**, 607–613.
14. L. I. Fessler, R. Timpl and J. H. Fessler, *J. Biol. Chem.*, 1981, **256**, 2531–2537.
15. N. J. Bulleid, J. A. Dalley and J. F. Lees, *EMBO J.*, 1997, **16**, 6694–6701.
16. J. F. Lees, M. Tasab and N. J. Bulleid, *EMBO J.*, 1997, **16**, 908–916.
17. N. J. Bulleid, R. Wilson and J. F. Lees, *Biochem. J.*, 1996, **317**, 195–202.
18. J. F. Lees and N. J. Bulleid, *J. Biol. Chem.*, 1994, **269**, 24354–24360.
19. P. Bruckner, H. P. Bächinger, R. Timpl and J. Engel, *Eur. J. Biochem.*, 1978, **90**, 595–603.
20. H. P. Bächinger, P. Bruckner, R. Timpl, D. J. Prockop and J. Engel, *Eur. J. Biochem.*, 1980, **106**, 619–632.
21. G. Beier and J. Engel, *Biochemistry*, 1966, **5**, 2744–2755.
22. W. F. Harrington and N. V. Rao, *Biochemistry*, 1970, **9**, 3714–3724.
23. J. Engel and D. J. Prockop, *Annu. Rev. Biophys. Biophys. Chem.*, 1991, **20**, 137–152.
24. R. W. Glanville, H. Allmann and P. P. Fietzek, *Hoppe-Seyler's Z. Physiol. Chem.*, 1976, **357**, 1663–1665.
25. H. P. Bächinger, P. Bruckner, R. Timpl and J. Engel, *Eur. J. Biochem.*, 1978, **90**, 605–613.
26. E. Heidemann and W. Roth, *Adv. Polym. Sci.*, 1982, **43**, 143–203.
27. M. Goodman, M. Bhumralkar, E. A. Jefferson, J. Kwak and E. Locardi, *Biopolymers*, 1998, **47**, 127–142.
28. L. Moroder, S. Fiori, R. Friedrich and J. Ottl, in *Synthetic Macromolecules with Higher Structural Order*, ACS Symposium Series, I. M. Khan, ed., ACS, Washington, DC, 2002, Vol. 812, pp. 103–116.
29. L. Moroder, H.-J. Musiol, M. Götz and C. Renner, *Biopolymers*, 2005, **80**, 85–97.
30. C. Boulègue, H.-J. Musiol, C. Renner and L. Moroder, *Antioxid. Redox Signal.*, 2008, **10**, 113–126.
31. D. E. Mechling and H. P. Bächinger, *J. Biol. Chem.*, 2000, **275**, 14532–14536.
32. K. Mann, D. E. Mechling, H. P. Bächinger, C. Eckerskorn, F. Gaill and R. Timpl, *J. Mol. Biol.*, 1996, **261**, 255–266.
33. S. Boudko, S. Frank, R. A. Kammerer, J. Stetefeld, T. Schulthess, R. Landwehr, A. Lustig, H. P. Bächinger and J. Engel, *J. Mol. Biol.*, 2002, **317**, 459–470.
34. S. Frank, S. Boudko, K. Mizuno, T. Schulthess, J. Engel and H. P. Bächinger, *J. Biol. Chem.*, 2003, **278**, 7747–7750.
35. S. P. Boudko and J. Engel, *J. Mol. Biol.*, 2004, **335**, 1289–1297.
36. D. Barth, O. Kyrieleis, S. Frank, C. Renner and L. Moroder, *Chem. Eur. J.*, 2003, **9**, 3703–3714.

37. S. Capasso, C. Mattia, L. Mazzarella and R. Puliti, *Acta. Crystallogr. B*, 1977, **33**, 2080–2083.
38. R. Chandrasekaran and R. Balasubramamian, *Biochim. Biophys. Acta*, 1969, **188**, 1–9.
39. C. J. Creighton, C. H. Reynolds, D. H. S. Lee, G. C. Leo and A. B. Reitz, *J. Am. Chem. Soc.*, 2001, **123**, 12664–12669.
40. C. Park and R. T. Raines, *Protein Eng.*, 2001, **14**, 939–942.
41. X. H. Wang, M. Connor, R. Smith, M. W. Maciejewski, M. E. H. Howden, G. M. Nicholson, M. J. Christie and G. F. King, *Nat. Struct. Biol.*, 2000, **7**, 505–513.
42. Z. X. Xia, Y. N. He, W. W. Dai, S. A. White, G. D. Boyd and F. S. Mathews, *Biochemistry*, 1999, **38**, 1214–1220.
43. O. Carugo, M. Čemažar, S. Zahariev, I. Hudáky, Z. Gáspári, A. Perczel and S. Pongor, *Protein Eng.*, 2003, **16**, 637–639.
44. C. Boulègue, A. G. Milbradt, C. Renner and L. Moroder, *J. Mol. Biol.*, 2006, **358**, 846–856.
45. M. Čemažar, S. Zahariev, J. J. Lopez, O. Carugo, J. A. Jones, P. J. Hore and S. Pongor, *Proc. Natl. Acad. Sci. USA*, 2003, **100**, 5754–5759.
46. L. Barth, E. K. Sinner, S. A. Cadamuro, C. Renner, D. Oesterhelt and L. Moroder, in *Peptides for Youth*, E. Escher, W. D. Lubell and S. Del Valle, eds., American Peptide Society, 2008, in press.
47. G. P. Lunstrum, A. M. McDonough, M. P. Marinkovich, D. R. Keene, N. P. Morris and R. E. Burgeson, *J. Biol. Chem.*, 1992, **267**, 20087–20092.
48. M. Mazzorana, H. Gruffat, A. Sergeant and M. van der Rest, *J. Biol. Chem.*, 1993, **268**, 3029–3032.
49. M. Mazzorana, C. Giry-Lozinguez and M. van der Rest, *Matrix Biol.*, 1995, **14**, 583–588.
50. M. Mazzorana, A. Snellman, K. I. Kivirikko, M. van der Rest and T. Pihlajaniemi, *J. Biol. Chem.*, 1996, **271**, 29003–29008.
51. M. Mazzorana, S. Cogne, D. Goldschmidt and E. Aubert-Foucher, *J. Biol. Chem.*, 2001, **276**, 27989–27998.
52. J. Jäälinoja, J. Ylöstalo, W. Beckett, D. J. S. Hulmes and L. Ala-Kokko, *Biochem. J.*, 2008, **409**, 545–554.
53. A. Lesage, F. Penin, C. Geourjon, D. Marion and M. van der Rest, *Biochemistry*, 1996, **35**, 9647–9660.
54. J. Engel, H. T. Chen, D. J. Prockop and H. Klump, *Biopolymers*, 1977, **16**, 601–622.
55. S. Sakakibara, K. Inouye, K. Shudo, Y. Kishida, Y. Kobayashi and D. J. Prockop, *Biochim. Biophys. Acta*, 1973, **303**, 198–202.
56. P. L. Privalov, *Adv. Prot. Chem.*, 1982, **35**, 1–104.
57. F. R. Brown, E. R. Blout, A. D. Corato and G. P. Lorenzi, *J. Mol. Biol.*, 1972, **63**, 85–99.
58. Y. B. Feng, G. Melacini, J. P. Taulane and M. Goodman, *J. Am. Chem. Soc.*, 1996, **118**, 10351–10358.
59. M. Goodman, Y. B. Feng, G. Melacini and J. P. Taulane, *J. Am. Chem. Soc.*, 1996, **118**, 5156–5157.

60. J. Ottl, R. Battistuta, M. Pieper, H. Tschesche, W. Bode, K. Kühn and L. Moroder, *FEBS Lett.*, 1996, **398**, 31–36.
61. J. Ottl and L. Moroder, *J. Am. Chem. Soc.*, 1999, **121**, 653–661.
62. B. Saccà and L. Moroder, *J. Pept. Sci.*, 2002, **8**, 192–204.
63. J. C. D. Müller, J. Ottl and L. Moroder, *Biochemistry*, 2000, **39**, 5111–5116.
64. D. Barth, H.-J. Musiol, M. Schütt, S. Fiori, A. G. Milbradt, C. Renner and L. Moroder, *Chem. Eur. J.*, 2003, **9**, 3692–3702.
65. F. W. Kotch and R. T. Raines, *Proc. Natl. Acad. Sci. USA*, 2006, **103**, 3028–3033.
66. T. Koide, D. L. Homma, S. Asada and K. Kitagawa, *Bioorg. Med. Chem. Lett.*, 2005, **15**, 5230–5233.
67. L. Labourdette and M. van der Rest, *FEBS Lett.*, 1993, **320**, 211–214.
68. D. E. Mechling, J. E. Gambee, N. P. Morris, L. Y. Sakai, D. R. Keene, R. Mayne and H. P. Bächinger, *J. Biol. Chem.*, 1996, **271**, 13781–13785.
69. V. Gauba and J. D. Hartgerink, *J. Am. Chem. Soc.*, 2007, **129**, 2683–2690.
70. J. Ottl, H.-J. Musiol and L. Moroder, *J. Pept. Sci.*, 1999, **5**, 103–110.
71. T. Koide, Y. Nishikawa and Y. Takahara, *Bioorg. Med. Chem. Lett.*, 2004, **14**, 125–128.
72. D. A. Slatter, L. A. Foley, A. R. Peachey, D. Nietlispach and R. W. Farndale, *J. Mol. Biol.*, 2006, **359**, 289–298.
73. B. Saccà, D. Barth, H.-J. Musiol and L. Moroder, *J. Pept. Sci.*, 2002, **8**, 205–210.
74. J. Ottl, D. Gabriel, G. Murphy, V. Knauper, Y. Tominaga, H. Nagase, M. Kröger, H. Tschesche, W. Bode and L. Moroder, *Chem. Biol.*, 2000, **7**, 119–132.
75. S. Fiori, B. Saccà and L. Moroder, *J. Mol. Biol.*, 2002, **319**, 1235–1242.
76. B. Saccà, C. Renner and L. Moroder, *J. Mol. Biol.*, 2002, **324**, 309–318.
77. B. Saccà, E. K. Sinner, J. Kaiser, C. Lübken, J. A. Eble and L. Moroder, *ChemBioChem*, 2002, **3**, 904–907.
78. T. Koide, S. Asada, Y. Takahara, Y. Nishikawa, K. Nagata and K. Kitagawa, *J. Biol. Chem.*, 2006, **281**, 3432–3438.
79. S. Frank, R. A. Kammerer, D. Mechling, T. Schulthess, R. Landwehr, J. Bann, Y. Guo, A. Lustig, H. P. Bächinger and J. Engel, *J. Mol. Biol.*, 2001, **308**, 1081–1089.

CHAPTER 7

Cystine-based Scaffolds for Functional Miniature Proteins

RUDOLF K ALLEMANN

School of Chemistry, Cardiff University, Main Building, Park Place, Cardiff CF10 3AT, UK

7.1 Introduction

Natural proteins rely for activity on large complex folds that hold in place a relatively small number of residues to bind to their targets or to catalyze reactions. The selectivity, specificity and in the case of enzymes, enormous catalytic efficiency has long tempted chemists to design synthetic peptides and proteins that rival the exquisite performance of natural proteins. Functional macromolecules can be created using two approaches. The first exploits a system that selects a few active biomolecules from a large pool of randomly generated (and mainly inactive) molecules; catalytic antibodies or target binding peptides selected by methods such as phage display are obtained in this way. The second approach involves the rational design of an active biomolecule that relies for activity on a predefined three-dimensional structure. True *de novo* design of functional proteins has proven difficult due to our incomplete understanding of the biophysical basis of protein folding and stability. The design of small peptides that display functional activities as a consequence of adopting structurally well-defined structures and bind targets such as DNA or proteins or show catalytic activity provides an attractive alternative.[1] Small peptides, however, only rarely adopt well-defined conformations with functional activities. One way to overcome this problem is to graft an array of pre-organized functional groups onto

RSC Biomolecular Sciences
Oxidative Folding of Peptides and Proteins
Edited by Johannes Buchner and Luis Moroder
© Royal Society of Chemistry 2009
Published by the Royal Society of Chemistry, www.rsc.org

stable, naturally occurring scaffolds thereby combining the high efficiency of natural proteins with the versatility and high stability of many small peptides.[2-14]

7.2 Pre-organization of Amino Acid Side-chains

The active sites of enzymes provide an organized environment in which the amino acid side-chains are optimally organized to bind and chaperone substrates along often complicated reaction paths. In many cases the remainder of the protein serves little function other than to provide a scaffold for the active site. A series of *de novo* designed oxaloacetate decarboxylases showed that it is possible to implement such active site geometries on the surface of small peptides without specialized binding pockets. Oxaldie-1 and -2 are 14-residue peptides composed of only three residues, namely leucine, lysine and alanine, and designed to fold into amphiphilic α-helices in a concentration-dependent manner.[15-17] Their structure is stabilized by aggregation into bundles of approximately four helical turns. Oxaldies-3 and -4, which were based on the natural peptide scaffold provided by the pancreatic polypeptides, catalyzed the decarboxylation even at concentrations as low as $2\,\mu M$.[6,7] The peptides showed Michaelis–Menten saturation kinetics and rate enhancements of 3–4 orders of magnitude relative to catalysis by simple amines; this is not insignificant when compared to 10^8-fold rate enhancement in natural (metal-dependent) oxaloacetate decarboxylases.[18] These experiments revealed for the first time that pre-organization of active site residues to bind a target and to catalyze a reaction can be achieved on the surface of relatively small peptides. Later work indicated that miniature proteins can also be used for the catalysis of more complex reactions such as the hydrolysis of esters. Based on the pancreatic polypeptide fold, Art-Est was designed to cata- lyze the hydrolysis of mono-*p*-nitrophenyl esters with bell-shaped pH depen- dence and through an acyl enzyme intermediate (Figure 7.1).[9]

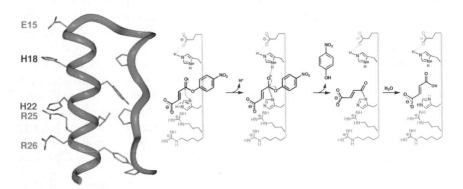

Figure 7.1   Representation of the three-dimensional structure of the designed esterase, Art-Est (amino acids involved in catalysis are highlighted) and the mechanism of the Art-Est catalyzed decarboxylation of mono-*p*-nitro- phenyl fumarate.[9]

These examples show that miniature proteins can display catalytic and stability properties similar to those of natural enzymes. For many applications, however, it is desirable to generate efficient catalysts that can function under non-physiological conditions such as high temperature or extreme pH values or show increased stability to chemical denaturants. Typical strategies to increase the stabilities of miniature proteins include the introduction of metal-binding sites or non-peptidic staples into natural polypeptides and the use of naturally occurring cystine-rich peptides.

7.3 Natural Linear Cystine-stabilized Peptides and Cyclotides

Many bioactive linear peptides of different sources and biological activities such as endothelins, sarafotoxins, bee and scorpion venom toxins, contain a common cysteine pattern of the type $Cys\text{-}(X)_1\text{-}Cys/Cys\text{-}(X)_3\text{-}Cys$ (Figure 7.2).[19-22] In the endothelin/sarafotoxin family of peptides two disulfide bridges to an extended β-type structure stabilize an α-helix with either a parallel or an antiparallel arrangement of the two elements of secondary structure (see also

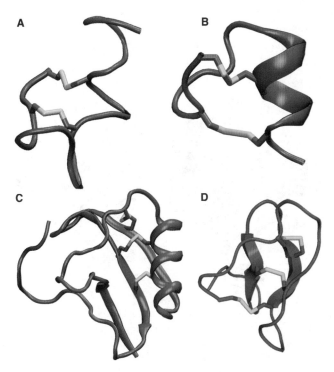

Figure 7.2 Representations of the disulfide stabilized peptides endothelin (A),[19] apamin (B),[21] scorpion toxin (C)[20] and kalata B1 (D).[29,30]

Section 6.3.2). In the case of the honey bee venom toxins, a globular overall fold that consists only of an α-helix containing the Cys-(X)$_3$-Cys portion is cross-linked to the β-strand with a crossed disulfide pattern. Scorpion toxins contain three disulfide bridges and are composed of an α-helix and a two- or three-stranded antiparallel β-sheet. In several growth factors a third cystine penetrates the Cys-(X)$_1$-Cys/Cys-(X)$_3$-Cys motif to form a cystine knot,[23] a rigid and stable structure that has been exploited for molecular engineering applications and for the development of lead compounds for drug design (see Section 6.3.4).[24,25]

Such disulfide-stabilized peptides show remarkable resistance to denaturation. For example, CD-spectroscopy of the 18-residue neurotoxic peptide apamin from *Apis melifica* indicated that the secondary structure of apamin was maintained for temperatures up to 75 °C.[8,28] Only above this temperature a gradual change in the CD-spectrum was observed, most likely indicating a gradual change of the overall structure rather than a loss of helicity. Similarly, apamin shows significant resistance to chemical denaturants and pH; its secondary structure is not affected by concentrations of guanidinium chloride as high as 6 M and by pH values as low as 2.[8,28] The remarkable structural robustness of the apamin fold was further underlined by the design, synthesis and characterization of NTH-18, an apamin-derived peptide, in which the arrangement of the elements of secondary structure were reverted relative to apamin; an N-terminal α-helix was connected through a reverse turn to a C-terminal extension of non-canonical secondary structure, which was linked to the helix through two disulfide bonds (Figure 7.3).[29] Air-oxidation of the fully reduced NTH-18 led to the production of two-disulfide linked peptides;

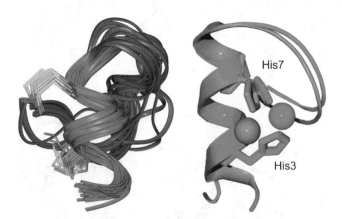

Figure 7.3 Diagrams of an ensemble of NMR-structures of NTH-18 (left), in which the order of the elements of secondary structure is reversed relative to the neurotoxic peptide apamin (Figure 7.2),[29] and of the X-ray-crystal structure of the Cu^{2+} complex of MINTS (backbone conformations and the orientations of the two histidine side-chains are shown for both peptide chains in the asymmetric unit).[30] The right panel of this figure was taken from reference 29. (Copyright ACS. Reproduced with permission.)

the major component (~60%) was characterized by a parallel pattern of di-sulfide bonds, while NTH-18 contained crossed disulfides and was stable over a wide range of pH and temperature as had been observed for apamin.

In the plant gene-encoded cyclotides, a cystine knot-like motif of three di-sulfide bonds is combined with a circular backbone to generate the so-called cyclic cystine knot (Figure 7.2).[26,27,31] The knotted disulfide arrangement combined with the circular backbone renders the cyclotides extremely stable (see Section 6.3.2).[32,33] They are resistant to thermal and enzymatic degrada-tion by proteases as shown by the retention of biological activity after boiling in medicinal applications (kalata B1, the first cyclotide to be discovered is used in native African medicine to induce contractions of the uterus; it is extracted by boiling the plant *Oldenladia affinis*). Studies of acyclic permutants of kalata B1 showed that while open-chain analogs maintained the basic three-dimensional structure, they are intrinsically less stable than the circular proteins. Clearly, the disulfides are central to the formation of tertiary structure of cyclotides, which are further stabilized through cyclization. The exceptional stabilities of cyclo-tides may in the future make them valuable for drug design applications despite potential problems with the degradation of such hyperstable drugs after action. While the potential of the cyclic cystine knot as a scaffold for the presentation of biologically active peptide epitopes has recently been demonstrated (see Section 6.3.3),[34,35] the discussion here will deal with miniature proteins derived from linear polypeptide scaffolds only.

7.4 Cystine-stabilized Miniature Proteins

The miniature decarboxylases and esterases described above indicate that small proteins can be designed that catalyze chemical reactions with transparent reaction mechanisms and catalytic activities that are not insignificant when compared to their much bigger natural counterparts. However, for many applications it is desirable to generate miniature proteins of high stability and with the ability to regulate their activity through external triggers. More than 10 years ago Claudio Vita described the first examples of the use of cystine-stabilized peptides as scaffolds for the generation of functional miniature proteins,[36,37] the activity of which could be controlled through changes in the redox-conditions. Other approaches for the creation of tunable miniature proteins can be based for instance on the introduction of cross-linkers[38–40] or metal-binding sites.

7.4.1 A Metal Ion Induced Helical Foldamer

Metal-binding sites can be used in miniature proteins to control structure formation in response to metal ion concentrations. This strategy is also used in many natural proteins where metal ions often stabilize α-helices. Based on the "reversed" apamin peptide NTH-18,[29] a high affinity Cu^{2+}-binding peptide (Metal Ion Induced N-terminally Stabilized Peptide, MINTS[41]) was designed through the introduction of two histidine residues in the solvent exposed face of the α-helix of NTH-18.[30] The presence of the two histidines was expected to reduce the α-helical character of the peptide. However, the spacing of the two histidines of approximately one helical turn apart implied that they might be ideally placed to promote helix formation through coordination of transition metal ions such as Cu^{2+}. As had been observed for NTH-18 and contrary to observations with apamin, air oxidation of the reduced peptide led to the formation of two products of identical mass corresponding to peptides with crossed and parallel arrangements of the disulfide bridges. The solution of the X-ray crystal structure of the Cu^{2+}-complex of the more helical of these peptides confirmed the presence of a crossed disulfide pattern (Figure 7.3).

CD-spectroscopy revealed that only 26% of the residues of MINTS adopted an α-helical conformation as compared to 44% observed for NTH-18 confirming the strong helix-destabilizing effect of the two histidines. This observation was supported by NMR-spectroscopy, which suggested that the metal-free peptide adopted a poorly defined structure (unpublished). However, the addition of certain transition metal ions allowed the reformation of a stable α-helix. CD-, NMR- and EPR-spectroscopy as well as MALDI-TOF mass spectrometry indicated that in solution MINTS bound to Cu^{2+} to form a 1:1 complex *via* the imidazoles of the two histidine side-chains with a dissociation constant of 5 nM at pH 8, which is the lowest reported value for a designed Cu^{2+}-binding peptide. MINTS displayed more than 100-fold selectivity for Cu^{2+} over Zn^{2+}, Ni^{2+} and Co^{2+}.

Metal-dependent foldamers like MINTS may have applications as oxidation-state sensitive affinity tags, for the generation of artificial models of active sites of metalloproteins, and as reagents to selectively target interfaces involved in protein-protein and protein-nucleic acid interactions that often rely on α-helices.

7.4.2   ApaMyoD: A Miniature DNA-binding Protein

DNA binding proteins have evolved to deal with a problem not normally encountered by enzymes, namely that the substrate, which is a specific DNA fragment, is immersed in a sea of other DNA sequences.[42] This sequence discrimination is often the basis of significant physiological differences. The production of the basic helix-loop-helix (bHLH) transcription factor MyoD in a wide variety of cell types, including fibroblasts and myoblasts, activates a cascade of genes eventually leading to cellular differentiation and the production of muscle cells.[43,44] However, in stark contrast to its high physiological specificity, MyoD displays only limited DNA binding specificity *in vitro*.[45,46] Similar to many other DNA binding reactions, the DNA binding reactions of bHLH proteins are characterized by a transition from a largely unfolded to a mainly α-helical conformation of the bHLH-domain.[45] The limited DNA binding specificity of bHLH proteins has been postulated to be a consequence of their conformational flexibility due to high solvent accessibility.[47] To test the hypothesis that increasing the stability of the DNA recognition helix of MyoD

MyoD ADRRKAATMERRRL...
apamin CNCKAPETALCARRCQQH

ApaMyoD CNCKAPETAACDRRCAATMERRRL...

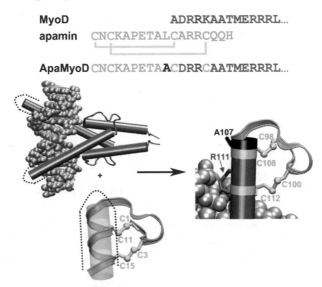

Figure 7.4 Construction of ApaMyoD as a hybrid between the DNA binding protein MyoD and the disulfide stabilized peptide apamin.[48] The recognition helix of MyoD is stabilized through two crossed disulfide bonds to an apamin-like N-terminal extension. To generate ApaMyoD, two amino acids in the solvent exposed face of the DNA recognition helix of MyoD were replaced with cysteines to allow the formation of disulfide bridges to the N-terminal apamin extension leading to a hybrid DNA binding protein where the recognition helix of MyoD was stabilized through two crossed disulfide bonds. Helices are depicted as tubes. This figure was taken from reference 48. (Copyright Elsevier. Reproduced with permission.)

enhances its DNA binding specificity, a hybrid protein, ApaMyoD, was designed in which a sequence based on the peptide apamin was fused to the N-terminus of the bHLH domain of MyoD (Figure 7.4).[48]

CD-spectroscopy had revealed previously that the bHLH-domain of MyoD was largely unfolded at concentrations below 5 µM.[45] At the same concentration, approximately 19 amino acid residues of oxidized apaMyoD corresponding to the whole DNA recognition helix were in an α-helical conformation. This value was slightly higher than would have been expected from the addition of a di-sulfide-stabilized apamin-like segment[21] and indicated that the apamin helix served as a nucleus for a helix that extended for a further 10 residues; the whole of the basic DNA recognition helix was apparently held in a predominately helical conformation by the apamin extension. Like bHLH-MyoD, reduced apaMyoD was largely unfolded in solution indicating that the stabilization of the secondary structure of apaMyoD was a direct result of the formation of the disulfide bridges.

Addition of DNA to oxidized apaMyoD induced a conformational change to a mostly folded conformation. This DNA complex underwent a cooperative thermal unfolding reaction with a midpoint of ~50 °C, 13 °C higher than the

Table 7.1   DNA binding parameters of MyoD–bHLH[46] and apaMyoD[48] in its oxidized and reduced forms.[48]

	MCK–S[a]		*NOE–Box[a]*		
	[P]$_{1/2}$ (nM)[b]	*K_D^c (M^2)*	*[P]$_{1/2}$[b] (nM)*	*K_D^c (M^2)*	*Specificity[d]*
MyoD–bHLH	28.3	8.0×10^{-16}	30.0	9.0×10^{-16}	1.13
ApaMyoD$_{Red}$	20.0	4.0×10^{-16}	27.7	7.7×10^{-16}	1.93
ApaMyoD$_{Ox}$	6.4	4.1×10^{-17}	21.2	4.5×10^{-16}	10.98

[a]MCK-S contains an E-box DNA sequence (CANNTG); NOE-Box is heterologous DNA.
[b]Protein concentration for which 50% of the DNA binding sites are filled.
[c]Apparent dissociation constants, $K_D = ([P]_{1/2})^2$.
[d]Specificity of the DNA binding reaction is defined as the ratio K_D(NOE-Box)/K_D(MCK–S).

melting temperature observed for the DNA complexes of the bHLH-domain of MyoD and of reduced apaMyoD, indicating that the addition of the apamin extension conferred significant additional stability to the DNA complexes of MyoD.

ApaMyoD bound to E-box (CANNTG) containing DNA sequences,[45,49] the targets of MyoD in the promoters and enhancers of genes, with significantly enhanced affinity relative to MyoD-bHLH (Table 7.1). Interestingly, the affinity of apaMyoD for heterologous DNA sequences was increased only approximately 2-fold. Limiting the number of accessible conformations of the recognition helix of apaMyoD appeared to stabilize the interaction with specific DNA and destabilize the complex with non-specific DNA leading to a 10-fold increase in the DNA-binding specificity of the hybrid protein through local stabilization of the DNA binding domain that results in a reduction of the conformational flexibility. The enhanced specificity was clearly a consequence of the disulfide-stabilized intramolecular interactions, as the reduced form of apaMyoD displayed specificity similar to that of MyoD-bHLH.

ApaMyoD is the first example of a designed transcription factor that binds to DNA with significantly increased DNA binding affinity and specificity as a consequence of disulfide-dependent intramolecular stabilization of the DNA recognition helix and suggests a possible resolution of the mechanistic paradox between the impressive physiological specificity displayed by bHLH proteins and their modest DNA-binding specificity *in vitro*. The intramolecular interactions between the N-terminal apamin-like extension and the basic region in apaMyoD can be seen as a model for the modulation of the DNA binding properties of transcriptional regulators through intermolecular interaction between their DNA recognition elements and other components of the transcription complex such as the members of the MEF-2 family. MEF-2C, a potent co-regulator of MyoD, could stabilize a specific conformation of the basic region of MyoD thereby altering its intrinsic DNA binding specificity.[50–52] Unlike protein-protein interactions, the properties of apaMyoD and hence of transcriptional activation can be controlled through changes in the redox conditions with many biotechnological and pharmacological applications.

7.4.3 Apoxaldie-1: A Miniature Oxaloacetate Decarboxylase

The rational redesign of enzymes is a difficult task due to the delicate balance between stability and catalytic activity and the intrinsic instability of many naturally occurring proteins. One way to overcome this problem is to create catalysts by grafting active site residues onto stable polypeptide scaffolds. A wide range of natural protein motifs have been used to create miniature enzymes, which present functional groups on the surface of specific elements of secondary structure. Examples include a $\beta\beta\alpha$ motif, α-helical bundles, mixed polyproline/α-helices and triple-stranded β-sheets.[6–7,13,17,53–58] Such scaffolds often require oligomerization for stability and functional activity. In addition, their stability with respect to heat, denaturants such as guanidinium hydrochloride or urea and extreme pH values is normally rather low. As discussed previously, the reduced thermal stability of miniature proteins is thought, at least in part, to result from their small size and hence low number of stabilizing interactions between different parts of the peptide chain.

Based on a general understanding of the amine-catalyzed decarboxylation of β-keto-acids, a reaction that proceeds through a protonatable imine intermediate,[59] the oxaldies were designed, a series of peptides that folded in solution and displayed catalytic activity as oxaloacetate decarboxylases.[6,7,17] These designs were successful in that imine formation, which is the slow step during the decarboxylation of β-keto-acids by simple amines, was no longer rate determining in reactions catalyzed by these peptides. However, the folding and the activity of these peptides were concentration dependent, suggesting that these scaffolds were stabilized by oligomerization and aggregation (Table 7.2). Their stability to denaturation by elevated temperature, denaturants or changes in pH was low.[6,7]

Hence based on our success with disulfide-stabilized transcription factors,[38–39,48] a miniature oxaloacetate decarboxylase (Apoxaldie-1) was designed based on the stable structure of apamin (Figure 7.5).[8] Three solvent exposed amino acids in close proximity in consecutive turns of the α-helix of apamin were replaced with lysine residues to provide an active site similar to that of oxaldies-2 to -4. As with apamin, intramolecular disulfides formed spontaneously on air oxidation and only the crossed disulfide pattern was observed.

Table 7.2 Kinetic parameters for the catalyzed decarboxylation of oxaloacetate.

Peptide	k_{cat} ($10^{-3} s^{-1}$)	K_M (mM)	k_{cat}/K_M ($M^{-1} s^{-1}$)
Oxaldie-2 (200 μM)[17]	7.5	48.0	0.16
Oxaldie-2 (100 μM)[17]	15.0	210.0	0.07
Oxaldie-3[6]	86.0	49.4	1.74
Oxaldie-4[7]	229.0	64.8	3.53
Apoxaldie (oxidized)[8]	66.0	29.8	2.22
Butylamine[8]	–	–	0.0005
Phe-OEt[8]	–	–	0.0061
Spontaneous[17]	0.013	–	–

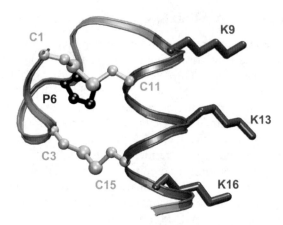

Figure 7.5 Apoxaldie-1 is a highly denaturation resistant oxaloacetate decarboxylase in which the "active site" is stabilized through two crossed disulfide bonds.[8] This figure was taken from reference 8. (Copyright Wiley-VCH Verlag GmbH & Co. KGaA. Reproduced with permission.)

The CD-spectrum of Apoxaldie-1 was concentration independent as would be expected for a monomeric protein and closely resembled that of unmodified apamin[28] with an α-helical content of ~35%, suggesting the formation of approximately two turns of an α-helix, a value that is lower than would have been expected from the structure of apamin (~44%). This suggested that the C-terminal end of Apoxaldie-1 might have been frayed; an interpretation that was supported by MD-simulations, which indicated an unraveling of the three N-terminal residues of Apoxaldie-1 either as a consequence of the three positively charged lysine residues on the solvent exposed face of the helix or due to the formation of a 3_{10}-helix rather than an α-helix. The structure of oxidized Apoxaldie-1 was dependent on the disulfide bonds since on reduction of the disulfide bonds, the peptide adopted a random coil-like structure.

In contrast to most natural oxaloacetate decarboxylases and the previously described Oxaldie-1 to -4,[6–7,17] Apoxaldie-1 was highly resistant to denaturation. Thermal unfolding experiments revealed that it maintained its secondary structure at temperatures in excess of 75 °C. Apoxaldie-1 also showed significant resistance to chemical denaturants; only a slight change in the structure was suggested by the CD-spectra for concentrations of guanidinium chloride in excess of 3 M (Figure 7.6). Apamin itself was resistant to denaturation by guanidinium chloride indicating a slightly reduced stability of Apoxaldie-1, most likely from the unfavorable interaction of the lysine side-chains. Both peptides maintained their structure to pH values as low as 2.2.

Despite its small size and the absence of a binding pocket, Apoxaldie-1 displayed saturation kinetics in its oxidized form (Figure 7.6) with k_{cat} of 0.0660 (±0.02) s^{-1} and a Michaelis constant K_M of 29.8 (±5) mM. It catalyzed the decarboxylation of oxaloacetate through a protonatable imine intermediate by almost four orders of magnitude thereby rivaling the performance of the best

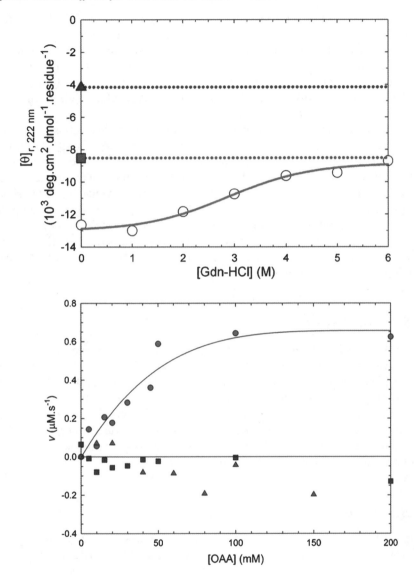

Figure 7.6 High stability of Apoxaldie-1 against chemical denaturants (top). Changes of mean residue ellipticity at 222 nm in the CD-spectrum of Apoxaldie-1 with increasing concentrations of guanidinium chloride (○). For comparison the mean residue ellipticities at 222 nm are also given for reduced (■) Apoxaldie-1 as well as of apamin (▲). Oxidized Apoxaldie-1 displays saturation kinetics for the decarboxylation of oxaloacetate (OAA) (●). For comparison, rate profiles for reduced Apoxaldie-1 (■) and for apamin (▲) are also indicated. This figure was adapted from reference 8. (Copyright Wiley-VCH Verlag GmbH & Co. KGaA. Reproduced with permission.)

synthetic oxaloacetate decarboxylases reported to date (Table 7.2). In contrast to the catalytic properties of the Oxaldies, the kinetic efficiency k_{cat}/K_M of Apoxaldie-1 was not dependent on the concentration of the peptide. Reduction of the cystines led to almost complete loss of catalytic activity (Figure 7.6) indicating that both the disulfide-stabilized, folded structure and the presence of the three lysines within the α-helix of oxidized Apoxaldie-1 were central to the catalytic activity of Apoxaldie-1.

Naturally occurring (metal-dependent) oxaloacetate decarboxylases speed up the decarboxylation reaction by approximately eight orders of magnitude relative to simple amines such as *n*-butylamine,[17,59] while Apoxaldie-1 enhanced the rate of decarboxylation by almost four orders of magnitude. Even relative to a catalyst such as phenylalanine-ethylester, which for catalysis at physiological pH has the nearly optimal pK_a of 7.2, Apoxaldie-1 speeded up the reaction almost 400-fold (Table 7.2). Despite having a 100-fold lower molecular weight, the catalytic performance of Apoxaldie-1 was equal to that of the catalytic antibody 38C2 that efficiently decarboxylates 2-(3′-(4″-acet-amidophenyl)propyl)acetoacetic acid.[60]

7.5   Conclusion

In this brief review, recent progress in the use of small cystine-stabilized protein scaffolds as templates in the design of metal sensors, DNA-binding proteins and enzymes has been described. Due to their small size and high stability such miniature proteins provide versatile agents for the study of fundamental bio-chemical processes like enzymatic catalysis, protein-nucleic acid and protein-protein interactions and for applications at high temperature, non-physiological pH-values and high concentrations of chemical denaturants, where natural proteins do not normally display good activity.

Acknowledgements

The author would like to thank Rhiannon Evans and Neil Young for generating Figure 7.2 and the UK's BBSRC and EPSRC for funding.

References

1. A. J. Nicoll and R. K. Allemann, *Recent Res. Devel. Chem.*, 2004, **2**, 227–243.
2. B. Imperiali and J. J. Ottesen, *Biopolymers*, 1998, **47**, 23–29.
3. B. C. Cunningham and J. A. Wells, *Curr. Opin. Struct. Biol.*, 1997, **7**, 457–462.
4. P. A. Nygren and M. Uhlen, *Curr. Opin. Struct. Biol.*, 1997, **7**, 463–469.
5. L. Martin, P. Barthe, O. Combes, C. Roumestand and C. Vita, *Tetra-hedron*, 2000, **56**, 9451–9460.

6. S. E. Taylor, T. J. Rutherford and R. K. Allemann, *Bioorg. Med. Chem. Lett.*, 2001, **11**, 2631–2635.
7. S. E. Taylor, T. J. Rutherford and R. K. Allemann, *J. Chem. Soc., Perkin Trans. 2*, 2002, 751–755.
8. C. J. Weston, C. H. Cureton, M. J. Calvert, O. S. Smart and R. K. Allemann, *ChemBioChem*, 2004, **5**, 1075–1080.
9. A. J. Nicoll and R. K. Allemann, *Org. Biomol. Chem.*, 2004, **2**, 2175–2180.
10. W. F. DeGrado, *FASEB J.*, 1999, **13**, A1431–A1431.
11. J. K. Montclare and A. Schepartz, *J. Am. Chem. Soc.*, 2003, **125**, 3416–3417.
12. J. W. Chin and A. Schepartz, *Angew. Chem. Int. Ed.*, 2001, **40**, 3806–3809.
13. J. W. Chin and A. Schepartz, *J. Am. Chem. Soc.*, 2001, **123**, 2929–2930.
14. N. J. Zondlo and A. Schepartz, *J. Am. Chem. Soc.*, 1999, **121**, 6938–6939.
15. R. K. Allemann, Evolutionary guidance as a tool in organic chemistry, Ph.D. thesis, ETH-Zürich, 1989.
16. K. Johnsson, Aspekte enzymatischer Katalyse in entworfenen Polypeptiden, Ph.D. thesis, ETH-Zürich, 1992.
17. K. Johnsson, R. K. Allemann, H. Widmer and S. A. Benner, *Nature*, 1993, **365**, 530–532.
18. J. R. Rozzell, *Immobilized Aminotransferase for Amino Acid Production*, Academic Press, New York, 1987, Vol. 136, pp. 479–493.
19. R. W. Janes, B. A. Peapus and B. A. Wallace, *Nat. Struct. Biol.*, 1994, **1**, 311–319.
20. D. Housset, C. Habersetzerrochat, J. P. Astier and J. C. Fontecillacamps, *J. Mol. Biol.*, 1994, **238**, 88–103.
21. J. H. B. Pease and D. E. Wemmer, *Biochemistry*, 1988, **27**, 8491–8498.
22. H. Tamaoki, R. Miura, M. Kusunoki, Y. Kyogoku, Y. Kobayashi and L. Moroder, *Prot. Engineer.*, 1998, **11**, 649–659.
23. M. P. Schlunegger and M. G. Grutter, *Nature*, 1992, **358**, 430–434.
24. S. Krause, H. U. Schmoldt, A. Wentzel, M. Ballmaier, K. Friedrich and H. Kolmar, *FEBS J.*, 2007, **274**, 86–95.
25. L. Chiche, A. Heitz, J. C. Gelly, J. Gracy, P. T. T. Chau, P. T. Ha, J. F. Hernandez and D. Le-Nguyen, *Curr. Prot. Pept. Sci.*, 2004, **5**, 341–349.
26. K. J. Rosengren, N. L. Daly, M. R. Plan, C. Waine and D. J. Craik, *J. Biol. Chem.*, 2003, **278**, 8606–8616.
27. O. Saether, D. J. Craik, I. D. Campbell, K. Sletten, J. Juul and D. G. Norman, *Biochemistry*, 1995, **34**, 4147–4158.
28. A. I. Miroshnikov, E. G. Elyakova, A. B. Kudelin and L. B. Senyavina, *Bioorg. Khim.*, 1978, **4**, 1022–1028.
29. A. J. Nicoll, C. J. Weston, C. Cureton, C. Ludwig, F. Dancea, N. Spencer, U. L. Günther, O. S. Smart and R. K. Allemann, *Org. Biomol. Chem.*, 2005, **3**, 4310–4315.
30. A. J. Nicoll, D. J. Miller, K. Fütterer, R. Ravelli and R. K. Allemann, *J. Am. Chem. Soc.*, 2006, **128**, 9187–9193.

31. D. J. Craik, N. L. Daly, J. Mulvenna, M. R. Plan and M. Trabi, *Curr. Prot. Pept. Sci.*, 2004, **5**, 297–315.
32. D. J. Craik, M. Čemažar and N. L. Daly, *Curr. Opin. Drug Discov. Develop.*, 2006, **9**, 251–260.
33. D. J. Craik, *Science*, 2006, **311**, 1563–1564.
34. R. J. Clark, N. L. Daly and D. J. Craik, *Biochem. J.*, 2006, **394**, 85–93.
35. C. I. Clark, R. C. Reid, R. P. McGeary, K. Schafer and D. P. Fairlie, *Biochem. Biophys. Res. Commun.*, 2000, **274**, 831–834.
36. E. Drakopoulou, J. Vizzavona, J. Neyton, V. Aniort, F. Bouet, H. Virelizier, A. Ménez and C. Vita, *Biochemistry*, 1998, **37**, 1292–1301.
37. C. Vita, C. Roumestand, F. Toma and A. Ménez, *Proc. Natl. Acad. Sci. USA*, 1995, **92**, 6404–6408.
38. L. Guerrero, O. S. Smart, G. A. Woolley and R. K. Allemann, *J. Am. Chem. Soc.*, 2005, **127**, 15624–15629.
39. L. Guerrero, O. S. Smart, C. J. Weston, D. C. Burns, G. A. Woolley and R. K. Allemann, *Angew. Chem. Int. Ed.*, 2005, **44**, 7778–7782.
40. L. D. Walensky, A. L. Kung, I. Escher, T. J. Malia, S. Barbuto, R. D. Wright, G. Wagner, G. L. Verdine and S. J. Korsmeyer, *Science*, 2004, **305**, 1466–1470.
41. A. J. Nicoll, C. J. Weston, C. H. Cureton, O. S. Smart and R. K. Allemann, *Abstr. Pap. Am. Chem. S.*, 228: 179-BIOL Part 1, Aug 22 2004.
42. R. K. Allemann, in *Encyclopedia of Molecular Biology*, T. E. Creighton, ed., Wiley, New York, 1999, pp. 743–744.
43. C. Murre, P. S. McCaw and D. Baltimore, *Cell*, 1989, **56**, 777–783.
44. J. D. Molkentin and E. N. Olson, *Curr. Opin. Gen. Develop.*, 1996, **6**, 445–453.
45. D. Meierhans, C. Elariss, M. Neuenschwander, M. Sieber, J. F. Stackhouse and R. K. Allemann, *Biochemistry*, 1995, **34**, 11026–11036.
46. A. G. E. Kunne, D. Meierhans and R. K. Allemann, *FEBS Lett.*, 1996, **391**, 79–83.
47. R. K. Allemann, in *Encyclopedia of Molecular Biology*, T. E. Creighton, ed., Wiley, New York, 1999, pp. 745–755.
48. E. C. Turner, C. H. Cureton, C. J. Weston, O. S. Smart and R. K. Allemann, *Chem. Biol.*, 2004, **11**, 69–77.
49. A. G. E. Künne, M. Sieber, D. Meierhans and R. K. Allemann, *Biochemistry*, 1998, **37**, 4217–4223.
50. E. N. Olson, M. Perry and R. A. Schulz, *Dev. Biol.*, 1995, **172**, 2–14.
51. Z. X. Mao and B. Nadal-Ginard, *J. Biol. Chem.*, 1996, **271**, 14371–14375.
52. J. D. Molkentin, B. L. Black, J. F. Martin and E. N. Olson, *Cell*, 1995, **83**, 1125–1136.
53. M. Allert and L. Baltzer, *Chem. Eur. J.*, 2002, **8**, 2549–2560.
54. M. D. Struthers, R. P. Cheng and B. Imperiali, *Science*, 1996, **271**, 342–345.
55. C. E. Schafmeister, S. L. LaPorte, L. J. W. Miercke and R. M. Stroud, *Nat. Struct. Biol.*, 1997, **4**, 1039–1046.

56. J. J. Skalicky, B. R. Gibney, F. Rabanal, R. J. B. Urbauer, P. L. Dutton and A. J. Wand, *J. Am. Chem. Soc.*, 1999, **121**, 4941–4951.
57. T. Kortemme, M. Ramirez-Alvarado and L. Serrano, *Science*, 1998, **281**, 253–256.
58. S. R. Griffiths-Jones and M. S. Searle, *J. Am. Chem. Soc.*, 2000, **122**, 8350–8356.
59. F. H. Westheimer, *Tetrahedron*, 1995, **51**, 3–20.
60. R. Björnestedt, G. F. Zhong, R. A. Lerner and C. F. Barbas, *J. Am. Chem. Soc.*, 1996, **118**, 11720–11724.

CHAPTER 8

Selenocystine Peptides – Synthesis, Folding and Applications

MARKUS MUTTENTHALER AND PAUL F. ALEWOOD

Institute for Molecular Bioscience, University of Queensland, 4072 Brisbane, Australia

8.1 Introduction

Sulfur incorporated in proteins in the form of cysteine, methionine and its various different oxidation states participates unlike any other element in an amazingly complex and diverse range of biologically essential reactions including structure formation and stabilization, redox pathways, exchange and radical reactions as well as atom-, electron- and hydride transfer reactions. Many sulfur-rich proteins and their function in the human body have been characterized and well studied. By contrast, the very similar, though distinct, family of selenopeptides and selenoproteins remains poorly understood and, as more selenoproteins are discovered and characterized, the more it becomes apparent that the emerging selenium biochemistry will be a fascinating way to explore this novel and still untouched field. Selenium incorporated in biomolecular structures leads to changes in catalysis, hydrolysis and redox properties[1–3] that can be utilized in a variety of applications (see Table 8.1).

RSC Biomolecular Sciences
Oxidative Folding of Peptides and Proteins
Edited by Johannes Buchner and Luis Moroder
© Royal Society of Chemistry 2009
Published by the Royal Society of Chemistry, www.rsc.org

Table 8.1   Applications involving selenocysteine.

Structural, functional and mechanistic probe	^{77}Se-NMR spectroscopy, X-ray crystallography, SAR (Structure Activity Relationships) studies, specific radiolabeling, PET (Positron Emission Tomography) studies
Robust drug scaffold design	Improvement of bioavailability of disulfide-bond-rich peptides in reducing environment
Directed peptide and protein folding	Induction of selective folding and examination of trapped intermediates
Enzyme function and kinetics	Change of specificity or function of enzymes by placing selenocysteine into the active site
Peptide conjugation	Introduction of dehydroalanine as a site-specific precursor for nucleophilic addition (*e.g.* for the preparation of lanthionines, glycopeptides or lipopeptides)

8.2   Selenium – Isosteric Replacement for Sulfur

8.2.1   Selenium

The non-metal selenium occurs in the earth's crust in the inorganic forms of selenide (Se^{2-}), selenate (SeO_4^{2-}) and selenite (SeO_3^-), and its abundance is about four magnitudes lower than that of sulfur, which is also reflected in the natural abundance of these elements in biological systems. A number of radionuclides and six stable isotopes exist for selenium with the most abundant isotopes ^{78}Se (23.52%) and ^{80}Se (49.82%) being responsible for the unique mass peak distribution (see Figure 8.1).

The element belongs to the group of chalcogens. Except for tellurium these elements are fundamental constituents of functional groups of amino acids and are important contributors to the chemistry and structure of peptides and proteins. Various oxidation states of selenium within proteins have been observed such as the reactive selenol, selenic acid, selenoxide, selenylsufide and the only just recently discovered diselenide bond.[4] The comparable physicochemical properties of selenium and sulfur indicate that similar effects should be expected from a substitution of these two elements. However, the question to what extent the chemical, electrochemical and pharmacological properties of biomacromolecules will change by such an interchange and how this could be of use to the scientific community captured the interest of many research groups over the years and will be a main focus in the following sections.

For a period selenium was considered a poison in biology especially when field research indicated that selenium poisoning was the leading cause of alkali and blind staggers, a disease which threatened livestock upon eating selenium accumulator plants of the genus *Astragalus* during periods of droughts in western USA and China.[5,6] Furthermore, laboratory studies led to declaring selenium a potential carcinogen.[7,8] Groundbreaking work of Schwarz and Foltz in 1957 changed that view significantly when they identified selenium as an essential trace element for bacteria, birds and mammals.[9] Later,

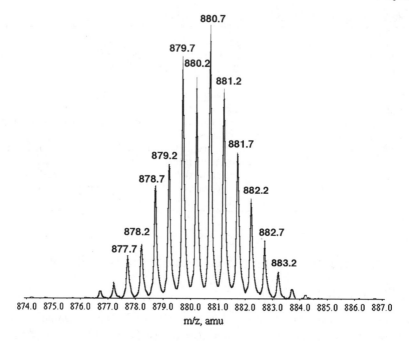

Figure 8.1 Characteristic isotopic selenium abundance seen in an MS spectrum of a peptide with one diselenide bond.

Flohe *et al.*[10] demonstrated in 1973 that selenium is an integral part of the active site of the mammalian glutathione peroxidase – covalently bound in stoichiometric quantities. This triggered a continuous increase in biomedical interest and the emergence of selenium biochemistry as a field of research.[11]

Many selenoproteins have been identified in all lineages of life,[12,13] the largest repertoire found in fish with 30 individual selenoproteins, followed by humans and rodents with 25 and 24, respectively.[14] Selenium has been established as a biologically essential element for cellular redox balance, immune responses, cancer prevention and inflammation protection.[15–17] Some of these seleno-proteins are already well characterized, such as glutathione peroxidase and thioredoxin reductase, though the precise function of many is still unknown leading to a growing interest in the synthesis and study of selenopeptides and -proteins.

8.2.2 Selenocysteine – the 21st Proteinogenic Amino Acid

Selenium is present in proteins of all three lines of descent, eukaryota, archaea and eubacteria.[18] It is predominantly present in the form of the naturally occurring amino acid selenocysteine (Sec), which Cone *et al.* showed for the first time in 1976, on analysis of the selenoprotein component of clostridial glycine reductase.[19] Selenomethionine is another important form in which selenium has been observed in proteins.[20,21] Selenocysteine is often found in enzymatic active sites, where its known function is acting as a nucleophile, a metal ligand or a redox element.[22,23] The importance of selenoproteins became very apparent when the selenocysteine-tRNA gene (necessary for the incorporation of Sec into proteins) knock-out experiment in mice resulted in early embryonic death.[24] Bioincorporation of selenocysteine is genetically controlled and occurs by a specific mechanism (see Section 5.2.1), and selenocysteine can therefore be referred to as the 21st proteinogenic amino acid.[25]

8.2.3 Selenocysteine as an Isosteric Replacement for Cysteine

The similarities between sulfur and selenium summarized in Table 5.1 of Chapter 5 indicate that a mutation of cysteine to selenocysteine represents a very conservative substitution. The largely isosteric character of selenocysteine has been exploited in the chemical syntheses of a wide range of bioactive peptides such as oxytocin,[26–29] somatostatin,[30] α-rat atrial natriuretic peptides,[31] endothelin-1,[32,33] apamin,[34] interleukin-8,[35] BPTI,[36] ribonuclease A,[37] glutaredoxin 3[38] and the α-conotoxins[39] either to elucidate structure activity relationships or to improve their stability in reducing environments.

All selenium analogs folded correctly, bioactivities were retained and structure analysis by NMR and CD spectroscopy further confirmed the isosteric character of the diselenide bond. Indeed, the substitution of a Cys residue with Sec has significant advantages over a substitution with other chemical moieties such as carba,[40] lactam,[41] thioether,[42] homocysteine or penicillamine,[35] which can all impart structural distortions that may compromise bioactivity and selectivity.

8.2.4 Selenocysteine and its Role as a Mechanistic Probe

Selenocysteine plays an important role in the elucidation of structural, mechanistic and functional features in many biomacromolecules. Functional information can be obtained by replacement of active-site Cys residues by Sec based upon the differences in redox properties and nucleophilicity demonstrated with selenosubtilisin,[1,43] metalloselenonein,[44] interleukin-8[35] and selenoglutaredoxin 3.[38] Incorporation of selenocysteine into proteins for the purpose of X-ray crystallography significantly facilitates the phasing problem, and the lengthy and problematic heavy-atom screening procedure can be avoided.[45–47] The difference in bond length between Se-C and S-C does not affect the properties of proteins or peptide analogs since the structures generally retain sufficient plasticity and flexibility to accommodate the selenium residues within the geometries of the wild type (see Figure 8.2).

Selenium has six isotopes ^{74}Se (0.87%), ^{76}Se (9.02%), ^{77}Se (7.58%), ^{78}Se (23.52%), ^{80}Se (49.82%) and ^{82}Se (9.19%). Only one of them, ^{77}Se, has a nuclear spin quantum number of $I = 1/2$ and can be employed in high-resolution NMR spectroscopy.[49,50] Uniform mutation of Cys residues by Sec should theoretically allow specific resonance assignment and

Figure 8.2 Superimposition of the crystal structures of PnIA and Sec[3,16]-[A10L]-PnIA illustrating the isosteric character of the diselenide bond in red and with the disulfide bond in yellow.[48]

conformational analysis of unknown disulfide-bond connectivities in Cys-rich peptides and proteins by ^{1}H-^{77}Se correlated NMR experiments.[51] However, ^{77}Se[^{1}H], ^{1}H-HMBC experiments performed on an oxidized diselenide gluta-redoxin fragment at natural ^{77}Se abundance did not allow for assignment of the diselenide connectivity[52] and reports on the use of this methodology are generally very rare. Recent advances on the rather expensive ^{77}Se-selenocysteine building blocks[51,53] should facilitate the labeling process, as ^{77}Se can now readily be incorporated into peptides and proteins by synthesis and native chemical ligation.

8.2.5 pK_a, Nucleophilicity and Reactivity

Despite the resemblance between the elements selenium and sulfur, the amino acids selenocysteine and cysteine exhibit significantly distinct chemical properties. Even though Sec has similar electronegativity to Cys, it is a stronger nucleophile[54–56] and a better leaving group than its sulfur analog.[57] Furthermore, in pK_a determination studies selenocysteine exhibited a much higher acidity than cysteine (p$K_{a(Sec)}$ = 5.24–5.63, p$K_{a(Cys)}$ = 8.25),[54,58,59] which means at physiological pH the Sec residue will be present largely in its reactive anionic form, the selenolate, while the cysteine residue would still be largely protonated. In pH-dependent titration studies it was also shown that selenocysteine reacted with iodoacetate or iodoacetamide at pH values much below the pK_a of the selenol.[54] Generally it is well established that exposed selenols are a very reactive species and readily oxidized by air.[60] ^{1}H-NMR studies of selenoenzyme selenosubtilisin revealed that the enzyme-bound selenol and seleninic acid have unusually low pK_a values when compared to typical selenium compounds and were found to be deprotonated at all accessible pHs.[61] Intriguingly, Sec residues in peptides are found to be fully oxidized directly upon deprotection *via* hydrogen fluoride (P. Alewood and M. Muttenthaler, unpublished results, and communication with P. Dawson, Scripps). Investigations towards pK_a determination of Sec residues incorporated in peptides and proteins though not trivial would certainly be of great value.

Considering the low pK_a and high reactivity of the selenolate at physiological pH the question arises as to which form of the catalytic selenium is present within a selenoenzyme. Crystallographic data on the selenoenzyme glutathione peroxidase showed that the well-conserved tryptophan and glutamine residues constitute a catalytic triad in which the selenolate is both stabilized and kept activated by hydrogen bonding with the imino group of the tryptophan and the amido group of the glutamine residue (see Figure 8.3).[62,63] In the case of selenosubtilisin the active selenolate is likewise stabilized by hydrogen bonding of a histidine and asparagine residue in close spatial proximity (see Figure 8.3).[64]

Most human selenoproteins have cysteine homologs, which are generally weaker catalysts. Mutations of a Sec residue to Cys in selenoenzymes confirm this, showing a 100- to 1000-fold decrease in catalytic activity.[57,65–67] Kinetic analysis of thioredoxin reduction by selenium analogs of glutaredoxin 3 suggested that it is the selenium's nucleophilicity that is the primary reason for the strong increase in activity rather than its role as the better leaving group.[38] This difference in pK_a and nucleophilicity can be used to discriminate between Sec and Cys, and is the underlying principle of the BESThio assay. BESThio (3′-(2,4-dinitrobenzenesulfonyl)-2′,7′-dimethylfluorescein) initially used as a fluorescent probe for thiols,[68] was shown not to react with thiols at pH values <7 (see Scheme 8.1). Consequently, when performed at a pH 5.8 in presence of DTT, this assay can be used for rapid identification and quantification of

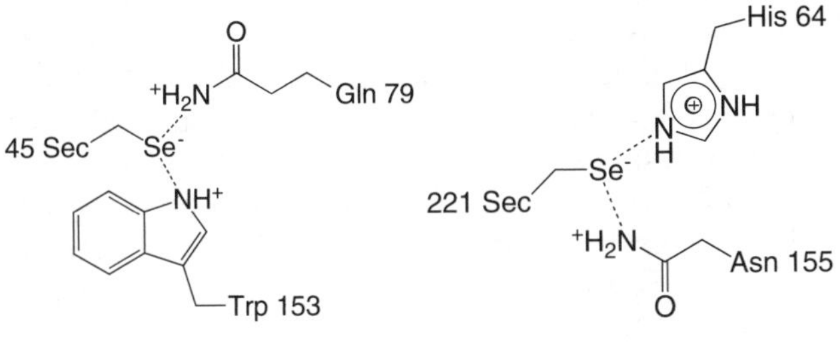

Figure 8.3 Left: Selenolate stabilization in the catalytic triad of glutathione peroxidase: Right: Selenolate stabilization in the catalytic triad of selenosubtilisin.

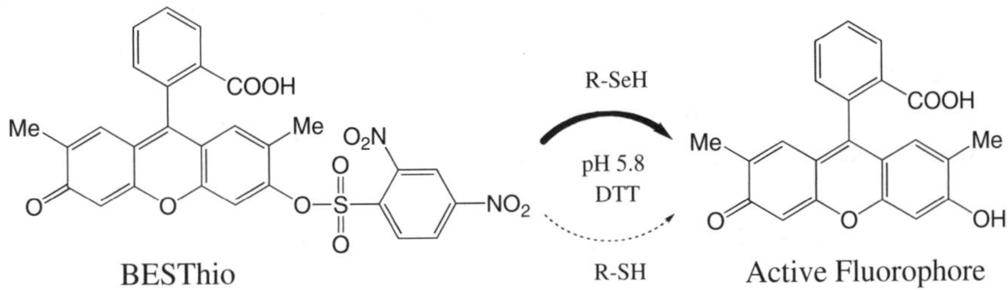

Scheme 8.1 Fluorogenic reaction of BESThio with selenols and thiols.

known and unknown selenoproteins, which was demonstrated on the seleno-proteins glutathione peroxidase and thioredoxin reductase.[69]

8.2.6 The Redox Potential of Selenocysteine

The redox potentials of thiols and selenols in peptides and proteins are usually extracted from the equilibrium constants of exchange reactions with reference redox systems such as DTT or glutathione. For this purpose apparent redox potentials of $-323\,\text{mV}$[70,71] and $-240\,\text{mV}$,[72] respectively, are generally used for calculating the apparent redox potentials by the Nernst equation (see Section 3.7). Studies on aliphatic and aromatic compounds as well as on peptides showed that the reduction of the diselenide bond is possible in the presence of excess irre-versibly reducing agents such as $NaBH_4$, tris(2-carboxyethyl)phosphine (TCEP) and tris-*n*-butylphosphine, or the strongly reducing DTT.[45,60,73–75,78] Excess of TCEP has been observed to lead to deselenation, similar to the well-established desulfurization reaction of disulfides in the presence of basic tris(ethylamine)-phosphine.[76,77] Cyclic voltammogram studies of cystine and selenocystine showed a striking difference in redox potential between them.[21] The highly negative redox potential of selenocystine ($-488\,\text{mV}$) compared to that of cystine ($-223\,\text{mV}$)[78] suggests a very different behavior in redox reactions, which has captured the interest of various research groups in recent years. Reinvestigation of the redox potential of selenocystine led to a value of $-386\,\text{mV}$ as electrode potential and $-388\,\text{mV}$ against DTT at pH 7,[79] which is almost identical to that determined for an unstructured linear bis(selenocysteinyl)peptide ($-381\,\text{mV}$ at pH 7.0 and 25 °C) using DTT ($-323\,\text{mV}$) as reference redox system[80] (see Table 8.2).

In general, the proximity of neighboring Cys and Sec residues and overall structural differences influence the redox potential significantly (see Table 8.2). In the case of Cys residues it can vary from approximately $-125\,\text{mV}$ for DsbA[81] to $-270\,\text{mV}$ for thioredoxin.[82] This diversity in potentials is also reflected in the various functions that these thiol/disulfide oxido-reductase proteins play *in vivo*, ranging from protein reduction to disulfide-bond for-mation see chapters in Section 1 of this book.[21]

Further comparative studies on diselenide, selenylsulfide and disulfide bonds in linear unconstrained glutaredoxin fragments as well as in folded gluta-redoxin 3 analogs were conducted, providing more relevant values for thiol/selenol oxidation.[38,80] The observed difference in redox potentials (111–166 mV) of the disulfide and the diselenide bond (see Table 8.2) in combination with the higher nucleophilicity of selenium suggest a highly favored diselenide or selenylsulfide bond formation over a disulfide bond. This hypothesis was experimentally confirmed on selenium analogs of native and non-native di-sulfide/diselenide analogs of the bee venom toxin apamin (see Section 5.3.1).[34]

8.2.7 Selenocystine in Reducing Environments

In stability studies, the globular selenium analogs of α-conotoxin ImI were exposed to various reducing conditions in several biological systems (blood

Table 8.2 Redox potentials of cysteine, related peptides and proteins and their selenocysteine analogs.

Compounds	Potential	Ref.	Structure/active motif
Cystine[21]	−233 mV	NHE	
Cystine[78]	−223 mV	GSSG[a]	
Selenocystine[21]	−488 mV	NHE	
Selenocystine[79]	−388 mV	DTT[b]	
Glutathione[72]	−240 mV	Lipoic acid	
Selenoglutathione[83]	−407 mV	DTT[c]	
Glutaredoxin 1 (Grx1)[84]	−233 mV	Trx[e]	-CPYC-
Glutaredoxin 3 (Grx3)[84]	−198 mV	Trx[e]	-CPYC-
[C[11],C[14]]-Grx-(10-17)[85]	−215 mV	GSSG[a]	Ac-GCPYCVRA-NH$_2$
[U[11],C[14],K[19]]-Grx-(10-17)[80]	−326 mV	DTT[d]	Ac-GUPYCVKA-NH$_2$
[U[11],U[14],K[19]]-Grx-(10-17)[80]	−381 mV	DTT[d]	Ac-GUPYUVKA-NH$_2$
[U[11],U[14]]-Grx3[38]	−309 mV	Trx[e]	-UPYU-
[U[11],C[14]]-Grx3[38]	−260 mV	Trx[e]	-UPYC-
[C[11],U[14]]-Grx3[38]	−275 mV	Trx[e]	-CPYU-

[a] Determined with $E_0' = -240$ mV for GSSG.
[b] With $E_0' = -332$ mV for DTT.[79]
[c] With $E_0' = -327$ mV for DTT.
[d] With $E_0' = -323$ mV for DTT.[71]
[e] With $E_0' = -270$ mV for Trx.[82]

plasma, glutathione, albumin, thioredoxin). While the *all*-Cys-peptide scrambled into a mixture of ribbon and globular isomers within 24 h in all of the conditions applied (see Scheme 8.2), the selenium analogs retained their three-dimensional structure and no diselenide/disulfide shuffling was detected. In addition, it seems that substitution of only one disulfide bond by a diselenide bond is sufficient to achieve complete structural and bioactive integrity in two-disulfide-bond containing peptides.[39]

8.3 Incorporation of Selenocysteine into Peptides and Proteins

Selenocysteine has been used as a mechanistic probe for structure-activity relationships since the beginning of selenium chemistry and there is more than one method to incorporate selenocysteine into peptides and proteins. This chapter focuses on purely chemical approaches. Incorporation of Sec through biosynthetic cell machinery such as transfection of eukaryotic cells and recombinant selenoprotein production techniques are reviewed in Section 5.2.

8.3.1 Peptide Synthesis

Chemical incorporation of Sec residues into peptides is largely achieved by solid phase peptide synthesis and in combination with native chemical ligation

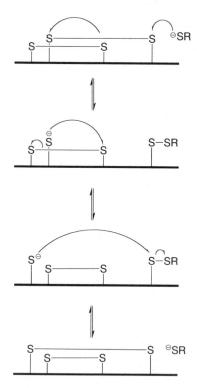

Scheme 8.2   Possible disulfide-bond scrambling in presence of free thiols.

it is regarded as a very effective tool in rational peptide/protein design. Selenocysteines are usually assembled in the form of the Fmoc-Sec(Mob)-OH derivative for Fmoc/tBu and as the Boc-Sec(MeBzl)-OH derivative for the Boc/Bzl strategy. These building blocks behave like their Cys analogs; coupling efficiency is high, chain assembly works smoothly and there are many examples in the literature where this method has been successfully applied.

The early syntheses of Sec-peptides were performed in solution and the benzyl (Bzl) group was used exclusively for the selenol protection, which was introduced predominantly as the Z-Sec(Bzl)-OH intermediate.[26–30,86–88] Its lability in alkaline media and the deprotection step with sodium in liquid ammonia provoked significant β-elimination of the phenylmethaneselenolate (BzlSe⁻) resulting in low yields. It is now well known that BzlSe⁻ is a much better leaving group than the corresponding thiolate.[89] An even better leaving group is the benzeneselenolate PhSe⁻, which is used as a precursor for the introduction of dehydroalanine through mild oxidative elimination (see Section 8.5.2).[90,91]

Following the development of Cys protection compatible with the Fmoc/tBu and Boc/Bzl chemistries the two protecting groups 4-methoxybenzyl (Mob) and 4-methylbenzyl (MeBzl) in the form of Fmoc-Sec(Mob)-OH[31,73] and Boc-Sec(MeBzl)-OH[44] were introduced. The Mob protection was used intensively by various groups,[31,32,34,36,56,73,80,91–94] but was shown to be prone to racemization during activation and coupling steps. Another drawback is the high tendency to deselenate *via* β-elimination during iterative piperidine-mediated Fmoc-deprotection steps resulting in dehydroalanine and consequently piperidyl adducts.[92,95,96] The deselenation and racemization could be largely suppressed by keeping the exposure to piperidine to a minimum and using pentafluorophenyl esters without the addition of base.[92] To avoid these side reactions completely, Boc/Bzl chemistry came back into focus, where deprotection, coupling and cleavage are carried out in acidic to neutral medium, *e.g.*, *via* the *in situ* neutralization protocol.[97] The optical purity of the Boc-Sec (MeBzl)-OH residue and its behavior during peptide synthesis was assessed showing that the stereochemical integrity was conserved and no major side reactions have been reported so far.[38,39]

8.3.2 Deprotection, Cleavage

The Mob group can be removed either in TFA in the presence of strong Lewis acids such as trimethyltrifluoromethane sulfonate[31,73] or with I_2 in acetic acid, which can cause complications and low yields if other intramolecular Sec or Cys residues are present, because of the selenolanthionine formation.[33,93,95]

Deprotection with mercuric acetate is unsuccessful as it leads to the formation of mercuric diselenide, which is stable to treatment with excess thiols for the displacement of the heavy metal.[92,95] The most applicable method to date is deprotection in TFA in the presence of DMSO, which leads directly to diselenide or selenylsulfide formation (see Scheme 8.3).[31,33,92,93,95]

Deprotection of Sec(Mob) can also be achieved under milder conditions *via* the addition of sub-stoichiometric amounts of 2,2′-dithiobis(5-nitropyridine) (DTNP) in TFA, where Sec(Mob) is converted to the 5-nitropyridylsulfenylselenide derivative, which can subsequently be cleaved by thiolysis.[98] This

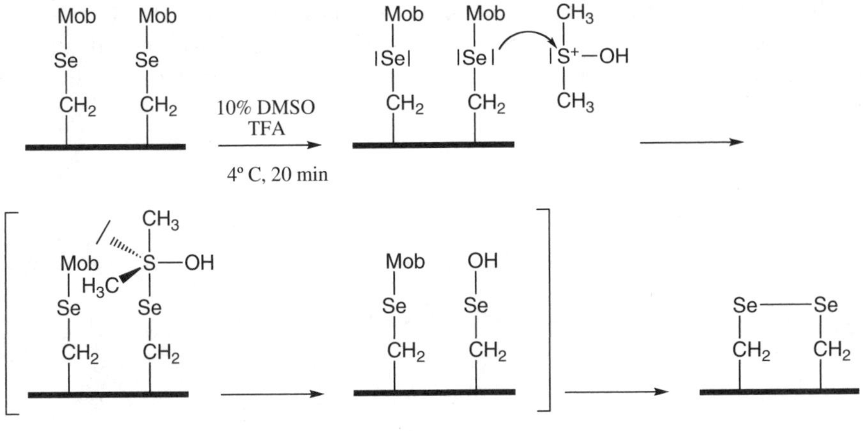

Scheme 8.3 Sec(Mob) deprotection with DMSO/TFA.

method can be used for desired selenylsulfide bond formation[99] and was also shown to produce vicinal selenylsulfide bonds.[100]

In the Boc/Bzl strategy the Sec(MeBzl) is deprotected by standard hydrogen fluoride methods at 0 °C within 1 h and without particular difficulties.[35,38,39,44] A very interesting feature is that upon deprotection the Sec residues are found to be already fully oxidized to the corresponding diselenide or mixed selenylsulfide bonds even at pH ~ 1. This observation has also been confirmed by Phil Dawson *et al.* (personal communication) in his work on selenogluta-redoxin 3[38] and is probably due to the high reactivity and low pK_a of the selenol within peptides and proteins as discussed in Section 8.2.5.

8.4   Synthesis of Selenocysteine Building Blocks

8.4.1   Overview

The increase in publications over recent years reflects the versatility of selenocysteine in its role in probing biological mechanisms and drug design. Quantitative amounts of Sec building blocks were required for the synthesis of analogs for structure-activity studies, which drove the development of novel and robust scaleable syntheses to yield optically pure compounds. This Section will give an overview of various synthetic approaches, discuss reaction key features and will compare yields, efficiency and applications.

Selenocystine was initially prepared by Fredga in 1936,[101] but it was Soda and co-workers who made the synthesis of selenocystine feasible by reacting excess (to prevent monoselenide formation) *in situ* generated disodiumdiselenide with β-chloro-alanine.[102–104] Stocking *et al.*[51] improved this method by reacting dilithiumdiselenide with the Boc-protected β-iodo-alanine-methylester yielding the Boc-protected selenocystine methylester, which upon deprotection could be converted into optically pure selenocystine (Scheme 8.4).

A different approach was introduced later by Siebum *et al.*[53] who took up the challenge of economically introducing the isotopes ^{13}C, ^{15}N, [17]O, [18]O and [77]Se to produce site-directed isotopomers for NMR studies. This method utilized the efficient Mitsunobu reaction to incorporate the expensive [77]Se into the building block (Scheme 8.5).

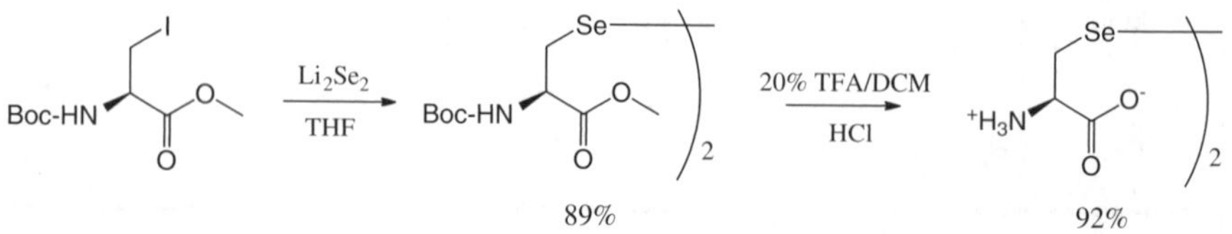

Scheme 8.4   Synthesis of selenocystine.[51]

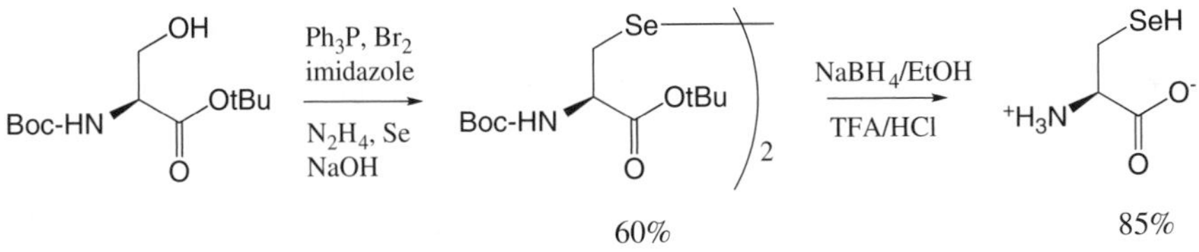

60%   85%

Scheme 8.5 Synthesis of selenocysteine.[53]

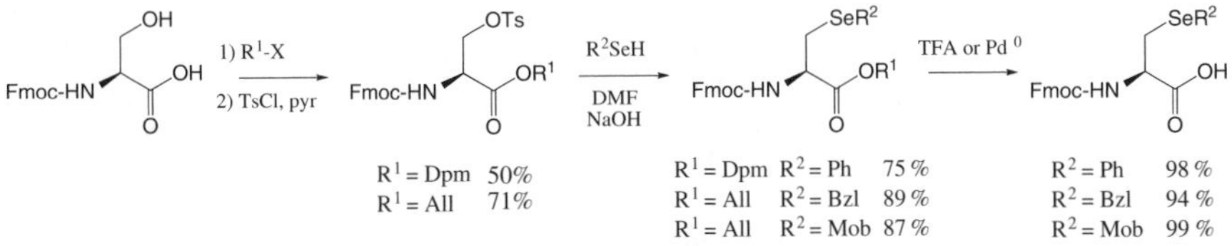

R^1 = Dpm	50%		R^1 = Dpm	R^2 = Ph	75%		R^2 = Ph	98%
R^1 = All	71%		R^1 = All	R^2 = Bzl	89%		R^2 = Bzl	94%
			R^1 = All	R^2 = Mob	87%		R^2 = Mob	99%

Scheme 8.6 Synthesis of *Se*-protected *N*-Fmoc-selenocysteine derivatives.[93]

8.4.2 Selenol Protection

Currently the most used selenol protections are the MeBzl and Mob groups, whereas the Sec(Ph) protection is now predominately used as a precursor for dehydroalanine and is incorporated either as Boc-Sec(Ph)-OH or Fmoc-Sec(Ph)-OH.[90,91]

8.4.3 Building Blocks for Fmoc/tBu Chemistry

The first optically pure Sec(Bzl) derivatives were reported in the late 1960s by Walter and co-workers.[88,105] The synthetic approach involved nucleophilic displacement of the *O*-tosyl moiety of L-serine derivatives by the benzyl selenolate anion (BzlSeNa). This key reaction was developed further by the group of van der Donk into a robust and scaleable synthesis to access Fmoc-Sec(Bzl)-OH, Fmoc-Sec(Ph)-OH and Fmoc-Sec(Mob)-OH in high yields and with simple recrystallization workup (Scheme 8.6).[93]

An alternative approach for the synthesis of phenylselenocysteine is the reduction of diphenyldiselenide with sodium metal followed by reaction with Boc-serine-β-lactone.[106,107] This procedure was slightly modified by Okeley *et al.*[91] performing the reduction with sodium trimethoxyborohydride (NaBH(OMe)$_3$) *in situ* and converting the Boc protecting group into the Fmoc derivative. Fmoc-Sec(Mob)-OH can also be obtained by reduction of selenocystine with NaBH$_4$ and *in situ* reaction with 4-methoxybenzyl chloride, followed by acylation with Fmoc *N*-hydroxy-succinimide.[73]

8.4.4 Building Blocks for Boc/Bzl Chemistry

Boc-Sec(MeBzl)-OH was synthesized by reaction of suitable β-halo-alanines with *in situ* generated disodium or dilithium,[51,53,104] leading to optical pure

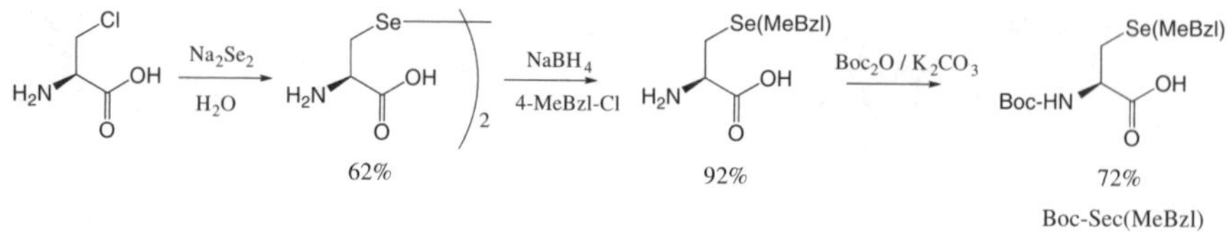

Scheme 8.7   Synthesis of *N*-Boc, *Se*-MeBzl protected selenocysteine.[39]

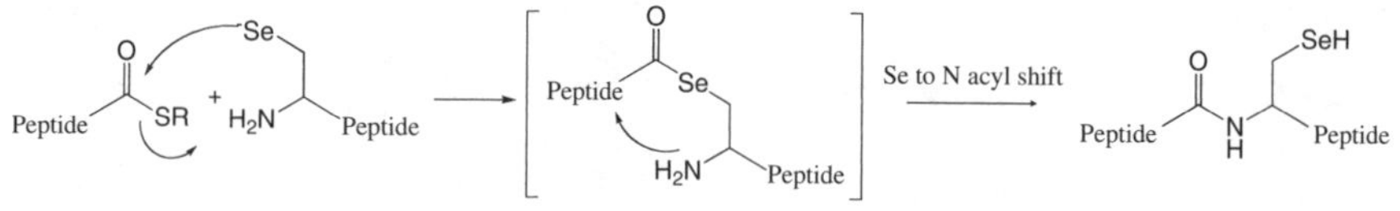

Scheme 8.8   Selenocysteine-mediated native chemical ligation.

selenocystine, which is used as a precursor for further building blocks.[44,102] This method was originally introduced by Oikawa *et al.*[44] but it was Alewood and co-workers who developed it into a fast and efficient synthesis for Boc-protected selenocysteine derivatives, amendable to scale-up with easy workups (Scheme 8.7).[39]

Alternatively Boc-Sec(MeBzl)-OH can be obtained analogous to the synthesis of the Fmoc building blocks, including significant changes such as the synthesis of the 4-methylbenzyl-diselenide, using elemental selenium under bubbling CO,[108] and the mild deprotection of the methyl ester using Me_3SnOH, also a very robust and scaleable synthetic route.[38,109]

8.5   Reactions with Selenocysteine

8.5.1   Selenocysteine-mediated Native Chemical Ligation

Native chemical ligation[110] has proven to be a very attractive approach to the synthesis and semisynthesis of a wide range of proteins and it is now possible to study selenoproteins by selenocysteine-mediated native chemical ligation. The native peptide bond is formed between a C-terminal peptide thioester and another peptide containing an unprotected Sec residue at its N terminus when mixed together in presence of a reducing agent. Without a reducing agent no ligation takes place supporting the mechanism of initial attack of the selenolate on the thioester to give a selenoester, followed by acyl migration from the Se- to the N-terminus to yield the thermodynamically more stable product (see Scheme 8.8).[36,94]

The first example of a Sec-mediated ligation was achieved with two peptides related to ribonucleotide reductase.[93] Although a faster and more efficient reaction was anticipated due to the higher nucleophilicity, the lower pK_a of selenocysteine and the faster aminolysis of the selenoester, the ligation only proceeded slowly and with low yields. The rate of the ligation seems to depend on the equilibrium between the reactive selenolate and its oxidized diselenide or selenylsulfide form. The use of only weakly reducing thiophenol is most likely the reason for slow reaction times as only little of the selenolate will be present for the nucleophilic attack due to higher stability of diselenide and selenylsulfide bonds. Indeed, in model studies on ligation of acetyl-glycine thioesters with Cys and Sec it was shown that once the selenolate was generated (in this case with the stronger and irreversibly reducing agent TCEP), the ligation proceeded faster than with cysteine. At a pH of 5 a 10^3-fold faster rate of product formation was observed, indicating that Sec-mediated chemical ligation can be chemoselective.[56] A drawback of using excess of TCEP can be deselenation, which leads to monoselenide formation, similar to the well-established desulfurization reaction of disulfides in the presence of basic tris-(ethylamine)phosphine.[76,77] An overview of native chemical ligation reactions incorporating a selenocysteine is given in Table 8.3.

Sec-mediated native chemical ligation was applied to the synthesis of a selenium analog of the three-disulfide bond BPTI. The peptide fragments BPTI(1-37)-thioester and [C38U]-BPTI(38-58) were synthesized by Fmoc/tBu synthesis on solid support and ligated in presence of TCEP (1 eq) and 3% (v/v) thiophenol to reduce the selenylsulfide bond and to initiate the ligation by generating the reactive selenolate.[36] [C38U]-BPTI was then subsequently folded to its fully active native structure. Another example is the synthesis of a 15-mer selenopeptide (15SeP) by ligating the fragment 1-14 thioester to selenocystine under reducing conditions yielding the 15SeP diselenide dimer.[112] Standard native chemical ligation in the presence of fully oxidized fragments (diselenide and selenylsulfide bond) was shown to be possible in the synthesis of the complete set of selenocysteine variants of glutaredoxin 3, which provided insights into the catalytic machinery of selenoenzymes.[38]

Intramolecular Sec-mediated ligation can be used for N-to-C terminal backbone cyclization for the formation of stable cyclic peptides. A linear 16-mer-diselenide dimer was cyclized in 0.1 M phosphate buffer at pH 7.5, containing 3% thiophenol (v/v). The reaction was complete after 3 hours, resulting in a mixture of the diselenide dimer and the mixed selenylsulfide with thiophenol (Scheme 8.9).[94]

8.5.2 Dehydroamino Acids – Versatile Precursors

The α,β-unsaturated amino acid dehydroalanine is often found in biological polypeptides and natural products,[113,114] and from a synthetic point of view represents a useful electrophilic precursor for preparation of peptide conjugates such as glycopeptides and lipopeptides. Introduction of dehydroamino acid can be achieved by various methodologies[115] with the most common ones being the activation and elimination of serine residues or Hoffmann elimination of 2,3-diaminopropionic acid.[116] These methods might be useful for selected peptides, but lack overall sufficient chemoselectivity. A more versatile approach is the

Table 8.3 Overview of native chemical ligation reactions incorporating selenocysteine.

Ligation reaction	Reactants	Yield
Ac-LVPSIQDDG-SBzl + UESGACKI → Ac-LVPSIQDDGUESFACKI	4% PhSH, 6 M GnHCl	60%[93]
Ac-LVPSIQDDG-SBzl + CESGAUKI → Ac-LVPSIQDDGCESFAUKI	4% PhSH, 6 M GnHCl	76%[111]
Ac-LVPSIQDDG-SBzl + UESGAUKI → Ac-LVPSIQDDGUESFAUKI	4% PhSH, 6 M GnHCl	48%[111]
LYRAG-SEt + Selenocystine → (LYRAGU)₂ diselenide dimer	3% PhSH, 1.7 eq TCEP, 6 M GnHCL	NA[36]
BPTI(1-37)-SEt + [C38U]-BPTI(38-58) → [C38U]-BPTI	3% thiophenol, 1 eq TCEP, 6 M GnHCL	NA[36]
[C11X]-[C14X]-Grx3(1-37)-MPAL-thioester + [A38C]-Grx3(38-82) → [C11X]-[C14X]-[A38C]-Grx3 with X = U/C	1.5% thiophenol, 3 mM peptide	40–50%[38]
14P-SEt + Selenocystine → 15SeP diselenide dimer	5% thiophenol, 6 M GnHCL	25%[112]
UYAVTGRGDSPAASSG-SEt → c[UYAVTGRGDSPAASSG] + {c[—UYAVTGRGDSPAASSG]}₂	3% thiophenol	Quant.[94]

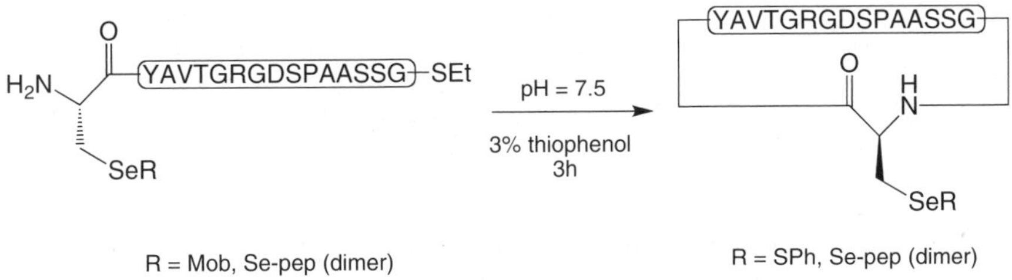

Scheme 8.9 Selenocysteine-mediated backbone cyclization.

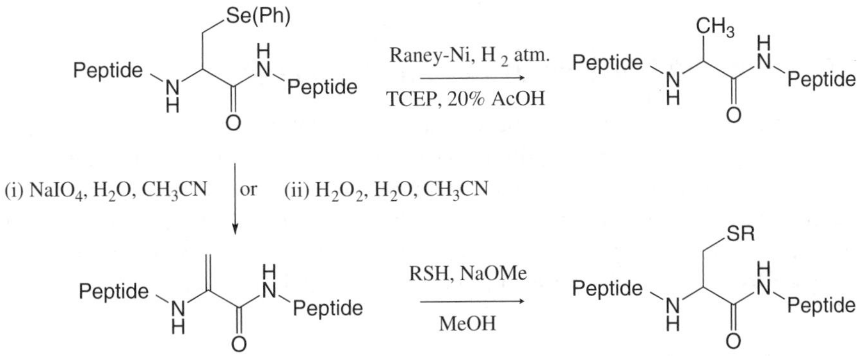

Scheme 8.10 Sec(Ph) – a versatile precursor.

oxidative elimination of cysteine derivatives, but that precludes other protected cysteine residues. With the selenocysteine residue being a better leaving group than the cysteine residue it enables selective access to dehydroalanine and other non-natural amino acids in a facile and convergent way. Selenocysteines can either be introduced site-specifically as phenylselenocysteine during the synthesis or by choosing the ligation site utilizing selenocysteine-mediated chemical ligation. It can then either be converted to dehydroalanine by mild oxidative elimination with hydrogen peroxide or sodium periodate,[90,91] or directly transformed into an alanine residue by reduction of the relatively weak Se–C bond under hydrogen atmosphere using Raney nickel and TCEP (see Scheme 8.10).[94]

Once electrophilic dehydroalanine is obtained, modification can take place by nucleophilic attack or other compatible chemoselective reactions such as Michael addition. The oxidation conditions tolerate functionalities such as tryptophan and methionine and are compatible with thiol protection such as the trityl and *tert*-butylthio group. One application for the electrophilic handle is the intramolecular Michael addition of a trityl-deprotected cysteine to generate the lanthionine (see Scheme 8.11). Interestingly the formation of only one single diastereoisomer was observed.[93]

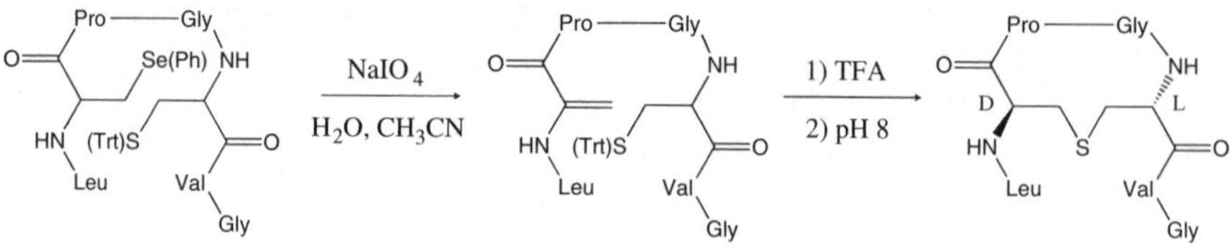

Scheme 8.11 Lanthionine formation *via* intramolecular Michael addition of a cysteine onto a dehydroalanine.

The incorporation of dehydroamino acid into peptides serves as a complementary chemoselective ligation approach and enables the access of highly complex biomolecules containing multiple site-specific modifications.

8.6 Concluding Remarks and Perspectives

Without a doubt selenium can be seen as the most conservative substitution to sulfur and its properties may be exploited in such roles as spectroscopic and mechanistic probes.[43,45,117] When incorporated into biomolecular structures, the resulting changes in catalysis, hydrolysis and redox properties[1–3] can be utilized in a variety of ways. Examples of future applications might be found in drug scaffolds, folding analysis and enzymatic reaction design.

8.6.1 Scaffold Design

Selenocysteine's isosteric character as a cysteine analog combined with the distinct physicochemical properties make it an interesting tool in rational drug design. The much lower redox potential and higher nucleophilicity can be used to direct peptide folding into desired isomers, which by design will have an overall more stable structure towards reducing environments. Scrambling of the disulfide-bond connectivity usually leads to structural changes that can lower the activity,[118] which can be suppressed by the presence of diselenide bonds. Hence from a drug developmental perspective using selenocysteines means not only having the advantage of an increase in bioavailability of a possible therapeutic peptide without losing bioactivity, but also having a more efficient oxidative folding procedure as many purification steps of the usually employed orthogonal protection strategies become unnecessary. A small disulfide/diselenide bond peptide could therefore be used as a robust scaffold with a bioactive drug motif incorporated in a suitable position. Head-to-tail cyclization and addition of conjugates and/or tags could further lead to improvement towards bioavailability or detection (see Figure 8.4).

8.6.2 Folding Pathways

With the exponentially growing number of possible isomers in Cys-rich polypeptides, correct oxidative folding is often not straightforward and ways to

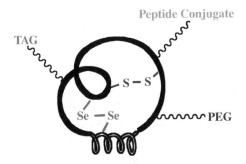

Figure 8.4 Rational scaffold design using selenocysteine.

increase the folding efficacy are sought. Selenocysteine may be a useful tool both to direct folding and to trap out specific intermediates, and therefore help to elucidate folding pathways. Care must be taken, due to the fact that diselenides do not scramble, which is often essential in the folding process. Nevertheless it should be possible to obtain important complementary information concerning folding pathways.

8.6.3 Tailoring of Enzymatic Reactions

Examples such as selenosubtilisin,[1,43,119] monoclonal antibodies[2,120] and phosphorylating glyceraldehyde 3-phosphate dehydrogenase (GAPDH)[3] show clearly that introduction of selenocysteine into the active site can change the specificity or function of an enzyme. This certainly is valid for many other enzymes, but the question is still open if it will eventually be possible to engineer enzymes with novel reaction mechanisms or functions.

References

1. Z. P. Wu and D. Hilvert, *J. Am. Chem. Soc.*, 1990, **112**, 5647–5648.
2. L. Ding, Z. Liu, Z. Zhu, G. Luo, D. Zhao and J. Ni, *Biochem. J.*, 1998, **332**, 251–255.
3. S. Boschi-Müller, S. Müller, A. Van Dorsselaer, A. Böck and G. Branlant, *FEBS Lett.*, 1998, **439**, 241–245.
4. V. A. Shchedrina, S. V. Novoselov, M. Y. Malinoski and V. N. Gladyshev, *Proc. Natl. Acad. Sci. USA*, 2007, **104**, 13919–13924, S/1–S/11.
5. A. L. Moxon, *S. Dak. Agr. Expt. Sta. Bull.*, 1937, **311**, 3–91.
6. S. F. Trelease and O. A. Beath, *Selenium. Its Geological Occurrence and its Biological Effects in Relation to Botany, Chemistry, Agriculture, Nutrition, and Medicine*, Champlain Printers, Burlington, Vt., 1949.
7. A. A. Nelson, O. G. Fitzhugh and H. O. Calvery, *Cancer Res.*, 1943, **3**, 230–236.
8. L. A. Cherkes, M. N. Volgarev and S. G. Aptekar, *Acta Unio Int. Contr.*, 1963, **19**, 632–633.

9. K. Schwarz and C. M. Foltz, *J. Am. Chem. Soc.*, 1957, **79**, 3292–3293.

10. L. Flohe, W. A. Günzler and H. H. Schock, *FEBS Lett.*, 1973, **32**, 132–134.

11. J. T. Rotruck, A. L. Pope, H. E. Ganther, A. B. Swanson, D. G. Hafeman and W. G. Hoekstra, *Science*, 1973, **179**, 588–590.

12. G. V. Kryukov, S. Castellano, S. V. Novoselov, A. V. Lobanov, O. Zehtab, R. Guigo and V. N. Gladyshev, *Science*, 2003, **300**, 1439–1443.

13. D. Behne, H. Pfeifer, D. Röthlein and A. Kyriakopoulos, in *Trace Elements in Man and Animals, Proceedings of the 10th International Symposium on Trace Elements in Man and Animals*, A. A. Roussel, R. A. Anderson and A. E. Flavier, eds., Plenum Press, New York, 2000, 29–33.

14. S. Castellano, A. V. Lobanov, C. Chapple, S. V. Novoselov, M. Albrecht, D. Hua, A. Lescure, T. Lengauer, A. Krol, V. N. Gladyshev and R. Guigo, *Proc. Natl. Acad. Sci. USA*, 2005, **102**, 16188–16193.

15. M. P. Rayman, *Lancet*, 2000, **356**, 233–241.

16. J. Chen and M. J. Berry, *J. Neurochem.*, 2003, **86**, 1–12.

17. L. Schomburg, U. Schweizer and J. Koehrle, *Cell. Mol. Life Sci.*, 2004, **61**, 1988–1995.

18. X.-M. Xu, B. A. Carlson, H. Mix, Y. Zhang, K. Saira, R. S. Glass, M. J. Berry, V. N. Gladyshev and D. L. Hatfield, *PLoS Biology*, 2007, **5**, 96–105.

19. J. E. Cone, R. Martin del Rio, J. N. Davis and T. C. Stadtman, *Proc. Natl. Acad. Sci. USA*, 1976, **73**, 2659–2663.

20. G. N. Schrauzer, *Adv. Food Nutr. Res.*, 2003, **47**, 73–112.

21. C. Jacob, G. I. Giles, N. M. Giles and H. Sies, *Angew. Chem., Int. Ed.*, 2003, **42**, 4742–4758.

22. T. C. Stadtman, *Annu. Rev. Biochem.*, 1996, **65**, 83–100.

23. L. Johansson, G. Gafvelin and E. S. J. Arner, *Biochim. Biophys. Acta*, 2005, **1726**, 1–13.

24. M. R. Bösl, K. Takaku, M. Oshima, S. Nishimura and M. M. Taketo, *Proc. Natl. Acad. Sci. USA*, 1997, **94**, 5531–5534.

25. A. Böck, K. Forchhamer, J. Heider, W. Leinfelder, G. Sawers, B. Veprek and F. Zinoni, *Mol. Microbiol.*, 1991, **5**, 515–520.

26. R. Walter and W.-Y. Chan, *J. Am. Chem. Soc.*, 1967, **89**, 3892–3898.

27. W. Frank, *Hoppe-Seylers Z. Physiol. Chem.*, 1964, **339**, 222–229.

28. R. Walter and V. du Vigneaud, *J. Am. Chem. Soc.*, 1966, **88**, 1331–1332.

29. R. Walter and V. Du Vigneaud, *J. Am. Chem. Soc.*, 1965, **87**, 4192–4193.

30. B. Hartrodt, K. Neubert, B. Bierwolf, W. Blech and H. D. Jakubke, *Tetrahedron Lett.*, 1980, **21**, 2393–2396.

31. T. Koide, H. Itoh, A. Otaka, M. Furuya, Y. Kitajima and N. Fujii, *Chem. Pharm. Bull.*, 1993, **41**, 1596–1600.

32. S. Pegoraro, S. Fiori, S. Rudolph-Böhner, T. X. Watanabe and L. Moroder, *J. Mol. Biol.*, 1998, **284**, 779–792.

33. D. Besse, N. Budisa, W. Karnbrock, C. Minks, H.-J. Musiol, S. Pegoraro, F. Siedler, E. Weyher and L. Moroder, *Biol. Chem.*, 1997, **378**, 211–218.

34. S. Fiori, S. Pegoraro, S. Rudolph-Böhner, J. Cramer and L. Moroder, *Biopolymers*, 2000, **53**, 550–564.
35. K. Rajarathnam, B. D. Sykes, B. Dewald, M. Baggiolini and I. Clark-Lewis, *Biochemistry*, 1999, **38**, 7653–7658.
36. R. Quaderer, A. Sewing and D. Hilvert, *Helv. Chim. Acta*, 2001, **84**, 1197–1206.
37. R. J. Hondal and R. T. Raines, *Methods Enzymol.*, 2002, **347**, 70–83.
38. N. Metanis, E. Keinan and P. E. Dawson, *J. Am. Chem. Soc.*, 2006, **128**, 16684–16691.
39. C. J. Armishaw, N. L. Daly, S. T. Nevin, D. J. Adams, D. J. Craik and P. F. Alewood, *J. Biol. Chem.*, 2006, **281**, 14136–14143.
40. R. F. Nutt, D. F. Veber and R. Saperstein, *J. Am. Chem. Soc.*, 1980, **102**, 6539–6545.
41. B. Hargittai, N. A. Sole, D. R. Groebe, S. N. Abramson and G. Barany, *J. Med. Chem.*, 2000, **43**, 4787–4792.
42. J. Bondebjerg, M. Grunnet, T. Jespersen and M. Meldal, *ChemBioChem*, 2003, **4**, 186–194.
43. Z. P. Wu and D. Hilvert, *J. Am. Chem. Soc.*, 1989, **111**, 4513–4514.
44. T. Oikawa, N. Esaki, H. Tanaka and K. Soda, *Proc. Natl. Acad. Sci. USA*, 1991, **88**, 3057–3059.
45. S. Müller, H. Senn, B. Gsell, W. Vetter, C. Baron and A. Böck, *Biochemistry*, 1994, **33**, 3404–3412.
46. J. F. Sanchez, F. Hoh, M. P. Strub, J. M. Strub, A. Van Dorsselaer, R. Lehrer, T. Ganz, A. Chavanieu, B. Calas, C. Dumas and A. Aumelas, *Acta Crystallogr. D: Biol. Crystallogr.*, 2001, **57**, 1677–1679.
47. M.-P. Strub, F. Hoh, J.-F. Sanchez, J. M. Strub, A. Böck, A. Aumelas and C. Dumas, *Structure*, 2003, **11**, 1359–1367.
48. S.-H. Hu, M. Muttenthaler and J. L. Martin, manuscript in preparation, 2008.
49. H. Duddeck, *Progr. Nucl. Magn. Reson.*, 1995, **27**, 1–323.
50. J. O. Boles, W. H. Tolleson, J. C. Schmidt, R. B. Dunlap and J. D. Odom, *J. Biol. Chem.*, 1992, **267**, 22217–22223.
51. E. M. Stocking, J. N. Schwarz, H. Senn, M. Salzmann and L. A. Silks, *J. Chem. Soc., Perkin Trans. 1*, 1997, **1**, 2443–2447.
52. D. Besse, Synthese and Redoxeigenschaften von Selenocystein peptiden, Ph.D. thesis, Technical University of München, 1997.
53. A. H. G. Siebum, W. S. Woo, J. Raap and J. Lugtenburg, *Eur. J. Org. Chem.*, 2004, **13**, 2905–2913.
54. R. E. Huber and R. S. Criddle, *Arch. Biochem. Biophys.*, 1967, **122**, 164–173.
55. R. G. Pearson, H. R. Sobel and J. Songstad, *J. Am. Chem. Soc.*, 1968, **90**, 319–326.
56. R. J. Hondal, B. L. Nilsson and R. T. Raines, *J. Am. Chem. Soc.*, 2001, **123**, 5140–5141.
57. L. Zhong, E. S. Arner and A. Holmgren, *Proc. Natl. Acad. Sci. USA*, 2000, **97**, 5854–5859.

58. B. Nygard, *Arkiv foer Kemi*, 1967, **27**, 341–361.
59. A. P. Arnold, K. S. Tan and D. L. Rabenstein, *Inorg. Chem.*, 1986, **25**, 2433–2437.
60. W. H. H. Guenther, *J. Org. Chem.*, 1967, **32**, 3931–3934.
61. K. L. House, A. R. Garber, R. B. Dunlap, J. D. Odom and D. Hilvert, *Biochemistry*, 1993, **32**, 3468–3473.
62. O. Epp, R. Ladenstein and A. Wendel, *Eur. J. Biochem.*, 1983, **133**, 51–69.
63. G.-M. Luo, X.-J. Ren, J.-Q. Liu, Y. Mu and J.-C. Shen, *Curr. Med. Chem.*, 2003, **10**, 1151–1183.
64. R. Syed, Z. P. Wu, J. M. Hogle and D. Hilvert, *Biochemistry*, 1993, **32**, 6157–6164.
65. S. R. Lee, S. Bar-Noy, J. Kwon, R. L. Levine, T. C. Stadtman and S. G. Rhee, *Proc. Natl. Acad. Sci. USA*, 2000, **97**, 2521–2526.
66. S. Gromer, L. Johansson, H. Bauer, L. D. Arscott, S. Rauch, D. P. Ballou, C. H. Williams Jr., R. H. Schirmer and E. S. J. Arner, *Proc. Natl. Acad. Sci. USA*, 2003, **100**, 12618–12623.
67. M. J. Axley, A. Böck and T. C. Stadtman, *Proc. Natl. Acad. Sci. USA*, 1991, **88**, 8450–8454.
68. H. Maeda, H. Matsuno, M. Ushida, K. Katayama, K. Saeki and N. Itoh, *Angew. Chem. Int. Ed.*, 2005, **44**, 2922–2925.
69. H. Maeda, K. Katayama, H. Matsuno and T. Uno, *Angew. Chem. Int. Ed.*, 2006, **45**, 1810–1813.
70. W. J. Lees and G. M. Whitesides, *J. Org. Chem.*, 1993, **58**, 642–647.
71. R. P. Szajewski and G. M. Whitesides, *J. Am. Chem. Soc.*, 1980, **102**, 2011–2026.
72. J. Rost and S. Rapoport, *Nature (London, UK)*, 1964, **201**, 185.
73. T. Koide, H. Itoh, A. Otaka, H. Yasui, M. Kuroda, N. Esaki, K. Soda and N. Fujii, *Chem. Pharm. Bull.*, 1993, **41**, 502–566.
74. R. C. Dickson and A. L. Tappel, *Arch. Biochem. Biophys.*, 1969, **130**, 547–550.
75. R. Singh and G. M. Whitesides, *J. Org. Chem.*, 1991, **56**, 6931–6933.
76. D. N. Harpp and J. G. Gleason, *J. Org. Chem.*, 1971, **36**, 73–80.
77. K. Fukase, M. Kitazawa, A. Sano, K. Shimbo, S. Horimoto, H. Fujita, A. Kubo, T. Wakamiya and T. Shiba, *Bull. Chem. Soc. Jpn.*, 1992, **65**, 2227–2240.
78. D. A. Keire, E. Strauss, W. Guo, B. Noszal and D. L. Rabenstein, *J. Org. Chem*, 1992, **57**, 123–127.
79. T. Nauser, S. Dockheer, R. Kissner and W. H. Koppenol, *Biochemistry*, 2006, **45**, 6038–6043.
80. D. Besse, F. Siedler, T. Diercks, H. Kessler and L. Moroder, *Angew. Chem., Int. Ed. Engl.*, 1997, **36**, 883–885.
81. J. Hennecke, A. Sillen, M. Huber-Wunderlich, Y. Engelborghs and R. Glockshuber, *Biochemistry*, 1997, **36**, 6391–6400.
82. E. Mössner, M. Huber-Wunderlich and R. Glockshuber, *Protein Sci.*, 1998, **7**, 1233–1244.

83. J. Beld, K. J. Woycechowsky and D. Hilvert, *Biochemistry*, 2007, **46**, 5382–5390.
84. F. Aslund, K. D. Berndt and A. Holmgren, *J. Biol. Chem.*, 1997, **272**, 30780–30786.
85. F. Siedler, S. Rudolph-Böhner, M. Doi, H. J. Musiol and L. Moroder, *Biochemistry*, 1993, **32**, 7488–7495.
86. W. Frank, *Hoppe-Seyler's Z. Physiol. Chem.*, 1964, **339**, 214–221.
87. D. Theodoropoulos, I. L. Schwartz and R. Walter, *Tetrahedron Lett.*, 1967, **8**, 2411–2414.
88. D. Theodoropoulos, I. L. Schwartz and R. Walter, *Biochemistry*, 1967, **6**, 3927–3932.
89. C. J. M. Stirling, *Acc. Chem. Res.*, 1979, **12**, 198–203.
90. R. Levengood Matthew and W. A. van der Donk, *Nat. Protoc.*, 2006, **1**, 3001–3010.
91. N. M. Okeley, Y. Zhu and W. A. van der Donk, *Org. Lett.*, 2000, **2**, 3603–3606.
92. D. Besse and L. Moroder, *J. Pept. Sci.*, 1997, **3**, 442–453.
93. M. D. Gieselman, L. Xie and W. A. van der Donk, *Org. Lett.*, 2001, **3**, 1331–1334.
94. R. Quaderer and D. Hilvert, *Chem. Commun.*, 2002, 2620–2621.
95. L. Moroder, D. Besse, H.-J. Musiol, S. Rudolph-Böhner and F. Sideler, *Biopolymers*, 1996, **40**, 207–234.
96. M. Rooseboom, N. P. E. Vermeulen, I. Andreadou and J. N. M. Commandeur, *J. Pharmacol. Exp. Ther.*, 2000, **294**, 762–769.
97. M. Schnölzer, P. Alewood, A. Jones, D. Alewood and S. B. H. Kent, *Int. J. Pept. Protein Res.*, 1992, **40**, 180–193.
98. K. M. Harris, S. Flemer Jr. and R. J. Hondal, *J. Pept. Sci.*, 2007, **13**, 81–93.
99. A. L. Schroll and R. J. Hondal, *Proc. 20th Am. Pep. Symp.*, 2008, in press.
100. S. Flemer Jr., B. M. Lacey and R. J. Hondal, *J. Pept. Sci.*, 2008, **14**, 637–647.
101. A. Fredga, *Svensk Kemisk Tidskrift*, 1936, **48**, 160–165.
102. P. Chocat, N. Esaki, H. Tanaka and K. Soda, *Anal. Biochem.*, 1985, **148**, 485–489.
103. D. L. Klayman and T. S. Griffin, *J. Am. Chem. Soc.*, 1973, **95**, 197–199.
104. H. Tanaka and K. Soda, *Methods Enzymol.*, 1987, **143**, 240–243.
105. J. Roy, W. Gordon, I. L. Schwartz and R. Walter, *J. Org. Chem.*, 1970, **35**, 510–513.
106. K. Hashimoto, M. Sakai, T. Okuno and H. Shirahama, *Chem. Commun.*, 1996, 1139–1140.
107. M. Sakai, K. Hashimoto and H. Shirahama, *Heterocycles*, 1997, **44**, 319–324.
108. F. Tian, Z. Yu and S. Lu, *J. Org. Chem.*, 2004, **69**, 4520–4523.
109. K. C. Nicolaou, A. A. Estrada, M. Zak, S. H. Lee and B. S. Safina, *Angew. Chem. Int. Ed.*, 2005, **44**, 1378–1382.

110. P. E. Dawson, T. W. Muir, I. Clark-Lewis and S. B. Kent, *Science*, 1994, **266**, 776–779.

111. M. D. Gieselman, Y. Zhu, H. Zhou, D. Galonic and W. A. van der Donk, *ChemBioChem*, 2002, **3**, 709–716.

112. G. Casi and D. Hilvert, *J. Biol. Chem.*, 2007, **282**, 30518–30522.

113. H.-G. Sahl, R. W. Jack and G. Bierbaum, *Eur. J. Biochem.*, 1995, **230**, 827–853.

114. R. C. M. Lau and K. L. Rinehart, *J. Antibiot.*, 1994, **47**, 1466–1472.

115. U. Schmidt, A. Lieberknecht and J. Wild, *Synthesis*, 1988, 159–172.

116. C. Blettner and M. Bradley, *Tetrahedron Lett.*, 1994, **35**, 467–470.

117. N. P. Luthra, R. C. Costello, J. D. Odom and R. B. Dunlap, *J. Biol. Chem.*, 1982, **257**, 1142–1144.

118. J. Gehrmann, P. F. Alewood and D. J. Craik, *J. Mol. Biol.*, 1998, **278**, 401–415.

119. I. M. Bell, M. L. Fisher, Z. P. Wu and D. Hilvert, *Biochemistry*, 1993, **32**, 3754–3762.

120. G. Luo, Z. Zhu, L. Ding, G. Gao, Q. Sun, Z. Liu, T. Yang and J. Shen, *Biochem. Biophys. Res. Commun.*, 1994, **198**, 1240–1247.

Subject Index